"Diane Vaughan's exhaustive new book, *The Challenger Launch Decision: Risky Technology, Culture, and Deviance at NASA* . . . explores the popular account of what happened leading up to the frigid morning of January 28, 1986, then revises the chronology by probing into a closed corporate world where anomalies are 'normalized' . . . Vaughan brings fresh perspective to what could have been a retread . . . Vaughan challenges conventional accounts of the accident and uncovers contradictions, especially during the teleconference the night before launch."

—*The Tampa Tribune-Times*

"*The Challenger Launch Decision: Risky Technology, Culture, and Deviance at NASA* sheds light on why NASA managers went ahead with the *Challenger* launch when, in retrospect, there was overwhelming evidence of the undue danger of doing so . . . Drawing on that untapped wealth of information, Ms. Vaughan was able to retrace what happened not only on the eve of the launch, but also years before."

—*The Chronicle of Higher Education*

"Interviews with space agency insiders, a review of evidence collected by the investigation, and exhaustive research by Boston College sociologist Diane Vaughan suggest that the fatal decision was not an isolated event."

—*Boston Globe*

"She looks beyond faulty o-rings to the decision-making processes and political structure at NASA and other organizations that deal in risky technology."

—*Science News*

"Sociologist Diane Vaughan provides a new level of detail and understanding about the accident, the agency, and the way people behave . . . The author provides minute detail to show that managers did not violate their safety procedures. Instead Vaughan says, they set and then followed bad precedents in a process she calls 'the normalization of deviance'. . . After *Challenger,* our dreams fell back to earth. Diane Vaughan believes that we ought to keep trying to understand what went wrong, and for that we should all thank her."

—Thomas Donlan, *Barron's*

"Her book is the result not only of in-depth, painstaking research in the vast collection of original documents related to the Challenger accident . . . but also of interviews conducted with many of those involved in the long chain of events that led up to the launch decision. . . [T]hose who follow her sophisticated line of reasoning will be rewarded with multiple insights into 'the hazards of living in this technological age'. . . Vaughan is writing as much for her academic colleagues as for a more general audience . . . In the end, however, the cumulative force of her argument and evidence is compelling."

—John Logsdon,
Chicago Sunday Tribune Books

"[Vaughan] demolishes the old theory that shuttle managers, under pressure from high-ranking officials, pushed ahead with the launch despite the strenuous objections from the engineers . . . Vaughan uses the launch decision as part of a broader context: the study of decision making, and making mistakes, in technological, bureaucratic organizations."

—*Spaceviews*

"Her unapologetic presentation of the technicalities is the book's main strength . . . it is the 'thick description' that demythologizes the retrospective account of the *Challenger* tragedy."

—Harry Collins, *Nature*

"Vaughan provides much original research, including detailed interviews with many of the key players . . . Her comments are a powerful warning to NASA as it embarks on ambitious projects in an era of sharply diminished funding."

—Frederic Golden,
San Francisco Chronicle Book Review

"This book is a tour de force, substantively, methodologically, and theoretically, based on solid work in traditional sociological inquiry."

—*Contemporary Sociology*

"Vaughan, a sociologist, offers a new explanation for the disaster. For a book that must combine techno-speak with social science jargon, it's well written and deserves a look . . . Vaughan's explanation of *Challenger* is more detailed and interesting than what has passed before."

—Bruce Berkowitz, *Air & Space*

"[Vaughan's] research and investigation present an illuminating explanation which dismisses managerial wrong-doing or any singular act of negligent conduct . . . Most striking is Dr. Vaughan's explanation of corporate misadventure through the sociology of mistake . . . The study demonstrates how mistake, mishap, and disaster are socially organized and systematically produced by social structures."

—*Design News*

"[Vaughan's] account opens whole new areas of research and inquiry into the social construction of risk . . . Vaughan's story is intellectually gripping. The conceptual tools she brings to bear . . . yield insights that are subtle, modulated, and clear."

—Gene Rochlin,
Issues in Science and Technology

THE CHALLENGER LAUNCH DECISION

The Challenger Launch Decision

Risky
Technology,
Culture, and
Deviance at
NASA

Diane Vaughan

The University of Chicago Press
Chicago and London

The University of Chicago Press, Chicago 60637
The University of Chicago Press, Ltd., London
© 1996 by The University of Chicago
All rights reserved. Published 1996
Paperback edition 1997
Printed in the United States of America

04 03 02 01 00 99 98   3 4 5

ISBN: 0-226-85175-3 (cloth)
ISBN: 0-226-85176-1 (paper)

Library of Congress Cataloging-in-Publication Data

The Challenger launch decision
    p. cm.
    1. Challenger (Spacecraft)—Accidents. 2. United States.
National Aeronautics and Space Administration—Management.
3. Aerospace industries—United States. 4. Organizational behavior—
Case studies. 5. Decision-making—Case studies.
TL867.C467   1996
363. 12′465—dc20   95-39858
                                                             CIP

♾ The paper used in this publication meets the
minimum requirements of the American National Standard for Information Sciences Permanence of Paper for
Printed Library Materials, ANSI Z39.48 1992.

For

Katherine, Zachary, and Sara Vaughan
Kristen, Lindsey, and Melissa Mortensen
Sophie and Cameron Nicoll

Contents

Figures and Tables

All figures are reproductions of documents appearing in *Report to the President of the Presidential Commission on the Space Shuttle Accident* (Washington, D.C.: Government Printing Office, 1986).

Figures

Tables

Preface

Some events are experienced by great numbers of people, diverse in interest, age, race, ethnicity, life style and life chances, gender, language, and place, who temporarily become bound together by a historic moment. The January 28, 1986, Space Shuttle *Challenger* disaster was such a moment. Collectively, the country grieved, and not for the first time. Many still vividly remember—and will quickly confess, when the subject comes up—exactly where they were, what they were doing, and how they felt when they heard about the tragedy. The initial shock was perpetuated by the television replays of the *Challenger*'s final seconds, the anguished faces of the astronauts' families and other onlookers huddled in disbelief on bleachers at the launch pad, by the news analyses, and then by the official investigation of the Presidential Commission.

Primarily, the disaster is remembered as a technical failure. The fault lay in the rubberlike O-rings. The primary O-ring and its backup, the secondary O-ring, were designed to seal a tiny gap created by pressure at ignition in the joints of the Solid Rocket Boosters. However, O-ring resiliency was impaired by the unprecedented cold temperature that prevailed the morning of the launch. Upon ignition, hot propellant gases impinged on the O-rings, creating a flame that penetrated first the aft field joint of the right Solid Rocket Booster, then the External Tank containing liquid hydrogen and oxygen. The image retained by the American public is a billowing cloud of smoke, from which emerged two white fingers tracing the diverging paths of the Solid Rocket Boosters across the sky.

Technology was not the only culprit, however. The NASA organization was implicated. The Presidential Commission created to investigate the disaster revealed that the O-ring problem had a well-documented history at the space agency. Earliest documentation appeared in 1977—nearly four years before the first shuttle flight in 1981. Moreover, the Commission learned of a midnight-hour tele-conference on the eve of the *Challenger* launch between NASA and Morton Thiokol in Utah, the contractor responsible for building the Solid Rocket Boosters. Worried Thiokol engineers argued against launching on the grounds that the O-rings were a threat to flight safe-ty. NASA managers decided to proceed.

Given this history of O-ring problems and a protest by engineers, why did NASA officials go forward with the launch? As the Com-mission, the House Committee on Science and Technology, whistle-blowers, and journalists probed and published accounts of the events at NASA leading up to the tragedy, they created a documentary record that became the basis for the historically accepted explanation of this historic event: production pressures and managerial wrongdo-ing. Published accounts agreed unanimously that NASA had been experiencing economic strain since the inception of the Space Shut-tle Program. Economic difficulties had focused the attention of NASA decision makers on the launch schedule, which became the means to scarce resources for the space agency. The die was cast for managerial wrongdoing as the explanation of the *Challenger* launch when the Presidential Commission identified a flawed decision-making process at NASA. The Commission found that NASA mid-dle managers had routinely violated safety rules requiring informa-tion about O-ring problems be passed up the launch decision chain to top technical decision makers, so that critical information was not forwarded up the hierarchy.

But a piece of the puzzle has always been missing. The Commis-sion's discovery explained why top administrators failed to act, thereby publicly exonerating them. The more significant question, however, was why the NASA managers, who not only had all the information on the eve of the launch but also were warned against it, decided to proceed. This question was never directly asked and there-fore never answered. In this vacuum, the conventional explanation was born and thrived: economic strain on the organization together with safety rule violations suggested that production pressures caused managers to suppress information about O-ring hazards, knowingly violating safety regulations in order to stick to the launch schedule.

No one has forgotten the astronauts, the incident, or the shape of those billowing clouds that recorded the final seconds of the *Chal-lenger's* flight. Nonetheless, the loss of the *Challenger* has receded into history, as the unfolding present, urgently demanding attention, replaces the past. The details of the investigations, the flaws in the NASA organization that the Presidential Commission and the House Committee illuminated, and the questions about the NASA managers who, informed of the O-ring problem, decided to proceed with the launch have dimmed. Yet for the generations who witnessed it, the tragedy is remembered as a technical failure to which the NASA organization contributed. The details may be forgotten, but for many, the idea of wrongdoing by NASA middle managers lingers.

This book contradicts conventional interpretations of the *Challenger* launch decision. It presents both a revisionist history and a sociological explanation. To achieve both these ends, the core of the book is bracketed by two versions of the eve-of-launch teleconference. Chapter 1 begins with a rendering that is drawn from—and typifies—conventional accounts, followed by a tracing of the discoveries that cast managerial wrongdoing as the historically accepted explanation. Then in Chapter 8, the Chapter 1 account is repeated, juxtaposed against another, more detailed telling that contradicts many aspects of the stereotyped version and revises the conventional explanation. Between these two chapters, the argument of the book unfolds. I present evidence refuting the traditional explanation that blames managerial wrongdoing. Then, relying primarily on archival data and interviews, I develop a historical ethnography that answers two questions: Why, in the years preceding the *Challenger* launch, did NASA continue launching with a design known to be flawed? And why did NASA launch the Challenger against the eve-of-launch objections of engineers?

To understand choices made in another time and place, we must circumvent the limits of retrospection. Thus, I have reconstructed the history of decision making about the Solid Rocket Boosters, chronologically tracing risk assessments and the development of meanings by insiders as the problem unfolded. This history portrays an incremental descent into poor judgment. It was typified by a pattern in which signals of potential danger—information that the booster joints were not operating as predicted—were repeatedly normalized in engineering risk assessments prior to 1986. The normalization of the boosters' technical deviation at NASA was shaped by social forces and environmental contingencies that impinged on and changed organizational structures and culture, routinely affecting the worldview that decision makers throughout the organization brought to their interpretation of technical information. Production pressures played a major role in the tragic outcome, but the role was very different from that conveyed in accepted explanations. Production pressures became institutionalized and thus a taken-for-granted aspect of the worldview that *all* participants brought to NASA decision-making venues. But to locate the cause of the tragedy only in production pressures is to oversimplify and, indeed, distort what happened. Other factors, which received less attention in posttragedy accounts and thus disappeared from public memory, figured importantly in the disaster.

The explanation presented in this book explicates the sociology of mistake. It shows how mistake, mishap, and disaster are socially organized and systematically produced by social structures. No extraordinary actions by individuals explain what happened: no intentional managerial wrongdoing, no rule violations, no conspiracy. The cause of disaster was a mistake embedded in the banality of organizational life and facilitated by an environment of scarcity and competition, elite bargaining, uncertain technology, incrementalism, patterns of information, routinization, organizational and interorganizational structures, and a complex culture. As this book revises historically accepted interpretations, it embraces broader themes. It describes how deviance in organizations is transformed into acceptable behavior. It discloses how production pressures, originating in the environment, become institutionalized in organizations, having nuanced, unacknowledged, pervasive effects on decision making. It opens the "black box" of engineering,[1] following the negotiation of risk and the production of technical knowledge by working engineers making the hands-on risk assessments. At the same time it shows the intersection of the social and the technical in the construction of risk, it gives insight into how scientific paradigms are created, the sources of their obduracy, and the circumstances in which they are—and are not—overturned.

This book also shows how accumulating history intertwines with structural and cultural factors to affect decisions in organizations. Illuminating how culture is created as people interact in work groups, it suggests how and why work group culture persists, and how it shapes future decisions. It reveals how the structure of competition becomes transformed into organizational mandates. It shows how signals of potential danger can be normalized, so that action becomes aligned with organizational goals. At a fundamental level, it exposes the incrementalism of most life in organizations and the way that incrementalism can contribute to extraordinary events that happen. In addition to these themes of general interest, the book demonstrates (1) the macro-micro connection, linking the new institutionalism in organizational analysis to individual action and thought,[2] (2) a historical ethnography, based on archival documents and interviews, that simultaneously builds on developments in social history and adds to an anthropology of modern complex organizations, and (3) theory elaboration as a methodological strategy for developing sociological theory.[3]

The practical lessons from the *Challenger* accident warn us about the hazards of living in this technological age. We learn that harmful outcomes can occur in organizations constructed to prevent them, as NASA was, and can occur when people follow all the rules, as NASA teleconference participants did. In contrast to conventional interpretations that focus on managerial wrongdoing, the book reveals a more complex picture that shifts our attention from individual causal explanations to the structure of power and the power of structure and culture—factors that are difficult to identify and untangle yet have great impact on decision making in organizations. For this reason, the revisionist history and sociological explanation presented here are more frightening than the historically accepted interpretation, for the invisible and unacknowledged tend to remain undiagnosed and therefore elude remedy. This case directs our attention to the relentless inevitability of mistake in organizations—and the irrevocable harm that can occur when those organizations deal in risky technology.

Chapter One

THE EVE OF THE LAUNCH

NASA's Space Shuttle *Challenger* originally was scheduled for launch January 22, 1986. A crew of seven was assigned. Commander Richard Scobee, Pilot Michael Smith, and Mission Specialists Ellison Onizuka, Judith Resnik, and Ronald McNair were astronauts. Gregory Jarvis, an aerospace engineer, and Christa McAuliffe, a teacher from New Hampshire, were payload specialists. McAuliffe's assignment—to teach elementary school students from space—gave the *Challenger* a special aura. Officially known as Space Transportation System (STS) mission 51-L, it became publicly identified as the "Teacher in Space" mission, despite the scientific and technical assignments of the other crew members. According to plan, *Challenger* would be the first launch of 1986. However, the launch date had to be "slipped" several times—first to January 23, then to January 25, then to January 26—because seven launch delays over a 25-day period postponed the December launch of *Columbia* (STS 61-C).[1] Setting a NASA record for false starts, STS 61-C was launched January 12.

Efforts for the January 26 *Challenger* launch from Kennedy Space Center, Cape Canaveral, Florida, were coordinated by the top technical managers and administrators in NASA's four-tiered launch decision chain. Among them were Jesse Moore, Associate Administrator for Space Flight, NASA Headquarters, Washington (Level I); Arnold Aldrich, Program Manager, Johnson Space Center, Houston, Texas (Level II); William Lucas, Director, Marshall Space Flight Center, Huntsville, Alabama; Stanley Reinartz, Manager, Shuttle Projects Office, Marshall; Lawrence Mulloy, Manager, Solid Rocket Booster Project, Marshall (Level III); and Allan McDonald, Director, Solid Rocket Motor Project, Morton Thiokol, Utah (Level IV). Following *Columbia*'s precedent for delay, early countdown activities were terminated because the forecast indicated that weather at Kennedy would be unacceptable throughout the launch window. NASA rescheduled for January 27. That day, countdown was proceeding normally when microswitch indicators showed that the exterior hatch-locking mechanism had not closed properly. By the time it was fixed, the wind velocity exceeded the Launch Commit Criteria for allowable crosswinds at the Kennedy Space Center runway used in case of a return-to-launch-site abort. The launch was scrubbed at 12:36 P.M. and rescheduled for January 28 at 9:38 A.M. EST.

NASA personnel at the Cape first became concerned about cold temperature at approximately 1:00 P.M. on January 27. The forecast for the eve of the launch predicted clear and uncharacteristically cold weather for Florida, with temperatures expected to be in the low 20s during the early hours of January 28. Marshall's Larry Wear, Solid Rocket Motor Manager, asked Morton Thiokol-Wasatch, in Utah, manufacturer of the Solid Rocket Motor (SRM), to have its engineers review the possible effects of the cold on the performance of the SRM. This was not the first time concerns about SRM performance had been raised between Marshall and Thiokol. On several previous launches, hot combustion gases produced when the propellant ignited at liftoff had charred and sometimes even eroded the surface of the rubberlike Viton O-rings designed to seal the joints between the SRM case segments of the Solid Rocket Boosters (SRBs) that help get the shuttle off the launch pad and into the sky (see figs. 1, 2, & 3).

In response to Wear's inquiry, Thiokol's Robert Ebeling, Ignition System Manager, convened a meeting at the Utah plant. The Thiokol engineers expressed concern that the cold would affect O-ring resiliency: the rings would harden to such an extent that they would not be able to seal the joints against the hot gases created at ignition, increasing the amount of erosion and threatening mission safety. A three-location telephone conference to discuss the situation was set for 5:45 P.M. EST. Participating were some of the managers and engineers associated with the SRB Project located at Thiokol, Marshall Space Flight Center, and Kennedy Space Center. During the teleconference, Thiokol expressed the opinion that the launch should be delayed until noon or after. A second teleconference was arranged for 8:15 P.M. EST so that more personnel at all three locations could be involved and Thiokol engineering data could be transmitted by fax to all parties.

Thirty-four engineers and managers from Marshall and Thiokol participated in the second teleconference (see appendix B, fig. B1). Thiokol engineers in Utah presented the charts they had faxed containing their technical analysis of the situation. They argued that O-ring ability to seal the booster joints at ignition would be slower at predicted temperatures than on the coldest launch to date: a January 1985 mission when the calculated O-ring temperature was 53°F. On that flight, hot propellant gases had penetrated a joint, eroding a primary O-ring to such an extent that it had not sealed at all, allowing the hot gases to "blow by" the primary toward the secondary O-ring.

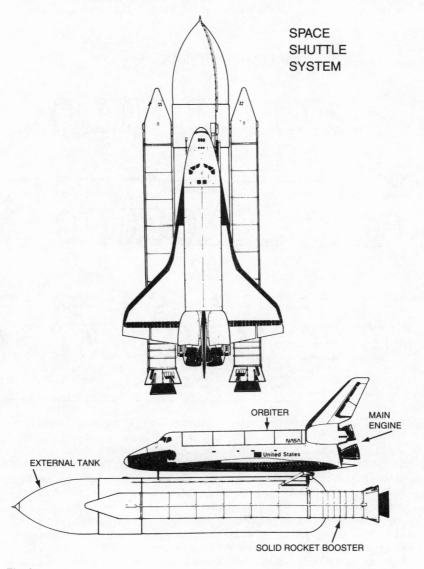

SPACE
SHUTTLE
SYSTEM

ORBITER

MAIN ENGINE

EXTERNAL TANK

SOLID ROCKET BOOSTER

Fig. 1

The secondary performed its intended function as a backup and successfully sealed the joint, but more extensive blow-by could jeopardize the secondary O-ring and thus mission safety. Thiokol Vice President of Engineering Robert Lund presented the Thiokol engineering
conclusion to teleconference participants: O-ring temperature must
be equal to or greater than 53°F at launch.

Marshall's Larry Mulloy asked Thiokol management for a recommendation. Thiokol Vice President Joe Kilminster responded that on

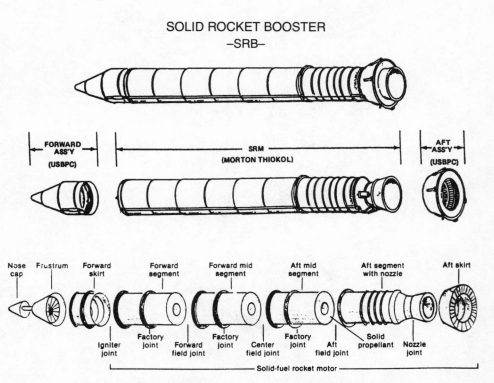

SOLID ROCKET BOOSTER
–SRB–

Fig. 2. The two solid-propellant rocket boosters (SRBs) contribute 80 percent of
the thrust at liftoff. They burn for two minutes, when, fuel exhausted, they separate from the Orbiter 24 miles downrange and drop by parachute to the sea.
The SRBs are approximately 149′ long and 12′ in diameter. The SRMs are composed of cylindrical segments joined by tang-and-clevis joints held together by
177 steel pins around the circumference of each joint. When assembled, they
form a 116′-long tube. The SRMs are partly assembled at Thiokol into four casting segments, into which propellant is poured (or cast). They are shipped by rail
as separate pieces to Kennedy, where they are connected when the booster segments are stacked for final assembly. "Factory joints" are those joints between
segments joined at the Thiokol factory; "field joints" are the ones between the
four casting segments that are joined at Kennedy (in the field).

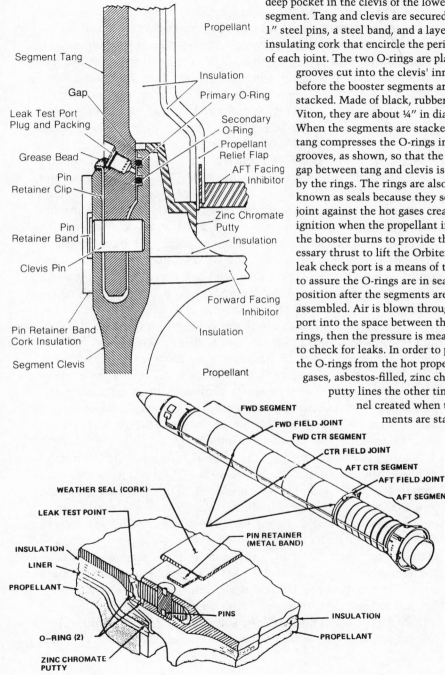

BOOSTER FIELD JOINT

Segment Tang

Gap

Leak Test Port
Plug and Packing

Grease Bead

Pin
Retainer Clip

Pin
Retainer Band

Clevis Pin

Pin Retainer Band
Cork Insulation

Segment Clevis

Propellant

Insulation

Primary O-Ring

Secondary
O-Ring

Propellant
Relief Flap

AFT Facing
Inhibitor

Zinc Chromate
Putty

Insulation

Forward Facing
Inhibitor

Insulation

Propellant

Fig. 3. Field joint cross section. The tang of the top booster segment fits into the 3¾"-deep pocket in the clevis of the lower segment. Tang and clevis are secured by 177 1" steel pins, a steel band, and a layer of insulating cork that encircle the perimeter of each joint. The two O-rings are placed in grooves cut into the clevis' inner edge before the booster segments are stacked. Made of black, rubberlike Viton, they are about ¼" in diameter. When the segments are stacked, the tang compresses the O-rings in their grooves, as shown, so that the tiny gap between tang and clevis is closed by the rings. The rings are also known as seals because they seal the joint against the hot gases created at ignition when the propellant inside the booster burns to provide the necessary thrust to lift the Orbiter. The leak check port is a means of testing to assure the O-rings are in sealing position after the segments are assembled. Air is blown through the port into the space between the two rings, then the pressure is measured to check for leaks. In order to protect the O-rings from the hot propellant gases, asbestos-filled, zinc chromate putty lines the other tiny channel created when the segments are stacked.

FWD SEGMENT

FWD FIELD JOINT

FWD CTR SEGMENT

CTR FIELD JOINT

AFT CTR SEGMENT

AFT FIELD JOINT

AFT SEGMENT

WEATHER SEAL (CORK)

LEAK TEST POINT

INSULATION

LINER

PROPELLANT

PIN RETAINER
(METAL BAND)

PINS

INSULATION

PROPELLANT

O-RING (2)

ZINC CHROMATE
PUTTY

the basis of the engineering conclusions, he could not recommend launch at any temperature below 53°F. Immediately, Marshall managers in Huntsville and at the Cape began challenging Thiokol engineers' interpretation of the data. Mulloy stated that since no Launch Commit Criteria had ever been set for booster joint temperature, what Thiokol was proposing to do was to create new Launch Commit Criteria on the eve of a launch. Mulloy then exclaimed, "My God, Thiokol, when do you want me to launch, next April?" Reinartz asked George Hardy, Marshall's Deputy Director of Science and Engineering, who was in Huntsville, for his view. Hardy responded that he was "appalled" at the Thiokol recommendation.

Reinartz then asked Kilminster for comments. Kilminster requested a five-minute off-line caucus for Thiokol managers and engineers in Utah. The caucus continued for about 30 minutes, during which all three sites were on mute. In Utah, after some discussion, Thiokol Senior Vice President Jerry Mason said, "We have to make a management decision," thus excluding Thiokol engineers from the decision making. Included were four senior Thiokol managers—among them Robert Lund, who had supported the engineering position. Engineers Arnie Thompson and Roger Boisjoly again tried to explain the engineering position. Thompson sketched the joint and discussed the effect of the cold on the O-rings. Showing photographs of the rings on the two flights with blow-by, Boisjoly argued for the correlation between low temperature and hot-gas blow-by. Getting no response, Thompson and Boisjoly returned to their seats. The Thiokol managers continued their discussion and proceeded to vote. Three voted in favor of launch; Lund hesitated. Mason asked him to "take off his engineering hat and put on his management hat." Lund voted with the rest.

The off-line caucus over, the second teleconference resumed. Kennedy and Marshall came back on line. Thiokol's Joe Kilminster announced that Thiokol had reconsidered. They reversed their first position, recommending launch. Kilminster read the revised engineering analysis supporting the recommendation. Mulloy requested that Kilminster fax a copy of Thiokol's flight-readiness rationale and launch recommendation to both Kennedy and Marshall. The teleconference ended at about 11:15 P.M. EST. At Kennedy, McDonald argued on behalf of the Thiokol engineering position. Subsequently, Mulloy and Reinartz telephoned Aldrich, discussing the ice that had formed on the launch pad and the status of the recovery ships. They did not inform Aldrich about the teleconference and Thiokol engineers' concerns about the effects of cold temperature on the O-rings. The Kennedy Space Center meeting broke up about midnight.

At 1:30 A.M. on January 28, the Ice/Frost Inspection Team assessed the ice on the launch pad. NASA alerted Rockwell International, prime contractor for the Orbiter in Downey, California, where specialists began investigating possible effects of ice on the Orbiter. Reporting for launch at 5:00 A.M., Mulloy and Reinartz told Marshall Director Bill Lucas and James Kingsbury, Director of Marshall's Science and Engineering Directorate, of Thiokol's concerns about temperature and the teleconference resolution. At approximately 7:00 A.M., the ice team made its second launch pad inspection. On the basis of their report, the launch time was slipped to permit a third ice inspection. At 8:30, the *Challenger* crew were strapped into their seats. At 9:00, the Mission Management Team met with all contractor and NASA representatives to assess launch readiness. They discussed the ice situation. Rockwell representatives expressed concern that the acoustics at ignition would create ice debris that would ricochet, possibly hitting the Orbiter or being aspirated into the SRBs. Rockwell's position was that they had not launched in conditions of that nature, they were dealing with the unknown, and they could not assure that it was safe to fly. The Mission Management Team discussed the situation and polled those present, who voted to proceed with the launch.

A little after 11:25 A.M., the terminal countdown began. STS 51-L was launched at 11:38 A.M. EST. The ambient temperature at the launch pad was 36°F. The mission ended 73 seconds later as a fireball erupted and the *Challenger* disappeared in a huge cloud of smoke. Fragments dropped toward the Atlantic, nine miles below. The two SRBs careened wildly out of the fireball and were destroyed by an Air Force range safety officer 110 seconds after launch. All seven crew members perished.

THE SEARCH FOR AN EXPLANATION: CREATING HISTORY

The grief and shock that encompassed the millions who witnessed the tragedy intensified in its dramatic aftermath. In early February, President Reagan created a commission, headed by former Attorney General William P. Rogers, to investigate the disaster. Within 120 days, the Presidential Commission (also called the "Rogers Commission") was to submit a report to the President that established the

probable cause or causes of the technical failure and was to make recommendations based on its findings.[2] NASA had already begun its
own investigation of the tragedy; the Presidential Commission saw
its role as one of collaborative oversight. Subsequently, however, the
Commission's search for an explanation uncovered evidence that
placed NASA managers and the launch decision at the center of controversy. Grief and shock were joined with public incredulity and outrage as the Commission discovered information revealing that the
disaster might have been avoided. Unraveling the history of decision
making between Marshall and Thiokol about the SRBs in its televised
hearings, the Commission incrementally laid the groundwork for
what became the historically accepted explanation of the *Challenger*
launch decision: production pressures and managerial wrongdoing.

The Commission's televised hearings began on Thursday, February 6, 1986. Witnesses testified under oath. The first business was to
bring the Commission up to date on NASA's in-house investigation.
Associate Administrator for Space Flight Jesse Moore showed photographs of *Challenger* in flight revealing what appeared to be a flame
coming from the right SRB at the aft segment. He said, "We don't
know for sure it is the SRB. I will caution you, it appears in that area,
but we are not ruling out anything at this point."[3] Neither he nor
others pointed definitively toward the O-rings as the probable cause
of the tragedy. In fact, Jud Lovingood, Deputy Manager, Shuttle Projects Office, Marshall, downplayed the O-rings as a potential cause.
He testified: "We have seen some evidence of what we call blow-by
of those seals, some erosion of those seals. The primary seal. We have
never seen any erosion of a secondary seal, but we have seen evidence
of soot in between the two seals."[4] However, Lovingood said that
problem had been "thoroughly worked" and offered to provide the
Commission with the paperwork documenting the engineering
analysis.[5] When asked whether the effect of the cold weather on the
SRBs was a concern, Lovingood testified: "We did have a meeting
with Thiokol. We had a telecon discussion with people in Huntsville,
people at the Wasatch division, and people at KSC [Kennedy Space
Center]. And the discussion centered around the integrity of the O-
rings under lower temperature. We had the Project Managers from
both Marshall and Thiokol in the discussion. We had the chief engineers from both places in the discussion. And Thiokol recommended to proceed on the launch, and so they did recommend the
launch."[6]

In the next session, Moore presented additional photographs and
the first results of flight data analysis indicating a time correlation

between the plume approximately at the aft field joint in the photographs and the telemetry data on the *Challenger*'s increasingly erratic flight performance. Both Moore and Aldrich acknowledged the possible significance of cold weather on the sealing capability of the O-rings. The weather, O-ring erosion, putty performance, and Rockwell's concern about the ice became a focus of the Commission's questions that day.[7]

Then, on February 9, the Sunday *New York Times* ran a front-page story headed, "NASA Had Warning of a Disaster Risk Posed by Booster."[8] Quoting from internal NASA documents released to the *Times* by an anonymous solid fuel rocket analyst knowledgeable about the SRBs, the article posed O-ring failure as the cause. Further, the article stated that "concerns about the O-rings had been expressed in agency documents at least as far back as 1982."[9] In July 1985, Richard C. Cook, a budget analyst at NASA who assessed the costs associated with the SRBs, wrote a memorandum warning that flight safety was being jeopardized by charring and erosion of the seals.[10] Moreover, a NASA internal memorandum stated that seal erosion had occurred 12 times during flights and contradicted Lovingood's testimony by noting that in one case the secondary O-ring showed erosion also.[11]

In response to the *Times* story, the Commission held a closed session on February 10. The Commission questioned personnel from Marshall Space Flight Center, the NASA center responsible for the shuttle's entire propulsion system, which included the Orbiter's Main Engine, the External Tank, and the SRB, and Morton Thiokol, the contractor responsible for the SRM (see appendix B, figs. B2.1–B2.5). It began to uncover the true extent of the SRB joint problems, which had not been disclosed by NASA administrators.[12] A dramatic turning point in the inquiry was the Commission's discovery of another critical bit of information that NASA top officials had failed to mention: during the teleconference, Morton Thiokol had initially recommended against the launch. The Commission immediately shifted its inquiry to the O-rings as the probable cause, probing extensively into the history of the O-ring problem and the midnight-hour teleconference. Two additional closed sessions were held, which included NASA and contractor participants in the teleconference on the eve of the launch. Subsequently, the Commission issued a statement that the decision-making process may have been flawed, requesting that no person involved in that process participate in the NASA internal investigation.[13] The Commission's stance toward NASA thus shifted from cooperative oversight to confrontation.

In June 1986, the Commission's report to the President identified the O-rings as the technical cause of the tragedy. Analysis of flight data, films, and the debris of the *Challenger* retrieved from the ocean floor indicated that the O-rings in the aft field joint of the right SRB did not do their job. Designed to seal the gap created by pressure at ignition, the O-rings' resiliency was impaired by the unprecedented cold temperature on the morning of the launch. Moreover, the zinc chromate putty designed to insulate the O-rings from hot combustion gases failed. Upon ignition, hot propellant gases penetrated the putty. Instantaneously, the gases moved through the putty-filled 3.25"-long, 0.25"-wide seam between segments of the SRB, impinging on the O-rings. O-rings, grease, and joint insulation began to burn. For a few seconds, the charred matter from burned material clogged the gap between the tang and clevis, preventing a launch pad debacle and allowing *Challenger* to begin its ascent.

At 58.788 seconds after ignition, a small flame emerged from the right SRB in the area of the aft field joint. As the leak grew, the flame developed into a continuous, well-defined plume. The thrust from the plume caused the struts attaching the right SRB to the External Tank to break free. The booster began swiveling, first smashing into the Orbiter's right wing, then hitting the External Tank. Deflected by the aerodynamic slipstream, the flame plume was directed onto the surface of the External Tank, which held the liquid hydrogen and oxygen fueling the ascent to orbit. The flame breached the External Tank at 64.6600 seconds into the flight. Nine seconds later, *Challenger* erupted into a ball of flame. Stop-action films showed the tiny crew compartment, intact and freed from its protective encapsulation in the Orbiter, tumbling through the smoke clouds as the vehicle broke apart.[14] The crew compartment fell to the sea in approximately 2.5 minutes, hitting the water at 200 miles per hour.

The Presidential Commission concluded that the design failure interacted with "the effects of temperature, physical dimensions, the character of the materials, the effects of reusability, processing, and the reaction of the joint to dynamic loading."[15] But there was more. The Commission concluded that the NASA organization contributed to the technical failure. In the report, the Commission documented the long history of problems at NASA with the SRB joints. NASA had had difficulties with the O-rings since 1977, nearly four years before the first shuttle flight in 1981, but the space agency continued to fly the shuttle while searching for solutions.[16] Moreover, the report described the midnight-hour teleconference, confirming that NASA managers had been warned by Thiokol engineers in Utah and at the Cape that O-ring performance was a threat to mission safety in the predicted cold weather. Yet NASA managers proceeded with the launch, apparently not relaying news of the teleconference controversy to upper-level NASA administrators.

The Commission attributed the controversial decision to launch to "a flawed decision-making process."[17] Rather than blaming the individuals responsible for the flawed decisions, the Commission pointed to weaknesses in the decision-making process itself: inadequate procedures for reporting problems and faulty information flows throughout the NASA organization. Behind these "faulty information flows" were Marshall middle-management failures to report O-ring problems up the hierarchy, in apparent violation of reporting rules. Yet communication problems were an inadequate explanation of the launch decision. They did not explain the behavior of Marshall management who had access to all the information about technical difficulties with the O-rings—despite procedural inadequacies and haphazard information flows—and decided, that critical night, to proceed with the launch despite the objections of Thiokol engineers.

The report of a bipartisan investigation by the U.S. House of Representatives Committee on Science and Technology, published in October 1986,[18] agreed with many of the findings of the Presidential Commission. But the House Committee offered a different explanation for the launch decision, placing the blame, not on the decision-making process, but on people making poor technical decisions about the O-rings over a period of years: top NASA and contractor personnel who "failed to act decisively to solve the increasingly serious anomalies in the Solid Rocket Booster joints."[19] The House Committee's finding blamed individuals for the launch (without naming names), suggesting (without stating directly) that perhaps these managers were unqualified for the positions they held—thus indicating the possibility that the decision to launch was the result of managerial incompetence.

But other facts contained in these same reports, the televised hearings, and the published accounts by journalists, historians, and whistle-blowers suggested an alternative interpretation that captured the attention of the public: NASA managers were subject to extraordinary pressures to launch the *Challenger*. All posttragedy accounts acknowledged that pressures on the NASA organization played a major role in the *Challenger* launch decision. Thus, pressure to launch became central to the conventional explanation of the tragedy.

PRESSURE TO LAUNCH: THE MEDIA, THE TEACHER IN SPACE, AND THE STATE OF THE UNION

After the disaster, analysts—in particular, journalists—first looked to circumstances peculiar to the *Challenger* launch for the source of these pressures, which varied from account to account.[20] The Presidential Commission followed up each possibility, never able to fully affirm or dismiss their effect on decision making.

Many posttragedy accounts speculated that media pressure was behind the launch decision.[21] The *Columbia* launch, immediately preceding *Challenger*, was delayed seven times over a 25-day period, setting a record for both false starts and wave-offs for landing.[22] Each delay had been greeted with jeers from the press. The *Challenger* launch had been delayed four times. Dan Rather's script for the evening news on Monday, January 27, ran, "Yet another costly, red-faces-all-around space-shuttle-launch-delay. This time a bad bolt on a hatch and a bad weather bolt from the blue are being blamed."[23] John Quinones, of ABC's "World News Tonight," opened with "Once again a flawless lift-off proved to be too much of a challenge for the *Challenger.*"[24] Seeking to improve its public image, the "media pressure" explanation went, NASA pushed ahead with the launch rather than be subjected to further ridicule. But the notion that a desire for a positive public image impelled Marshall managers to push for a launch under conditions that might lead to the worst of all possible public images—loss of crew and mission—was counterintuitive. Also, since attention from the media accompanied every launch, that NASA would push ahead with the *Challenger* launch to win favor from the media seemed to have little credibility. Indeed, media pressure had become routine at NASA. Why would NASA managers respond in January 1986 by pressing ahead with a launch argued to be risky?

Other accounts suggested that the launch decision was the result of external pressure of another sort: political pressure from the Reagan administration.[25] The *Challenger* mission appeared to have more political ramifications than usual for shuttle flights. The launch's Teacher in Space Project and the subsequent vicarious inclusion of schoolchildren across the country in the flight had created an extraordinary amount of interest. The successful launch of a mission including an "average citizen"—a teacher—was a major statement about the reliability of space travel and the achievement of the administration (and the country) responsible. Moreover, the President's State of the Union message was scheduled for the evening of

January 28, the first night of *Challenger's* orbit, and many sources reported that it was to include a mention of the first teacher in space.[26] A launch timed so that the administration could crow about this "space first" on prime-time television would be politically advantageous at home and abroad, thus (the argument went) motivating the administration to press for *Challenger* to be up in time for the speech.

The Presidential Commission investigated the report that NASA officials were subjected to outside pressures to launch in time for the Teacher in Space Project to be mentioned in the State of the Union message.[27] Specifically, posttragedy speculation from some quarters was that a White House official gave orders for *Challenger* to lift off.[28] In its investigation, the Commission questioned individuals who were involved to a greater or lesser extent in the launch decision. The major figures at NASA testified that there had been no outside intervention or pressure of any kind leading up to the launch.[29] In addition, 28 sworn affidavits were taken by experienced investigators supporting Commission activities in which statements to similar effect were made. The Commission reported finding no evidence that confirmed the allegation that the White House intervened.

Following this line of inquiry nonetheless, the Commission also investigated the report that plans had been made for a live communication hookup with the STS 51-L crew during the State of the Union message. All persons were interviewed who would have been involved in a hookup had one been planned. Although three live telecasts of crew activity during the flight were scheduled, the first was not to begin until the fourth day of flight. Apparently, none were to occur in conjunction with the State of the Union message.[30] The Commission concluded that "the decision to launch the *Challenger* was made solely by the appropriate NASA officials without any outside intervention or pressure."[31] This conclusion was vigorously affirmed by White House staff. But it was as vigorously challenged by Senator Ernest F. Hollings (D-S.C.) and Representative James R. Traficant, Jr. (D-Ohio).[32] Commissioner Richard P. Feynman learned that a hookup did not need to be scheduled in advance: the telemetry people could connect the shuttle to the White House, Capitol Hill, or any place, for that matter, with no more advance planning than a three-minute warning.[33]

No evidence surfaced either fully confirming or dismissing the possibility of pressure exerted by administration officials to launch in time for the State of the Union address, but it seemed unlikely, considering the following. Although communication hookups between orbiting flight crews and presidents had occurred in the past, neither President Reagan nor Vice President Bush had spoken live to a space crew since STS 51-A, launched November 8, 1984. According to White House Press Secretary Larry Speakes in a press briefing held February 26, 1986, the President had not called the subsequent 10 shuttle crews because the flights had become such a regular occurrence.[34] This hiatus would not necessarily preclude the possibility of a space communication during the *Challenger* flight, but a perhaps more significant factor was a procedural inhibition: NASA does not schedule communication hookups for at least 48 hours after a launch in order to give the crew time to settle into orbit and become oriented.[35] If the *Challenger* had been launched January 22 as scheduled, or even Sunday, January 26, a hookup timed to coincide with the speech would have been possible, for the requisite 48-hour grace period would have elapsed. But by the time of the teleconference on the evening of Monday, January 27, the 48-hour rule would have placed a hookup with the orbiting crew after the speech.

The Commission found another reason to be skeptical about the theory that the administration had a sufficiently keen interest in the timing of the *Challenger* launch to intervene in NASA internal affairs. Vice President Bush was to have attended the launch had it occurred on Sunday as planned.[36] The presence of the Vice President, traditionally the administration's representative for space policy, would supply a great photo opportunity to an administration known for its skill at taking advantage of such situations. But neither Bush nor any other upper-level representative of the administration was present for Tuesday's launch. If this launch was important enough that special pressure was exerted to see that it got off Tuesday, in time for the speech, as alleged, surely the administration would have sent a representative to the Cape to grace newspaper and television coverage of the event across America. In fact, a successful mission would notify the world that America had mastered space to such an extent that average citizens could travel, thus accomplishing the administration's political goals—regardless of the mission's inclusion or exclusion in the President's speech, regardless of the presence or absence of an administration luminary at the Cape.

Other accounts speculated that it was NASA, not the White House, that sought to launch in time for a favorable mention in the State of the Union message.[37] The staff of Shirley Green, then NASA's Director of Public Affairs and liaison with the White House Press Office, had drafted a statement for potential inclusion in the State of the Union message in early January. Concerning this draft,

Green stated, "We had suggested that the President restate his commitment to the completion of the Space Station by 1994; mention the initiation of the Aerospace Plane Research program; acknowledge the importance of space science; and comment on the flight that was expected to be in orbit at the time he was speaking."[38] The draft emphasized Christa McAuliffe's role as the first citizen-passenger on the Space Shuttle. When Green arrived at the Cape prior to the launch, she called the White House Press Office, asking about NASA's inclusion in the President's speech: "Did NASA make the State of the Union? And I was told, 'Yes.' And I said, 'Good. Did we also get the recommendation on the Space Station?' He said, 'Yes, you did, also Space Science and the Aerospace Plane.' And I said, 'You know, that's terrific. Did they tie it to the Teacher in Space being up?' And he said, 'No, there's no mention of the Teacher.' And I said, 'I'm sorry to hear that because it's going to be a big media event, I think.'"[39]

The draft, her phone call, and the above statement indicate Green's interest in the administration's acknowledgment and program approval in the speech. As Public Affairs Director, this was her official responsibility. No evidence surfaced indicating whether Green or other NASA officials subsequently took action to assure the Teacher in Space Project was mentioned. However, the speech still might have been amended by the independent initiative of a speechwriter or President Reagan himself if the *Challenger* were launched Tuesday, January 28, as planned. So the inclusion of the Teacher in Space Project was still a possibility. But even discounting the possibility of intervention by the White House, a communication hookup, and/or a mention in the State of the Union address, NASA management's mere awareness that the speech was going to happen might have increased determination to launch on Tuesday.

Would a launch delay that missed the speech have affected NASA negatively? Even without the Teacher in Space mention, NASA's desired programmatic areas were to be included and endorsed in the speech. NASA's public relations representative was informed of this before the launch. While the Commission discovered no evidence concerning the pattern and extent of communication of this news at NASA, surely if managers *were* deeply concerned about a NASA mention in the State of the Union message, word of NASA's inclusion and program endorsement would have reached the relevant administrators. Adding to the puzzle, Christa McAuliffe's broadcasts of lessons from space were likely to capture not only the attention of schoolchildren everywhere but the imagination of the nation. Even if the launch were delayed, this general attention and acclaim would reinforce NASA's image and hoped-for executive approval, more than compensating for any lack of mention in the address.

Some accounts identified scheduling conditions necessary for the Teacher in Space Project as a possible source of pressure to launch. According to this explanation of the launch decision, NASA had to launch *Challenger* on January 28 because if they did not, Christa McAuliffe's lessons would have to be canceled. The mission was established with a return on day 7. When the launch date changed from January 22 to January 26, the live space lessons were changed from day 6 to day 4 (a Friday) to keep them on a weekday, when children were in school.[40] The launch had to go on January 28 (went posttragedy speculation) because a delay would eliminate the lessons; the next school day (a Monday) was the day of scheduled mission return. However, the Commission learned that alternative dates for deorbit and landing at Kennedy were always planned for one and two days later in case of weather or other problems that prevented deorbit on the date designated for end of mission.[41] Further, the Commission learned that the return of orbiting flights was often delayed. If the *Challenger* launch had been deferred, reentry could have been delayed until Tuesday so that the lessons could be taught on Monday.

Posttragedy accounts that initially focused the search for an explanation on pressures surrounding the circumstances of this particular launch—the media, the Teacher in Space Project, and the State of the Union message—eventually expanded the focus to pressures on the Space Shuttle Program as a whole. These pressures were lodged in the institutional history of the NASA organization and agency economic struggles deeply embedded in the structure of the national and international political scene. The "media pressure" explanation, "Teacher in Space" explanation, and "State of the Union" explanation all pointed to favorable publicity as a NASA goal. Image-related pressures, it turned out, were the tip of the iceberg. For NASA, the desire for a positive public image was simply an outward manifestation of a continuous struggle for funding to support the Shuttle Program. Every account of the disaster verified that NASA's Shuttle Program was beset by economic problems throughout its history and that these difficulties had influenced decisions at NASA since the Shuttle Program's inception. R. Jeffrey Smith, writing in *Science* in March 1986, reported "widespread concern that over a long period, budget constraints and schedule pressures have been allowed to exert undue influence on the process of decision making."[42]

The evidence that economic constraints governed decision making at NASA throughout the Shuttle Program was extensive. Moreover, many sources acknowledged that this strain translated into "production pressure": adhering to the launch schedule had become the means to secure the funds essential to the program.[43] At the time of the *Challenger* mission, NASA was competing with rival agencies for the attention of the President in a year when massive budget cuts were predicted.[44] Administration program endorsement would enhance the space agency's chances for budget appropriations from Congress. Commissioner Feynman, concluding that there was no pressure from the White House to launch the *Challenger*, asserted "that the people in a big system like NASA know what has to be done—without being told. There was already a big pressure to keep the shuttle flying. NASA had a flight schedule they were trying to meet, just to show the capabilities of NASA—never mind whether the President was going to give a speech that night or not."[45]

The institutional history of how political power struggles, space science, and production pressures became intertwined at NASA is fascinating in its own right.[46] But it became critically important to the historically accepted explanation of the launch decision. It suggested a causal connection between the space agency's environment of competition and scarce resources and NASA managers' actions on the eve of the launch. The history that follows, which I reconstructed from posttragedy accounts and their major sources, shows how NASA's political environment forced the space agency to compete for scarce resources. It demonstrates production pressures and a history of decision making characterized by cost/safety trade-offs that endangered shuttle safety from the beginning of the program. This history, revealed to the public after the disaster in a near-deluge of published accounts, was taken by many as irrefutable evidence that budgetary considerations and the ever-present production pressures weighed heavily on NASA managers, influencing their decision to launch.

THE STRUCTURAL ORIGINS OF DISASTER: COMPETITION, SCARCE RESOURCES, AND THE SPACE SHUTTLE PROGRAM

From the space agency's inception in 1958, NASA's military and space science goals played a role in U.S. competition for international supremacy. With the launch of the Soviet *Sputnik* in 1957, space became the "new frontier": the United States and the Soviet Union competed for technological supremacy through the exploration and military mastery of space. Not only did the individual achievements in these areas matter in their own right, but they also conveyed an image of power and leadership with the potential to secure and maintain allies and influence international negotiations. Although NASA certainly was not the only means to political influence for a given administration, the government's dependence on NASA for both military and scientific advances in space lasted until the mid-1980s. So central was the space program to international competition, first with the Soviet Union, then with Europe, India, Japan, and China as well, that to narrow the question of "pressure to launch" by the administration to a desire to crow in the State of the Union message is to minimize a concern about the accomplishments of the space program that had been fundamental to every administration since NASA's inception.

Although administration concern about how the space agency's accomplishments affected the national interest was continuous, historically its willingness to fund the agency waxed and waned in rhythm to national and international events and political swings. President John F. Kennedy's mandate to land a man on the moon by the end of the 1960s was a response to a national crisis evolving from the political consequences of Laos, the Congo, the Bay of Pigs, and the Soviet launch of *Sputnik*. Widespread American concern about these events created an environment of institutional consensus for the Apollo Program.[47] Space technology was to become the vehicle for national and international prestige.[48] The result was that Apollo received abundant resources. With the triple luxuries of adequate funding, institutional consensus, and a clear goal, NASA created an innovative system and rapidly progressed toward accomplishing a mission right out of science fiction: in 1969, astronauts planted an American flag on the bleak terrain of the moon.

In the mid-1960s, however, international and domestic factors caused uncertainty about the future direction of the space program and a consequent decline in congressional appropriations for NASA.[49] Vietnam and economic issues displaced the space program as a national priority.[50] The Shuttle Program was born in the midst of this decline, funding struggles its birthright.[51] NASA, which had for so long enjoyed budgetary certainty grounded in consensus about its mission, suddenly experienced the uncertainty of other agencies. During this period, the space agency was developing its post-Apollo plans. After the 1969 Apollo moon landing, the NASA organization needed a new, exciting, and nationally important mission in order to justify its continued existence and centrality in U.S. affairs. In 1969, the space agency proposed the permanent location of people in space as its new long-range goal. Plans for the manned space program included three projects: a mission to Mars, a space station in earth orbit, and a space shuttle to transport people and materials to the space station and to move objects around in space.[52] These plans were ideal, for they would promote national interests and at the same time secure NASA's position as a powerful agency.

But their actualization depended on the national purse. Nixon was opposed to replicating the costs of the Apollo Program because concern about the costs of the Vietnam War was widespread.[53] Funding cuts were reducing NASA personnel by 1,000 employees each year, from an all-time high of nearly 34,000 in 1969.[54] Cost objections caused NASA to back off from the proposed mission to Mars and the space station and to settle for the Space Transportation System: a fleet of reusable spacecraft designed to reduce the cost of putting objects in orbit. The shuttle was intended to be the only launch vehicle for scientific satellites and space probes.[55] Although plans for the space station had been deferred, the shuttle was conceived as the eventual service vehicle for the station, and thus fundamental to the manned space program and all its future missions.[56] NASA's proposed fleet of space "buses" would be the first objective in a long-range, multifaceted plan. The Space Transportation System was an idea whose time had come: not only was it visionary, but the shuttle's reusability promised cost effectiveness.

SCARCE RESOURCES AND COMPROMISED EXCELLENCE

But NASA's environment of institutional consensus had changed to one requiring political allegiances to secure funding.[57] Although originally conceived as an independent civilian space program dedicated to space science, NASA's early analysis indicated that a link with military interests would justify the shuttle economically to Congress and the President.[58] The shuttle could be used to launch all military and intelligence payloads (aptly named, for NASA would be paid for every "load" it carried) during the 1980s, estimated to represent over 30 percent of future space traffic.[59] This seemed a promising arrangement: the Air Force wanted the shuttle to launch space missions essential to national security—especially reconnaissance satellites.[60] They made a deal: NASA would design the shuttle to meet military requirements; in return, the Air Force agreed not to develop any new launch vehicles of its own and endorsed the Shuttle Program to the White House and Congress. Hence, from the beginning, the program was inextricably linked to the Department of Defense.[61] Military goals were added to NASA's original goal of excellence in space science.

Budgetary constraints also plagued shuttle development. NASA's original plans called for a fully reusable shuttle with estimated development costs of $10 billion and an annual budget of approximately $2 billion.[62] Surprisingly, the principal cost at the outset was predicted to be, not the necessary hardware, but the people required to do the job.[63] Putting the 31,000 heat tiles on the *Columbia* to protect it during reentry, for example, would take 670,000 man-hours.[64] Salaries at Kennedy Space Center, at Mission Control in the Johnson Space Center, at tracking and telemetry stations around the world, and at all the other NASA facilities were estimated at the outset at nearly $1 billion per year. Nonetheless, the Office of Management and Budget (OMB), in an effort to moderate NASA's planning for what critics called "a Cadillac of space shuttles," severely limited NASA's annual budget, with no possibility of budget increases during the succeeding five years.[65]

But the Air Force required a large, high-performance shuttle and insisted on certain costly features for which it refused to pay. Since Air Force endorsement was absolutely essential to congressional appropriation of funds, in 1971 NASA called in a think tank, Mathematica, Inc., that would analyze these difficulties. The Mathematica finding justified the Shuttle Program on a cost-effectiveness basis. Using data supplied by prospective shuttle contractors, Mathematica reported that the payload capacity would allow the vehicle to pay for itself provided it had a launch rate of more than 30 flights per year, which was considered a conservative estimate in 1971.[66] This finding strengthened NASA's case for development of the shuttle, but OMB's funding restrictions meant NASA could not develop the shuttle it had been planning. The allocation was half of what NASA had requested.[67]

Design Trade-Offs: From Cadillac to Camel

Historically, economic conditions, social relations, and politics have shaped technological products, undermining their quality.[68] What happened next resulted in "one of the most imaginative innovations in the history of political compromise."[69] The power struggles between NASA, OMB, Congress, and the administration directly affected shuttle design. Compromise was necessary to get the program going. Alternative designs were sought that would still meet military needs but also fit NASA's drastically cut budget.[70] The entire design process was characterized by trade-offs between development costs and future operating costs.[71] Development costs had to be low in order to secure the funding to get the program started, but in many cases, an economical design decision would result in higher operating costs. Because the shuttle also was expected to earn money by bringing in payloads from commercial satellite customers, shifting the economic burden to the operational phase seemed to make good sense. This logic was behind design decisions about the SRBs. John Logsdon, Space Policy Institute, wrote: "Most ideas for lowering development costs involved substituting some form of expendable booster for the manned first-stage booster. This did not appear technically possible, however, given the large size of the Orbiter. A June 1971 design breakthrough solved this problem."[72]

The breakthrough was a smaller Orbiter that could be launched with boosters that expended their fuel then dropped into the sea. NASA had little experience with expendable SRBs. Nonetheless, the decision was for expendable SRBs that would be attached to the Orbiter's external tank. These SRBs cut costs by eliminating the need for the crew and additional fuel necessary to return a manned first-stage booster to the launch site. And they could be retrieved from the sea and reused. Reusability was an important budgetary plus. Both Liquid and Solid Rocket Boosters could be recovered and reused,[73] but Liquid Rocket Boosters had high development costs and lower operating costs. When announcing the decision, then-NASA Administrator James F. Fletcher stated candidly that the SRB choice turned on a "trade-off between future benefits and earlier savings in the immediate years ahead: liquid boosters have lower potential operating costs, while solid boosters have lower development costs."[74] But production and schedule concerns also played a role in this design decision. Liquid Rocket Boosters dropped by parachute into the sea after launch required more refurbishing before reuse than did SRBs jettisoned in the same manner.[75] SRBs could be ready to fly again in two weeks, fitting with the rapid turnaround time necessary to maintain the launch schedule Mathematica predicted.

NASA's restricted budget entailed further design trade-offs, some apparently swapping safety for cost benefits. Weight had been a consideration since the earliest planning stages. The payload requirements necessary to break even added to the vehicle's weight. When NASA began incorporating military requirements into the design, the weight of the vehicle increased still more.[76] The Orbiter weight would put a strain on the shuttle's propulsion system,[77] so design decisions were made to lighten the vehicle when possible. Ominous in retrospect, two escape rockets on the Orbiter were scrapped to reduce weight, leaving the astronauts locked onto the launch vehicle during liftoff.[78] Malcolm McConnell, author of *Challenger: A Major Malfunction*, wrote: "All previous manned American spacecraft had carried small, solid-fuel escape rockets that would extract the crew capsule from a failed booster during the critically dangerous early moments of a mission. But once more, weight restrictions forced NASA designers to abandon the prudence they had shown in the moon landing program."[79]

The final design was far from NASA's original conceptualization. Although NASA was given the responsibility for meeting the nation's goals for space exploration, the ability to accomplish them was constrained by other organizations in NASA's environment, each seeking to fulfill its own political mandate. Alex Roland, space historian, observed: "Congress, OMB, the Air Force, and NASA had all pulled in different directions: Congress toward cost recovery, OMB toward low development costs, the Air Force toward operational capabilities, and NASA toward a future of manned spaceflight. Instead of a horse, NASA got a camel—better than no transportation at all and indeed well suited for certain jobs, but hardly the steed it would have chosen".[80]

A Bitter Bargain

The result was ironic. NASA's post-Apollo survival required the space agency to negotiate for congressional approval of a shuttle that was not the vehicle it desired. Nonetheless, NASA top officials argued hard for the compromised shuttle design, extolling its scientific merits, technological utility, and cost effectiveness. They claimed it would supplant all existing expendable launch vehicles. They promised that at least four flights annually would carry spacecraft designed to explore other planets.[81] Most important in response to tight congressional purse strings, they argued that the shuttle would be self-supporting. NASA officials stated that by the early 1980s, 60 flights per year would be launched, each flight conveying 65,000 pounds of payload into low earth orbit at a cost of $100 per pound.[82] They argued that this price was low enough to attract commercial customers with payloads to be launched into space. The flight rate and the pounds of payload for each flight would bring in enough income to compensate for the high operating costs.

In January 1972, President Nixon authorized the development of the shuttle. The future of the agency and its space centers was secured. NASA divided managerial responsibility for the program as follows: Johnson Space Center in Houston was assigned management of the Orbiter; Marshall was made responsible for the propulsion system, consisting of the Orbiter's Main Engine, the External Tank, and the SRBs; Kennedy was given the job of assembling the Space Shuttle components, checking them out, and conducting launches.[83] NASA got its project, but the victory was secured at the potential expense of future users. Program survival would depend on the number of launches each year and the income from commercial cargo. The early design decisions to keep development costs low in order to gain program approval would automatically increase the future operating costs of the shuttle. Hence, the commercial goal became absolutely essential to all other operations.[84] The transportation of people and materials into orbit and back again to achieve military and scientific goals would depend on the shuttle's success as a business.

The production pressure that was thus built into the program increased as design implementation was met with deepening economic concerns. NASA officials believed OMB was committed to a level budget during the shuttle's development phase, but NASA did not get all the funds it requested.[85] The desired fleet of five Orbiters had to be reduced to four. This reduction was a major threat to NASA's ability to meet the flight rate projected by Mathematica. As four shuttles were now to do the work of five, any technical problems causing extensive delays in refurbishing a craft for its next flight could reduce the number of flights possible in a given year. Moreover, NASA was already experiencing schedule slippage. Initial orbital test flights would be delayed more than two years.[86]

In another move to economize, NASA cut spending on safety testing and other development work for the shuttle components.[87] This strategy of limited testing diverged from that of the Apollo Program, in which extensive and redundant testing programs were carried out early in equipment development. Criticized by government auditors, NASA officials responded that the space agency's greater experience with spaceflight meant that fewer tests were needed than during the Apollo Program, when American space exploration was in its infancy. Yet, in February 1979, a National Academy of Engineering panel headed by Eugene Covert of the Aeronautics and Astronautics Department of MIT (and later member of the Presidential Commission) was created to investigate problems with the shuttle's Main Engine. The Covert report concluded that key components may not have been tested sufficiently and that certification of components for flight required more time than NASA was allocating. In response, NASA altered Main Engine testing procedures.[88]

OPERATIONAL SPACEFLIGHT: THE SHUTTLE AS BUSINESS

NASA's bitter bargain required that the space agency live up to its promise to be self-supporting, so top administrators concentrated on getting the economic resources essential for agency survival. From the beginning, the shuttle was depicted as the space vehicle that would make space transportation "routine and economical."[89] The key to attracting customers was to match this image. NASA's efforts to do so took two directions.

First, NASA committed itself to moving the shuttle from its "developmental" stage into an "operational" stage as soon as possible.[90] An operational system is one that is mature and ready for routine use.[91] The written guidelines for achieving operational status were established at the program's beginning.[92] The first four flights (STS-1 to STS-4) would be test flights. If the test series went well, the Space Transportation System would be considered operational. The fourth orbital test flight was completed with a flawless countdown with no delays and a dramatic on-time landing at Edwards Air Force Base on July 4, 1982. A red, white, and blue, Fourth of July, pull-out-all-the-stops triumph was announced. NASA officials declared the Space Shuttle operational.[93] President Reagan addressed the American people, proclaiming that the shuttle was the "primary space launch system for both national security and civil government missions," stressing its future cost effectiveness in "providing routine access to space."[94] Together, NASA's announcement and the President's speech created an image that the shuttle was reliable, safe, and ready for routine use.

Second, NASA emphasized a flight schedule that would attract both commercial payloads and continued congressional approval.[95] But the shuttle never paid its own way. In 1976, NASA forecast 49 flights in fiscal year 1984 and 58 in 1985. But by the early 1980s, NASA's projection of 60 flights per year, each flight carrying 65,000 pounds of payload at $100 per pound, seemed pure fantasy.[96] Science experts charged that NASA officials had made "wildly overoptimistic estimates of cost effectiveness," grossly overestimating space traffic by the shuttle and grossly underestimating developmental costs.[97] The 1971 Mathematica study justifying the shuttle on a cost-effectiveness basis forced NASA to maintain before Congress and the general public (not only in the early 1970s but throughout the Shuttle Program) that the shuttle was a good investment on economic grounds.[98] But Mathematica used data furnished by contractors who hoped to receive shuttle contracts, so the estimates they gave for payload and launch vehicle costs were "optimistic."[99] In addition, the overly optimistic estimates were shaped by the very real difficulty of estimating costs and income from an innovative, untried technology when the projections must be made years in advance.

To add to NASA's problems, Mathematica's argument that the shuttle could be self-supporting, which had gained approval for the program, was based on flawed analysis. Increasing the flight schedule reduced the cost per flight for customers and permitted more commercial payloads, as believed. But an increased flight rate also increased the overall cost of the program.[100] The Shuttle Program had a large fixed cost of $1.2 billion per year, which covered the workforce and the costs of maintaining launch facilities. But each flight added approximately $60 million in marginal costs. As the number of flights increased, the fixed cost was distributed over more flights, but the total cost of the program increased as more flights were launched each year.

Not only were cost estimates off in the beginning, but they rose in unpredictable ways. Inflation had its effect. The true launching cost turned out to be 20 times greater than the original estimate.[101] Moreover, the Main Engines never achieved the 109 percent of thrust NASA predicted.[102] This "thrust shortfall" restricted the payload capacity to 47,000 pounds instead of the projected 65,000 pounds on which the Mathematica calculations were based, so each launch would bring in less income than originally predicted.[103] Worse, the launch turnaround time necessary to break even was never achieved. NASA originally anticipated two weeks between launches by 1985. Instead, turnaround time limited NASA to nine missions in 1985— a far cry from the 60 flights predicted for that year at the shuttle's inception.[104] Add to these difficulties the development of new commercial satellite services in other countries. Arianespace, the commercial satellite launch consortium organized by the European Space Agency, began successfully competing with NASA for space business, offering low rates and winning commercial satellite customers.[105] NASA had to keep payload cost per pound down in order to compete.[106] If the price went up, NASA ultimately lost business, increasing the cost of each flight.

To make matters worse, NASA also faced possible loss of major customers in the United States. The National Oceanic and Atmospheric Administration began to pursue independent launch capability.[107] NASA's early ally, the Air Force, threatened to withdraw. In 1984, the Air Force requested money from Congress to develop expendable launch vehicles in order that it might have its own capability to launch reconnaissance satellites "on short notice, with strong assurance of success."[108] This action was based on Air Force dissatisfaction with NASA's floundering operational capabilities. The intelligence community's dependence on the shuttle for reconnaissance had increased with the 1979 SALT II treaty requiring verification of Soviet compliance.[109] Later, the shuttle became a critical component of Strategic Defense Initiative (SDI) research.[110] But many missions were late; five were scrubbed altogether. Some satellites launched from the shuttle were either lost or placed in the wrong orbit.[111]

In an effort to maintain the positive image that would continue to secure resources necessary for the Shuttle Program, in 1984 NASA committed to spectacular retrieval missions to accommodate existing paying customers and demonstrate to potential customers the service the astronauts and shuttle could perform.[112] The agency carried out an extraordinary mission to repair a communications satellite in space in one instance and, in another, captured two failed satellites, escorted them into the payload bay, and returned them to earth.[113] Both spectacular missions were accomplished with much excitement and acclaim, but with a great drain on resources necessary to the fundamental aspects of the program.[114] In another attempt at "impression management," NASA offered "prestigious payload specialist assignments to the agency's best—and most cooperative—customers."[115] Once the shuttle had been declared operational, an assortment of nonastronauts became part of the flight crew: a Saudi prince, a Mexican engineer, aerospace contractor engineers, a French pilot, and a Canadian Navy officer.[116]

The idea of nonastronaut passengers was first discussed at NASA Headquarters in 1976. It was consistent with the original program goals of "permanent occupancy of space with manned systems" and an "ever-broadening base of participation by individuals from all walks of life."[117] It was also consistent with NASA's need to win attention and resources, even at that early date. A study team drafted a preliminary report titled "Unique Personality for Space Shuttle Flights," suggesting a VIP candidate be drawn from

> categories which might yield either a unique personality in his own right or by virtue of his occupation or organizational affiliation: News Media, Popular Science, Eminent Scientist, Politician, Prominent Statesman, "Layman," Space Science Student, Promi-

nent Transportation Figure, Artist (Poet, Painter, Writer), Human-
itarian, Entertainer. The general ground rules were that the individual and/or his activity in connection with flight should produce
something of high public interest as well as contribute to the
widest practicable dissemination of information concerning spaceflight activity and results thereof. The latter consideration is consistent with Section 203(a)(3) of the Space Act and would provide
the legal basis for such activity in the event a non–space professional were chosen.[118]

The consensus was for the general area of popular science, with a
writer-photographer from the National Geographic Society the
team's first choice. Second was a news media personality. Flying a
member of Congress was considered, but the study team concluded
that the choice was "fraught with undesirable difficulties involving
partisan politics."[119] NASA Administrator Fletcher agreed that the
idea of taking civilians into space deserved further study. Subsequently, NASA circulated the possibility among leading figures in
the arts, humanities, and politics and representatives of the media
and space community. It received widespread endorsement. No further action was taken, however. In 1982, the year the shuttle was
declared operational, then–NASA Administrator James Beggs asked
the NASA Advisory Council to determine how the agency might
select civilians to fly.[120] A task force was formed. Then the 1984 election campaign intervened. Following a National Education Association endorsement of Walter Mondale and a Gallup poll indicating
President Reagan's vulnerability on the issue of education, August 27
was declared Education Day in the Reagan reelection campaign.[121]
President Reagan announced that a teacher would be chosen as the
first private citizen to fly on the Space Shuttle. "When that shuttle
lifts off," stated the President, "all of America will be reminded of
the crucial role that teachers and education play in the life of our
nation."[122] Eleven thousand teachers applied.[123]
    Meanwhile, contrary to the recommendation of the 1976 study
team, the first "unique personalities" to fly as "payload specialists"
were Senator Jake Garn (R-Utah), who flew on STS 51-D in April
1985, and Congressman William Nelson (D-Fla.), then-Chair of the
House Space Science and Applications Subcommittee, which
approved the NASA budget, who flew on STS 61-C, launched two
weeks before *Challenger*. Both these elected officials were late additions to their respective flights, causing manpower repercussions for
other flights.[124] One effect was of little note at the time but of great
consequence: late scheduling of Congressman Nelson to STS 61-C

resulted in the reassignment of Payload Specialist Greg Jarvis and his experiments, originally scheduled for that flight, to the *Challenger* launch.[125]

THE WIDENING GAP BETWEEN GOALS AND MEANS

In 1985, having completed only nine missions that year, NASA published a projection calling for the annual rate to increase to 24 flights per year by 1990.[126] But this goal was sought without a change absolutely essential to meeting it: effective conversion to an operational mode. Though the shuttle had been declared operational in 1982, the mature system envisioned never emerged. This failure occurred, first, because two ingredients essential for the conversion—expertise on the part of individuals designing the changes and the time for them to do the job—were diverted from the long-range planning to the more immediate problems of the flight schedule and, second, because the budget that controlled the development of needed facilities and equipment did not support the level of capability required for a mature system.[127]

NASA was required to make system changes in order to support a reusable fleet, and it was slow to develop these capabilities.[128] When the shuttle was declared operational, the emphasis shifted from applying resources to a single flight, which was the case during its developmental stages, to applying them to several flights concurrently. Human and material resources devoted to any single flight were diluted. The attempt to reach 24 missions per year was limited by lack of spare parts and resulted in compression of training programs. Resources became concentrated around short-term, not long-term, problems. The production of flight software, flight trajectory information, and crew training materials was difficult with the accelerated program. Inadequate replacements existed for parts that were damaged in flight, resulting in the "cannibalization of spare parts" from one vehicle to another in order to launch.[129] Not only did these parts transfers take extra personnel time, but they were wildly disruptive and created opportunities for component damage.

Resources were further strained by changes in flight manifests (the schedule of planned operations and elements for a given flight).[130] For example, commercial customers sometimes changed the scheduled launch dates for their payloads because of problems with development, finances, or market conditions.[131] With over 50 percent of the work for launch preparation occurring in the last three months before launch, and with typically 20 or more flights in the works at a given time, last-minute changes were disruptive and costly.[132] When a change occurred, the effects on budget, cost, and man-power were significant, for one change precipitated others. Shifting a payload from one shuttle flight to another often meant bumping another scheduled payload to make room. In addition to finding a space for the payload that was bumped, payload specialist and crew member assignments had to be altered. The changes resulting from one such customer request affected several flights. When flight manifest changes occurred late in the planning and preparation cycle, the effect was absorbed by all the subsequent steps in the prelaunch process. The burden fell on the last elements in the chain: training of the flight crew and the ground launch crew.[133]

In a move that under the circumstances strains credulity, in July 1985 the House cut 5 percent ($375 million) from NASA's fiscal year 1986 budget, freezing the agency's appropriations at the 1985 level of $7.51 billion.[134] From the all-time funding high during the Apollo Program in the early 1960s, appropriations had dropped in the mid-1960s by "a factor of three in constant dollars" and had remained at the same level into the 1980s.[135] The number of NASA employees had dropped from Apollo's high of 34,000, bottoming out around 22,000. Marshall, with responsibility for three elements of the shuttle system, had gone from nearly 7,000 employees in 1968 to 3,500 in 1985.[136] Moreover, the sustained scarcity had a deepening effect on the quality of the program. It was the July 1985 House budget cut that created the spare parts crisis, for example. Inventory of spare parts had been consistent with plans until the second quarter of fiscal year 1985.[137] At that time, NASA managers altered plans to meet budget restrictions, shaving the spare parts budget from $285.3 million to $83.3 million in October 1985. As a result, the inventory necessary to meet the 1986 flight schedule was jeopardized.[138] In another July 1985 action that had dramatic consequences for 1986, Vice President Bush announced at a White House ceremony that Christa McAuliffe and Barbara Morgan had been selected as primary and backup candidates for NASA's Teacher in Space Project.[139]

JANUARY 1986: PRODUCTION PRESSURES AND A DISASTROUS DECISION

This institutional history, assembled and made public in the aftermath of the *Challenger* tragedy by numerous analysts, revealed the dark underbelly of NASA's glittering record of public accomplishments. Evidence of production pressure, cost/safety trade-offs in decision making, and the effects of economic strain on the operation of the Space Shuttle Program served not only to indict the past but to fuel public demands for an explanation of the *Challenger* launch decision. Threatened by competition and scarcity, NASA was a system under stress at the time of the *Challenger* launch. The difficulties began not in 1986, but in 1972 when NASA received political endorsement of the Shuttle Program and its mission without the political commitment necessary to provide resources adequate to meet program goals. Instead of the exciting challenge of responding to the Soviet launch of *Sputnik* that justified the Apollo missions, cost effectiveness became the public justification for the Space Shuttle. NASA sold the shuttle on the basis of economics; thus economics became the criterion by which shuttle performance was to be judged in budget requests year after year. The original, inflated Mathematica estimates created performance expectations that became more and more difficult to meet. Throughout the shuttle's history, the NASA organization experienced environmentally produced production pressures with a limited capacity to meet those demands.[140] In response to these pressures, NASA raised performance goals— apparently without assessing available resources and capabilities— and these performance goals were not reduced to accommodate subsequent workforce reductions.[141] Consequently, externally generated pressures on the organization were met with internally generated ones, increasing system stress.

Posttragedy accounts indicated that production pressures were at an all-time high on the eve of the *Challenger* launch. All 1986 missions would greatly influence the future well-being of the space program. Fifteen flights were scheduled—six more than in 1985.[142] In January, activity at NASA was frenetic: workers and managers were working on three missions simultaneously.[143] All the strains on the NASA system fell on a reduced workforce. Retirements, hiring freezes, and transfers to other programs like the space station affected the organization's ability to meet these increased demands. On January 1, 1986, when a new contract consolidated the entire contractor workforce under a single company, a number of workers in important areas chose not to change contractors. Not only was the workforce reduced, but training programs were required for new personnel, further draining the system as it geared up to meet the projected 1986 flight rate.[144] The Teacher in Space flight was the first scheduled flight of the year. Seven launch delays for *Columbia* (originally scheduled for December 18, 1985), however, gave the honor (albeit dubiously) to that flight, launched January 12. Originally scheduled for January 22, similar extensive delays for *Challenger* would have affected the schedule of every subsequent launch.

NASA's 1986 launch schedule included three important scientific missions. The first of these, *Columbia*'s Astro-1 flight in March, was America's scheduled examination of Halley's comet.[145] Two missions in May would be major probes of the solar system involving "cryogenic Centaur upper-stage boosters and a plutonium nuclear-electric generator in the payload bay for the first time."[146] The March Astro mission was of vital interest to NASA officials. The scheduled launch date was March 6. The Soviet Union's Vega 2 robot probe was to send computer images back to earth beginning March 9. If Astro's launch date could be met and live films sent around the world by American astronauts, NASA's image and the country's international space reputation would be boosted. Launch delays prior to this mission might jeopardize the exquisitely close timing necessary to scoop the Soviets.

To fail in the Astro mission would be public confirmation of charges by American space scientists that the shuttle was an expensive attention getter that was draining resources from space science experiments. The astronomer James Van Allen, a leading space science advocate, published an article in *Scientific American* in January 1986 (on newsstands several weeks before the *Challenger* launch) criticizing NASA's emphasis on the development of manned space-flight and military objectives and lamenting the scientific achievements delayed or lost because of NASA's direction.[147] To successfully pull off the Halley's comet launch would convincingly argue for the interdependence of manned missions and space science, silencing critics and (perhaps) converting some of them, building a coalition of interests that would bring the space program greater resources.

It was, posttragedy analysts agreed, a disaster waiting to happen. The extensive evidence concerning economic scarcity, competition, and production pressure in the Space Shuttle Program were fundamental to what became the historically accepted explanation of the decision to launch the *Challenger:* production pressure and managerial wrongdoing. The Presidential Commission stated that these production pressures took priority over safety: "Pressures developed because of the need to meet customer commitments, which translated into a requirement to launch a certain number of flights per year and to launch them on time. Such considerations may occasionally have obscured engineering concerns."[148] The House Committee, in a statement that more specifically connected the production pressures to the *Challenger* launch decision, concluded "that NASA's drive to achieve a launch schedule of 24 flights per year created pressure throughout the agency that directly contributed to unsafe launch operations. The Committee believes that the pressure to push for an unrealistic number of flights continues to exist in some sectors of NASA and jeopardizes the promotion of a 'safety first' attitude throughout the Shuttle program. . . . There is no doubt that operating pressures created an atmosphere which allowed the accident on 51-L to happen. Without operating pressures the program might have been stopped months before the accident to redesign or at least understand the SRB joint. Without operating pressure the flight could have been stopped the night of January 27."[149]

Building on the revelations of the official investigations, other analysts sustained the importance of production pressure, as their published accounts added to the accumulating recorded history. Many questions remained unanswered, as we will see in the next chapter, but in essence, the conventional explanation of the launch decision was a trickle-down theory. Competition, scarce resources, and production pressure, originating in the national and international political environment, were the structural origins of the disaster. They permeated the NASA organization, affecting decision making on the eve of the launch. Far from being incompetent, those responsible for the decision to launch were experienced managers who understood thoroughly both the technical and managerial issues involved. Understanding, they made a choice. They knowingly violated safety requirements, risking lives in order to achieve the projected launch schedule. In the interest of securing resources for the organization's well-being and the country's competitive goals, these NASA managers took a calculated risk—and lost.

Chapter Two

LEARNING CULTURE, REVISING HISTORY

Like millions of others around the world, I followed as the Presidential Commission's televised hearings unraveled in excruciating detail the events at NASA that preceded the destruction of the *Challenger* on January 28, 1986. Like everyone else, I wondered how NASA could have launched when they had known about the O-ring problem for years and were warned the night before against launching. Why had it happened? Removed and remote from the event, my early understanding of why NASA proceeded with the launch was based on the hearings of the Presidential Commission, as mediated by the press and television. The circumstances surrounding the launch decision that I described in the previous chapter are the ones that led me, along with many other people, to assume that some sort of wrongdoing was behind the decision to launch the Space Shuttle *Challenger*.

But I may have begun this project with even greater suspicions about wrongdoing than other citizens. Sociologists view the social world through a critical lens created by a research tradition that penetrates both the interactions of everyday life and the major societal institutions that are the centers of power, ideology, and wealth. In my case, this general critical orientation was joined to a professional experience that further shaped my interpretation of the *Challenger* launch decision. I had been conducting research on misconduct by organizations for many years. As the Commission's televised hearings on the accident progressed and newspaper accounts accumulated, the *Challenger* disaster looked more and more like an example of organizational misconduct to me.

First and paramount, the Presidential Commission indicated that NASA personnel had violated both industry rules and internal NASA rules designed to assure safety.[1] People committing violations to achieve the goals and interests of the organization for which they work is what makes this misconduct "organizational." Second, the unfolding media drama revealed the three factors that scholars traditionally associated with misconduct by organizations.[2] *Competitive pressures and resource scarcity* were widely acknowledged to have created unrelenting production pressure, the safety rule violations suggesting that NASA decision makers intentionally risked astronaut safety in order to keep the shuttle flying and meet the production goal. Further, certain NASA *organizational characteristics* appeared to have facilitated wrongdoing in much the same way as in other well-known cases of organizational misconduct: the sprawling NASA-contractor structure obscured what was happening about the O-rings from insiders and outsiders alike; an organizational culture described as "success oriented" and "can do" rewarded managers for achievement and may have driven them to meet the launch schedule by correcting the joint rather than halting flights to redesign it; a transaction system for the exchange of information became a mechanism for concealing rather than revealing technical problems. As a final parallel with other instances of organizational misconduct, postaccident accounts documented *regulatory ineffectiveness:* the system regulating the safety of the Space Shuttle Program appeared to have suffered a massive breakdown.

Shocking perhaps that NASA, with its red, white, and blue, "Right Stuff" image, might have been so preoccupied with the importance of production goals that the space agency would intentionally violate safety rules in order to stick to the launch schedule and accept the possibility of human tragedy as a by-product. But we know that risk taking is routine in most organizations.[3] Moreover, giving organizational interests priority over human life also has its precedent. Charles Perrow's classic, *Normal Accidents: Living with High-Risk Technologies,* chronicles incident after incident in which production concerns overrode adherence to safety regulations, contributing to incidents and accidents in complex, tightly coupled technical systems.[4] In *Workers at Risk,* Dorothy Nelkin and Michael Brown chillingly portray the devastating physical consequences for employees when organizations skimp on safety equipment and regulations to save money.[5] Clearly, other managers, in other organizations, have weighed the trade-offs and risked lives in the name of competitive and economic success. An internal memo documenting such calculations surfaced in the investigation of the Ford Motor Company's decision to manufacture a flawed Pinto. It showed Ford officials calculated the cost of redesigning the Pinto to eliminate the flaw and the cost of retooling to manufacture a new, safe design. Then they calculated the probability of accidents and loss of life, estimating the value of a human life at $200,000.00. The cost of redesign far outweighed the cost of accidents. Production went forward.[6]

Management decisions in the business world that value competitive and economic success more highly than the well-being of workers, consumers, or the general public so often have come to public attention that today's most widely accepted model of corporate criminality portrays the business firm as an amoral, profit-seeking organization whose actions are motivated by rational calculation of costs and opportunities.[7] In this model, managers are "amoral calculators." Driven by pressures from the competitive environment, they will violate the law to attain desired organizational goals unless the anticipated legal penalties (the expected costs weighed against the probability of delaying or avoiding them) exceed additional profits the firm could make by violation.[8] The imagery of amoral calculation reflects the logic of rational choice, with an important distinction. When decision makers' calculations of costs and benefits are tainted by self-interest, economics, or politics so that *intentional* wrongdoing and/or harm result, their calculation becomes amoral.[9]

Undeniably, NASA was experiencing serious economic strain prior to the *Challenger* launch. The result was production pressure that manifested itself internally in attention to the launch schedule. The rule violations suggested blatant and intentional disregard of formalized safety precautions, and thus amoral calculation. Clearly, the *Challenger* incident does not fall into the category of corporate criminality because NASA is not a private enterprise. But consider the following: although it is a government organization rather than a corporate profit seeker, NASA is subject to the same difficulties. In order to survive in an environment populated by competitors, suppliers, customers, and controllers,[10] all organizations must compete for scarce resources—regardless of their size, wealth, age, experience, or previous record.[11] We know too well the extensive social costs. Competition for scarce resources encourages misconduct by a wide variety of organizations: research institutions falsify data in order to win grants and prestige; universities violate NCAA recruiting regulations in order to secure the stuff of winning athletic teams; police departments violate the law to make arrests that bring recognition and funding; political parties and governments commit illegalities to secure national and international power.

What was behind the launch decision? This was, for me, a particularly compelling question because it joined my personal desire to know why the astronauts had died with a long-standing riddle of my profession. One of the enduring puzzles in the search for the causes of organizational misconduct is the question of "good" people and "dirty" work:[12] how is it that people who are employed, have opportunities and education, and appear to have a lifetime record as law-abiding citizens can and do violate rules, regulations, and laws, doing harm in furtherance of organizational goals? We do not have good answers to this question. Although the amoral calculator hypothesis has been the leading explanation of decision making when organizations engage in misconduct, empirical support for it is fragile indeed. Support comes primarily from research showing an association between competition, scarce resources, and corporate violations. However, in these studies the link between environmentally produced economic strain and decision making has not been traced.[13]

This gap is not the result of any oversight: the data have been hard to come by. Typically, the power, boundaries, and secrecy of organizations have either restricted or completely prevented researchers' access to information about decision making in instances of misconduct. When information did become available, it was at best a partial record. Like historians, sociologists studying misconduct have been constrained by missing data: critical conversations never recorded; documents undiscovered, distorted, destroyed. Case studies have sometimes allowed us to view internal organizational activities, but even in the few case studies in which we do have evidence of a relationship between economic strain and choice in organizations,[14] we seldom have information about how those decisions were formulated. In *Moral Mazes*, Robert Jackall brilliantly describes the pressures, trade-offs, and moral ambiguity that typify managerial cultures.[15] But are violations a result of rational calculation of costs and benefits of some harmful act? Peter Grabosky, in his research on illegality in the public sector, only found evidence of such careful assessment and weighing of costs and benefits in possibly two of seventeen cases.[16] Regardless of the general support for an association between economic strain and organizational misconduct, the amoral calculator hypothesis has remained just that—a hypothesis, often invoked or assumed as an explanation, but seldom tested.

There is good reason for skepticism. We know that not all organizations experiencing economic strain violate laws and rules to attain goals.[17] Further, some organizations violate for other reasons—incompetence; misunderstanding of or improper attention to regulatory requirements; or even principled disagreement with the form, content, or ideological underpinnings of a particular rule or regulation.[18] A violation might occur under circumstances of economic strain, but amoral calculation may not be the explanation. Also, the way managers actually assess risks in the workplace is far from the systematic calculation the amoral calculator hypothesis implies.[19] They do not weigh all possible outcomes but instead rely on a few key values. The magnitude of possible bad outcomes is more salient, so that there is less risk taking when greater stakes are involved. But even then, they do not quantify and calculate: they "feel it," because quantifying costs and benefits is not easy. Finally, decision making does not occur in a vacuum. To understand decision making in any organization, we must look at individual action within its layered context: individual, organization, and environment as a system of action.[20] When all are taken into account, we find that they modify the notion of rational choice that undergirds the amoral calculator hypothesis. The weighing of costs and benefits occurs, but individual choice is constrained by institutional and organizational forces.

Norms—cultural beliefs and conventions originating in the environment—create unreflective, routine, taken-for-granted scripts that become part of individual worldview. Invisible and unacknowledged rules for behavior, they penetrate the organization as categories of structure, thought, and action that shape choice in some directions rather than others.[21] DiMaggio and Powell note that "institutional arrangements constrain individual behavior by rendering some choices unviable, precluding particular courses of action . . . [but] institutions do not just constrain options: they establish the very criteria by which people discover their preferences."[22]

These constraints on choice are reinforced in organizations as executives set the premises for decision making through organizational routines that reduce uncertainty.[23] Decision making is more an example of rule following than of calculation of costs and benefits.[24] Rationality is constrained further: the organization has limited abilities to search for information and solutions to problems; individuals have limited knowledge of alternatives, access to and ability to absorb and understand information, and computational capacity; the decision-making process is influenced by deadlines, limited participation, and the number of problems under consideration. Rather than a model of perfect rationality, decision making in organizations is characterized by "bounded rationality"; performance is described as "satisficing" rather than optimizing.[25] In fact, some scholars argue that often the only rationality that might be credited to the process is imposed retrospectively by participants in order to justify a particular decision.[26] The notion of individual rationality has become so circumscribed and discredited that some organization theorists even have described the decision process by the "garbage can model" and characterized managers as "muddling through"—a far cry from the imagery of cool, calculated managerial capability suggested by the ideology of rational choice.[27]

Had NASA managers acted as amoral calculators, knowingly violating safety rules in order to stick to the launch schedule? To solve the puzzle of "good" people and "dirty" work, I would have to explore what is known in my trade as the "macro-micro" connection: how macrostructural forces—in this case, the competition, scarce resources, and production pressures originating in the environment—affected decision making about the SRBs. The investigations of the Presidential Commission, the media, and others immediately following the tragedy were producing the kind of information that made it possible to do exactly that. The institutional history establishing the structural origins of the disaster was extremely well documented, as demonstrated in chapter 1; data on NASA decision making—the sort coveted but too often elusive in instances of organizational misconduct—were abundant. It was a rare research opportunity, not to be missed. Thus I embarked on what turned out to be a much longer, more complicated, more surprising research journey than any I had undertaken. It consisted of layers of learning: deeper and deeper immersion in the world of engineers, the technology of the SRB Project, and the NASA culture. As I went beyond the readily available published accounts and into original documents and less accessible archival data, I became familiar with the culture of the organization I was studying. I made discovery after discovery that challenged key aspects of the case on which my beginning assumptions were based.

My understanding of why the *Challenger* was launched is now different from what I believed in the first year or so of this research. In order to clear the way for the explanation presented in this book, which, in many aspects, contradicts conventional explanations, I now retrace the steps of my research and the discoveries I made that contradicted the worldview I held when I began—the one that many people still hold. I use my research experiences as a device to dissuade readers, as I was dissuaded, from accepting the conventional view that couples economic strain on the organization with managerial wrongdoing as the cause of the *Challenger* tragedy. First, I will present the evidence that initially persuaded me that the environmental pressures NASA experienced resulted in amoral calcula-

tion, building an argument that coincides with conventional interpretations. Then, I will gradually dismantle that straw man, showing
how still deeper immersion in the documentary record led me to a
different understanding.

EARLY AFFIRMATION OF AMORAL CALCULATION

Initially, I did no interviewing. Before I could interview, I had to master the event, the cast of characters, and the technical and bureaucratic language essential to any semi-intelligent conversation about
NASA and the *Challenger*. In the first months, I relied on newspapers, magazines, and scientific journals, but when volume 1 of the
Presidential Commission's report was published in June 1986, it
became my primer. A detailed 256-page summary of the findings and
recommendations, it was written for a general audience. Still, the lay
reader had to persist, for in places the text was tough going. Even
with the fascinating, carefully written explanations of shuttle design
and technology and the accompanying technical diagrams, charts,
and photographs, volume 1 was a challenge. My two auto repair
courses (circa 1976) scarcely had trained me for the world I was entering, a predominantly male world typically inhabited by aerospace
engineers, astronauts, military personnel, and technicians.

As I began studying the event in more detail, I soon learned that
some of my most basic assumptions about the technical cause of the
disaster were wrong. For example, the shuttle did not explode, as I
thought. According to engineers, there was a fireball and a structural breakup, but no explosion. And the much-publicized and dramatic demonstration that maverick Presidential Commissioner Feynman made during the televised Commission hearings, showing the
loss of resiliency that occurred when he dipped a piece of an O-ring
into ice water, greatly simplified the issues on the table on the eve of
the launch.[28] Managers and engineers alike knew that when rubber
gets cold, it gets hard. The issues discussed during the teleconference
were about much more complicated interaction effects: joint dynamics that involved the timing (in milliseconds) of ignition pressure,
rotation, resiliency, and redundancy.

While learning more details enabled me to rectify some of my mis-
taken notions, I developed new ones. Throughout this project I felt
like a detective, but my detective work never had the unerring linear
clarity of a Sherlock Holmes investigation. For longer than I care to
admit, for example, I was convinced by the cross-sectional diagrams
accompanying the text (figs. 3 and 5) that an O-ring was like a Nerf

ball, expanding or contracting as needed to fill its groove. Because the boosters were enormous (12' in diameter and 149' high—only 2' shorter than the Statue of Liberty), at first I simply could not grasp the minuscule size of the joint: the infamous gap between tang and clevis was 0.004"; and when it opened up during joint rotation at ignition that "enlarged gap" that so worried everyone was estimated to be 0.042"–0.060" at maximum! Add to this the unbelievable brevity of joint rotation: six-tenths of a second (0.600) from start to finish, which Marshall engineer Leon Ray later told me "is, as you know, a lifetime in this business."[29]

When I grasped the temporal and physical dimensions of the joint, I understood that an O-ring is like a huge length of licorice—same color, same diameter (0.280")—joined at the ends so it forms a circle 12' across. It lies between the booster segments much like the rubber ring between a Mason jar and its lid. Thus face to face with my "Nerf ball" error (with considerable embarrassment, as I had discussed the *Challenger* launch in a colleague's class and used the Nerf ball analogy to misinform them on the technical cause), I realized I would need to consult people who understood the technical information. But I believed that with sufficient immersion in the case materials and by consulting technical experts, I could sufficiently master the technical details necessary to get at the sociological questions. It was, after all, human behavior I wanted to explain, and I was trained to do that.

In contrast to my first forays into the mysteries of the technology, my early reading continually verified the influence of the competitive environment, the organizational characteristics, and regulatory failure—the three factors commonly associated with organizational misconduct. Most fascinating, the following details of the case seemed to support a causal connection between the structural origins of the disaster, as described in chapter 1, and managerial amoral calculation in the launch decision:

NASA management actions on the eve of the launch. The Commission's report confirmed that Marshall managers played a major role in the reversal of Thiokol's initial recommendation to delay launch. Though Marshall managers denied that they "pressured" Thiokol, they did vigorously challenge Thiokol's position (recall George Hardy's statement about being "appalled" by the contractor's no-launch recommendation and Larry Mulloy's "When do you expect me to launch, next April?"). When asked by Chairman Rogers why he seemed to change his mind when he changed his hat, Lund responded:

> We have dealt with Marshall for a long time and have always been in the position of defending our position to make sure that we were ready to fly, and I guess I didn't realize until after that meeting and after several days that we had absolutely changed our position from what we had been before. But that evening I guess I had never had those kinds of things come from the people at Marshall that we had to prove to them that we weren't ready . . . and so we got ourselves in the thought process that we were trying to find some way to prove to them it wouldn't work, and we were unable to do that. We couldn't prove absolutely that that motor wouldn't work.[30]

Moreover, Thiokol engineers' testified that *they* perceived Marshall management's response to the no-launch recommendation as pressure. Roger Boisjoly testified: "One of my colleagues that was in the meeting summed it up best. This was a meeting where the determination was to launch, and it was up to us to prove beyond a shadow of a doubt that it was not safe to do so. This is in total reverse to what the position usually is in a preflight conversation of a Flight Readiness Review. It is usually exactly opposite of that."[31]

Also controversial, at the conclusion of the teleconference, Marshall managers requested that Thiokol representatives transmit their launch recommendation in writing to Kennedy Space Center. Several of the engineers who protested the launch said this was an unusual request. It also suggested possible amoral calculation and wrongdoing: perhaps Marshall managers wanted proof that they had proceeded upon advice from Thiokol. Then, should anything go wrong with the O-rings, Thiokol would bear the brunt of the responsibility.

Furthermore, the Commission found that procedural rules instituted to assure safety were violated in the 24 hours preceding the *Challenger* launch. These rules were violated by the same Marshall management personnel who proceeded with the launch despite the arguments and data presented by Thiokol engineers. The violated rules required NASA managers to transmit information about technical problems up the hierarchy. But NASA top administrators in the launch decision chain were not informed that the teleconference had taken place. By alerting additional NASA administrators to the O-ring problems, these procedural rules had the potential to alter the final decision on the *Challenger* launch by subjecting the O-ring controversy to further review. The failure to pass word of Thiokol engineers' concerns up the hierarchy appeared to be a blatant attempt to conceal O-ring problems. If these rules were intentionally violated in order to assure that the launch would proceed on schedule, then the amoral calculator hypothesis gains credibility.

Previous cost/safety trade-offs. NASA had a history of decision making in which economic interests appeared to take priority over safety considerations. By inference, this pattern suggested an organizational culture in which amoral calculation was the norm, suggesting that the *Challenger* launch decision may have been influenced by this cultural predisposition. Particularly damning was evidence that the design of shuttle elements was often affected by considerations of both cost and schedule.[32] Two incidents stood out: (1) the decision made during the design phase to eliminate the escape rockets, thereby reducing the weight of the vehicle in order to increase payload capacity (see chap. 1), and (2) the decision to award the SRB contract to Morton Thiokol. Four firms submitted proposals for the design in 1973. One of Utah's leading industries was the Wasatch Division of Thiokol Chemical Corporation in Brigham City, north of Salt Lake City. Its parent company was Morton Thiokol, Inc., Chicago, which sold specialty chemicals and salt. At the time, Thiokol-Wasatch was a medium-sized contractor with experience in munitions, military pyrotechnics, and missile solid rocket manufacture. They had worked on the Minuteman and Trident missile SRMs, but they were not viewed as a leader in the field.[33]

Thiokol won the approximately $800 million contract to manufacture the SRBs despite its engineering design being ranked lowest of the four bids. Then–NASA Administrator James C. Fletcher, publicly announcing this decision, stated that the rationale behind this choice was cost: Thiokol's bid was about $100 million lower than the competitors.[34] After the *Challenger* accident, Fletcher was discovered to have had deep political and personal ties in Utah at the time the contract was awarded.[35] The payoff for Thiokol was big. In 1985, Thiokol-Wasatch employed 6,500 workers, nearly twice as many as Marshall Space Flight Center. Its space-related sales, most of which were tied to the Shuttle Program, accounted for about $380 million, or 44 percent, of the Chicago-based company's aerospace division revenues in that fiscal year.[36]

NASA had started out doing the shuttle "on the cheap," apparently compromising safety in the process. Later, after the SRB joint problems started occurring, NASA's method of dealing with them also appeared to be governed by cost rather than safety considerations.

The space agency kept tinkering with the existing design rather than redesigning the joint. Clearly, time was a consideration in decision making. And time was money, given the advanced scheduling of commercial payloads. Redesign could take two years. If the shuttle did not fly during the design process, projected income from payload contracts would be deferred, or perhaps permanently lost. Correcting the existing design conserved resources in two ways: modifications were less expensive than redesign, and the flight schedule could proceed with minimum alteration—if any. NASA tended to choose the solution that would keep the shuttle flying by fixing the joint rather than the one that would eliminate the problem.[37]

Cover-up. Adding to the impression of wrongdoing at the space agency, NASA apparently withheld information from the public after the accident.[38] The possibility of "cover-up" at NASA fueled speculation about managerial amoral calculation.[39] All data related to the *Challenger* and the accident were promptly impounded at the space agency. Inquiring journalists learned nothing. NASA began its own investigation, conducting it in "great secrecy."[40] In the first days after the disaster, NASA officials vacillated about the possible role of the O-rings, suggesting other explanations. Although they mentioned the teleconference, they neglected to tell the Presidential Commission about Thiokol's initial no-launch recommendation. When the Commission unearthed the history of problems with the O-rings, NASA's initial reluctance to admit the extent of the problem seemed willful and suspicious.[41]

Conjecture about a cover-up was also stimulated by NASA's reluctance to release information about the astronauts' deaths.[42] At first, it was assumed that the astronauts had died instantly, without suffering, vaporized in the fireball. Naval operations to recover the *Challenger* wreckage from the ocean floor were shrouded in secrecy.[43] In March 1986, nearly six weeks after the tragedy, debris from the crew compartment was finally recovered from the ocean floor. The media reported that a Navy salvage ship returned to port after nightfall with darkened decks, carrying a flag-draped coffin, presumably containing the remains of the crew. Queried by journalists, NASA officials made no comment. Questions were raised about why recovery took so long when evidence of crew equipment had floated to the surface a month before.[44] Again, no comment.

On July 17, NASA made public the transcript of tapes from the flight-deck tape recorder salvaged from the ocean floor. The presumption that the astronauts had died instantly, unaware of impending disaster, was reinforced by the last words heard on the tape, Commander Francis Scobee's "Roger, go at throttle-up." But a NASA report made public 11 days later contradicted this conclusion.[45] Astronaut Joseph P. Kerwin, M.D., the scientist who headed the investigation into the astronauts' deaths, stated unequivocally that the astronauts survived the breakup of the launch vehicle. Kerwin reported that the last words on the tape were not Scobee's, but Pilot Michael Smith's "Uh-oh," indicating that the crew was aware something was wrong.[46] Moreover, analysis of the four emergency oxygen packs recovered from the ocean indicated it was possible that some were alive until the crew compartment hit the water.[47] Film of the *Challenger* breakup analyzed at NASA immediately after the accident showed the crew compartment intact and falling free of the rest of the vehicle.[48] Hence, NASA's long silence about the details of the astronauts' deaths and slow release of information was taken by some as intentional concealment in an effort to limit public relations damage and the financial impact of insurance claims filed by astronaut families.[49] But beyond that, it suggested that the "Wrong Stuff" was guiding decision making at NASA. By implication, the "Wrong Stuff" may also have been behind the decision to launch the Space Shuttle *Challenger*.

Young, Boisjoly, Cook, and Feynman. The public actions and statements of veteran astronaut John W. Young, Thiokol engineer Roger Boisjoly, NASA Resource Analyst Richard Cook, and Presidential Commissioner Richard Feynman after the *Challenger* tragedy gave further credibility to amoral calculation as an explanation of the launch decision. Cook, employed by NASA's Comptroller's Office in the summer of 1985, was responsible for assessing the budgetary implications of the External Tank, the SRBs, and the Centaur upper stage.[50] One of his first assignments was to write a background memo on the SRBs. Written in July 1985, Cook's memo stated: "There is little question, however, that flight safety has been and is still being compromised by potential failure of the seals, and it is acknowledged that failure during launch would certainly be catastrophic. There is also indication that staff personnel knew of this problem sometime in advance of management's becoming apprised of what was going on."[51] The memo supports amoral calculation: personnel knew of the problem and also knew failure would result in catastrophe.

John Young, Chief of NASA's Astronaut Office, specifically charged in a March 4, 1986, internal NASA memo that the space agency had compromised flight safety to meet launch schedule pressures and that some shuttle crews were lucky to be alive.[52] Referring to the potentially catastrophic result of a failure of seals in the shuttle's booster rockets, Young wrote, "There is only one driving reason that such a potentially dangerous system would ever be allowed to fly—launch-schedule pressure." With his memo was an enclosure listing other potentially dangerous aspects of shuttle components and procedures that he argued were accepted for the same reason.[53] He concluded by noting that safety and production pressures were conflicting goals at NASA—and in the conflict, production pressure consistently won out: "People being responsible for making flight safety first when the launch schedule is first cannot possibly make flight safety first no matter what they say. The enclosure shows that these goals have always been opposite ones. It also shows overall flight safety does not win in these cases."[54]

Roger Boisjoly brought formal charges against Thiokol. No longer employed, Boisjoly filed two suits against the company on January 28, 1987, one year to the day after the tragedy. In a personal suit, he filed for punitive and compensatory damages for personal and mental health injuries, accusing Thiokol of "criminal homicide" in the death of crew members.[55] NASA was implicated in the second suit, which Boisjoly filed on behalf of taxpayers under the Federal False Claims Act (31 U.S.C. 3729-3733). Under this act's provisions, private parties may sue on behalf of the government and the government has the right to join the action. In this suit, Boisjoly charged that Thiokol knowingly supplied NASA with defective SRMs for use in the Space Shuttle Program and unlawfully filed "false and fraudulent claims for payment or approval" with the government and, further, that the company had falsely certified the safety of the *Challenger* launch.[56]

Although Morton Thiokol, Inc., was the defendant in his complaint, Boisjoly also included charges against NASA. Boisjoly charged NASA officials with collusive action based on economic interests. These two sections of his complaint imply that both Thiokol and NASA officials were amoral calculators, knowingly withholding information about the O-ring problems from others in order to satisfy production concerns:

> Sec. 24. The senior NASA officials collusively participated in the fraudulent and fatal decision to launch 51-L as part of a long term pattern of fraudulent reports to the Congress in regard to the safety of the Shuttle in order to induce the Congress to continue funding the program.

Sec. 25. Despite the fact that close to a score of engineers familiar with the SRM's [sic] and its seals had unanimously recommended against the flight on January 28, 1986, NASA and Thiokol intentionally and fraudulently withheld information of this known danger from the seven astronauts comprising the flight crew, the owners of the cargoes on board the orbiter *Challenger* and the insurance companies who had insured various elements of the flight.[57]

During the hearings, Commissioner Feynman's remarks also seemed to support the idea that managerial amoral calculation prevailed at NASA. He stated that NASA officials were playing "Russian roulette," going forward with each launch despite O-ring erosion because they got away with the last one: "It is a flight review, and so you decide what risks to accept. I read all of these (pre-flight) reviews, and they agonize whether they can go even though they had some (O-ring erosion). . . . And they decide yes. Then (the Shuttle) flies and nothing happens. Then it is suggested, therefore, that risk is no longer so high. For the next flight we can lower our standards a little bit because we got away with it last time. . . . It is a kind of Russian roulette. You got away with it, and it was a risk."[58]

On the surface, the public statements and actions of Young, Boisjoly, Cook, and Feynman, the information about economic strain, rule violations, previous decisions where economic goals took priority over safety at NASA, and NASA management actions both the night before the launch and after the disaster all seemed to affirm that intentional managerial wrongdoing—the kind both explicit and implicit in theories of amoral calculation and organizational misconduct—was behind the *Challenger* launch decision. But there were many unanswered questions.

PUZZLES AND CONTRADICTIONS

To establish that NASA managers knowingly violated safety rules, risking the lives of the astronauts in order to stick to the launch schedule, I would have to understand how the environment of competition and resource scarcity affected these managers' actions. But in the extensive hearings of the Presidential Commission, I found that specific questions about the *connection* between economic strain at NASA, rule violations, and decision making about the SRB joint were never asked and therefore never resolved.[59] Because of this omission, the evidence was still circumstantial as far as I was concerned: the link between economic pressure on the organization and managerial actions on the eve of the *Challenger* launch had not been traced. Furthermore, as I immersed myself more deeply in the case, the details seemed either to completely contradict or, at least, to throw doubt on some of the statements made in posttragedy accounts that led many (including me) to believe that misconduct was behind the decision to launch. Although not all of them shed light on the *Challenger* launch decision per se, these discoveries did suggest the extent to which the documents I was relying on had systematically distorted public impressions of NASA actions and thus of history. What appeared in publications soon after the disaster as black and white became more gray.

NASA management actions on the eve of the launch. Simple logic challenged the notion that NASA managers had acted as amoral calculators on the eve of the launch. The basic assumption underlying the amoral calculator hypothesis—that the launch was a cost/safety trade-off intended to sustain the Shuttle Program's economic viability—did not make sense. To explain the launch decision as a cost/safety trade-off posited safety and economic success as competing organizational goals: to pursue one was to interfere with or undercut the other. In many management decisions at NASA, safety and economic interests competed. For example, reducing the number of safety personnel frees budget allocations for spare parts necessary to keep the shuttle flying but weakens safety regulation; redesigning the shuttle's cargo area to reduce the space available to the crew increases available space for additional payloads but limits in-flight crew mobility, sometimes precious to safety. But does the same assumption that safety and economic success are competing goals apply to a launch decision? No.

In the decision to launch the *Challenger*—or in any of the preceding 24 launch decisions, for that matter—safety and economic success were interdependent rather than separate conflicting goals. Both mattered, the one inextricably tied to the other. Included in the amoral calculator's weighing of the costs and benefits is the probability of *detection*. If there is a low probability of being caught, rewards may be greater than costs. But if detection is not only possible but probable, the costs must be taken into account.[60] For NASA, an SRB failure during a mission would be irretrievably and globally public. Detection was a certainty. Moreover, the costs were readily calculable: they would be high and instantaneous, with both short-run and long-run repercussions. Neither the astronauts nor the vehicle would survive an SRB failure. The lost lives and the subsequent impact on the astronauts' families, the country, and NASA employees would be an enormous personal and social cost. Already in perilous condition, the manned spaceflight program might not recover. Technical failure would be accompanied by financial consequences so extensive that they could threaten the existence of the Shuttle Program. The flight schedule would be interrupted, and income from payload contracts would stop. NASA's inability to fulfill aerospace industry contracts would threaten the industry, perhaps causing contractors to become more independent of NASA and space projects. In addition to the impact on the space agency, U.S. military and scientific goals would suffer serious setbacks. Such costs do not require pencil and paper calculation.

It defied my understanding. If NASA managers launched *Challenger* because of production pressures, knowingly violating safety rules in the attempt to sustain the program's economic viability, how could they fail to take into account the extensive costs of an O-ring failure? If NASA managers were so calculating—and also preoccupied with convincing potential customers, the President, Congress, and the world that spaceflight was routine and economical—surely they would not knowingly risk failure on a mission carrying Christa McAuliffe, a nonastronaut whose presence was intended to symbolize the routinization and safety of spaceflight. Why would they not scrupulously follow every safety rule? After the accident, the *Wall Street Journal* stated, "Most critical decisions—whether to begin countdown at all—whether to interrupt it—have to be weighed against mounting demands from military and civilian customers under tight schedules to keep their payloads aloft."[61] If NASA managers were concerned about meeting the demands of military and civilian customers to keep their payloads aloft, would they knowingly violate rules designed to assure the safety of those precious income-producing payloads? Completing missions was essential to obtaining the resources necessary for continuing the program. To argue that the *Challenger* launch decision was amoral—blatant and intentional disregard of formal safety requirements in the interest of production pressures and economic necessity—is to argue that decision makers acted in direct opposition to NASA's competitive interests, rather than in concert with them. On the basis of logic alone, the amoral calculator hypothesis appeared to lack validity.

Add to these puzzles and contradictions another point: the issue that night was the weather. Engineers were objecting to launching at the predicted temperature of 26°F—an unusually cold temperature for Florida. What got lost in posttragedy accounts was that they were not calling for a launch cancellation; they were recommending delay.[62] Weather changes, and the forecast predicted that the temperature would rise during the day. True, the *Challenger* launch had already been delayed four times. And delay has costs, as discussed in chapter 1. We know that workers experiencing production pressure sometimes take risks to avoid costs rather than to make gains.[63] But whatever the costs of further delay of this particular launch, they had to be low in comparison with the costs of a catastrophe. In the words of George Hardy, then NASA's Deputy Director, Science and Engineering Directorate, Marshall Space Flight Center, "I would hope that simple logic would suggest that no one in their right mind would knowingly accept increased flight risk for a few hours of schedule."[64]

Previous cost/safety trade-offs. The examples of previous cost/safety trade-offs identified by the press, the official investigations, and Astronaut John Young cannot serve as automatic verification of an amoral calculator explanation of the *Challenger* launch. We must beware of the retrospective fallacy. These examples were raised after the tragedy: analysts sorted through NASA's decision-making history, citing certain documents and specific technical and procedural decisions as illustrative of cost/safety trade-offs.[65] Then these examples claimed to describe a pattern that, by further inference, explained the launch decision. Caution is in order when examples are retrospectively selected to explain some event. Cost probably influenced many if not most of the decisions related to the Space Shuttle Program. The tragic outcome of the *Challenger* launch made it easy, with the luxury of hindsight, to see the risk associated with certain decisions and retrospectively construct an argument that safety had been traded for the economic viability of the program— particularly when such a decision pertained to the components responsible for the *Challenger's* technical failure.

The disaster made certain examples salient and others less so. Contradicting the emphasis on previous cost/safety trade-offs in postaccident accounts, I found that the Presidential Commission report was full of examples of decisions that supported a diametrically opposed point: NASA administrators also had a history of making costly decisions *in the interest of safety.* The intricate system of checks and balances that NASA instituted to assure safety, the extensive program for crew training, the number of procedural and technical requirements, the four volumes of items routinely checked in each prelaunch countdown, and even the midnight-hour teleconference on the eve of the *Challenger* launch itself are examples of NASA's formal and informal measures in the interest of safety. Moreover, other launches had been delayed when technical and weather problems jeopardized mission safety, concern about production pressures and schedule slippage notwithstanding. Posttragedy arguments that NASA's "can do" attitude create a gendered "macho-techno" risk-taking culture that forced them to push ahead, no matter what, are contradicted by NASA's many launch delays (see table 1). These delays were for safety reasons. These counterexamples demonstrate how the selective use of examples can lead to different conclusions. Determining that NASA had a pattern of giving cost priority over safety in previous decisions would require (1) looking at all NASA decisions and (2) understanding how the safety and economic issues were evaluated by management *at the time* the choices were made. Even if such an assessment were possible and verified a pattern of cost/safety trade-offs, the *Challenger* launch decision would still have to be examined to see whether it fit the pattern.

More from curiosity about past choices than from any real hope of shedding light on the launch decision, I investigated the two decisions that after the disaster became perhaps the most controversial (and sensationalized) as examples of cost/safety trade-offs: (1) the early design decision to jettison the escape rockets and (2) the controversial awarding of the SRB contract to Thiokol. (Because of its length, my analysis appears in appendix A.) My search revealed that the published accounts oversimplified and hence grossly misrepresented the complex factors that went into these two decisions. In the case of the escape rocket decision, to conclude that it was a cost/safety trade-off (and thus a representative example of amoral calculation) was absolutely and unambiguously incorrect. Engineering opinion held that if the SRB joint failed, the result would be a catastrophic loss of crew, vehicle, and mission. Such a failure was most likely to happen on the launch pad within the first six-tenths of a second after ignition, making it impossible to save the astronauts because there would be no warning. Even in the unlikely case that a joint failed later (which happened on STS 51-L) or some other life-threatening failure happened during the brief, two-minute trajectory powered by the SRBs, no abort procedures could save the crew.[66] In the case of the Thiokol contract award decision, I concluded that the decision to go with Thiokol was not amoral calculation, but a typical example of satisficing: cost and schedule played a role, but based on the technical knowledge available at the time, NASA engineers believed that the Thiokol design would do the job and do it safely.[67]

Table 1. SHUTTLE LAUNCH DELAYS

Launch Date	Mission	Comment
4/12/1981	STS-1	2-day launch delay because of computer problems
11/12/1981	STS-2	Delayed from November 4 to November 12 because of a problem with the auxiliary power units.
3/22/1982	STS-3	Launched one hour late
6/27/1982	STS-4	On schedule
11/11/1982	STS-5	On schedule
4/4/1983	STS-6	Launch postponed from January 20 to April 4 because of engine problems
6/18/1983	STS-7	On schedule
8/30/1983	STS-8	Launch delayed 8 minutes due to bad weather
11/28/1983	STS-9[a]	Rescheduled from October 28 to November 28 because of Orbiter and Solid Rocket Booster nozzle problems
	STS-10	Cancelled because of problems with the inertial upper stage (IUS)
2/3/1984	STS 41-B	On schedule
	STS-12	Cancelled because of IUS problems
4/6/1984	STS 41-C	On schedule
8/30/1984	STS 41-D	Delayed from June 25 to August 30 because of fuel valve malfunctions and computer problems; STS 41-D carried the combined payloads of STS 41-D and 41-F
	STS 41-E	Cancelled due to IUS problems
10/5/1984	STS 41-G	On schedule
11/8/1984	STS 51-A	Delayed one day by weather (wind shear)
1/24/1985	STS 51-C[b]	Delayed one day by weather (freezing temperature)
4/12/1985	STS 51-D	Delayed 45 minutes by weather; STS 51-E combined with 51-D due to IUS problems
4/29/1985	STS 51-B	On schedule
6/17/1985	STS 51-G	Delayed from June 14 to June 17 due to satellite antenna problems
7/29/1985	STS 51-F	July 9 launch delayed at 3 seconds before launch due to engine valve problem
8/27/1985	STS 51-I	Delayed one day by weather, then two days due to computer problems
10/3/1985	STS 51-J	On schedule
10/30/1985	STS 61-A	On schedule
11/26/1985	STS 61-B	On schedule

Table 1. SHUTTLE LAUNCH DELAYS *(continued)*

Launch Date	Mission	Comment
1/12/1986	STS 61-C[c]	Delayed December 12 due to malfunction in booster rocket, delayed January 4 due to computer problem, delayed January 6 due to valve malfunction.
1/28/1986	STS 51-L	Delayed January 22 due to need to update crew training, delayed January 25–27 due to weather plus technical difficulties (hatch handle and inaccurate reading from launch pad fire detection system).

Source: Transcription of "Shuttle Launch Delays," memorandum on launch delays of more than 24 hours prepared at request of U.S. Congress, House, Committee on Energy and Commerce, Subcommittee on Energy Conservation and Power, by Patricia Humphlett, Science Policy Research Division, Congressional Research Service, Library of Congress, Washington, D.C., 25 February 1986.

[a] For an explanation of the booster problem, see chap. 4.—AU.

[b] The ambient temperature at Cape Canaveral must be 31°F or higher for launch to proceed. When the temperature falls below that level, as it did in the case of STS 51-C, the Launch Director at the Cape stops the launch until the temperature rises to the appropriate level. No temperature concerns were raised about the SRB joints by Thiokol prior to this launch. See chap. 5.—AU.

[c] The booster problem cited here was due to an out-of-round segment. See chap. 6.—AU.

Cover-up. The "cover-up" did not constitute evidence about the *Challenger* launch decision in any sense. Actions chronologically subsequent to an incident cannot contribute to its cause. Thus, the *Challenger* launch decision could not be understood in light of the actions of administrators after the tragedy. Furthermore, all organizations (as well as individuals) attempt to control their public image.[68] This general tendency is increased when we are under public scrutiny. Whether NASA's actions after the disaster were part of some broad, organizationally based plan to conceal wrongdoing had never been established. Here again, I found information that challenged conventional interpretations. I mention it to show that this matter, too, was one that deeper inquiry transformed from black and white to gray.

Speculation about a cover-up was grounded in NASA's reluctance to release information afterward. NASA's postdisaster information disclosure was limited for several different reasons, not all of which appeared to contribute to a concerted cover-up. The allegation that NASA's instant impounding of *Challenger* flight data was part of a cover-up was incorrect. A computer-activated program was initiated at 1 minute 30 seconds after loss of signal from *Challenger* in order to capture and record as much data as possible in the Mission Control Center in Houston.[69] Impounding data after an accident was standard operating procedure throughout the aerospace industry because careful postflight analysis of the technical cause was essential to preventing a recurrence. NASA's program to secure data for postflight analysis had been formalized long before the *Challenger* disaster. NASA had conducted its own in-house investigation after the 1967 Apollo fire (without the appointment of a special accident investigation commission) and expected to be solely responsible this time. Securing all available data was a preprogrammed response for NASA's own investigation, which began without delay.

Another NASA action that led to speculation about a cover-up was NASA senior administrators' testimony in the early days of the Presidential Commission investigation. They minimized the concern at NASA about the O-rings both in the years before January 27 and during the teleconference. Further, when asked about the technical cause of the accident, they mentioned the SRB joints as a possible cause but presented them as one among many that NASA was investigating. Later, it was revealed that stop-action films available within two days of the disaster showed the smoke puff coming from the aft field joint of the right SRB.[70] Yet in the hearings, the top NASA administrators vacillated about the technical cause. While cover-up is certainly a possible explanation, there are two others. First, Jack Macidull, a Federal Aviation Administration (FAA) investigator who worked with the Commission, said that at first top agency administrators did not understand enough about shuttle technology to know what happened.[71] Second, engineering accident investigations are conducted using "fault trees" to determine the technical cause of a failure. A fault tree is a diagram. Listing all the elements of a total technical system, it includes "branches" for possible physical and environmental failure modes for each element. Thus, a fault tree contains every possible hypothesis about how a single element might fail.

A fault tree's explanatory potential is limited by what is known, however. In addition, it does not include interaction effects between components, identified by Perrow in *Normal Accidents* as the cause of many failures in complex technical systems.[72] Although observations, such as analysis of damaged components and films, can lead to a strong working hypothesis about the technical cause of a failure, usually no formal conclusions or accident report are issued until the fault tree analysis is completed, eliminating all false hypotheses and verifying the technical cause.[73] The apparent cause is not always borne out by the fault tree analysis, so administrators hesitate to identify cause publicly before the analysis is complete to avoid the embarrassment of correcting the record each time "apparent cause" changes.

Further speculation about cover-up resulted from journalists' difficulty in getting other kinds of information out of NASA. Many news organizations were forced to file through the Freedom of Information Act (FOIA) to get details. Even so, the information was exceedingly slow in coming. Was this stonewalling in order to cover-up wrongdoing in the launch decision? Although NASA was well prepared to secure computerized flight data in case of disaster, Presidential Commissioner Robert B. Hotz reported that NASA had no other coordinated response planned for a shuttle failure, and absent a plan, chaos reigned at the space agency.[74] Research on organizational response to other disasters supports his contention. When tragedy first strikes, confusion is more commonplace than orderly response—even when contingency plans are in effect, communication channels are preplanned, and equipment available and ready.[75] Complicating matters at the time of the disaster, NASA was essentially leaderless. Top NASA Administrator James Beggs was on a leave of absence, and William Graham, a policy analyst with no experience to prepare him for running a technical organization like NASA, had been named Acting Administrator.[76] Some of what seemed to be concerted cover-up may have been simply the unrelated actions of many separate individuals scattered throughout the organization responding ad hoc to the postdisaster environment.

Years after the Commission's investigation, NASA FOIA officer Bunda L. Dean supported an "organizational chaos" explanation of NASA's information release after the disaster.[77] *Challenger*-related documents currently occupy a two-story warehouse. These same documents were scattered throughout the NASA system in January 1986. After the tragedy, requests for information were at an all-time high. Information could not be released until it was found, sent to the Johnson Space Center FOIA office in Houston, and entered in a computerized record and index. The disaster increased the workload for thousands of NASA employees, and people were slow to send relevant documents to Houston. As late as June 1992, two hundred boxes of materials pertaining to the tragedy had just been received and were being cataloged and filed in the warehouse where the *Challenger* paper trail is permanently stored.

NASA's reluctance to release information about the astronauts does support the theory of a concerted, organizationwide attempt to conceal information. A key controversy centered on NASA officials' unwillingness to make available to the press the flight-deck tape recordings of *Challenger* astronaut voices that rescue vehicles retrieved from the ocean floor (transcripts only were released), photographs of the debris of the *Challenger* crew cabin, and the details of financial settlements with crew members' survivors.[78] NASA officially explained these refusals as an attempt to protect the privacy of the astronauts' families.[79] NASA's argument was that details of the astronauts' deaths and the settlements would be followed by sensational press coverage that would cause the astronauts' families renewed suffering. The space agency argued (and the Justice Department concurred) that the law and the public interest had been served by the information already released.[80]

Obviously, suppression would also benefit NASA by preventing further negative publicity and another round of public shock and outrage directed at the space agency. Consistently after the disaster, the space agency's actions concerning the fate of the *Challenger* crew appeared to be governed not by chaos, but by a well-organized, politically sensitive response designed to conceal rather than reveal. But, as noted above, there were a number of reasons why information was slow in coming. To generalize the term "cover-up" to all the questionable incidents was inappropriate, for each appeared to have a different explanation.

Young, Cook, Boisjoly, and Feynman. Concluding this list of puzzles and contradictions, I found that no one accused any of the NASA managers associated with the launch decision of being an amoral calculator. Although the Presidential Commission report extensively documented and decried the production pressures under which the Shuttle Program operated, no individuals were confirmed or even alleged to have placed economic interests over safety in the decision to launch the Space Shuttle *Challenger*. For the Commission to acknowledge production pressures and simultaneously fail to connect economic interests and individual actions is, prima facie, extremely suspect. But NASA's most outspoken critics—Astronaut John Young, Morton Thiokol engineers Al McDonald and Roger Boisjoly, NASA Resource Analyst Richard Cook, and Presidential Commissioner Richard Feynman, who frequently aired their opinions to the media—did not accuse anyone of knowingly violating safety rules, risking lives on the night of January 27 and morning of January 28 to meet a schedule commitment.

Even Roger Boisjoly (who in speaking engagements across the country after the tragedy continued in his role of whistle-blower by telling his compelling story of the *Challenger* launch and the impact of the disaster on him and his family) had no ready answer on this point. When I asked him why Marshall managers would respond to production pressures by proceeding with a launch that had the potential to halt production altogether, he was stymied. Boisjoly thought for a while, then responded, "I don't know."[81] In fact, Boisjoly and others who for years had worked with Mulloy and Hardy expressed nothing but the greatest respect for Mulloy's and Hardy's previously demonstrated ability, knowledge, and concerns and actions in the interests of SRB safety. The evidence that NASA management violated rules, launching the *Challenger* for the sake of the Shuttle Program's continued economic viability was not very convincing. Hardy's statement "No one in their right mind would knowingly accept increased flight risk for a few hours of schedule" rang true.

RULE VIOLATIONS

The puzzles and contradictions I had found left me with many questions by the end of 1986. Walton makes the point that cases are always hypotheses.[82] They are selected on the basis of "typological distinctions" grounded in prior assumptions about what defines a case. But those assumptions can be faulty. Consequently, the first goal of a case study is to find out exactly what we have a case of. To find out whether this was a case of misconduct or something else, I decided on the following strategy. Rule violations were essential to misconduct as I was defining it. I would examine management rule violations in relation to the booster joints, looking specifically at environmental and organizational context to see what influenced these decisions. I chose two that were the most controversial and had an extensive paper trail:

The Criticality 1 waiver. In 1982, Marshall assigned the SRBs a "Criticality 1" (C 1) status, a NASA designation indicating that the SRB joints were not redundant, as the design intended: if the primary O-ring failed, the crew and vehicle would be lost. But soon after the C 1 status was imposed, Marshall's SRB Project Manager Lawrence Mulloy waived the C 1 status, just in time for the next shuttle launch, allowing it to proceed.[83]

The Launch Constraint waivers. In 1985, after extensive in-flight O-ring erosion, a "Launch Constraint" was imposed on the boosters. A Launch Constraint is an official status at NASA imposed in response to a flight safety issue that is serious enough to justify a decision not to launch. But Marshall's Mulloy waived the Launch Constraint prior to each of the next five shuttle flights, allowing each to proceed without the O-ring problem being fixed.[84]

The C 1 status and the Launch Constraint were formal administrative actions indicating that management knew the SRB problem was serious. Waiving the requirements suggested amoral calculation: knowingly violating safety rules in order to keep the shuttle flying.

Looking for explanations of these violations led me from volume 1 of the Commission's report into the hearing transcripts in volumes 4 and 5 and NASA documents describing the procedural requirements. From them, I gained a deeper understanding of daily operations at NASA than I had before. I realized I had made another mistake. This time it was not a technical error (like the "Nerf ball" mistake), but a misunderstanding of NASA's bureaucratic language and procedures. It was not my last mistake, but it was perhaps the most significant because upon this error was founded a crucial assumption that had guided my entire inquiry to this point. These two controversial management decisions were not "rule violations." I had assumed that they were from the language used in volume 1 of the Presidential Commission report and the responses and reactions of the Presidential Commissioners (in the televised hearings and in the verbatim hearing transcripts in volumes 4 and 5). Repeatedly, the Commission used the word "waive" as a verb (e.g., "Mulloy waived . . .") to explain what happened to both the C 1 status and the Launch Constraint.

But I discovered that "waiver" is a noun in NASA's bureaucratic language. *A waiver is a formalized procedure that, upon approval of the technical rationale behind the request, allows an exception to some internal rule.*[85] Waivers were a standard managerial option at NASA. The verb, to waive, refers to the implementation of this formal procedure. Also, I discovered that waivers were frequent at NASA, used in all parts of the shuttle system. Further, waivers were recorded. If amoral calculation was operating at NASA, and waiving requirements was an indication of it, would not people try to cover up decisions deviating from prescribed standards, rather than record their misconduct? Finally, I learned that the "Criticality 1 waiver" and the "Launch Constraint waiver" were two different procedures. Indeed, the term "waiver" was inappropriate to describe management response to a Launch Constraint. Although the Presidential Commission had repeatedly used the term "waiver" in connection with the Launch Constraint, this usage did not appear in Marshall documents describing Launch Constraint procedures, nor did it

appear in the documents in which NASA recorded both the imposition of Launch Constraints and their removal. These records used the
word "lifted," an indication of the difference between the two procedures. I learned that, like C 1 waivers, lifting Launch Constraints was
a legitimate managerial option, widely used in the program, and not
a rule violation.[86]

Now here was something really interesting. Apparently, NASA
had rules that allowed them to circumvent other rules: in particular,
safety regulations. The C 1 waivers and lifting Launch Constraints
were clearly in the interest of organizational goals—by virtue of
these actions, the shuttle continued to fly—suggesting the presence
of competitive pressures. But what I, as an outsider, thought of as
rule violations not only were behaviors conforming to formalized
NASA rules but also appeared to be the norm, or "standard operating
procedure," at NASA. It was now February 1987, and I had been
involved in the project for a year. My hypothesis that the decision to
launch was a case of organizational misconduct hinged on NASA's
rule violations, and I had not yet found any. Where I expected rule
violations, I found conformity. The tidy introduction to the chapter
"Rule Violations" that I had written went into the wastebasket. As
is so often the case when we begin to learn the complexities of a situation, some of the issues that had seemed very clear at the outset
had become more confused. Only much later would I fully understand the extent to which oversimplification obfuscates and complexity brings understanding.

FROM CURRENT EVENT TO
HISTORICAL ETHNOGRAPHY

My discovery of this mistake transformed my research. This much
was clear: the two controversial NASA actions that outsiders defined
as deviant after the tragedy were not defined as deviant by insiders at
the time the actions took place. Deviance refers to behavior that violates the norms of some group. No behavior is inherently deviant;
rather it becomes so in relation to particular norms.[87] Because norms
vary—between groups, over time, across societies—the same behavior can be seen by some as deviant and by others as conforming.
Deviance is socially defined: to a large extent, it depends on some
questionable activity or quality being noticed by others, who react to
it by publicly labeling it as deviant. The Presidential Commission's
public hearings and pronouncements were the major source of outsider definitions of events at NASA.

The Commission's revelation of rule violations was central to the historically accepted explanation of the tragedy. However, the Commission had made some mistakes of fact and interpretation of fact regarding the "rule violations" mentioned in volume 1, which I (and other analysts relying on this volume of the official report) had repeated. It seemed to me, as a result of studying the verbatim hearing transcripts and appendices in the other volumes, that the Commission had not thoroughly understood these NASA procedures and how they were used in daily decision making. Consequently, the Commission failed to distinguish them as different kinds of procedures and mistakenly identified both as violations of safety requirements. Their mistake indicated that the Commission had an inadequate grasp of this aspect of NASA culture. The meaning of waivers and Launch Constraints only became clear to me from extensive examination of a full set of documents and complete testimony, not available to the Commission while the information was still accumulating. In addition, I was systematically comparing SRB decisions with decisions about other shuttle elements, while the Commissioners were examining SRB decisions only. From this insight, I concluded that I needed to know more about NASA culture prior to the disaster.

Typically, culture is studied by ethnographers who participate in the culture they want to understand. I could learn something about NASA culture by interviewing, but I was, for the time being at least, restricted to archival data. Suits brought by the astronauts' families against NASA and Thiokol in the fall of 1986 cut off information from employees at the two organizations.[88] Interviewing would have to wait until the suits were settled. Although relying on documents was a strategy I assumed out of necessity, I decided it was the better strategy. Even if people had been available for interviews, I doubted that they would tell me anything that diverged from what they had said in the Commission hearings. They had all been sworn in. Their testimony had been recorded, transcribed, and televised. As far as I was concerned, it was etched in stone. Perhaps I would learn *more* (especially from whistle-blowers), but I was not likely to learn anything *different*. Only if I asked them different questions than the Commission had—questions that were grounded in my own independent analysis and understanding—would interviewing benefit me. Most important, I had to be sufficiently well informed about both the organization and the technology in order to evaluate the validity of their responses. My plan was to interview when my analysis was more mature.[89]

In the meantime, the five-volume Commission report gave me access to NASA culture prior to the *Challenger* launch. It was filled with technical diagrams and documents, memos and letters, NASA formal language and everyday expressions, rules and procedures, and both verbatim descriptions and a paper trail of how people in the space agency behaved in the 1970s and 1980s. Using "rule violations" as a point of entry into decision making about the SRB joint had proved to be a good way to learn about NASA culture while learning about "misconduct." According to the Presidential Commission report, industry standards had been violated by those making decisions about the SRB joint in the early development period, before the shuttle even started flying. Were *these* true rule violations? And what was the role of competitive pressures emanating from the environment, filtering through the organization, and shaping not only the response to rules but perhaps the rules themselves? In order to examine both the environmental and organizational influences on these (what now I had to call) "alleged" SRB-related rule violations that had begun in the early design stage and occurred throughout the program, I shifted my research focus from the *Challenger* launch decision to the past.

By this time, the House Committee on Science and Technology had published the findings of its investigation; moreover, all the Commission's documents, including transcripts of interviews conducted by investigators working for the Commission, were available at the National Archives. Relying on this expanded archival record, I traced the alleged rule violations in the history of decision making. Repeatedly, I was struck by the difference between the meaning of actions to insiders as the problem unfolded and interpretations by outsiders after the disaster. As I gained an understanding of cultural context and chronological sequence, many actions, much publicized and controversial, took on new meaning. Incidents that when abstracted from context contributed to an overall picture of managerial wrongdoing became ordinary and noncontroversial. For example, after writing a 1978 memo objecting to the joint design, Marshall engineer Leon Ray helped develop corrections to the joint that assuaged his concerns, leading him to believe the design was an acceptable flight risk. Ray's memo became part of the official record, creating the impression that managers had been overriding engineering objections for years; his subsequent actions did not. Now alert to the importance of relocating controversial actions in the context of prior and subsequent actions, I expanded my focus beyond its restricted attention to rule violations. I began reconstructing a chronology of all decision making about the SRB joint, for the period 1977–85.

Thus the research became a historical ethnography: an attempt to elicit organizational structure and culture from the documents created prior to an event. The ethnographic historian studies the way ordinary people in other times and cultures made sense of things by "passing from text to context and back again, until (s)he has cleared a way through a foreign mental world."[90] My work was in harmony with the work of many social historians and anthropologists who use documents to examine how cultures shape ways of thinking. Yet the research setting was decidedly modern: a complex organization in which the technology for producing records and the process of record keeping were valued, thus creating the artifacts for its own analysis. Moreover, interviews were possible.

As I reconstructed the decision-making history, I found many things that contradicted conventional interpretations. But the most significant discoveries were these:

1. *A five-step decision-making sequence in which the technical deviation of the SRB joint from performance predictions was redefined as an acceptable risk in official decisions.* This sequence was repeated, becoming a pattern. Each time tests or flight experience produced anomalies that were signals of potential danger, the risk of the SRBs was negotiated between working engineers at NASA and Thiokol. Each time, the outcome was to accept the risk of the SRBs. This pattern indicated the existence of a *work group culture* in which the managers and engineers working most closely on the SRB problems constructed beliefs and procedural responses that became routinized. The dominant belief, grounded in a three-factor technical rationale, was that the SRB joints were an acceptable risk: therefore it was safe to fly. From the early development period until the eve of the *Challenger* launch, the work group's cultural construction of the risk of the O-ring problem became institutionalized. Despite dissensus about what should be done about the problem, consensus existed that it was safe to fly.

2. *Decision making in the SRB work group conformed to NASA rules.*[91] The general impression left by many posttragedy accounts was that managers ignored engineers and violated safety rules in the years preceding the *Challenger* launch. Instead, I found that managers depended on engineers for risk assessments. Behind all those launches were determinations of risk acceptability originating with Thiokol working engineers at the bottom of the launch decision chain. And rather than managers violating rules about passing information on O-ring problems up the hierarchy to their

superiors, I found conformity to NASA rules prescribing the information that should be passed on.

Discovery of these two patterns led to an explanation of the string of controversial launch decisions at NASA prior to 1986. The attention paid to managers and rule violations after the disaster deflected attention from the compelling fact that, in the years preceding the *Challenger* launch, engineers and managers together developed a definition of the situation that allowed them to carry on as if nothing was wrong when they continually faced evidence that something *was* wrong. This is the problem of the normalization of deviance. Three factors, with which we will be occupied throughout this book, explain the normalization of deviance: the production of a work group culture, the culture of production, and structural secrecy.

RISKY DECISIONS AND THE NORMALIZATION OF DEVIANCE AT NASA

Risk is not a fixed attribute of some object, but constructed by individuals from past experience and present circumstance and conferred upon the object or situation. Individuals assess risk as they assess everything else—through the filtering lens of individual worldview. A butcher at work, for example, does not see the same immediate danger in the tools of that trade as does a parent catching a preschooler (who likely sees no danger at all) pulling a carving knife out of a kitchen drawer. Risk is in the eyes of the beholder; it can be present or absent in the same situation or object, or, if present, present to a greater or lesser extent.[92]

How is this variety possible? Each person—the butcher, the parent, the child—occupies a different position in the world, which leads to a unique set of experiences, assumptions, and expectations about the situations and objects she or he encounters. From integrated sets of assumptions, expectations, and experience, individuals construct a worldview, or frame of reference, that shapes their interpretations of objects and experiences.[93] Everything is perceived, chosen, or rejected on the basis of this framework. The framework becomes self-confirming because, whenever they can, people tend to impose it on experiences and events, creating incidents and relationships that conform to it. And they tend to ignore, misperceive, or deny events that do not fit.[94] As a consequence, this frame of reference generally leads people to what they expect to find. Worldview is not easily altered or dismantled because individuals tend ultimately to disavow knowledge that contradicts it.[95] They ward off information in order to preserve the status quo, avoid a difficult choice, or avoid a threatening situation. They may puzzle over contradictory evidence but usually succeed in pushing it aside—until they come across a piece of evidence too fascinating to ignore, too clear to misperceive, too painful to deny, which makes vivid still other signals they do not want to see, forcing them to alter and surrender the worldview they have so meticulously constructed.[96]

The diverse experiences, assumptions, and expectations associated with different positions can manifest themselves in distinctive worldviews that often result in quite disparate assessments of the same thing. Because worldview affects the social construction of risk, a particular object or situation with a readily ascertainable capacity for harm (like the knife) can be interpreted differently by different people. When risk is no longer an immediately knowable attribute of the object and the possible harm associated with it depends on other, less knowable factors, we move into the realm of uncertainty and probabilities. Under conditions of uncertainty, the potential for variation in interpretation of risk is even greater. Technological innovations, at least for part of their sometimes brief lifespans, tend to have uncertain or unpredictable qualities. Organizations that create potentially hazardous technical products often have a formal responsibility to determine not only whether risk is present or absent but, when present, how much hazard is present. As experience accumulates, the formal assessment of risk will vary with changes in the technology, changes in knowledge about its operation, or comparison with other similar technologies. The transient character of the product necessitates that risk assessment become an ongoing process, subject to continual negotiation and renegotiation by the experts whose job it is.

Implied in the term "expert" is some technical skill, gained either by experience, by professional training, or by both, that differentiates the professional from the lay assessment of risk. Also implied is that professionalism will somehow result in a more "objective" assessment than that of the amateur. But professional training is not a control against the imposition of particularistic worldviews on the interpretation of information.[97] To the contrary, the consequence of professional training and experience is itself a particularistic worldview comprising certain assumptions, expectations, and experiences that become integrated with the person's sense of the world. The result is that highly trained individuals, their scientific and bureaucratic procedures giving them false confidence in their own objectivity, can have their interpretation of information framed in subtle, powerful, and often unacknowledged ways.[98]

THE PRODUCTION OF CULTURE, THE CULTURE OF PRODUCTION, AND STRUCTURAL SECRECY

In the history of decision making about the SRB joints from 1977 to 1985, a cultural construction of risk developed that became a part of the worldview that many participants brought to the January 27, 1986, teleconference. A culture is a set of solutions produced by a group of people to meet specific problems posed by the situations that they face in common.[99] These solutions become institutionalized, remembered and passed on as the rules, rituals, and values of the group. Culture is sometimes falsely assumed to be a characteristic peculiar to a formal organization as a whole. This may be the case when the organization is small and has a simple structure. But most organizations are segmented and potentially have as many cultures as subunits.[100] It cannot be assumed that these subunits are linked culturally, in some uniform way, to the larger organization or to each other.[101] For example, it cannot be presumed that a corporation's accounting department has a culture identical to its marketing, research and development, or legal departments, or that the zoology department of a university has the same culture as fine arts, the management school, or sociology—or, for that matter, that students have the same culture as faculty. Although some uniform cultural values and patterned ways of behaving and believing may be shared throughout a given organization, the degree to which they permeate a particular division, department, or subunit remains a matter of inquiry.

The idea of organizational culture must be refined even more, however, for rules, rituals, and beliefs can evolve that are unique to work groups. People in an organization who interact because they have a central task in common constitute a work group. The common task draws them together, and in the doing of that task lies the potential for development of a culture unique to that particular *task*. A work group is not constrained by the organization's formal structure, for people scattered in positions that span the hierarchy can take on a project together. So an organization simultaneously may have many work groups that crosscut its internal structure, persisting while the task persists, dissolving when the task is completed or terminated. In interaction, work groups create norms, beliefs, and procedures that are unique to their particular task. Although often far from harmonious, work groups nonetheless do develop certain ways of proceeding and certain definitions of the situation that are

shared and persist. These collectively constructed realities constitute the work group culture. The work group culture contributes to
decision making about the group's designated project by becoming a
part of the worldview that the individuals in the work group bring to
the interpretation of information.

Early in the Shuttle Program, an unexpected problem developed in
the SRB. Test results indicated that the SRB joint deviated from the
performance predicted by the design. Because no precedent existed
for the joint design, no precedent existed for responding to the problem. The shuttle was not yet flying, thus the test evidence presented
no immediate threat to flight safety. However, in conformance to
NASA procedures, it was treated as a signal of potential danger. The
SRB work group—composed of the Level III and IV Marshall and
Thiokol engineers responsible for the SRB joints—assessed the consequences of the new technical information. My analysis of their
first negotiation of risk in the history of decision making showed this
sequence of events:

1. Signals of potential danger
2. Official act acknowledging escalated risk
3. Review of evidence
4. Official act indicating the normalization of deviance:
   accepting risk
5. Shuttle launch

As the SRB work group negotiated the risk of the SRB joints in this
five-step decision sequence, the work group normalized the deviant
performance of the SRB joint. By "normalized," I mean that behavior
the work group first identified as technical deviation was subsequently reinterpreted as within the norm for acceptable joint performance, then finally officially labeled an acceptable risk. They redefined evidence that deviated from an acceptable standard so that it
became the standard. Once this first challenge to field joint integrity was resolved, management's definition of the seriousness of the
problem *and* the method of responding to it (correct rather than
redesign) became a collectively constructed cultural reality, incorporated into the worldview of the work group. The work group brought
their construction of risk and their method of responding to problems with the SRB joints to the next incident when signals of potential danger again challenged the prevailing construction of risk. Risk
had to be renegotiated. The past—past problem definition, past
method of responding to the problem—became part of the social
context of decision making.

The significance of the above sequence of events lies not in its initial occurrence, but in its repetition. Decision making became patterned. Many times in the shuttle's history, information indicated that the O-rings deviated from performance expectations, thus constituting a signal of potential danger. Each time, the above decision sequence occurred. The connection between some incident in the past and the present is demonstrated when it is repeated. Patterns of the past—in this case, decision-making patterns pertaining to technical components—constitute part of the social context of decision making in the present. This decision-making pattern indicates the development of norms, procedures, and beliefs that characterized the work group culture.

A culture is generally typified by a dominant worldview, or ideology. A belief in redundancy prevailed from 1977 to 1985. Engineering determinations of acceptable risk were based on a three-factor technical rationale confirming that, erosion by hot gases notwithstanding, the primary O-ring would seal the joint, and if, under a rare and unexpected worst case condition, it should not, the secondary O-ring would act as a backup.[102] Certainly there was dissent. Any culture encompasses many ways of thinking, feeling, and behaving that often conflict. Engineering controversy is normal, especially when a technological innovation is involved. And between NASA and Thiokol working engineers, the "hands on" people evaluating and constructing risk each time a shuttle returned from space, disagreement about the dynamics of the joint began in the earliest phases of the Shuttle Program. But that dissent was about the mechanics of joint operation; it did not alter the work group's collective belief that the SRB joints were an acceptable flight risk in the years preceding the *Challenger* launch. This belief prevailed and became institutionalized. It survived many challenges. Even in 1985, when two signals of potential danger finally attracted sufficient attention that both Thiokol and NASA began funneling resources into a design modification, the work group—and this included the Thiokol engineers who, on January 27, 1986, voiced their objections to the *Challenger* launch—continued to recommend shuttle launches, based on the three-factor technical rationale that supported their decision to proceed with flight.

The production of culture shows how the work group normalized the joint's technical deviation. We also want to know *why* this normalization happened. We know from research on organizational learning that organizations do identify problems and correct and eliminate them, thereby avoiding mistake, mishap, and disaster. The many launch delays in response to signals of potential danger at NASA tell us that NASA also learned by doing, identifying potential problems and correcting them. In the case of the SRB joints, why did the work group consistently normalize the technical deviation in the years preceding the *Challenger* launch? The answer is in how the culture of production and structural secrecy affected the development of meanings in the work group. The culture of production was a key environmental contingency that was part of their worldview. The culture of production included norms and beliefs originating in the aerospace industry, the engineering profession, and the NASA organization, then uniquely expressed in the culture of Marshall Space Flight Center. It legitimated work group decision making, which was acceptable and nondeviant within that context. These taken-for-granted assumptions shaped choices in response to NASA's institutional history of competition and scarcity, contributing to the normalization of deviance. The second factor reinforcing work group decision making was structural secrecy: patterns of information, the organizational structure, and the structure of regulatory relations perpetuated the normalization of deviance. Decision making in organizations is always affected by how information is sent and received, the characteristics of that information, and how it is interpreted by the individuals who send and receive it. In the case of the SRB joints, structural secrecy concealed the seriousness of the O-ring problems, leaving the official definition of the situation unaltered.

When signals of potential danger occurred on the eve of the *Challenger* launch, the patterns that shaped decision making in the past—the production of culture, the culture of production, and structural secrecy—were reproduced in interaction, to devastating effect. The norms, beliefs, and procedures that affirmed risk acceptability in the work group were a part of the worldview that many brought to the teleconference discussion. That night, the work group culture fractured. What had previously been an acceptable risk was no longer acceptable to many Thiokol engineers. That night, the five-step decision sequence was initiated a final—and fatal—time, with one alteration. It was followed not by a successful launch, but by a failure: that "piece of evidence too fascinating to ignore, too clear to misperceive, too painful to deny, that makes vivid still other signals they do not want to see, forcing them to alter and surrender the worldview they have so meticulously constructed." One of the great ironies of this event is the paradoxical power and powerlessness of the working engineers who were the lower-level participants in the NASA hierarchy. Having created and repeatedly affirmed a construction of risk that became the dominant ideology of the work group culture in preceding years, they themselves could not overturn it. Despite a fractured work group culture, the belief in redundancy survived as the official construction of risk, as once again the factors that had contributed to the normalization of deviance all along were played out.

Amorally calculating managers intentionally violating rules to achieve organization goals does not explain the *Challenger* disaster. Managers were, in fact, quite moral and rule abiding as they calculated risk. Production pressures exerted a powerful influence on decision making, but not in the way many posttragedy accounts led us to believe. Originating in the environment of competition and the space agency's struggle for survival, production pressures became institutionalized. They permeated the organization, operating forcefully but insidiously in a prerational manner, influencing decision making by managers and engineers without requiring any conscious calculus. Moreover, production pressures were just one of many factors in a complex causal system. These larger social forces shaped the outcome, not individual misconduct or malintent. To a great extent, when individual actions are embedded in an organizational context, evil becomes irrelevant as an explanation because the meaning of one's actions and the locus of responsibility for one's deeds become transformed, as this book will show.[103] Most certainly, managers and engineers alike considered the costs and benefits of launching as the teleconference unfolded and the decision was made that night. No rules were violated. But they emerged from the teleconference having accepted increased risk once more. Following rules, doing their jobs, they made a mistake. With all procedural systems for risk assessment in place, they made a disastrous decision.

RETROSPECTION AND HISTORY

Organizations no doubt have failures more frequently than we realize. Only when these failures lead to harmful consequences that then publicly become defined as failures (an important distinction) do outsiders have the opportunity to consider the cause. Public understanding of this historic event developed, not from immediate experience, but from accounts created by the official investigations, media, and other analysts. The *Challenger* disaster generated an enormous amount of archival information as well as much conflicting public discourse, making possible analysis of a complex technical case not possible otherwise. But this advantage had an accompanying disadvantage. Dorothy Smith reminds us: "Our knowledge of contemporary society is to large extent mediated to us by documents of various kinds. Very little of our knowledge of people, events, social relations and powers arises directly from our own experience. Factual elements in documentary form, whether as news, data, information or the like, stand in for an actuality which is not directly accessible. Socially organized practices of reporting and recording work upon what actually happens or has happened to create a reality in documentary form, and though they are decisive to its character, their practices are not visible in it."[104]

In many instances, the accounts published after the *Challenger* incident that constitute the historic record were distorted by retrospective analyses and hindsight. Starbuck and Milliken point out that when observers who know the results of organizational actions try to make sense of them, they tend to see two kinds of analytic sequences.[105] Starting from the bad result and seeking to explain it, observers seek the incorrect actions, the flawed analyses, and the inaccurate perceptions that led to the result. Nearly all explanations of crisis, disaster, or organizational failure single out how managers "failed to spot major environmental threats or opportunities, failed to heed well-founded warnings, assessed risks improperly, or adhered to outdated goals and beliefs."[106] In contrast, analyses of success celebrate accurate managerial vision, wise risk taking, well-conceived goals, and diligent, intelligent persistence, despite scarce resources and other obstacles. These two analytic sequences lead to a selective focus that oversimplifies what happened. First, they focus attention on individual decision makers, positing managerial perceptions as the cause of all organizational outcomes. Second, they obliterate the complexity and ambiguity of the task environments that people once faced.

Turner, in *Man-made Disasters*, notes the tendency for a problem that was ill structured in an organization to become a well-structured problem after a disaster, as people look back and reinterpret information ignored or minimized at the time, that afterward takes on new significance as signals of danger.[107] After the *Challenger* incident, the SRB joint problem became a well-structured problem as the tragedy made salient and selectively focused attention on the NASA decisions that seemed to lead inexorably to it. The information then was strung together in posttragedy accounts that presented a coherent set of signals of potential danger that was not characteristic of the situation as it existed for NASA and Thiokol managers and engineers in the work group prior to the tragedy. Furthermore, even the most detailed of these postdisaster accounts extracted actions from their historical and organizational context in a stream of actions, the sequence of events and structures of which they were a part. Robbed of the social and cultural context that gave them meaning, many became hard to understand, controversial, and, in some cases, incriminating. The result was a systematic distortion of history that obscured the meaning of events and actions *as it existed and changed for the participants in the situation at the time the events and actions occurred.*

Not that this was intentional. Each account reconstructed the *Challenger* accident or some aspect of it to inform others. Each analyst was faced with overwhelming detail about technical issues, NASA organizational structure and procedures, and organizational and human history. Each published account had to be abbreviated because, first and obviously, the full account could never be known. Second, limits on the time available to gather, understand, and absorb information, on the time and space available in which to tell the story, and on the audience's ability to wade through the details made it imperative that each analyst shorten and simplify. So facts were presented; individual actions were described, but of necessity excised from the stream of actions that gave them their essential meaning. This excision of action from historical context also was true of the Presidential Commission's report, which was by far the most systematic, thorough, detailed attempt to construct a historical chronology that took into account both the organization and its context.

The Commission's investigation had particularly significant consequences for public understanding of this event. Both the televised hearings and volume 1 of the Commission's final report were the basis for many, if not most, public definitions of the situation. Receiving near-universal praise for what was the most extensive and expensive public investigation to date, the Commission successfully identified many problems at NASA that contributed to the loss of the *Challenger* crew. In volume 1, however, we repeatedly read that "NASA" did this or "NASA" did that—language that simultaneously blames everyone and no one, obscuring who in the huge bureaucracy was actually doing what. Top administrators who made policy, middle managers, and the engineers at the bottom of the launch decision chain who made the risk assessments were indistinguishable. The Commission's report did not officially blame any specific individuals. But their questioning of NASA managers in the televised hearings, the erroneous finding that NASA middle managers violated rules in both the history of decision making and on the eve of the launch, and the Commission report conclusion that the "decision making was flawed" focused public attention in one place. Obscuring the roles of elites who slashed NASA budgets, created production goals, and put a teacher on the shuttle as well as the roles of engineers who diagnosed this uncertain and risky technology, the public drama created the impression that NASA middle managers alone were responsible for the tragedy. This impression was reflected in an FBI agent's joking comment to Marshall SRB Project Manager Lawrence Mulloy as he left the hearings following testimony one day, "After they get you for this, Mulloy, they're going to get you for Jimmy Hoffa."[108]

Other aspects of the Commission investigation added to the public perception of managerial wrongdoing. For example, the Commission's inability to fully grasp NASA reporting rules resulted in many published accounts that had Mulloy suppressing information when he presented the engineering assessments to top NASA administrators. But a late-breaking video recording of Mulloy briefing top administrators prior to STS 51-E (the flight following the January 1985 cold-weather launch) showed a full oral review of the O-ring problem that was consistent, both in form and content, with guidelines for NASA Level I reporting requirements.[109] The existence of this video was not known to the NASA or Commission investigations at the time of the hearings.[110] Therefore, neither Mulloy's full briefing nor the video of it were acknowledged in volume 1 of the Presidential Commission's report. Because the video was discovered after volume 1 had gone to press (it was published months before the other volumes), it was mentioned and transcribed only in volume 2, which contains nine appendices of densely detailed technical reports in small print.[111] Had this video surfaced during the Commission hearings and its existence and transcription been reported in volume 1—or had a press conference been called when it surfaced—the historic record about managerial wrongdoing and "who knew what when" at NASA might have been different.

Another important distortion resulted from the selection of witnesses to testify. For the televised hearings, nearly 80 witnesses were called. Working engineers were underrepresented. The seven asked had all opposed the launch. But fifteen working engineers participated in the teleconference. The National Archives interview transcripts show that not all engineers who participated in the teleconference agreed with the position that Boisjoly and Thompson represented so vigorously. The engineers who differed did not testify, so these diverse views never became part of posttragedy accounts. Furthermore, since the emphasis of the questioning was on the launch decision, the fact was also lost that the same working engineers who opposed the *Challenger* launch originated the official risk assessments and recommendations for all booster launch decisions in preceding years.

The Commission findings were extensive and complex, covering every possible point of inquiry. But the televised hearings and the summary volume 1 represented only a small portion of the findings contained in volumes 2–5. And those volumes were further reduced from the several hundred thousand pages of data the Commission gathered, now stored at the National Archives. The media, working from the televised hearings and volume 1, seized on some of the discoveries, covering them extensively, while ignoring others. From the published and televised accounts, the public made further reductions. Unable to retain all the details, they latched on to whatever for them explained the unexplainable and went on. Most of those individually held theories were, to a greater or lesser extent, a form of the amoral calculator hypothesis. The House Committee on Science and Technology investigation report was not published until October 1986. Benefiting from the data collected by the Commission and beginning with a review of the Commission report, the House Committee went on to write a report that contradicted some of the Commission's main conclusions, among them the finding of rule violations by middle managers. But the House Committee report, coming later, never received the same media attention. Consequently, many people, at the time witnessing the event through the media, today believe some version of what they believed when the Presidential Commission completed its work: wrongdoing by middle management was behind the launch decision.

A REVISIONIST ACCOUNT

This book challenges conventional interpretations. I address two questions: Why, from 1977 to 1985, did NASA continue to accept the risk of a design known to be flawed, accepting more and more booster joint damage? And why did NASA launch the *Challenger* when the engineers who worked most closely on the joint were opposed? In answering these questions, I look at deviance from two perspectives: (1) the statistical deviation of booster technology from expected performance, and how it was normalized, and (2) the contrast between outsider interpretations of NASA actions as deviant after the disaster and insider definitions of these same actions as normal and acceptable at the time they occurred. Both perspectives on deviance are illuminated by a historical ethnography that shows the development of meanings by insiders as the problem unfolded.[112] Barbara Tuchman, historian and writer extraordinaire, emphasized that understanding history rests on being able to step into the past: "According to Emerson's rule, every scripture is entitled to be read in the light of the circumstances that brought it forth. To understand the choices open to people of another time, one must limit oneself to what they knew; see the past in its own clothes, as it were, not in ours."[113] My explanation rests on the presentation of details necessary to elicit cultural patterns. It does this by (1) presenting a chronological account of the history of decision making, resituating controversial actions in the stream of actions in which they occurred, (2) restoring the launch decision to its position as one decision in a stream of SRB joint decisions, and (3) examining the connection between the environment, organization, and individual choice. As a result, many controversial actions that defied understanding after the accident—including the launch decision—become understandable in terms of the meaning they had for participants at the time those actions were taken.

The architecture of this book is designed to take into account the worldview that many readers are likely to bring to the interpretation of the information in it: the tragedy was the result of wrongdoing by success-blinded, production-oriented middle managers. For this reason, I have organized the book around two versions of the eve of the launch, building my explanation incrementally between the two. The chapter 1 version was a bare bones description, intendedly stereotypical of accounts published in the first months after the disaster. I constructed it by identifying the key turning points of the event that were reported and repeated in numerous sources without variation. This stereotyped representation reminded everyone what happened, created a common starting point for those whose memories have dimmed or who were too young to pay attention in 1986, and introduced some basic information about the technology, the organization, and the cast of characters. Then, again drawing on other accounts, I raised all the rumored explanations that circulated after the accident, established NASA's environment of competition and scarcity as the origin of disaster, and ended with an argument for managerial wrongdoing as an explanation. In this chapter, I challenged the historically accepted explanation of production pressures and managerial wrongdoing by revisiting my research discoveries in order to clear the way for the explanation presented in the remaining chapters.

Chapters 3 to 7 introduce the three elements of the theory of the normalization of deviance: the production of culture, the culture of production, and structural secrecy. These chapters revise history, contradicting conventional interpretations by virtue of ethnographic thick description that puts controversial NASA actions back into chronological and social context. Necessarily they also teach culture so that readers have the tools that the Presidential Commission did not have at the outset of its investigation: aspects of the technology, engineering methods, NASA language, and organizational procedures that are as essential to understanding the *Challenger* launch decision as is the institutional history. Technical diagrams and documents, themselves cultural artifacts, are included to instruct readers and substantiate the arguments.

Chapters 3, 4, and 5 show the production of culture in the work group. They reconstruct the history of decision making about the SRB joints from 1977 to 1985, demonstrating how the statistical deviation of joint performance was officially normalized. We see how risk was negotiated and interpreted by the people who worked closely on the SRBs, and how their definition of the situation was conveyed to the top of the organization, becoming a collective construction of risk. These chapters expose the routine actions, daily decisions, and incremental production of technical knowledge that laid the foundation for the disaster. Chapters 6 and 7 illuminate how the institutional history of competition and scarcity created structural effects that impinged on decision making. They show the effects of the culture of production and structural secrecy, respectively.

Then, having laid out the explanation for past decision making, we return to the eve of the launch. Chapter 8, by its position in the book, reinserts the event in chronological sequence so that readers see the launch decision as a choice shaped by previous choices. Many of the details of that night, excluded from most previously published accounts, are restored. The abbreviated, stereotypical description that begins chapter 1 is repeated in boldface; new information is added in the standard font. A reconstruction in thick description, the chapter calls into question conventional interpretations. Chapter 9 explains the launch decision, demonstrating how the production of culture, the culture of production, and structural secrecy resulted, not in rule violations and misconduct, but in conformity and mistake: a decision to go forward once again despite signals of potential danger.

Chapter 10 moves beyond the case, examining its implications for what we know about culture; decision making in organizations; risk, science, and technology; and deviance. Then it examines the theory of the normalization of deviance as it applies to decision making in a variety of settings. Normalizing signals of potential danger is not a problem over which NASA had exclusive domain. The normalization of deviance is common to organizations and individuals alike, resulting in mistake, mishap, and misconduct, too often with disastrous consequences. Although the issue at NASA was an engineering controversy about a technical object and the deviance they accepted was statistical deviation, the case uncovers general processes that account for the normalization of deviance in other situations. Overall, the book forces us to confront the social organization of mistake. It reminds us of the power of political elites, environmental contingencies, history, organizational structure, and culture as well as the impact of incrementalism, routines, information flows, and taken-for-granted assumptions in shaping choice in organizations.

A final note: The effect of worldview on the interpretation of information is well documented in scientific research, despite the introduction of tools and techniques to control these very human tendencies.[114] This tendency does not discredit science as a useless enterprise, but instead alerts us to the importance of understanding the position—historical, political, cultural, gendered, organizational, and professional—of the analyst and thus something of the worldview she or he brings to the interpretation of information. I am situated in a position external to NASA and historically subsequent to the tragedy, bringing previous experience and training as a sociologist to bear on information that I have found. This book is affected by the same constraints on gathering, presenting, and interpreting information as other accounts of the incident. Although I assert that my analysis differs because it presents more of the social context of SRB decision making, the NASA and Thiokol employees whose actions form the basis of this book no doubt will view it as an overly simple, overly condensed version of events. I restore much of the context of decision making, but I, too, have shortened and simplified. In following up Dorothy Smith's concerns about the practices that are brought to bear on the construction of documentary reality, it is important to consider not only the information that was the formal basis for this book but also the more elusive but equally influential practices that I brought to bear on its selective use and interpretation.

The theory of this book—that structure and culture affected worldview, and thus affected the interpretation of information at

NASA—is exemplified by my own research experience. As a professional and an outsider, I started from a set of typological distinctions that suggested that the case was an example of organizational misconduct. The data dragged me to a different position. In this chapter, I described how I made the transition to the conclusion that the launch decision was rational calculation but not amoral, that it was a mistake but not misconduct. Here and in appendix C, I have tried to make visible to you some of the practices that have gone into my construction of this account. Still, much of my process and practice remains invisible. Lack of space as well as common sense about what sort of, perhaps academically pleasurable, tedium will bore the general reader makes it impossible for me to relate it all.

The public has been confronted with many alternative accounts, written by people positioned differently in relation to the tragedy, bringing a variety of worldviews and socially organized practices of reporting and recording to their interpretation of this same event. Each author has sorted and sifted through information to create what is by definition a particularistic account of what surely must be one of the best-documented technological failures of modern times. Massive documentation notwithstanding, we do not have and probably never will have all the information about the *Challenger* disaster. Yet it is important to record and analyze this event because of its historic significance and for whatever lessons can be learned about organizational failures. While such understanding is relevant for all organizations, it is absolutely essential for organizations that design, manufacture, produce, and use high-risk technological products.

Chapter Three

RISK, WORK GROUP CULTURE, AND THE NORMALIZATION OF DEVIANCE

In its report, the Presidential Commission summarized decision making about the SRB joints from 1977 through 1985, describing a series of acts of omission and commission that, in the Commission's view, were deviant:

> The Space Shuttle's Solid Rocket Booster problem began with the faulty design of its joint and increased as both NASA and contractor management first failed to recognize it as a problem, then failed to fix it and finally treated it as an acceptable flight risk.
>
> Morton Thiokol, Inc., the contractor, did not accept the implication of tests early in the program that the design had a serious and unanticipated flaw. NASA did not accept the judgment of its engineers that the design was unacceptable, and as the joint problems grew in number and severity NASA minimized them in management briefings and reports. Thiokol's stated position was that "the condition is not desirable but is acceptable."
>
> Neither Thiokol nor NASA expected the rubber O-rings sealing the joints to be touched by hot gases of motor ignition, much less to be partially burned. However, as tests and then flights confirmed damage to the sealing rings, the reaction by both NASA and Thiokol was to increase the amount of damage considered "acceptable." At no time did management either recommend a redesign of the joint or call for the Shuttle's grounding until the problem was solved.[1]

We want to know how this extraordinary sequence of events could happen. The explanation begins in the microcosmic world of daily decisions made by the SRB work group in the years preceding the *Challenger* tragedy. The ethnographic pursuit is a search for the "native view": what interpretation do people in a particular situation give to their experience? What is the formal and informal logic of their everyday life? The native view can be found by examining culture. Since culture itself is not visible, one way we can know it is by inspecting events to see how people interacting in groups identify, define, and transform previously known cultural items and create a distinctive culture of their own.[2] It is true that much of culture is tacit understandings that are hard to uncover.[3] But we can approximate it by examining what people do and say. Then, to convey another culture, the ethnographer must write in "thick description" that retrieves and interprets what was said and done at the microcosmic level of everyday life.[4] To this end, chapters 3, 4, and 5 reconstruct the history of decision making, relocating controversial NASA actions in the stream of actions in which they occurred. From these chapters, we can infer the existence of culture through the patterned behaviors, we can identify common understandings, we can discover inconsistencies among group members and witness how this within-in-group variation was resolved. These chapters allow us to see (1) how the work group normalized the statistical deviation of booster technology in official launch decision making and (2) the contrast between outsider interpretations of NASA actions as deviant after the disaster and insider definitions of these same actions as normal and acceptable at the time they occurred.

The work group developed an understanding of the situation through the members' interaction with one another, but it was an understanding in which the structure of power and key environmental contingencies loomed large. Structural factors were very important to decision making about the SRB joints and, as we shall see in chapters 6 and 7, explain *why* the work group always normalized technical deviation. For now, however, we restrict ourselves to the process of culture creation as the work group interacted about its task and witness how that culture, once created, affected subsequent choices. This chapter shows the production of culture in the work group in the early developmental period of the shuttle, from the mid-1970s through the first launch in 1981. It shows how Marshall and Thiokol Level III and IV managers and engineers initiated the norms, beliefs, and procedures that would shape future decisions. We see how the initial construction of risk of the SRB joints was negotiated when the first signals of potential danger occurred, and how the work group's belief that the design was safe to fly was sustained when first challenged. We find the initial occurrence of the five-step decision sequence that would occur again and again, becoming a pattern in which the deviant performance of the SRB joint continued to be normalized. Finally, we examine two controversial actions that led to speculation about misconduct and amoral calculation at

NASA after the tragedy: (1) NASA's failure to redesign the shuttle after two memoranda from Marshall engineers stating the SRB joint design was unacceptable and (2) NASA's certification of the SRB design as flightworthy, despite its being in violation of three industry standards.

Clifford Geertz writes, "Understanding a people's culture exposes their normalness without reducing their particularity. . . . It renders them accessible: setting them in the frame of their own banalities, it dissolves their opacity."[5] Thus, we begin with the aspects of the NASA culture that are crucial to understanding how decisions about risk were made: the assumptions, rules, and language of risk assessment and how risky decisions were negotiated. From this primer, we learn that NASA work groups were calculating risk under circumstances that made risk fundamentally incalculable, thereby routinely putting people in great danger. To do so was written into formal procedures, normal in NASA culture, and a taken-for-granted aspect of their jobs.

RESIDUAL RISK AND WORK GROUP DECISION MAKING

Technological uncertainty was the shifting terrain on which the decisions of the work group were built. Never before imagined was a vehicle designed to serve as a sort of space "bus," routinely hauling people and objects back and forth from earth to a station permanently fixed in space. To accomplish this goal, NASA and contractors developed one of the true phenomena of modern times: a large-scale technical system that encompassed shuttle components, launch equipment, and command and control networks that spanned the globe.[6] The shuttle design was unprecedented. Although it was a composite of existing aerospace technology, the various ideas had never been put together in this particular way before. Many parts were designed by different manufacturers; many were standard sizes, designed for general use, and had to be adapted to the unique environment of the shuttle. Problems getting them to fit together and work were expected, but no one knew what they would be. Also, the shuttle was designed to be reusable. It would not emerge unscathed from the extreme dynamic loads, wind shear, and other environmental forces during launch, orbit, and reentry, however. Although NASA engineers could calculate and predict performance, run lab tests, and test under simulated flight conditions, they could not verify the shuttle's capability under actual ascent, orbit, and descent dynamics. Therefore, anomalies—NASA's term for any deviation from design specifications—were expected on every mission.

The uniqueness of the shuttle design, its components, and its mission made calculating risk a fact of life. From the beginning of the Space Shuttle Program, the assumption was that risk could not be eliminated. It could be calculated, it could be predicted, it could be regulated, but after everything that could be done had been done, certain residual risks would remain. The risk of shuttle flight was acknowledged early in the planning stage of the program when NASA weighed whether to have an astronaut crew on the first orbital mission. The agreed-upon plan was that the success of the first mission would be enhanced by (1) taking a conservative design approach, (2) subjecting the vehicle to the most benign environment possible, and (3) flying a manned mission in which corrective action by a crew would be possible.[7] The purpose of the crew was to reduce mission vulnerability by adding human intelligence to cope with the uncertainties of the machinery. Sending people up in a vehicle with residual risks was not unique to the Shuttle Program, but had characterized NASA's efforts since the earliest days of spaceflight. And like those other space missions, decisions had to be made about whether it was safe to fly.

Risk assessments were the outcome of engineering analysis done by what I am calling work groups: managers and engineers at the bottom of the launch decision chain whose primary daily responsibility was technical decision making for the shuttle component to which they were assigned. Work groups were headed by Project Managers responsible for each of the main shuttle elements—the Main Engine, External Tank, Orbiter, and SRBs. Project Managers worked closely with the managers and engineers at their space center and with the prime contractor's managers and engineers assigned to that particular element. In these work groups, the distinction between managers and engineers was not as clear as the dichotomy suggests. All managers in work groups were trained engineers. But in the NASA culture, the hands-on work was done by "working engineers," or "the guys back in the engineering department that get to spend their 40 hours a week working on engineering problems not interfacing with the outside world."[8] Most of the working engineers also had management responsibilities, but they managed subunits of engineers and did not "interface" with the outside world.[9]

Work group risk assessment went on within a complicated set of rules. Two are central to the *Challenger* story: the Acceptable Risk Process and Flight Readiness Review.

Acceptable Risk Process. The Acceptable Risk Process was the basis for all technical decision making at NASA, from daily decision making to the formalized, final decision process immediately prior to a launch known as Flight Readiness Review. Starting from the assumption that all shuttle components are risky, engineers had to determine whether the risk of each item was acceptable. "Acceptable Risk" was a formal status conferred on a component by following a prescribed NASA procedure. When engineers discovered an anomaly or questionable condition, it was treated as a signal of potential danger. The shuttle could not fly unless the hazard was eliminated or controlled by some corrective action and/or engineering calculations and tests establishing that the condition was not a threat to flight safety. It could be classified as an Acceptable Risk only on the basis of a documented engineering risk rationale (in hearings testimony and interview transcripts of work group participants often called simply a "technical rationale," or "rationale"; another usage was "we rationalized it"). A rationale was an analysis of the problem, the probability of its recurrence, and data supporting a conclusion of acceptable risk.

The Acceptable Risk Process was set forth in the *Space Shuttle Safety Assessment Report,* issued prior to the first shuttle launch in April 1981. The document stated the parameters of hazard assessment for the first shuttle flight, STS-1, and all subsequent missions. For us, it is also a good introduction to NASA language used in risk assessment:

> *Acceptable Risk Process.* Hazard analyses were performed during the development phase to identify potential hazards associated with the design and operations of the Space Shuttle. The hazards identified were then subjected to a formal hazard reduction precedence sequence as described in NHB 5300.4 (1D-2).
>
> In instances where the hazard could not be eliminated or controlled, the hazards were highlighted and were subjected to a formal risk assessment. The responsible technical personnel determined the credibility and relative probability of the hazardous condition occurring and provided the technical rationale for retaining the existing design or procedure. The risk assessment was presented to the Senior Safety Review Board for evaluation.
>
> The hazard was classified as an accepted risk only after (a) all reasonable risk avoidance measures were identified, studied, and documented, and (b) project/program management made a decision to accept the risk on the basis of documented risk rationale. Documented accepted risks are continually evaluated by project, program, and safety management. . . .

As in all previous space programs, certain residual risks have been accepted by program management. These residual or accepted risks, which remain after all feasible corrective actions have been taken, have been thoroughly reviewed by technical personnel and received the concurrence of project and program level safety organizations.

The conclusion of this review is that there is no single hazard nor combination of hazards that should be considered a constraint to the STS-1 launch. All phases of Shuttle development and operations are continually being monitored to insure that *the aggregate risk remains acceptable.* (Emphasis added)[10]

This document is important not only because it acknowledges that the shuttle had aggregate risk (the accumulated risk of each component) and that monitoring risk acceptability was a continuous process but because it explains a postaccident controversy by establishing the meaning of the term "acceptable risk" in the NASA culture. The Acceptable Risk Process and the language of acceptable risk apparently were unknown to many investigators. Most vocal among them was Presidential Commissioner and physicist Richard Feynman, who publicly expressed astonishment at the discovery of the words "acceptable risk" and "acceptable erosion" in documents recording technical decisions about the SRB joints.[11] But flying with acceptable risks was normative in NASA culture. The five-step decision sequence I found that characterized work group decision making about the SRB joints was nothing less than the work group conforming to NASA's procedure for hazard analysis, as described in the above document.[12] Flight Readiness Review documents for other technical components routinely used the term "acceptable risk."[13] In fact, a listing and description of the acceptable risks on the Space Shuttle prior to the first launch in April 1981 filled six volumes.[14]

Flight Readiness Review. The day-to-day negotiation of risk acceptability took on special formal characteristics prior to a launch. Each mission is preceded by 15 months of mission-specific activity, following a generic schedule.[15] Flight Readiness Review (FRR) is the final, formal review in an intricate process of launch preparation and decision making involving thousands of people and thousands of hours of engineering work. NASA's Associate Administrator for Space Flight sets the date for FRR so it will be completed about two weeks before launch.[16] The goal is to determine that the shuttle is ready to fly and to fly safely. In order for launch to proceed, the aggregate risk of the shuttle must be found acceptable by all parts of NASA's complex, sprawling, decentralized organizational system, which includes NASA Headquarters in Washington; Johnson Space Center in Houston, Texas; Kennedy Space Center in Florida; Marshall Space Flight Center in Huntsville, Alabama; and all contractors (see appendix B, figs. B2.1–B2.5). FRR pulls together all of these organizations, which are incorporated into a hierarchical launch decision chain (see fig. 4). NASA's Safety, Reliability, and Quality Assurance staff participate at all levels.

FLIGHT READINESS REVIEW AND
SHUTTLE PROGRAM MANAGEMENT STRUCTURE

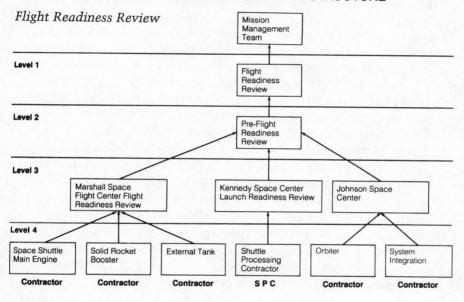

Flight Readiness Review

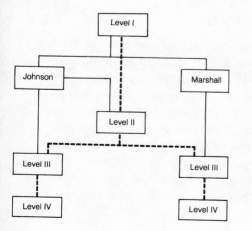

Shuttle Program Management Structure

_____ Institutional Chain
--------- Program Chain

Level I: The associate administrator for Space Flight. Oversees budgets for Johnson, Marshall and Kennedy. Responsible for policy, budgetary and top-level technical matters for Shuttle program.

Level II: Manager, National Space Transportation Program. Responsible for Shuttle program baseline and requirements. Provides technical oversight on behalf of Level I.

Level III: Program managers for Orbiter, Solid Rocket Booster, External Tank and Space Shuttle Main Engine. Responsible for development, testing and delivery of hardware to launch site.

Level IV: Contractors for Shuttle elements. Responsible for design and production of hardware.

Fig. 4

FRR proceeds from Level IV through Level I (the last known as "*the* Flight Readiness Review").[17] Level IV encompasses contractors responsible for design and production of shuttle hardware. Contractor engineers first air their data analyses and conclusions for each item with their own managers, who review all items in order to certify in writing the flight readiness of their component. The contractors' risk assessments then are reviewed with Level IV Marshall personnel. Subsequently, the contractor presents these data and findings about risk acceptability to Level III NASA Project Managers responsible for development, testing, and delivery of the Orbiter, SRBs, External Tank, and Main Engine to the launch site. The Project Managers then present the engineering analysis and conclusions to the Level II Program Manager at Johnson. Each Project Manager certifies in writing the satisfactory completion of the manufacture, assembly, tests, and checkout of the elements for which he is responsible.

Finally, the Project Managers make presentations to Level I. The Level I FRR assays all previous stages, merging all parts of the NASA system. Participants are Chief of the Astronaut Office; Director of Safety, Reliability, and Quality Assurance; NASA Associate Administrator for Space Flight; NASA Chief Engineer; NASA Program Manager; Center Directors (Johnson, Marshall, and Kennedy), Shuttle Project Managers; and Senior Contractor Representatives. This review, held approximately two weeks before a launch, does not terminate the review process, however. A Mission Management Team, headed by the Level II Program Manager, assumes responsibility for each mission's readiness beginning at the conclusion of the Level I review and extending through the Orbiter landing. This team is on call throughout the entire period, reporting to Level I.

NEGOTIATING RISK: COOPERATIVE ADVERSARIES

Both the daily determinations of acceptable risk and the prelaunch decision making of FRR were characterized by engineering disagreements that had to be negotiated. They originated, first, because the shuttle design was unprecedented and its operation was extremely complicated, as we shall see. But disagreement also was built into the organizational arrangements of the government/contractor relationship, NASA in-house subunits, and the mandates of FRR. First, we will examine how risk was negotiated from day to day, then how it was negotiated in FRR. The description that follows will focus on Marshall and Thiokol, although the process described applies to risk assessments throughout the NASA-contractor system.

ROUTINE DECISIONS

In the daily implementation of the Acceptable Risk Process, one source of disagreement was the government/contractor relationship. Marshall and Thiokol had a contractual relationship for the SRM that bound the two engineering communities together in symbiosis: their fortunes were linked, rising or falling together with the successes and failures of the Shuttle Program.[18] Cooperation was built into the relationship because of their shared goals, but their different role responsibilities also created disagreements, making the two engineering communities cooperative adversaries. Jim Smith, Marshall's Chief Engineer, SRB Project, stated: "We worked with those people every day and you have a working relationship with them. It was not buddy-buddy, because it was business. We deal with a lot of money with these contractors. We expect a certain performance from them, for the dollar. Even though we deal with them, it's not like, 'Oh, my best friend here, you know, well, my best friend didn't do this or that and I'll kind of let him off,' you know. It's a business. But it was a cordial-type working relationship."[19]

Marshall's position with Thiokol was one of oversight responsibility. The government was not responsible for the everyday, step-by-step operations at the Thiokol plant, but Marshall Level IV and III engineers and managers actively directed contractor activities on the SRM's technology, "defining what was important, not important, approving certain things, delegating certain things to the contractor: yes, this is not as important, you can go ahead and do that, don't tell us about it. Some [technical issues] you penetrate quite deeply; others you don't penetrate as deeply."[20] The day-to-day oversight responsibility fell to engineers in Marshall's Science and Engineering Directorate (S&E). They monitored contractor engineers during the design review and approval cycle to make sure that the SRM matched the design specifications. Once flight began, these S&E engineers talked to Thiokol engineers every day. They participated in and monitored Thiokol risk assessments. They worked closely with Thiokol on prelaunch booster assembly at Kennedy and on its disassembly and postflight analysis.

In its monitoring role, Marshall had to contend with geographic distance. Since the SRMs were designed and manufactured at Thiokol in Utah, a number of Marshall people were located at the prime contractor's plant permanently and a number of people from Huntsville visited the plant regularly. Morton Thiokol contractor representatives also traveled to Huntsville. Communication by telephone was on a daily basis, usually many times a day. Telephone conference calls—"telecons"—were integral to communication. Marshall's Larry Wear, SRM Program Manager, noted:

> We ran on telecons. Telecons were a way of life. You faxed charts in and you talked to each other. The conference rooms were all wired for that. The Shuttle Program wired them with four telephone hook-ups. Each of the contractors had them. All of that was set up. Special switchboards and so forth. . . . Telecons were a way of getting the job done without spending the one day travel to [Morton Thiokol in] Utah and the one day travel back. It meant three days for every review, whether they came here or whatever, you lost three days work. So telecons were a way of life."[21]

In addition, the working engineers from both Marshall and Thiokol had biweekly reviews. Known as "data dumps," these meetings were in-person gatherings at Thiokol or Marshall where the "nuts and bolts people told each other everything they knew."[22] They also met in more formal monthly and quarterly project reviews. These two and a half day meetings were divided into two sessions: an executive session lasting half a day, combining working engineers and management from Level IV and Level III, and a two-day data dump between the "nuts and bolts" people. At these periodic data dumps, the engineers talked about technical issues only. In the executive sessions, technical issues were discussed, taking into account cost and schedule. Also, Marshall's Mulloy presented the schedule and got engineering feedback on its viability. The S&E engineers discussed the engineering changes they wanted: the technical reasons for the change; its impact (how much improvement would it produce and its effect on other systems); and its effect on cost and schedule. Changes were common and were effected by Engineering Change Requests (ECR). Initiated by working engineers, ECRs were reviewed and approved or disapproved by the Configuration Control Board, bypassing Project Managers.[23]

Negotiating Daily Disputes

Engineering disagreements and their negotiation were also structured into the work relationship of Marshall and Thiokol engineers in the work group. The S&E engineers were responsible for "keeping the contractor honest."[24] Key players in the *Challenger* story, they were known by some as "bad news guys," for their job was to be critical of all Thiokol conclusions about technical issues, negotiating with the contractor to assure that the SRM design and its operation conformed to standards. According to former S&E engineer Keith

Coates, now Chief Engineer of Marshall's Special Projects Office: "Our laboratory people tend to be critical of Thiokol, and I think there is a philosophy adopted, our role being more of finding fault, of finding mistakes. I mean, our role is not to just go out there and rubber stamp things, so the things [memos or reports] that I always see are comments relative to what's wrong. So you can develop an impression that the Thiokol people are always wrong because that's all you ever hear."[25]

Disagreements also were a natural product of testing arrangements. Marshall had its own testing facilities and labs, where the S&E engineers were working on the SRM just like the Thiokol engineers. However, Marshall's tests and testing equipment were not identical to Thiokol's, and in fact, Marshall engineers considered them superior. Because of the uncertain technology, in combination with different tests and testing facilities, Marshall and Thiokol outcomes and interpretations of the same phenomenon often conflicted. Even the same results could be interpreted differently.[26] Sometimes disagreements between the two communities were hard to settle because, as one long-time Marshall S&E representative put it, contractor working engineers tended to be "defensive about their design" because they believed in their own methods and analysis. In addition, a disagreement with Marshall's S&E "bad news guys" generally cost the contractor. Thiokol engineer Roger Boisjoly stated: "Thiokol management hated [Marshall S&E engineers] Ben Powers and Leon Ray. Thought they were muckrakers. Everything Leon found was not in Thiokol's interest, bad news. Leon would find it, tell his management, his management would call Thiokol, and there would be trouble."[27]

To settle these engineering disputes, Project Managers usually asked for further data comparison or more tests. In common, government and contractor work group members endorsed risk assessment based on scientific methods. Positivistic science and quantitative data settled disagreements between the two engineering communities. To resolve disagreement based on controversies about types of test and/or equipment differences between Thiokol and Marshall, the same result confirmed by multiple methods, multiple tests, and the most rigorous engineering analysis won the day. Consensus was the goal. If the available data did not create consensus, the Marshall SRB Chief Engineer made a recommendation, Safety, Reliability, and Quality Assurance representatives made a recommendation, and the Project Manager made the decision, based on the information and recommendations available. If the technical evidence still was divided and choice was not clear, the Project Manager could request a "referee test": the problem was given to another manufacturer to do independent testing to obtain an outside assessment, providing both engineering communities agreed to abide by the results.

A second organizational arrangement that routinely generated engineering disagreements was divisions within Marshall. Project Management and the S&E Directorate had differences in orientation that grew out of differing organizational roles. The S&E Directorate and Project Management were viewed as "different sides of the house." They, too, were symbiotic. Certainly, as NASA subunits, they held in common the goal of a successful Shuttle Program, but like the Marshall-Thiokol relationship, they were cooperative adversaries. Marshall's S&E engineers were responsible for the quality of the design and its operation; cost and schedule were not their responsibility. Thus, they were regarded by themselves and others as engineering "purists." They reported deviations and recommended the *best* of all possible resolutions. Usually, this meant increased cost in time and money. The Project Management, to whom they reported, considered not only technical issues and safety but also cost and schedule. Project Managers were responsible for the administrative management of their groups, which entailed not only managing the day-to-day activities but also acting as interface with the rest of the organization on the work group's behalf. By definition, they were more informed about and more directly responsible for seeing that the technical work was within budget and met the production schedule than working engineers.

Marshall's Keith Coates, noting the different responsibilities of Project Managers and S&E engineers, remarked: "I'm on the technical side of the house, and I have said that one of the easiest jobs that there is is to be an engineer in a laboratory because you can take the most conservative position in the world, and you don't have to make the decision. The decision gets bounced up to somebody who has to consider all factors and take the risks."[28] John Q. Miller, Technical Assistant, Marshall SRM Project: "I've never been involved in anything of the schedule type situations, 'I need to do this or I need to do that' because of schedule. I don't know of any schedule pressure that I have ever seen because I don't work schedules, and I did most of my work on the engineering side, and an engineer doesn't pay much attention to the Program Manager's schedule. I have never known one to. They are going to do it as best they can and as quick as they can do it, but they are not driven ever, that I know of, by schedule."[29]

The ability for S&E engineers to disagree with Project Manage-

ment was protected by NASA's matrix system. Like many traditional bureaucracies that are R&D organizations, the Project Management side of the house drew resources from other divisions.
Although Project Managers had their own small staff, they also had
people who were on loan, or in NASA-speak "matrixed," from other
parts of the organization. The S&E engineers assigned to the SRB Project were evaluated, paid, and promoted by their own division, which
gave them a degree of autonomy in their risk negotiations. But other
conditions mediated potential strain between the Project Managers
and S&E engineers.[30] First, each needed the support and products of
the other to accomplish its goals. Managers needed the knowledge,
techniques, and hands-on work of the working engineers; engineers
needed managers to secure the facilities, personnel, and financial
resources necessary to do their work. Second, much of the administrative load of NASA Project Managers belonged to business administrators, allowing managers to stay more closely connected to their
engineering staff. Problems with resources and procedures were
blamed on the business administrators, allowing Project Managers to
hold the respect of working engineers on both technical and managerial grounds.

When the inevitable controversies about risk occurred, in most
cases they did not need to be settled by managerial fiat. Scientific
positivism reigned; it was dispute resolution by numbers. Like disputes between NASA and contractor engineers, the parties provided
data to which each had special access. Quantitative data sufficient to
constitute a scientific proof of engineering logic would convert disagreements to consensus.[31] S&E engineer Leon Ray stated that S&E
often won: the Project Manager, who controlled his own budget,
found more money or changed priorities.[32] But the purists did not
always get what they wanted. In case of an unresolvable conflict
between the design needs, as defined by the purists, insufficient data
to create consensus, and the cost and schedule concerns of the Project Manager, a "management risk decision" was made. In NASA
culture, this term was not pejorative, but prosaic, referring to the
routine weighing of cost, schedule, and safety in management decisions when the data were not sufficiently compelling to create consensus. But the purists were not powerless: the working engineers
could submit an ECR that bypassed their Project Manager. Marshall's
Larry Wear described it:

> Anybody can write an ECR. It has to be formally dispositioned.
> There is no ducking it. Thiokol has a similar system. All the aero-
> space contractors have this approach. Any junior engineer who has

been hired fresh off the street can write an ECR, and there is no choice on anybody's part but to consider it straight up through the formal chain of command and disposition it. Now they may disapprove it, or they many approve it, that is a call. But my point is . . . there is no way you can duck the issue. You can't lose the paper. You can't hide the paper. You have to do something with it. So that avenue is open.[33]

LAUNCH DECISIONS

Engineering disagreements also were endemic to FRR. First, the Level III and Level IV managers and engineers in the work group who talked to each other every day carried their disagreements into FRR. The technology did not obediently hold still so that the launch decision chain could operate. FRR for each launch cut into what was a continuous process of assessing risk. Although the main purpose was to certify that technical issues had been resolved,[34] many were still being "worked" as FRR got underway. Testing and analysis were always ongoing, as were changes to improve shuttle performance. The result was an FRR agenda that included "open" (unresolved) technical issues, open hazard analysis, and other checks and balances that needed to be completed in order to qualify items as flight-ready, "acceptable risks." Second, disagreement was built into the formal procedure. By design, FRR was adversarial at each level: the engineering presentations were "probed" and "challenged" to identify any weaknesses in the engineering analyses that might conceal a flight safety issue. The sessions were open. The matrix system brought in many people from outside the work groups. The contractor engineers' initial construction of risk became the basis for discussion at each level of the FRR hierarchy, where it had to be negotiated with people from a variety of specializations who were positioned differently in the NASA-contractor system.

As the review moved up the hierarchy, more people were required to attend and more interested others showed up. Although a board formally was responsible for assessing and testing the engineering recommendations and conclusions at each level, all attendees were to bring their expertise to bear on the technical issues by adversarial challenges to the data analysis and conclusions. The certification process bound all levels to the final outcome; the questioning was therefore rigorous, as the ethnographic history will demonstrate. New technical issues came up at each level, calling for further engineering testing, analysis, or corrective actions. The goal was to "close" open items during the FRR, completing all qualification requirements, and to reach consensus on technical issues before a launch. Thus, engineering disagreements had to be settled. Disputes in FRR were settled by the same methods as daily disputes between the two engineering communities or between Project Management and S&E personnel: the methods of positivistic science, emphasis on quantitative analysis, and multiple tests with data covering every known condition were the means by which uncertainty was converted to certainty. To resolve them, Project Managers assigned "Action Items" directing that more research, more tests, more data analysis be done to settle these issues. "Delta" reviews were additional, follow-up meetings called to check out remaining open technical issues and review results of Action Items. The problem, of course, was that for the Space Shuttle, all conditions could never be known.

FRR Negotiation: Levels III and IV

FRR operated from the bottom up. The Level IV FRR was the linchpin that held the entire process together. It began with an internal review at Thiokol. Thiokol engineers prepared a package of charts, following prescribed guidelines, and an oral presentation for the Level IV FRR. For each part of their component, they included their (1) technical analysis, (2) conclusion about risk acceptability, and (3) recommendation about proceeding with the next launch. Thiokol engineers then met with their own management (known as the SRM Pre-Board), with Marshall representatives sitting in. The engineers presented their FRR "package." Since work always remained to be completed, the usual practice was that the Thiokol board certified the SRM as flight-ready, "pending the closure of open work and Action Items."

Thiokol then reported on SRM flight readiness to Level IV at Marshall. The format for the charts and discussion were prescribed by Marshall requirements.[35] This review was known as the SRM Board and was headed by Marshall SRM Project Manager Larry Wear. Ordinarily these meetings were held at Marshall, but when the launch schedule picked up, the Level IV FRR tended to be handled by teleconference because time could not be spared for travel. Wear noted: "The formality and the rigor or whatever of the review doesn't change whether its a face-to-face or a telecon. Some of the most intense meetings I was involved in were telecons and some of the most laid-back ones were in person. It just depended on the schedule or people's time. If you were pressed for time you did a telecon. If you weren't pressed for time, they (Thiokol) came in (to Huntsville). So a telecon could be the most serious meeting in the world and an in-person meeting could be the most mundane thing. It varied."[36]

The presentation was made by a Thiokol management representative—usually SRM Project Director Al McDonald, or sometimes Joe Kilminster, Vice President of the Booster Program—and the working engineers designated as "support personnel." The idea of "support" is that the hands-on people, who presumably possessed the most recent, in-depth information, participate fully and contribute what they know to the conversation. Engineers were chosen who had expertise on the technical issues on the agenda. Thiokol's charts were distributed in advance. Marshall's Jim Smith and the S&E working engineers also participated in what often was an all-day question-and-answer session. Marshall's Mulloy "sat in," although he was not required to do so. Smith described the Level IV review as "an engineering roundtable discussion of facts: understanding what we have, and do we completely understand it, do we partially understand it, and can we defend this position? And even today, we get in situations where we at Marshall may have one opinion, our contractor has another opinion. Being engineers, even those here at this Center, it is very difficult to get a whole group of engineers together and reach the same conclusion on something that's difficult."[37]

Yet at both Level IV and Level III, the two engineering communities had to negotiate differences and agree on the risk acceptability of numerous items because they had to present a consensus position at the next levels. In order to explore and settle differences, the Marshall managers and S&E engineers, in their oversight role, would probe Thiokol's analyses and conclusions. Thiokol's McDonald describing these lower level Marshall reviews: "On every launch, any time we had an anomaly, we had to get up and explain what that anomaly was, what we knew about it, why we felt that anomaly was acceptable to fly with if it should occur again, and what we had done to fix it. And that is the way every launch was conducted."[38]

Weak analysis or unresolved issues resulted in Action Items requiring Thiokol to "go home and do more homework." Thiokol's response to these Action Items would be discussed in teleconference Delta reviews with Level IV participants. At the conclusion of the Level IV FRR, Marshall's Wear and Thiokol's management representative signed a Certificate of Flight Readiness, "pending closure of defined open items." Thiokol's original package of engineering analysis would be reduced for each subsequent review, as problems got resolved and the "pending" qualification work got completed. Larry Wear noted: "The first review [Level IV] would be very broad, very general. All the soup and nuts of everything would be in the first review, but as you go up that ladder, things that you discuss get more and more limited and more and more focused. Discussion becomes problem-oriented as you go up the line."[39]

At the Level III FRR (known as the SRB Board), McDonald and his engineering support personnel reported in person to Mulloy, Smith, Safety, Reliability, and Quality Assurance representatives, and representatives from each of the engineering disciplines in the laboratories: electronics, materials processing, propulsion, and dynamics. As usual, the package of charts was distributed in advance. Although they followed a rigidly detailed, prescribed format, FRR discussions at both Level III and Level IV were relatively informal, primarily because participants dealt with each other every day. Informality notwithstanding, the Level III environment was much harsher and more adversarial. As Project Manager, Mulloy was responsible for presenting these risk assessments and defending them to all succeeding FRR levels. Mulloy described himself as the "neck of the hourglass" in the FRR process: all the information from the engineering communities came to him; then he was responsible for presenting it at the remaining reviews.[40] Mulloy was held accountable for all work on the SRB Project. Wear noted: "Other people exist, but they are faces in a crowd. You know, as you look down the organizational structure, you can only see so far, and the Project Manager is the one who is held accountable, and that's Level III."[41]

Mulloy subjected Thiokol's analysis to rigorous scrutiny to be sure that the hazard analyses would withstand the extensive examination yet to come. Mulloy noted: "As the Project Manager, I am in the position of getting data from Morton Thiokol and having the data assessed and they present the data to me along with a recommendation. I test the data and make my own judgment whether or not the data support the recommendation being made."[42]

The outside specialists, too, were extremely critical in their questioning and observations. At the conclusion of the Level III review, the SRB Board signed a Certificate of Flight Readiness that formalized the construction of risk developed by the work group. It was the consensus work group position on all items that Mulloy forwarded up the hierarchy.

From the Work Group to the Top

Subsequent reviews were more formal, but remained adversarial. Because Marshall had three of the shuttle's main elements—the Main Engine, the External Tank, and the SRBs—it had two additional reviews (not required by NASA and known as Level "2 1/2") that preceded the Level II and Level I FRRs. The first was the Shuttle Projects Office, which oversaw and coordinated the three projects. The second was the Marshall Center Board FRR, chaired by Director William Lucas. The Shuttle Projects Office Board review was attended by few people; the Marshall Center Board review was held in a large conference room, sometimes with over 100 people attending. Mulloy was one of three Project Managers presenting. He gave both boards the same documents (a package further compressed from Level III and distributed in advance) and oral presentation.

By the time of Level II and Level I review, most of the work was done: items qualified according to guidelines, technical issues from the previous flight analyzed, corrective actions implemented, predictions made for performance on this launch, open problems closed. Therefore, the package of charts distributed in advance was smaller. Significantly fewer issues were discussed than at Marshall, where the prescribed format required broad issue coverage. Always some technical issues remained open. However, only selected "topics of substance" were presented in these final reviews, in order to inform the upper administrators of major issues in the launch decision process. What qualified as a topic of substance was not left to the discretion of Project Managers, however. Rules prescribed the information that was to be carried forward to Levels II and I. These will be discussed in detail as the chronology unfolds. One very important one, however, needs to be mentioned here.

Level II and Level I FRR reporting was governed by a NASA policy that insiders informally called "management by exception."[43] Formally, it was known as Delta review—"delta" meaning change. Delta review had two meanings in NASA culture. The first, discussed earlier, simply referred to an additional meeting, a follow-up review on some pending issue. The second meaning prescribed the information Level III Project Managers presented at Level II and I FRR.[44] In flight-readiness reporting, NASA's "Delta review concept" required that any change, or deviation, from what was previously understood or done had to be reported. Changes required further analysis and revised technical rationale for risk acceptability; thus senior management needed to be informed.[45] For example, if an aluminum wire had been used on the previous flight and a copper wire was being used on this flight, that change and its technical rationale had to be presented up the hierarchy. Also, any new problem or new manifestation of an old problem had to be reported with engineering analysis and conclusions about risk acceptability. If a test or flight showed data that deviated from existing technical standards for acceptable risk, that had to be reported at all levels.[46] The converse was true: if the preexisting technical rationale still covered the condition, it was not discussed at upper-level reviews because the preexisting engineering rationale for risk acceptability still applied. The Delta review concept was intended to streamline FRR reporting at upper levels, which brought together all parts of the system.

At the final Certification of Flight Readiness at Level I, Project Managers reported these technical issues, indicating that they had been reviewed and understood and that they were ready to fly. After yet another adversarial session, the Level I signatures on the Certificate of Flight Readiness were the final approval for launch. When the Mission Management Team took over prelaunch activities (reporting back to Level I as necessary), their agenda included "closeout of any open work, a closeout of any FRR Action Items, a discussion of new or continuing anomalies," and weather briefings. Action Items would often delay and extend FRR, but rarely was a launch delayed because of a technical issue that turned up during FRR. Most launch delays occurred because of problems found during countdown. Marshall S&E engineer Ben Powers said: "Most of them are stopped at the Cape. I don't know of one that's stopped in the FRR process, because normally when you get to the FRR process your hardware's already been cleaned and it's ready to go. When you get it at the Cape, that's typically where you're going to stop a launch, right there. That's when problems show up."[47]

"Gut Check"

From the bottom-up negotiation in FRR, an official construction of risk resulted, certifying all shuttle components as flight-ready, acceptable risks. However, as countdown proceeded to its conclusion, the engineers most familiar with the technology were well aware of foibles of the technology and the negotiated nature of the aggregate residual risk of the shuttle.[48] Larry Wear said:

> We knew that going to space is a challenging thing. The weight margins have to be as low as you can possibly make them otherwise you aren't going to achieve orbit. The performance that you are requiring of everything in that vehicle is top notch. . . . Everything that's designed is more like a fighter plane than a commercial airplane. There are risks and there are hazards. It is a high performance flying machine. It's a high performance re-entry body. It's a high performance landing device. Everything in it is a high performance tool and it has elements of risk all the way through. So

you know, none of it is easy. There is not a great deal of what we thought of as design margins in anything. . . . It was a white knuckle event just to run the main engine on a test stand, let alone on a launch pad with people on top of it. High performance equipment is, by nature, unforgiving.[49]

Many said a prayer as each shuttle was about to be launched;[50] for Jim Smith:

It doesn't matter how many you go through, they're always a gut check for you, gut check meaning you are never totally comfortable. Your stomach's tight, it's always anxiety, it's not just routine business, not like hopping on a plane. There is always an uneasy feeling. You can feel the emotions of us all at the time, so we call it gut check. You are dealing with explosives. You are dealing with large structures. You've got engines with tremendous amount of force. It's a complicated vehicle. So you always have that kind of tight stomach. There is a big sigh of relief once it's over. . . . As confident as you are, you have every reason to believe it's going to be successful, but there is a fear, an anxiety of the unknown, something you don't have any control over. You know, you've done everything you can to the best of your ability, and then something could happen that you don't have any control over."[51]

From this primer on the culture of risky decision making, now we turn to the decision making itself and the first of the two questions this book addresses: Why, in the years preceding the *Challenger* tragedy, did NASA proceed with launches with a design known to be flawed? In the remainder of this chapter and the next two, we examine the production of a work group culture that repeatedly normalized the technical deviation of the joint from performance predictions. We focus on the native view: the interpretive work of the technical experts at the bottom of the launch decision chain as they grappled with booster anomalies, beginning with the early development period.

THE NORMALIZATION OF DEVIANCE: THE EARLY DEVELOPMENT PERIOD

The beginning construction of risk was that the SRB joints were failsafe. This belief was grounded in the working engineers' faith in the Thiokol design.[52] It was based on the Air Force's Titan III solid rocket, widely regarded in the aerospace industry as one of the most reliable ever produced.[53] The Titan had one O-ring, the shuttle design had two—a primary and a secondary to back up the first. Both Thiokol and NASA believed that the second O-ring provided a redundancy that further increased the safety and reliability of the original Titan design.[54] If, for any reason, the primary ring did not perform as expected, the secondary O-ring would seal the SRB joint. Marshall S&E engineer Leon Ray was a member of the Source Evaluation Board that evaluated the designs originally submitted by contractors who bid for the SRB contract. Ray remarked that at the time Marshall had no questions about the way the joint or seal in the Thiokol design would operate: "We all went into it with the very best knowledge that we could gather, and we at that time thought it was an acceptable design. We thought from the technical standpoint that probably Lockheed had a little better design, but that's yet to be proven also. It's strictly speculation."[55] Neither Marshall nor Thiokol expected problems.[56]

Despite basic similarities, however, many differences existed between the two designs. In order to meet the distinctive needs of the shuttle, the Titan design was altered. The second O-ring resulted in a longer tang on each segment of the SRB, creating a greater tendency for it to bend under combustion pressure; the fit of insulation between segments of the shuttle's SRB was not as gas-tight as the Titan, resulting in reliance on greater amounts of putty to fill the space; pressure of combustion in the shuttle's booster joint was one-third higher than the Titan's; and finally, unlike the Titan, the shuttle's SRBs were designed to be recovered after flight and reused, perhaps up to 20 times.[57] These design changes—especially the elongated tang necessary to support two rings—altered the functioning of the joint. In reality, the SRB joints were a different, and untried, design.

JOINT ROTATION

In the mid-1970s, even before Thiokol began producing the SRM segments, uncertainty developed about how the joint would operate. Marshall S&E engineers Ben Powers and Leon Ray had calculations indicating that at ignition the SRB joint momentarily would rotate open: the tang would move away from the clevis, creating a gap (see fig. 5). Joint rotation was a well-known phenomenon in the aerospace industry, and so were gaps: O-ring manufacturers' books contained industry standards listing gap sizes acceptable for all standard O-ring sizes. Consequently, the SRB joint rotation was not necessarily a signal of potential danger. Nonetheless, this was the first indication that the joint might not operate as expected. Thiokol calculations, however, indicated the opposite: the joints would close at ignition.[58] The disparity between Marshall and Thiokol calculations about joint rotation was the first of the engineering controversies that would characterize the history of decision making about the SRB joints. Ironically, in these early disagreements, the positions of Marshall and Thiokol representatives were the reverse of those they held the night before the *Challenger* launch: Marshall was expressing reservations—sometimes vigorously—about the functioning of the SRB joints; Thiokol was expressing a belief that the joints would perform as the design predicted.

The rotation controversy was settled when a September 1977 Thiokol test, called a hydroburst test, confirmed the Marshall engi-

JOINT ROTATION

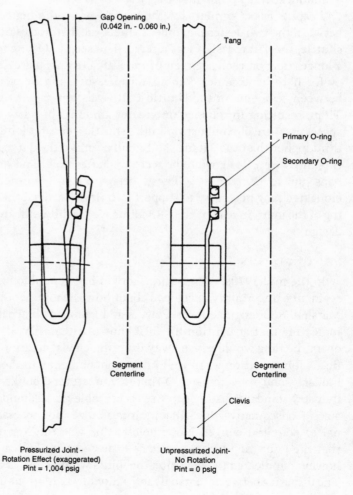

Fig. 5

neers' calculations: pressure at ignition forced the joint tang and cle-vis to move away from each other. As the joint rotated open, a gap occurred, reducing rather than increasing pressure on the O-rings in the milliseconds after ignition.[59] Now Thiokol agreed with Marshall that the joint opened, but they differed on the implications of this finding for joint redundancy. Thiokol's engineers did not believe joint rotation would cause problems, so they scheduled no further tests to explore it.[60] But Marshall engineers, in particular Leon Ray, dis-agreed with Thiokol's conclusions. By his own account, Ray was "way, way down in the organization."[61] An S&E engineer, his job was to pursue every possible problem with the SRB design. Many people respectfully described him as a sort of organizational gadfly: he frequently stood up in FRRs and challenged his superior's opin-ions. And he kept after them, sometimes picking up the phone when he had something to say. Based on the hydroburst test results, Ray was concerned that joint rotation might be a threat to redundancy.

Thiokol engineer Howard McIntosh told why Thiokol engineers disagreed. First, according to McIntosh, "He (Ray) perceived that the rings blew out, but he didn't take into account the fact that it was a test of more than a single (ignition) cycle."[62] The hydroburst test put a single O-ring through 20 cycles that simulated pressure at ignition to see whether the rings would hold. In actual use, an O-ring experi-ences ignition pressure once, not 20 times. The leakage that occurred on the test was from an O-ring worn and weakened from 20 tests. The rings had worked perfectly for 8 of the 20 trials. In Thiokol engi-neering opinion, the hydroburst was a successful test, showing that the rings were operating as intended.[63]

Second, McIntosh stated that Ray's concern was an example of the conservative engineering philosophy that typified S&E engineers at Marshall, and McIntosh disagreed with it:

> You take the worst, worst, worst, worst, worst case, and that's what you have to design for. And that's not practical. There are a number of things that go into making up whether the joint seals or doesn't seal or how much O-ring squeeze you have, and they took the max and the min which would give you the worst particular case on everything that went into that calculation, such as the groove depth, the tang thickness, the gap of the clevis, the O-ring dimensions . . . the shims, they took the worst on that one . . . all those worsts were all put together, and they said you've got to design so that you can withstand all of that on initial pressuriza-tion, and you just can't do that or else you couldn't put the part together.[64]

Third, Thiokol engineers objected because the hydroburst test was a horizontal test that produced more extreme conditions than would occur on the vehicle when it was vertical on the launch pad.[65] Gravity, Thiokol believed, was distorting the test results. Ray believed the results were influenced by ignition pressure, not gravity. George Hardy, Manager of the SRB Project at Marshall from 1974 to 1982, remarked: "There was quite some discussion between the two engineering communities for some time over the appropriateness of the data that was measured there as to whether it was truly measuring joint rotation or whether it was not doing that, and it had to do with the fact that the test was run with the portion of the Solid Rocket Motor in the horizontal position versus [i.e., instead of] the vertical position and the fact that there would be some effect of the sag of the motor."[66]

In response to the hydroburst test data, in October 1977 Ray wrote a memo, the first of two he authored that after the *Challenger* disaster raised questions about managerial actions. Contrary to some accounts reporting that Ray's memo called the design unacceptable, Ray stated that "no change" in the Thiokol design was "unacceptable."[67] Then he listed a number of "design options" for dealing with rotation and initial compression that would make it acceptable. Redesign, Ray wrote, was "the best option for a long-term fix." In NASA language, "long-term" and "short-term" fixes refer to implementation times, not the durability of the corrections. Once a new design was worked out, new hardware would have to be manufactured, so Ray knew that a new design was years away.[68] Consequently, he recommended shimming the joint as an "acceptable short-term fix" to increase the initial compression. A shim is a piece of metal that fits between the outside edge of the tang and the clevis, tightening up the joint. Shims could be used to reduce gap size and increase the initial O-ring compression, or "squeeze." Although shimming would have no effect on the gap created at rotation, it was "highly desirable" because of "potential defects in the sealing surfaces and in the O-ring from contamination and so forth."[69] Ray recommended that shimming be discontinued when the existing hardware was phased out.

Continued tests to resolve the disagreement simply added to it. Marshall and Thiokol engineers disagreed about the size of the gap that resulted when the joint rotated. Although joint rotation was not necessarily a problem, the size of the gap mattered: if a gap were sufficiently large, it could affect the sealing capability of the rings. But tests to determine gap size continued to produce conflicting results:

Marshall's data indicated a larger gap (0.060″) than Thiokol's (0.042″). On such seemingly small fractional differences rested major differences in the perception of risk. Thiokol's measurement led them to believe that *the secondary O-ring would always be in position to seal and that the joint thus was redundant for the full ignition cycle.* Marshall's larger figure created *uncertainty about the availability of the secondary to back up the primary.* Both sides ran test after test, but the disagreement went unresolved. As Leon Ray commented, "We had sort of a gentleman's running battle argument, if you want to call it, our people and Thiokol, over the years about this situation."[70]

In January 1978, Ray authored his second controversial memo, written in the name of John Q. Miller, Ray's superior and chief of the SRM branch at Marshall.[71] The memo identified two other SRB joint problems: (1) O-ring quality was unacceptable because the calcium material was not being ground fine enough by the manufacturer, Parker Seal Co., and (2) O-ring compression before ignition was in violation of the minimum 15 percent industry standard.[72] This memo, known after the *Challenger* tragedy as the Ray/Miller memorandum, brought to Marshall management's attention the first rule violation in the history of the O-ring problem: the joint operation violated an industry standard. As before, Ray advocated redesign to bring initial compression up to industry standard and advocated improved shimming and O-rings as an acceptable short-term measure. The Ray/Miller memo concluded: "In summary, we believe that the facts presented in the preceding paragraphs should receive your most urgent attention. Proper shim sizing and high quality O-rings are mandatory to prevent hot gas leaks and resulting catastrophic failure."[73]

Both concerns in this 1978 memo were resolved. Parker Seal corrected the calcium flaws in the O-ring composition. Acknowledging that "you can't produce a perfect O-ring 456 inches long, you just can't do it," Ray nonetheless felt that a "very high quality O-ring" resulted after the corrections.[74] He was satisfied that it was the best possible. In addition, Ray and other engineers at Marshall and Thiokol began work on shimming the joint to resolve the initial compression problem. Meanwhile, Marshall began an important four-month series of tests called a Structural Test Article (STA-1). This test simulates the pressure loads the shuttle experiences during launch. Since engineers could not see what was happening in the joint during this test, electrical devices were inserted to measure joint rotation at ignition. Marshall data affirmed Ray's fears: the size of the gap was excessive. As a result, Ray believed that both rings would be out of position during ignition, when sealing occurred. Figure 5 shows the appropriate position for sealing in the right panel. Note that both rings touch both tang and clevis. At ignition, the gap size predicted by the design would allow both rings to maintain contact with both surfaces during sealing. The left panel shows the reasons for Ray's concern. The recent measurements showing excessive gap size indicated that at joint rotation, the rings would no longer be wedged between both surfaces. Under that circumstance, the primary would seal by hopping off its "seat" in the groove and pushing up into the gap. After the primary sealed, the secondary O-ring would be left "floating" in its groove, unable to seal the joint if the primary failed.[75]

Subsequently, Ray and Miller wrote another memorandum stating, "We find the Thiokol position regarding design adequacy of the clevis joint to be completely unacceptable."[76] They gave the following reasons: (1) The joint was violating yet a second industry standard: joint rotation caused the primary O-ring to "extrude into the gap, forcing the seal to function in a way that violates industry and Government O-ring application practices."[77] The SRB primary ring sealed, but by an unapproved mechanism: extrusion, not compression. Although sealing by extrusion was common in industry, it was not recommended.[78] (2) Joint rotation created a gap sufficiently large that the secondary could not be counted on to seal. The possible unavailability of the secondary was in violation of the design specification that required dual (i.e., redundant) seals.

But Thiokol engineers thought Marshall was wrong. They believed that the electrical devices used on STA-1 did not perform as they were supposed to.[79] The measurements of joint rotation were "off the charts" compared to physical measurements of rotation during STA-1, leading Thiokol engineers to believe the calibration was off.[80] The physical measurement confirmed Thiokol's own test data indicating a smaller gap size, so they believed that the secondary *was* in sealing position. To try to resolve the debate between the two engineering communities, Marshall authorized Leon Ray to visit two O-ring manufacturers in February 1979, seeking advice about the joint rotation problem. Ray presented Marshall's data on the joint opening. Since the SRB design was the only one of its kind, neither O-ring manufacturer had comparable data. Examining manufacturers' books listing industry standards for gap sizes acceptable for particular O-ring sizes, they informed Ray and Robert Eudy, Chief Engineer, SRM Project Office, that the SRB joint gap size was larger than the recommended industry standard. (This violation of industry standards constituted the third for the booster joint.) Both manufacturers said that the O-ring was being asked to perform beyond its intended design. They also expressed concern about the appropriateness of the tests that were producing these data. Both manufacturers recommended that "tests which more closely simulate actual conditions should be done."[81] As Ray summarized the result:

> I made the presentation to their technical folks, and we told them, the joints are opening up, and here's how much. What do we do? We've got the hardware built, we're going to be flying before long, what do you think we ought to do? And those guys said, well, that's a tough one, a real tough one. I feel sorry for you guys, I don't know what you're going to do, but we will study it over and give you an answer. . . . Later on they both wrote letters back to us and said, hey, you've got to go with what [hardware] you've got. Do enough tests to make sure the thing's going to work. Now that's their recommendation. Go on and do some tests, and go with what you've had. That's all we can tell you. So we did.[82]

Following the advice of the O-ring manufacturers to "make sure the thing's going to work," Marshall and Thiokol engineers continued their tests. To determine whether the design was an acceptable risk, the working engineers analyzed SRB joint performance in the three areas in which their design was violating recommended industry standards: (1) sealing by extrusion, not compression, (2) gap size, and (3) initial compression level. They did "extrusion tests" to examine the primary's ability to seal with gaps much larger than either the Marshall or Thiokol figures on gap size. With a gap of 0.125" (an eighth of an inch—about twice Marshall's assessment) and using 5,000 psi (pounds per square inch; five times the pressure per square inch experienced during launch), they had no problem with the seal. Said Ray, "We opened the gap up that far (0.125") and pressurized up to 5,000 psi and everything was okay."[83] Both Marshall and Thiokol now felt confident in the primary's ability to seal. *They found that despite its being in violation of the industry guidelines for gap size and sealing mechanism, the primary ring sealed under conditions far more severe than anything likely to be experienced during a launch. Once the primary sealed, and the ignition transient completed, ending joint rotation, the primary would remain tightly wedged in the now-closed gap through the two-minute period until the boosters were jettisoned into the sea.* Then they checked to see what would happen under a worst-on-worst (WOW) condition (rare circumstances in which a number of negative conditions occur simultaneously) in which the primary failed at initial pressurization. Would the secondary still act as an effective backup? Thiokol engineers tested with a purposely failed primary, grinding the ring down (simulating the effect of erosion, although at this point they were not expecting it) so it would not be able to seal. *With a purposely failed primary, they found pressure at ignition activated the secondary, which then sealed the joint, fulfilling its redundant function.*[84] This was completely acceptable. As Marshall's Ben Powers explained: "The whole concept of redundancy is that if the primary system doesn't function, the secondary system will. Therefore, if the primary system fails and the secondary system backs it up, the system is working as expected. Building in redundancy is expensive. You don't build it in unless you expect to use it."[85]

Finally, they turned to the issue of initial compression. All the O-ring manufacturers' design manuals and a report by Aerojet Solid Propulsion Company stated that the joint *would seal with compression values lower than* the industry standard.[86] By altering the shims, reducing the joint metal tolerances, and increasing O-ring size, Ray and his colleagues were able to upgrade initial joint compression to 7.54 percent, which still was only half the 15 percent industry standard. The result? When the joint was tested, Ray reported: "We find that it works great, it works great. You can't tell any difference. You don't leak at 7 1/2 percent."[87] The changes worked.[88] Performance in both subscale and full-scale hardware tests indicated that the industry standard for initial compression could be violated to no ill effect.[89]

But this decision, too, was the outcome of conflict and negotiation between Marshall and Thiokol engineers. Roger Boisjoly had joined Thiokol in July 1980. He began working on the joints and seals in 1981. His first assignment on the SRB joint was to work closely with Leon Ray on shimming the joint to resolve this initial compression problem. Thiokol's Arnie Thompson, Supervisor of the SRM Cases and Boisjoly's boss, reported: "He [Roger] had a real good experience, hard, tough, head-knocking experience at first with Leon [Ray], because both happen to be vigorous people, and both of them, you know, stand for what they believe, and after, oh, gee, I guess it would be seven or eight months of difficult conversations, Roger and Leon came to an agreement on 0.020 . . . which is about 7.5 percent squeeze [initial compression] . . . that was negotiated, you know, that we would never go below 0.020 squeeze, and we would in fact select hardware [shims] so that that would not happen."[90]

At this juncture, engineers at Marshall and Thiokol unanimously agreed that although the joint performance deviated from design expectations, it was an acceptable risk. Jack Kapp, who supervised Arnie Thompson, Roger Boisjoly, and the other working engineers at Morton Thiokol doing the hands-on work, stated:

> I think it's fair to say that the consensus was that if the primary O-ring seals, we've got it made. We had not seen erosion, remember. If the primary O-ring, for whatever reason, fails or seals initially, we have got it made. If the primary O-ring fails to seal initially, we have a redundant seal. Now we all agreed to the fact that if for some reason the primary O-ring fails at a high pressure [not initially, but milliseconds later—a WOW condition], then yes, there is a finite possibility that the secondary O-ring will not be in a position to seal. But again, the extrusion tests that were done at that time told us that there was a relatively high factor of safety on the primary O-ring once it got in there.[91]

Marshall's Leon Ray concurred, based on tests and the Titan experience: "We had faith in the tests. The data said the primary would always push into the joint and seal. The Titan had flown all those years with only one O-ring. And if we didn't have a primary seal in a worst case scenario, we had faith in the secondary."[92] Ray had gotten no response to his 1977 report recommending redesign, and he remained concerned. However, the testing and corrective actions that occurred after he wrote that memo convinced him that it was safe to fly. He stated: "Our recommendation after we talked to these O-ring vendors was not to stop the program, but to shim up, and then all new hardware coming down the pike we redesign. And we did not have a new design at that time. We had a lot of ideas that were tossed in. None were ever finalized. But I guess I was quite sure we'd come up with something, right or wrong, one way or the other, had we been given the go-ahead."[93]

The "go-ahead" would not be given until 1985. A redesign decision was contingent on consensus among the working engineers. Marshall's Maurice Parker, Manager of Marshall Space Flight Center Operations at Thiokol, who retired two years before the *Challenger* incident, remarked:

> There was lack of unanimity on the seriousness of the situation and whether it required something be done. Under pressure the gap which was being sealed by the O-ring was substantially larger than we had intended initially. The O-ring was having to perform beyond what we had originally intended, but the question was, is this a serious problem? One of the tests we ran to find out whether it was serious was this test where the O-ring withstood a 0.125" gap: 1,600 and 5,000 psi still did not prevent it from doing its job. . . .

To redesign at that point and stiffen that joint sufficiently to mate-
rially improve this condition [was out of the question because] first
of all, we didn't know for sure that it [the problem] was real: the
extent, the quantification of it, was very, very, uncertain. There was
really no serious thought of stopping the program and saying, hey,
we have to redesign this. As an old O-ring man, I was not pleased
either that it didn't work out the way we originally designed it, but
as far as its ability to do its job, it was not a concern.[94]

Both engineering communities agreed the primary would seal;
both asserted that the joint was redundant—but they were operating
on two very different understandings about redundancy. Leon Ray
and some others at Marshall believed *redundancy was limited to certain conditions, in violation of the design specification for dual
seals.* Data on gap size indicated that the secondary O-ring provided
redundancy at a crucial moment: if the primary failed at initial pressurization. But they believed that the gap would be sufficiently large
that the secondary would be unavailable in a WOW condition causing the primary to fail late in the ignition transient. At Thiokol, however, Roger Boisjoly's own measurement of gap size led him and others to believe that the secondary O-ring would be in position to seal
throughout full rotation, thus *the joint was fully redundant, meeting
the specifications for dual seals.* Boisjoly remarked: "In all honesty,
the engineering people, namely Leon Ray and Ben Powers and
myself, always had a running battle in the Flight Readiness Reviews
because I would use 0.042" [gap size] and they were telling me I was
using a number too low, and I would retort back and say no, the horizontal number [Marshall's 0.060" gap size] doesn't apply because we
really don't fly in a horizontal position."[95]

The two engineering communities were at the first turning point
in the normalization of deviant joint performance. From very different definitions of redundancy, both came to the same conclusion: the
design, although deviating from performance expectations, was an
acceptable risk. Further, fixing the joint by improving the shims and
the O-rings was accepted as an appropriate response to the problem.
These two early conclusions—accept the risk and proceed with
flight; correct rather than redesign—would become norms guiding
subsequent decision making, characterizing the work group culture
in the years to come.

FORMALIZING THE CONSTRUCTION OF RISK

In 1980 and 1981, three events transformed the construction of
risk negotiated by working engineers into an official NASA organi-

zational construction of risk: certification of the SRBs as flightworthy, assignment of a Criticality category, and FRR for the first shuttle flight, STS-1. The formalization of the construction of risk was
another turning point in the normalization of deviant joint performance at NASA.

NASA's Verification and Certification Committee was responsible for certifying the flightworthiness of each element of the shuttle
system prior to the first scheduled launch in 1981. The Committee
paid particular attention to the joint design and operation. Marshall
S&E engineers told them that NASA specialists had reviewed the
field joint design, updated with larger O-rings and thicker shims, and
found the safety factors adequate for the current design. In the wake
of the *Challenger* tragedy, some analysts interpreted NASA's "certification of a flawed design" as one of the early incidents attesting to
the priority of schedule over safety. It is important to note that the
Committee's decision was based on the findings and recommendations of Ray, Boisjoly, and other engineers in the SRB work group. At
the time, the working engineers believed they understood SRB joint
dynamics and that the boosters were an acceptable risk. NASA's
report to the Committee stated "the joint has been sufficiently verified with the testing accomplished to date."[96] The SRB was certified
as flightworthy by the Committee in September 1980.

The second action formalizing the construction of risk was the
assignment to the component of an official status on NASA's Critical Items List (CIL).[97] Criticality categories were formal labels
assigned to each shuttle component, identifying the "failure consequences" should the component fail. These designations were conservative because their assignment was based on failure analysis
under WOW conditions.[98] Here is the labeling system:[99]

CRITICALITY 1: Loss of life or vehicle if the component fails
CRITICALITY 2: Loss of mission
CRITICALITY 3: All others
CRITICALITY 1R: Redundant components, the failure of both could
    cause loss of life or vehicle
CRITICALITY 2R: Redundant components, the failure of both could
    cause loss of mission

The CIL entry for each item told why each component was an
acceptable risk. Since all items were risky, the entries described the
data and actions taken that the engineers believed precluded catastrophic failure. The "Rationale for Retention" included the "technical rationale," based on evidence from tests and flight experience,

that stated *why the design should be retained* for the critical item. Written by contractor engineers working closely with the S&E engineers at the NASA center responsible for the element, it presented all the data analysis to date, documenting risk acceptability.[100] Engineering disagreements between the two engineering communities had to be negotiated to get the document written and approved because government, contractor, and Safety engineers had to sign it. Marshall's Mulloy described the process of negotiation between working engineers of the two organizations:

> Sometimes it's hotly debated, and the wording gets changed a lot in those things. Thiokol may come in with something when its first submitted and the government rejects it saying, "That won't fly; that's not good enough." That causes Thiokol to go back and do more analysis and more testing before that rationale is finally accepted. . . . It's not adversarial from the standpoint of the government saying, "Guess again, that won't do." The responsibility rests with the prime contractor, but we always have a number of people at the prime contractor's plant, and a number of people visiting the prime contractor's plant, and the prime contractor comes to Huntsville a lot. So in the end, by the time something is agreed upon it's been through the "boiler room" of opinion, if you want to call it that, and an agreed upon rationale is derived.[101]

In November 1980, the SRB joint was classified as Criticality 1R (C 1R) on the CIL, the "R" acknowledging the redundancy of the joint (see appendix B, figs. B3.1 and B3.2). The document accurately depicts the disagreement between Marshall and Thiokol engineers about the timing of redundancy. In the Rationale for Retention, Thiokol engineers report Marshall's concern about the secondary's ability to seal if the primary failed late in the ignition cycle (see the fourth paragraph of section A): ". . . redundancy of the secondary field joint seal cannot be verified after motor case pressure reaches approximately 40% of MEOP [maximum expected operating pressure]. It is known that joint rotation occurring at this pressure level with a resulting enlarged extrusion gap causes the secondary O-ring to lose compression as a seal. It is not known if the secondary O-ring would successfully re-seal if the primary O-ring should fail after motor case pressure reaches or exceed 40% MEOP."[102]

The next three sections constituted the technical rationale for retaining the design: the tests, inspection procedures, and failure history that convinced both engineering communities that the design was an acceptable risk.[103] The many test results affirmed the ability of both the primary and the secondary to seal early in the ignition transient. The Thiokol conclusion of full redundancy from STA-1 was included. The CIL entry concluded: "Failure History: No known record of failure due to case joint seal leakage on segmented 156" or Titan IIIC motors. No failures in the four development and three qualification SRM motor test firings."[104]

The third event transforming the construction of risk by working engineers into an official NASA organizational construction of risk was the March 1981 FRR for *Columbia* (STS-1), the first of NASA's four scheduled test flights in the developmental period. The construction of risk negotiated and confirmed by Marshall and Thiokol working engineers was presented for review and approved by Level IV, Level III, Level II, and Level I. Certified at all levels, the SRBs were officially defined as an acceptable risk, as were the Orbiter, Main Engine, and External Tank.

On April 12, 1981, the first Space Shuttle was launched from Kennedy Space Center. Orbiting the earth 36 times, Astronauts John W. Young and Robert L. Crippen brought *Columbia* down at Edwards Air Force Base in California's Mojave Desert two days later. After the boosters were recovered from the sea, they were disassembled and examined. The field joints were taken apart at the landing site. But the nozzle joints (which followed a different design from the field joints and were "factory joints," welded together at the Thiokol plant in Utah) had to be shipped back to Utah by train to be taken apart, so disassembly was always delayed. No anomalies were found in either field or nozzle joints: they had performed as Marshall and Thiokol engineers had predicted. For engineers, a design is a hypothesis to be tested.[105] But tests only approximate reality. The proof is in the performance. For the shuttle, flight was the ultimate test. STS-1 affirmed the official construction of risk of the Space Shuttle but not just because the mission was completed. The affirmation was based on engineering analysis of all components done from in-flight data transmission and inspection after the vehicle landed. Milton Silveira, Chief Engineer, NASA Headquarters, later wrote: "The first flight of the shuttle proved a number of concepts that were under research and development for years. The concept of a winged recoverable and reusable spaceflight vehicle had been under study since the early 1950s. A new reusable surface insulator [tile] thermal protection system was used successfully for the first time. A new dual-cycle, high-specific-impulse, main liquid propulsion system, which is bottleable and reusable, was flown for the first time. . . . This represented a state-of-the-art system. Many other new technology systems were used, and *the first flight represented a proof of the design concept"* (emphasis added).[106]

At Marshall and Thiokol, the postflight analysis of the STS-1 booster joints affirmed the decisions the work group had made before the launch. They had developed a technical rationale, based on many tests, that the SRB joint was an acceptable risk. The determination of acceptable risk had two significant procedural implications: (1) the design was not a threat to flight safety, so flight could proceed, and (2) the appropriate corrective action was to fix, not redesign. STS-1 affirmed these choices and what was to become the dominant ideology of the work group in the years preceding the *Challenger* tragedy: the belief in redundancy.

This chronology of the early development phase of the shuttle shows the early turning points in the production of a work group culture at NASA that was created by the people assigned to the booster joints. We see how the beginning construction of risk—that the design was a redundant system—survived the first challenges that the joint was not working as expected. In the decision making leading up to the launch of STS-1, we see the first occurrence of the five-step decision sequence revealing the initial choices that would become a pattern in the history of decision making pertaining to the O-ring problem. This decision sequence allows us to follow the negotiation of risk as the first evidence showing that the behavior of the technical component was deviating from predictions was normalized.

Beginning construction of risk: a redundant system. Both Marshall and Thiokol believed that the original Titan design, respected in the industry for its reliability, was improved further by the addition of a second O-ring to back up the first and that redundancy was thus designed into the SRB joint. No problems were expected.

Signals of potential danger. Test performance deviated from design predictions, creating uncertainty. Marshall engineers discovered that the SRB joint rotated open at ignition, creating a gap between the tang and the clevis. According to Marshall data, the gap size was sufficiently large that the secondary would not be in position to seal if the primary ring failed late in the ignition cycle. This finding indicated a threat to mission safety, thus challenging the initial construction of risk.

Official act acknowledging escalated risk. No official action was taken. In the developmental phase of an innovative technology, engineering controversies such as the ones occurring between Marshall and Thiokol engineers are common as kinks are worked out of the system. Yet joint rotation was taken seriously, as indicated by Marshall's authorization of Ray's trip to manufacturing firms to discuss SRB joint problems and the resource allocation that initiated more extensive testing.

Review of the evidence. Although the two engineering communities continued to disagree about the size of the gap, test results convinced both that the O-rings were an acceptable risk. Based on many tests and seven static motor firings, they agreed that the primary would seal at initial compression, and if it failed initially (which analysis indicated was likely only under rare WOW conditions), the secondary would seal the joint, performing its redundant function. In combination, the tests and corrective action (improved shimming and larger O-rings) resolved work group concerns about the O-rings. Even Ray, a consistent voice of concern in the developmental stage of the program, found the joints an acceptable risk.

Official act indicating the normalization of deviance: accepting risk. The construction of risk negotiated by the Level IV and III Marshall and Thiokol engineers and managers in the work group was formalized by three mechanisms: the joint was officially certified as flightworthy by NASA's Verification and Certification Committee; it received the official designation of C 1R—the "R" a formal symbol indicating the joint's redundancy and the official construction of risk; it was officially certified at each level of FRR for STS-1. In the process of officially declaring that the joint was an acceptable risk, the joint's deviation from design predictions became known, accepted, and expected as an aspect of normal joint performance throughout the NASA hierarchy. The construction of risk developed by the work group became the official NASA organizational construction of risk.

Shuttle launch. The completion of this first mission, with no problems found in the SRB joints in postflight analysis, confirmed the official construction of risk and the technical rationale that convinced the working engineers that the SRB joint was safe for flight. The postflight analysis also confirmed the method of responding to the problem in the prelaunch period: correcting the joint rather than redesigning it; flying despite data that the joint deviated from expected performance.

NORMS, RULES, AND DEVIANCE
IN THE WORK GROUP

As we will see, these early decisions were precedent setting, in the full literal meaning of the term. This decision-making sequence summarized above—and its important outcome, the normalization of technical deviation and the belief that the SRB joint was an acceptable risk—was to be repeated many times. Emerson notes the importance of the first decision for subsequent decisions.[107] The first decision establishes a precedent that becomes a normative standard for future decisions in similar cases, paving the way for development of a pattern. Why does it become a precedent, a normative standard, and thus an aspect of culture?

Some answers can be found in the history of decision making in the early developmental period of the Shuttle Program. Many decisions were made prior to the first flight, but for now let us simplify, treating the "first decision" broadly as the decision to fly with a joint that deviated from design expectations. As Thiokol and Marshall engineers and managers responded to the first signals of potential danger, they committed themselves to certain engineering methods, theories, and procedures. The decision process itself generates commitment to the group stance.[108] But in addition, the certification of the boosters, the CIL assignment of C 1R status, and FRRs for the first launch made public the methods, theories, procedures, and decisions of the work group. Going public with our opinions and actions commits us to our positions because we create an audience with expectations about our future behavior based on our present stand. To behave contrary to these expectations requires us to back down from previous positions, sometimes causing us to lose face. As the initial decision—that the SRB joint was an acceptable risk, and thus it was safe to fly—became more public through the certification process and the first FRR and became encoded in documents, the commitment to the engineering methods, theories, and procedures that went into it were reinforced. Marshall's Keith Coates observed that "when you get acceptance of that configuration and its known lack of perfection . . . by the higher level, that tends to be a little bit of an umbrella for subsequent decisions."[109]

Performance feedback cemented work group decisions as normative standards. In the aerospace industry, flight experience makes an important contribution to risk assessment because engineering tests and analysis can only simulate and predict what might happen in flight. Hence, the postflight analysis of STS-1 SRB joints was additional confirmation, creating a predisposition to use these same methods, theories, and procedures in future decision making. This tendency was reinforced by NASA's FRR procedures, which *required* that the most recent flight experience be assessed against previously established standards. With the completion of the first shuttle launch, the Level III and IV Marshall and Thiokol managers and engineers had established the beginnings of a normative structure that would characterize the work group culture, guiding future decisions.

In addition to the normalization of technical deviation, this chapter shows the contrast between retrospective outsider interpretations and work group understandings at the time decisions were made. Two decisions made by NASA in the early development period of the Shuttle Program became controversial after the loss of the *Challenger:* NASA's failure to redesign after the Ray memorandum stated that for there to be "no change" in the design was "unacceptable" and NASA's certification of the SRB design as flightworthy despite violating three industry standards. These actions were interpreted by analysts as evidence that NASA managers were engaged in deviance and rule violations early in the program, acting as amoral calculators who knowingly gave economic goals priority over safety even in the earliest phases of the program. The purpose of this history of decision making is to restore actions to their position in a chronological sequence in order to see them in terms of the meanings they had for participants at the time. The history shows that, within the NASA culture, these actions were acceptable and not deviant.

Ray's memos took on a sense of urgency and ominousness after the tragedy that did not match his position on the issue at the time. Their dramatic quality is lessened by knowing that as an S&E engineer, Ray wrote memos containing "bad news" for every work assignment. Although Ray believed the best course of action was to redesign the joint, he believed that fixing the joint was a satisfactory measure (and stated that in the memos), even though redesign was probably years away. He remained concerned, but on the basis of O-ring corrections and tests in which he participated after those memos were written, he believed that it was safe to fly with the existing design. There was consensus between Marshall and Thiokol working engineers that the design was an acceptable risk and that correction, rather than redesign, was an appropriate hazard reduction strategy. And it was their analysis and conclusions about risk acceptability that was the basis for all management decision making during that period. The Verification and Certification Committee's conclusions were based on the work group's technical rationale that the boosters were an acceptable risk. As we will see again and again in the history of decision making, the social construction of risk at lower levels shaped the procedural response to a problem.

MACRO-MICRO CONNECTIONS

The premise of this book is that individual behavior cannot be understood without taking into account the organizational and environmental context of that behavior. Although in these early history chapters we focus on the microcosmic world of routine decisions to show the process of culture production and *how* the work group normalized technical deviation, the very obvious influences from the external environment and the NASA organization warrant some discussion now. They sensitize us to *why* the normalization of deviance occurred. I identify some of these structural influences briefly in order to establish them as a context for the next two chapters and to set the agenda for a more thorough discussion in chapter 6.

During its formative phase, the work group culture was influenced by and conformed to norms and cultural beliefs originating in the NASA organization, the engineering profession, and the aerospace industry. Because their actions conformed to these environmental contingencies, the decisions they made were legitimate, acceptable, and not deviant, in their view. In the NASA organizational culture, flying with known flaws was not deviant. What post-tragedy analysts referred to as "known flaws" in the SRB joint were the "residual risks" that NASA acknowledged in the 1981 document describing the Acceptable Risk Process. Flying with residual risks was the norm at NASA, as the six volumes of Acceptable Risks compiled prior to the first shuttle flight attest. Because of an innovative shuttle design without precedent, "technical issues," "anomalies," and "open problems" were taken-for-granted aspects of NASA culture. Most certainly, these decisions were calculated risk taking, because that was how NASA made every technical decision.

Flying with a design that violated industry standards also was not deviant in the work group culture. In fact, deviation from industry standards is accepted practice in the aerospace industry. Carol Heimer makes the distinction between universalistic and particularistic rules: the former being very general proscriptions, limited because not all applications are foreseeable; the latter developing as conditions become known through practice.[110] Industry standards are understood not as hard and fast rules that must be adhered to, but as general *guides* for engineers to use. Manufacturers produce many products that are uniform (thus "sold off the shelf," like O-rings) and used in a variety of environments: ships, airplanes, and rockets. These standard items are desirable because they are cheaper. However, because many designs are innovative and unprecedented, these items have not been tested (or "qualified") in order to find the optimal circumstances for use in specific designs. When engineers incorporate one of these standard devices into their particular system, they test it, analyze it, and come up with the in-house guidelines—particularistic rules—that qualify it for their specific use environment. Development of in-house standards for decision making when formal rules did not apply was normative, not deviant, in aerospace technology. As Larry Wear noted:

> Any airplane designer, automobile designer, rocket designer would say that [O-ring] seals have to seal. They would all agree on that. But to what degree do they have to seal? There are no perfect, zero-leak seals. All seals leak some. It's a rare seal that doesn't leak at all. So then you get into the realm of, "What's a leaking seal?" From one technical industry to another, the severity of it and the degree that's permissible would change, you know, all within the same definition of seals. Generally speaking, from the outset, from a design standpoint, you have to establish some standards for when you design. Some of them are relatively easy to establish. Some of them are quite arbitrary. How much is acceptable? Well, that gets to be very subjective, as well as empirical. You've got to have some experience with the things to see what you can really live with, and that's why you run ground tests. And from that you establish limits that are acceptable and unacceptable. So legal limits were established empirically on all these [shuttle elements], in particular, the Solid Rocket Motor.[111]

Ray noted that deviating from formal industry standards was "common" but not "recommended" in the industry. However, the practice of deviating from industry standards *was* recommended in this case. The power of norms is that they influence behavior without specific articulation. These industry norms, however, were conveyed in conversation, allowing us to glimpse the intersection of institutional forces and microlevel decisions. In the face of uncertainty (unique design and unique data), both firms that Ray visited encouraged Marshall and Thiokol engineers to test in order to find the limits of their own design, explicitly encouraging deviation from the industry standards. The Aerojet report and O-ring manufacturers' manuals further legitimized deviation from industry standards. The industry standards were not perceived as relevant to the SRB design, so NASA engineers created their own standards for the SRB joint design, or, as Wear called them, "legal limits," qualifying the assembled parts to their unique environment.

The response of these two manufacturers also acknowledges the constraining influence of costs and schedules in the industry—a major concern at NASA, even as early as the developmental period. This incident demonstrates the institutionalization of production pressures and how they affected decision making. To redesign the joint to conform to industry standards would cost millions and delay the schedule. When Leon Ray visited the two firms in the industry to seek help on the SRB joint problem in 1979, the SRB hardware was already in production at Thiokol. The first shuttle flight was scheduled for spring 1981. He posed both hardware and schedule as constraints: "We've got the hardware built, we're going to be flying before long, what do you think we ought to do?" Advised of these circumstances, representatives of both firms later wrote letters encouraging Marshall and Thiokol engineers to "do enough tests to make sure the thing's going to work," then to "go with what [hardware] you've got." Marshall engineers received support from others in their industry for testing and correction rather than redesign.

Engineering decisions are biased toward making existing hardware and designs work, as opposed to scrapping it and coming up with a better design. But safety concerns also contribute to this bias. In the engineering profession, the belief that "change is bad" is widely held. In the short run, a new design brings new uncertainties, not greater predictability. Because designs never work exactly as the drawings predict, the learning process must start all over again. A change introduced in one part of a system may have unpredicted ramifications for other parts. In the interest of safety, the tendency is to choose the evils known rather than the evils unknown.

Note also that the manufacturers that the NASA engineers consulted also mentioned safety: they told Ray to "make sure the thing's going to work." All the evidence suggests that based on the tests they did following these visits, the working engineers believed that safety had not been compromised. Although these decisions and the certification of the design that followed were defined as deviant in accounts published after the *Challenger* tragedy, for key participants these actions were in keeping with engineering norms that encompassed cost, schedule, and safety in the development of innovative technological products. The decision to accept risk and proceed with flight exemplifies what Simon referred to as a tendency for organizations to "satisfice" rather than optimize when making decisions.[112] Trying to balance a number of different goals, organizations reach decisions that minimally meet competing needs. Marshall's John Schell, a retired (1981) engineer who specialized in nonmetallic materials, described this instance of satisficing as both typical and not threatening to flight safety:

> In every program we've ever had where O-rings were involved, and it will be this way in the future, there's always been pressures to relax specifications on O-rings and pressures to tighten. The structures proportion people, the people who were working on that side in O-rings, always wanted them tightened. Sure. I want something better. I mean, that finish of this glass is good, but it could be better. So everybody wants the best.
>
> You've got another side of the house, the flip side of the coin is, you've got people who've got to fabricate that item, you've got people who've got to manufacture it, you've got to inspect it, and they're saying that we just can't make it that way. And in this type product, there's always been a trade-off of somewhere between what you want and what you can get.
>
> The fact that something is not perfect doesn't mean that it isn't good. It means that in many cases, that's the best with the state of the art technology that can be done. I would like to have seen perfect O-rings, every one of them. I think everyone would have. But at the same time, that is just not a practical thing to look for.[113]

Most certainly, cost, schedule, and safety were weighed in the work group in the early developmental period. Robert Eudy testified:

> MR. EUDY: Yes, the production line was running, and there was considerable data base that we had with test results from those motor cases.
>
> COMMISSIONER RUMMEL: Do you know, had you made a decision at that time or had NASA made a decision at that time to redesign the joint and gone to some different configuration, what would the impact have been on the flight schedule? Can you say approximately?
>
> MR. EUDY: . . . [pause] I guess we would have been down probably two years. We would have been back to the billet stage, all the way through the pipeline.
>
> COMMISSIONER RUMMEL: I assume that was one of the factors you took into account and decided to continue with the unsatisfactory joint, is that correct?
>
> MR. EUDY: I didn't consider it unsatisfactory. This [design] was flying well in Titan . . . and all of the ground programs we had with our own joints looked very good. We had absolutely no test failure history of any kind.[114]

The engineering analysis did not support a redesign decision. Jim Smith framed it as an intersection of cost, schedule, and safety:

Engineering-wise it was not the best design, we thought, but still no one was standing up saying, "Hey, we got a totally unsafe vehicle." With cost and schedule, you've got to have obviously a strong reason to go in and redesign something, because like everything else, it costs dollars and schedule. You have to be able to show you've got a technical issue that is unsafe to fly. And that really just was not on the table that I recall by any of the parties, either at Marshall or Thiokol. We just couldn't go out and redesign something because some engineer would like to do it a different way. There had to be a safety reason to redesign it.[115]

Cost, schedule, and safety were weighed in the decision making, and compromises were made. But I find no indication of amoral calculation and organizational misconduct in the decisions made in the SRB work group. Testimony and interviews in which working engineers stated they believed it was safe to fly are supported by a paper trail of test results, engineering calculations, internal memos, and letters created at the time. Even the statements of postaccident whistle-blowers affirm that this was the work group's definition of the situation in the early development years of the shuttle. Marshall's Leon Ray was a key participant in the hands-on engineering work, with a reputation as a gadfly and the most conservative engineering voice during this early development period, and Thiokol's Roger Boisjoly, like Ray, became a hero after the *Challenger* tragedy for his integrity and willingness to stand up for what he believed—both men believed the design was an acceptable risk. Based on the test results and flight experience available to the work group at the time, the work group concluded that the joint design, warts and all, was not a threat to flight safety. To proceed with flight, to correct rather than redesign, was not a deviant action within the work group culture.

Chapter Four

THE NORMALIZATION OF DEVIANCE, 1981–84

For NASA, 1981–84 was an extraordinary time. The developmental period of the Space Shuttle Program gave way to the operational phase, thus apparently moving the space agency closer to the goal of a "space bus" that would routinely carry people and equipment back and forth to a yet-to-materialize space station. In glaring contrast to this public image, the engineers and technicians conducting risk assessments prior to each launch grappled with an uncertain technology that continuously produced anomalies—evidence that the shuttle was developmental, not operational. In this chapter, we trace the production of culture by the work group into the operational phase of the Shuttle Program. We examine the major turning points in the history of decision making to show (1) how the work group continued to normalize technical deviation as flight data accumulated and (2) the contrast between posttragedy interpretations of these events and the meanings these events had for insiders at the time. By immersing ourselves in work group routines, we see the incrementalism behind this historic event. We witness the gradual accrual of information, action, and definitions that shaped the work group's cultural construction of risk. We are reminded of how repetition, seemingly small choices, and the banality of daily decisions in organizational life—indeed, in most social life—can camouflage from the participants a cumulative directionality that too often is discernible only in hindsight.[1]

During these years, the work group continued to learn about the dynamics of their unique joint design. They were surprised by new signals of potential danger: among them, the first occurrence of "impingement erosion" on STS-2 and the first evidence of "blow-by" on STS 41-D. These incidents did not alter the work group's cultural construction of risk, however. To the contrary, their definition of the

119

situation received repeated affirmation, becoming institutionalized. In response to new signals of potential danger, the five-step decision-making sequence was repeated. The work group calculated and tested to find the limits and capabilities of joint performance. Each time, evidence initially interpreted as a deviation from expected performance was reinterpreted as within the bounds of acceptable risk. By the end of 1984, the work group had, in a problem-driven, incremental fashion, developed a three-factor technical rationale consisting of a "safety margin," the "experience base," and the "self-limiting" aspects of joint dynamics that convinced them that impingement erosion and blow-by were within the bounds of acceptable performance. And in FRRs, the work group's official definition of the situation was conveyed up the hierarchy, creating at NASA a uniform cultural construction of risk for the booster joints.

As ethnographic history, this chapter necessarily reveals aspects of NASA decision rules that work group members took into account as they assessed risk. In so doing, it revises conventional posttragedy interpretations of controversial NASA actions: among them, continuing to fly with the existing design after the first—and extensive—occurrence of erosion on STS-2; publicly declaring the Space Shuttle "operational," despite an SRB joint that was not performing as expected; waiving the newly imposed C 1 status of the joint just before the flight of STS-6; and failing to report information about SRB joint performance to upper-level NASA administrators during FRRs. As we relocate these controversial actions in the stream of actions of which they were a part, we see the native view: the meaning they had for participants when they occurred. Again we find the effects of cultural understandings on choice at NASA: actions that analysts defined as deviant after the disaster were acceptable and nondeviant within the NASA culture.

EROSION

The second developmental flight, STS-2, flew in November 1981. Thiokol always sent a team of four or five engineers and a photographer to Kennedy Space Center to inspect the boosters when they were returned from the sea. When Thiokol engineers disassembled the booster segments, they found the first in-flight O-ring anomaly. Thiokol's Jack Buchanan, permanently assigned to Kennedy Space Center as manager of Thiokol operations, was one of the first to see the damage: "At first we didn't know what we were looking at, so we got it to the lab. They were very surprised at what we had found."[2]

Hot motor gases had eroded 0.053" of the primary O-ring in the right SRB's aft field joint.[3] Although erosion was a known phenomenon in the aerospace industry, neither Marshall nor Thiokol expected it because Titan tests showed none. Also, in Thiokol's static firings of four developmental motors and three qualifying motors and in the flight experience of STS-1, the rings had not eroded.

Thiokol engineers began reviewing the evidence. They discovered that "blowholes" in the zinc chromate putty that lined the space between the booster segments had caused the erosion. This putty, which after the disaster appeared to some to be a Band-Aid, on-the-cheap correction (e.g., when I talked about my research in progress to colleagues, invariably someone would comment derisively, "You mean they tried to patch it with putty?"), had two functions: (1) to protect the O-rings from direct exposure to the hot motor gases coming from the propellant at the center of the booster and (2) to act like a piston at ignition to compress the air in the joint and thus seal the primary O-ring.[4] Unknown to the engineers, however, tiny bubbles had formed in the putty when the booster segments were stacked. These bubbles left weak spots. At ignition, the hot propellant gases moved through these weak spots, blowing holes in the putty and eating away, or "impinging" on, portions of the O-ring. Since erosion occurred on only one of the 16 O-rings on the two boosters, the cause appeared to be a deficiency in the putty in only that location. As Marshall S&E engineer Ben Powers explained, "the putty was creating a localized high temperature jet which was drilling a hole right into the O-ring."[5]

Once they had established cause, the Acceptable Risk Process required the work group to determine the "relative probability of the hazardous condition occurring." Despite 0.053" erosion, they found that the SRB field joint primary O-ring had sealed the gap created at ignition between the tang and the clevis. They calculated the "safety margin"—the maximum impingement erosion that could occur under an in-flight WOW condition—finding it was 0.090".[6] Then they performed tests to find conditions under which the primary would both seal and fail. Cutting pieces out of a primary O-ring to simulate an erosion depth of 0.095", they found it would still seal the joint under pressure of 3,000 psi—three times the amount the rings would experience at the peak of ignition pressure during a launch. They concluded that the joint was an acceptable risk because the STS-2 primary erosion was 0.053"—well within these parameters. This safety margin was an important precedent for future risk assessment: a numerical boundary was established defining the parameters of normal and/or deviant joint performance. It was the first of three in-house standards that would, by the end of 1984, constitute the full technical rationale that would be used repeatedly in assessing the risk of O-ring erosion.

Thiokol's Jack Kapp described the definition of the situation at the time:

> We didn't like that erosion, but we still had a couple of mitigating circumstances. First of all, it occurred very early on in firing, within a couple hundred milliseconds, and if the primary O-ring was burnt right through that early in the ignition sequence . . . the secondary should be in a good position to catch it. In addition to that, we came home and we took our little extrusion test rig and we sliced off some rather significant amounts of the O-ring up to about 90,000ths and reran some extrusion tests. We found out that even with major portions of the O-ring missing, once it gets up into that gap in a sealing position it is perfectly capable of sealing at high pressures—2–3,000 psi.
>
> So that gives us another little comfort zone in that, from the standpoint of O-ring material removal, what we were seeing was fairly minor compared to what would be critical for the O-ring. There was no question in our minds, as a matter of fact, and I want to state it again, that even if the primary O-ring failed to seal initially we felt that there was a high probability under those circumstances that the secondary would pick it up.[7]

The Acceptable Risk Process also required that the work group "implement hazard reduction": a corrective action that would reduce risk. Because the problem was defined as a "putty problem," not a joint rotation problem, the working engineers immediately began running tests on the putty composition and its ability to insulate the joint.[8] They altered the putty composition and the putty "lay-up"—the way it was applied to the booster segments—in order to eliminate the bubbles that formed when the segments were stacked. Thiokol engineer Howard McIntosh, stated that at the time impingement erosion was viewed as a "solvable problem."[9] Engineering analysis and tests would continue; redesign was not considered. Neither Marshall nor Thiokol working engineers viewed erosion as a constraint to flight. It had not occurred on STS-1. This was the only occurrence; they thought they correctly understood it and had taken necessary actions to control it. The construction of risk affected the procedural response: correct and fly.

This erosion was the most extensive prior to the fatal *Challenger* flight, but it was not discussed in FRRs for the next launch, STS-3, nor was it reported in the Marshall Problem Assessment System

(MPAS), a computer system for tracking serious problems. After the *Challenger* tragedy, this reporting failure was interpreted as the first of many attempts by Level III Marshall managers to keep bad news about the joint from top NASA officials. Indeed, the STS-2 erosion was not discussed in FRR until erosion occurred again nearly three years later. However, it was working engineers, not managers, who were responsible for the failure to report. Their inaction might appropriately be attributed to what Arthur Stinchcombe labeled "the liability of newness."[10] Although Stinchcombe had in mind difficulties that beset new organizations, new programs in existing organizations also face obstacles. Thiokol engineer Roger Boisjoly explained that the emphasis on the first four flights was on "R&D."[11] Consequently, so many physical instruments were on each flight to measure performance and feed the data into the system that the workload between flights was unusually heavy during the developmental period. Boisjoly remarked, "It was a major task just to turn around that data and assimilate it into the system."[12] The heavy instrumentation revealed many anomalies. These were important data, for they provided the first opportunity to compare engineering design predictions and test results with performance.

When Thiokol engineers spotted the STS-2 erosion immediately upon postflight disassembly at Kennedy, much analysis and discussion occurred between engineers; tests began. However, after analysis and putty corrections the engineers considered it a resolved problem. Marshall's Leon Ray said that the erosion discovery "generated quite a bit of excitement at the time," but it was not reported because only the generic problems—the things that could happen on the next flight—were mentioned in FRR.[13] Overwhelmed by the need to process and report the masses of data each developmental flight was producing, the engineers did not write up the erosion discovery and analysis for their Level IV FRR presentation for STS-3, and so it was not carried up the hierarchy.[14] The belief in redundancy was fundamental to their decision not to report. Marshall's George Hardy, then SRB Project Manager, explained: "On STS-2, we had no secondary erosion. Therefore, the joint performed as it was supposed to. The system was working exactly the way it was supposed to. You don't build in redundancy and never expect to use the back-up. If you never use your back-up, you're wasting money."[15]

Hardy made additional comments that support a "liability of newness" explanation. Procedures were also in a developmental phase. Not only was the work group doing baseline learning about how the technology operated during flight, but they were also trying out the rules for FRR presentations for the first time. What should and should not be reported in Level IV FRR may not have been clear to all the working engineers.[16] They did record the incident, however. Catching up on its paperwork much later, Thiokol completed a post-flight report on STS-2 that, unlike an FRR analysis, includes a complete rundown on all flight data.[17]

Accepting erosion was a major turning point in the work group's normalization of technical deviation. The engineers expanded the boundaries of acceptable risk from test results that deviated from predictions to include deviation from the in-flight performance expectation. The five-step decision sequence shows the pattern characterizing their interpretive work, indicating the work group's production of culture.

Signals of potential danger. Postflight analysis of the SRB joints showed that the primary ring on the aft field joint of the right SRB had 0.053″ of its surface eroded by hot motor gases. The appearance of erosion deviated from design expectations, creating uncertainty about the future performance of the joints.

Official act acknowledging escalated risk. The work group initiated no administrative action mandating special treatment when erosion first was discovered. They were concerned, but erosion was not as remarkable to the work group at the time as it was to outsiders after the *Challenger* disaster. Although erosion had not been predicted by the SRB design, deviations from design expectations are common during the developmental phase of an innovative technology. The context mattered. At Marshall, erosion was one of many flight anomalies that occurred on STS-2 and required risk assessment.

Review of the evidence. Marshall and Thiokol working engineers examined the aft field joint, measuring erosion and trying to pinpoint the cause of the problem. They identified it as an idiosyncratic incident: a localized deficiency in the putty allowed hot gases to impinge on the primary O-ring. Following guidelines of the Acceptable Risk Process, they calculated a safety margin that quantified the amount of erosion that could occur without interfering with the ability of the primary O-ring to seal the joint. Their analysis affirmed their belief in redundancy. The safety margin was a yardstick against which the STS-2 erosion was measured and assessed. They altered the putty and its lay-up in order to increase the margin of safety, concluding that the design was an acceptable risk.

Official act indicating the normalization of deviance: accepting risk. In FRR for STS-3, the working engineers' analysis, conclusions, and recommendation that the SRBs were an acceptable risk were presented and discussed. The SRBs were certified as flight-ready at each review level. As a result of the "liability of newness," erosion was one of many resolved anomalies not included in engineers' FRR presentations. Contrary to the outsider interpretation that this was intentional managerial concealment of a highly dangerous situation, the engineers' failure to report reflected the prosaic meaning the problem held for them after their risk analysis.

Shuttle launch. STS-3 was launched in March 1982. Since no erosion occurred, the postflight analysis convinced the work group that the alterations to the putty had done the trick. It confirmed the belief that the SRB joint was an acceptable risk. The postflight analysis also affirmed the procedural responses to the problem: correcting the joint rather than redesigning it and flying despite data that the joint deviated from expected performance.

THE BEGINNING OF OPERATIONAL SPACEFLIGHT

According to NASA plans formulated before flights were begun, after four successful test flights the developmental period would end and the operational period would begin. Launched June 27, 1982, STS-4 was the last of the four test flights. After a July 4 landing scheduled for maximum publicity and drama, NASA and President Reagan publicly declared the shuttle "operational"—ready for routine use.[18] Undoubtedly, the celebratory public announcements of the beginning of operational spaceflight were intended to alter the construction of risk of the Space Shuttle for important political audiences, primary among them potential payload customers, holders of the congressional purse strings, the administration, voters, and foreign nations. But political manipulation notwithstanding, the public announcements reflected a change in the construction of risk at NASA too.

In the aerospace industry, developmental technology is inherently risky; operational technology is considered sufficiently tested to qualify for routine use. Thus, by definition, operational technology is less risky than developmental technology. In the industry, less testing and fewer procedural constraints were required in any operational system. Tests were expensive, and both were cumbersome to a streamlined operational phase. The shuttle now had flown successfully four times, the final STS-4 flight beginning with a flawless countdown and ending with a perfectly timed landing. Following the announcement of the operational phase of shuttle flight, NASA instituted certain planned procedural changes that reflected top administrators' altered perceptions of risk. Testing of the vehicle and its components was reduced; procedures related to processing the Orbiter and boosters were reduced; requirements for reporting problems were reduced (and, in some cases, eliminated).[19] Moving forward with the shuttle's operational phase had serious structural consequences that affected the work group's decision making. These ramifications will be pursued in depth starting in chapter 6. For now, however, I want to emphasize one point: while top NASA administrators were busy publicizing the operational phase and converting to an operational system, ongoing engineering work on the SRBs was quite obviously still developmental in character.

In May 1982, NASA awarded a contract to Hercules, Inc., to develop a new kind of booster segment that included an innovation designed to control the joint rotation problem: the "capture" feature. A year earlier, Marshall engineers had already been at work on two new booster rocket designs that were thousands of pounds lighter.[20] This great weight reduction would be particularly advantageous in light of future NASA plans to launch vehicles from Vandenberg Air Force Base in California. From that launch site, shuttles had to achieve a more difficult, polar orbit, rather than the equatorial orbit possible from Florida: the lighter the vehicle, the greater the ease of achieving orbit. The working engineers were concerned, however, that lighter cases might increase joint rotation. They learned that Frederick Policelli, a Hercules engineer, had designed a "capture lip" to control rotation problems at Hercules.[21] The beauty of the design was that the entire joint could be machined out of a single piece of steel in such a way that the joint would not open when rotation occurred. This piece could then be bonded to the end of a booster segment.

Thiokol had considered incorporating a capture feature when they originally designed the SRM. They decided against it because it created another kind of risk. Thiokol's Jack Kapp reflected:

It was another surface that had to fit up as the joint comes together. At that time, all of the static firings were going to be done in the horizontal mode. We knew of no place in the free world where a large segmented clevis joint with non-tapered pins had ever been put together in the horizontal mode, and we were concerned about it because we knew it would be hard to maintain the motors' round as we put them together in the horizontal mode.

If we, in fact, developed a motor, loaded it [with propellant] and were ready to statically fire it and found out that we couldn't even assemble it, that represented about a two-year slip. Because of our analysis of the joint and the success of Titan without the capture feature, the decision was made that it wasn't needed. . . . We knew that the analysis was somewhat nebulous, just because of the complexity of the area, and so the question was asked, should we put the capture feature on there to help us if things go out of round or if some unknown cropped up. The decision was made that it would be nice to have but the risk for getting this thing together horizontally was just too great.[22]

Later, after they learned about joint rotation, Marshall's Leon Ray examined the Hercules capture lip idea for the lightweight case and championed it.[23] By May 1982, Marshall was considering adding a capture feature by altering one end of a booster segment so that the capture feature could be fitted and machined on. Marshall S&E engineers did not believe that a radical redesign (starting over from scratch with a new booster design) was necessary. Even after the discovery of joint rotation, they believed the design was an acceptable risk. The capture feature was viewed as a possible means of improving the safety margin by modifying the existing design. But there were good reasons to test thoroughly before going ahead: (1) the capture feature would add 600 pounds to each SRB; (2) assembly of booster segments would be more difficult; (3) the change would require 27 months to implement; (4) no one was certain that the capture feature design, which looked good on paper, would solve the joint rotation problem in practice; (5) incorporating the capture feature might cause new, unpredicted problems. Ray commented: "We wanted to be sure it worked. That joint looks simple, but it's very complex. You don't want to accidentally make it worse."[24] The engineering maxim "Change is bad" applied. Marshall engineers decided to make no new hardware recommendation until test devices could be developed and the capture feature thoroughly tested. These tests were scheduled for mid-1984.

Soon after the public announcement of the operational phase, in November 1982, Lawrence Mulloy succeeded George Hardy as SRB Project Manager. Although Mulloy had been with NASA since 1960, this was his first involvement with the SRB Project.[25] When Mulloy assumed his position, a significant aspect of his cultural inheritance was that the operational phase of the program already was underway.

The prevailing construction of risk was that the shuttle was sufficiently developed that it could be offered for routine spaceflight. Another aspect of his cultural legacy was that technical problems were a taken-for-granted aspect of the operational phase of the Space Shuttle Program. An ironic example was STS-4, publicly touted as the beginning of the operational phase despite a major glitch. A minor change had caused the spent boosters to separate from their parachutes before hitting the water: supposedly reusable, both sank in the Atlantic, never to be recovered.[26] But this unpredicted event did not alter the construction of risk. The operational phase still could begin because the loss of the boosters was not a threat to crew or to mission completion; moreover, in accordance with the Acceptable Risk Process, the cause of the problem was subsequently identified and corrected.

NASA's declaration of the shuttle's operational phase also brought the flight schedule to the fore of Mulloy's managerial responsibility, for routinized spaceflight was a critical goal of the operational phase of the program. Mulloy noted that in the transition period between Hardy and himself, the "hot problem at that time was the booster assembly facility (planned for construction at Kennedy Space Center) for getting to a rate of 24 (launches) per year."[27] The "hot problem" was definitely not O-ring erosion, for another significant cultural understanding Mulloy inherited was the work group's opinion of the SRBs. Mulloy was briefed on the SRB Project by Hardy, the Booster Assembly Project people, engineers, and the SRM Project Manager at NASA.[28] He visited Thiokol facilities, where he was briefed by contractor managers and engineers about SRB history.

In these briefings, Mulloy was instructed in the tenets that constituted the work group's risk assessment in November 1982. Two anomalies that deviated from design expectations had been discovered: joint rotation and primary O-ring erosion. These anomalies had been analyzed, corrections implemented, and the design defined as acceptable risk. Mulloy also inherited two significant procedural precedents: flying with joint performance deviating from design predictions, and an ongoing capture feature test program to improve the performance of the joint, rather than one to redesign it. In mid-November 1982, days after Mulloy assumed his new responsibilities, STS-5 was launched. This first flight of the operational phase of the shuttle carried the first payloads, two commercial satellites. No O-ring anomalies occurred. Thus, STS-5 affirmed the work group's cultural construction of risk and the engineering decisions made so far.

FROM CRITICALITY 1R TO CRITICALITY 1

Soon after Mulloy took charge, he changed the SRB joint Criticality status from C 1R (redundant) to C 1. Any shuttle component that did not meet the fail-safe design requirement of a redundant system had to be designated a C 1 item.[29] "The redundancy requirements for all flight vehicle subsystems . . . shall be established on an individual subsystems basis, but shall not be less than fail-safe. 'Fail-safe' is defined as the ability to sustain a failure and retain the capability to successfully terminate the mission."[30] C 1 items also were known, more picturesquely perhaps, as "single-point failures" because they had no backup. In Marshall's view, however, the booster joint did not cleanly meet the requirement for either the C 1 or the C 1R category because its status varied from one to the other during the 600 milliseconds after ignition during which the joint rotated. Mulloy was convinced by Marshall's engineering data that the secondary O-ring was not available as a redundant seal if the primary failed late in the ignition cycle. However, Marshall engineers and managers agreed that the possibility of joint failure in that period had low probability, depending on a rare coincidence of physical and temporal circumstances that created a WOW condition.[31] Consequently, switching to the C 1 designation was a conservative step. Although the worst-case circumstances were the exception and redundancy was the rule, Mulloy initiated the change to C 1 status in order to accurately reflect total experience with the joint.[32]

But Thiokol engineers still disagreed with Marshall engineers about when the joint was redundant and when it was not, which resulted in a dispute about the change of status. Thiokol thought it should remain C 1R because their measurements showed that the secondary was available *throughout* the ignition cycle; therefore, it was fully redundant.[33] According to Thiokol's Howard McIntosh, switching to C 1 was another example of Marshall's extreme conservatism in all engineering judgments. He thought the criticality change corresponded with the Marshall philosophy that "you take the worst, worst, worst, worst, worst case."[34] Thiokol's Al McDonald, Director, SRM Project, stated: "I had always considered it a Criticality 1R. And our paperwork in the plant today [19 March 1986] carries it as a Criticality 1R, Thiokol does. The reason for that is that we never agreed with Marshall when they made that change, because [of] the data they changed it on. . . . And we never agreed with that data. . . . We still had a 1R condition and our records still show that, and we never flew one after that that didn't have that condition."[35]

The official label had changed, but the SRB work group's construction of risk had not. Despite disagreement about the size of the gap—the basis for the opposed views on redundancy—both engineering communities agreed that the SRB joint was an acceptable risk. Marshall's Maurice Parker, who managed the Marshall Space Flight Center Operations Office located at Thiokol in Utah before his retirement prior to the *Challenger* tragedy, remarked:

> I personally wasn't concerned whether it was called a C 1 or a C 1R. The condition was the same as far as I was concerned, so I didn't have any great concern as to whether it got reclassified or not. I did feel that we had a safe situation, and I felt that the critical time for the joint to be redundant was at the start [of the ignition transient], and I felt we had a redundant joint at the start. Having had some previous experience with O-rings, I personally always felt that once the sealing job is accomplished by either the primary or the secondary, only one O-ring was doing the job anyway, so the question was, when does it have to be redundant?[36]

Leon Ray, who pushed for the C 1 change, agreed with the logic of risk acceptability at that time. As an S&E "cop," Ray participated in all engineering tests, corrections, and risk assessments. Ray affirmed the work group's belief in redundancy: "Unless we had a primary failure in a worst-case scenario, we had faith in the secondary."[37] Concerned that the C 1 change implied that they were not doing their job (i.e., the design specified a redundant joint and the C 1 symbol announced that it was not redundant), Thiokol engineers continued to test in order to convince Marshall to change the designation back to C 1R.[38] Their renewed testing continued to support 0.042" as maximum gap size, data critical to their interpretation. Marshall argued from their own data on gap size, and the status remained C 1. In an example of uncertainty writ large, Mulloy eventually sent this engineering disagreement to an outside "referee" to be settled. The results were not available until after the *Challenger* tragedy.[39]

The CIL document changing the C 1R designation to C 1 tells why the working engineers thought the design was an acceptable risk at that time (appendix B, figs. B4.1 and B4.2). This entry accurately reflects areas of work group consensus and disagreement. Note the section headed "Rationale for Retention." It gives the engineering logic for continuing to fly with the existing design—mandatory for all C 1 items on the CIL. The Rationale for Retention (or, as Mulloy put it, "the rationale for living with that situation")[40] was based on test results, flight experience, and corrective actions. By analogy, Marshall's Wiley C. Bunn, Director of Reliability and Quality Assurance, describes the engineering work and data that constitute the Rationale for Retention, necessary for all C 1 items:

> Let's take your car. Right front spindle, no back-up. Single point failure: if the spindle breaks, the wheel rolls off. It's got to be a Criticality 1, because even if you're only going 10 miles an hour, it might be through a school zone and you might hit a child. Now the Risk Retention Rationale then reads: the way you drive your car with this Criticality 1 failure [possibility] is you're going to absolutely know what that spindle is made of; you're going to absolutely know for certain that it was heat treated this way; you're going to absolutely know that it was made to these dimensions; and you're going to absolutely know that in the handling and the installation of that thing you didn't nick it, ding it, and put a stress riser on it.
>
> Now, that would be the kinds of things that Quality [engineering] would have to do in order to let you drive your car with that single failure point. Quality engineering makes that assessment of the Risk Retention Rationale and says: hey, we need to tie down A, B, C, and D. And then out of that logic, there still may be another thing that says, don't drive over 100 miles an hour. So if it said that, then that would lead you into another loop that would say, when you get over there to the accelerator, make sure it won't let him go over 100 miles an hour.
>
> So it's that kind of engineering, quality engineering, reliability engineering, that has to go into every one of those single point failure modes. Then they have to come up with the worst case: here's the design; what's the worst thing that can happen? And you do [calculate and test] worst cases. You just sit there and the guys go through that thing—worst case, worst case, worst case. And then quality engineering takes that and puts the inspection [requirement] in it.[41]

The Rationale for Retention for the SRB joint design was a team effort by Thiokol and Marshall working engineers.[42] It begins with a statement of the problem:

> Full redundancy exists at the moment of initial pressurization. However, test data shows that a phenomenon called joint rotation occurs as the pressure rises, opening up the O-ring extrusion gap and permitting the energized ring to protrude into the gap. *This condition has been shown by test to be well within that required for safe primary O-ring sealing.* This gap may, however, *in some cases,* increase sufficiently to cause the unenergized secondary O-ring to lose compression, *raising questions as to its ability to energize and seal if called upon to do so by primary seal failure.* Since, under this latter condition only the single O-ring is sealing, a rationale for retention is provided for the simplex mode [where only one O-ring is acting].[43] (Emphasis added)

The Rationale for Retention then tells why, given these conditions, the design was acceptable: (1) tests showed the joint was redundant except in a rare, worst-case scenario, (2) the Titan had no failures flying with a single seal, and (3) corrective actions had been taken to improve joint performance. Finally, it reported all test results and flight experience with the joints, concluding:[44] "Failure History: No failures have been experienced in the static firing of three qualification motors, five development motors and ten flight motors."

In interpreting this last statement, it is important to know that in the work group culture at that time, "erosion" and "joint failure" were two different phenomena. Joint failure was a possible consequence of joint rotation. Joint failure could only occur if the primary failed and the secondary was not available to fulfill its redundant function. Erosion had occurred, but did not—and, the engineers believed, would not—interfere with redundancy. Notice the phrase "after motor pressurization" in the note to the "Failure Mode & Causes" section of the CIL entry. Timing—when in the ignition cycle the joint would seal or fail—was behind the work group's belief in acceptable risk. In testimony, Marshall's Mulloy tried to explain the meaning of the CIL entry to the Presidential Commission, understandably struggling to interpret this incredibly opaque but essential document:

> I can see your interpretation of the words in the CIL. I would acknowledge that the wording could have been clearer. The intent was to show what is the physical phenomenon. The physical phenomenon is that after motor pressurization [i.e., late in the ignition cycle, after the primary O-ring seals] under "worst case conditions," the secondary O-ring may not be in a position to seal if called upon to do so by failure of the primary O-ring. . . . After motor pressurization—that is the key. Those words are in there, and the sentence structure is probably very poor, but the fact is that on all flights that we have flown to date . . . after motor pressurization, the secondary O-ring still has positive squeeze on it, and is indeed a redundant seal. We relied first on the test and analysis that said that the primary O-ring erosion would not cause failure of the primary O-ring to seal. That is the first thing. The second thing, then, is, one has redundancy.[45]

THE CRITICALITY 1 WAIVER

C 1 status on a component was, in NASA language, a "constraint to flight." But Marshall's Mulloy waived the C 1 status, allowing the next shuttle flight, STS-6, and all subsequent launches to proceed.

The waiver was rushed through without convening the appropriate board, receiving final approval less than one week before STS-6 was launched on April 4, 1983. The C 1 waiver was among the most controversial NASA management actions revealed by the official investigations after the *Challenger* disaster. Flying with disregard for the "constraint to flight" imperative for C 1 items seemed to be a violation of NASA rules for responding to safety-critical items. The waiver and the flight of STS-6 occurred in a sufficiently narrow time frame that the official investigators were concerned about the effect of production pressures and Marshall managers' decisions even at this early juncture in the shuttle's history. The failure to convene the appropriate board added an air of secrecy and stealth to the launch of STS-6. Were these actions rule violations indicative of organizational misconduct and amoral calculation?

In initiating the waiver, Mulloy was conforming to NASA rules, not violating them. To understand this controversial action, we have to explore further the meaning of C 1 status in the NASA culture. Any component without a backup did not meet the fail-safe design requirement of redundancy and therefore required the C 1 designation. However, for many shuttle components a backup was out of the question. Neither Orbiter wing, for example, had a backup, and thus each was a C 1 item. Thus, C 1 was a commonly used designation: hundreds of shuttle components were assigned to the C 1 category.[46] Because they were not redundant, all C 1 items had the same "failure consequences"—loss of crew, vehicle, and mission. *But not all had the same probability of failure, so among C 1 items, risk varied.* A wing, for example, was not likely to be disabled. And, according to the Rationale for Retention on the CIL, neither was the SRB joint. But was the probability of failure equal for the two components? Also, C 1 items had different possibilities for hazard reduction. For the Orbiter wing, no backup could be provided: it could never meet the fail-safe requirement. Wiley Bunn said that, comparatively speaking, the booster joints were "less risky": "Now, there are many Criticality 1 items that you don't have redundancy at any time, so having the period of time that this joint is redundant gives you some measure of comfort in accepting that C 1 for flight."[47]

The "constraint to flight" that accompanied C 1 status did not mean flight was halted. C 1 status was assigned to items that did not meet the fail-safe requirement in order to assure that they received extra attention. The waiver system was one form that extra attention took. With the waiver system, NASA established a formal procedure by which *the configuration requirement for redundancy could be bypassed when the outcome of engineering analysis was a decision to accept risk.*[48] Once engineers determined that a component with C 1 status was an acceptable flight risk, NASA *required* a waiver of the redundancy requirement before the component could be used in flight.[49] A waiver meant that the official designation C 1 remained, but that, on the basis of a documented engineering rationale that the component was an acceptable risk, the "constraint against launch" was waived to allow all subsequent flights to proceed, bypassing the redundancy requirements.[50] To obtain a waiver to the redundancy requirement not only conformed to NASA rules, it was the norm at NASA. As Mulloy testified:

> My understanding of the waiver is [that] the design goal on the shuttle was to have redundant systems. That design goal is not met in all systems. There are some 829 Criticality 1 waivers on the shuttle system. There are 213 Criticality 1 waivers on the Solid Rocket Booster. This particular waiver is one of 18 on the Solid Rocket Motor. . . .
>
> [With a C 1 designation] we have to do one of two things. We either have to redesign the joint so that it's redundant all the time or we have to get a waiver to the requirement. We chose to get a waiver to the requirement [for the SRB joint] and had good rationale [i.e., Rationale for Retention], which is documented, for that. . . . And with the waiver, I mean we didn't do anything more or certainly nothing less. We just continued to look at the hardware and the joints. *The waiver was merely to put the configuration requirements in consonance with the actual performance of the joint.*[51] (Emphasis added)

The method of processing the waiver, controversial after the tragedy, was also defined as nondeviant and acceptable within the NASA culture. Although most reports after the disaster noted that "Mulloy waived" C 1 status, a C 1 waiver could not be accomplished by one person acting alone. The Level III manager had to submit an ECR, written by engineers and presenting engineering data backing the request, to the Marshall Configuration Control Board for approval.[52] Mulloy conformed to these requirements, convening that board in January 1983, far in advance of the April STS-6 launch date.[53] It was Glen Lunney, STS Program Manager at Johnson at the time and chair of the Level II Program Requirements Control Board, who approved the waiver on March 2, nearly two months later, without convening the appropriate board. Each person on Lunney's board received the usual "change package" that contained all the relevant data, then later someone hand-carried the approval form around to get signatures.[54] Belying posttragedy interpretations of this as a stealthy act, the waiver approval document itself contains a clear statement that the waiver was processed "outside" the board.[55] Lunney explained: "Outside the board means that we did not have a meeting where it was discussed at the meeting. Sometimes that is done when the responses to the change are such that there is not any disagreement and there isn't any known issue that wants to be debated on the subject. In that case, a change would be processed outside the board. I would not say that means that it's outside the review process, which it was well within, but rather it was not dealt with formally at a board with people sitting in the room looking at view-graphs at the same time."[56]

Although Lunney violated the NASA requirement to assemble the Program Requirements Control Board, his action was not deviant in the NASA culture, for "walking documents through" was the norm when the engineering analysis contained no controversial issues. In March 1983, working engineers unanimously agreed that the design was an acceptable risk. The sort of debate of technical issues Lunney mentioned above that would, in his opinion, require a formal board meeting, did not exist. They could not launch with a C 1 item without waiver approval at Levels II and I. Concerned about getting the waiver approved and the paperwork completed and into the system in time, Lunney had the waiver walked through so that the necessary paperwork would be completed before the next launch date.[57] Mulloy, who initiated the waiver far in advance of STS-6 and with a vested interest in meeting the deadline because the C 1 waiver was on his hardware, stated: "[We were] trying to expedite it through so it'd be clear on the next flight, legal on the next flight. . . . I was very adamant that we get that waiver approved or attempt to get it approved . . . and get the paper in before we flew STS-6."[58]

For these NASA administrators, processing the waiver outside the board was an act of conformity: conformity to the Acceptable Risk Process, conformity to norms about what sorts of issues required a formal board meeting, conformity to procedural rules about C 1 items prior to launch. The flight safety issue had been settled. The issue was not production schedule versus safety, but conforming to a bureaucratic regulation about processing paperwork to meet the deadline. Michael Weeks, Deputy Associate Administrator, Office of Space Flight, the Level I official who signed the SRB joint C 1 waiver, explained: "We felt at the time—all of the people in the program, I think, felt that this Solid Rocket Motor in particular more than the SRB was probably one of the least worrisome things we had in the program."[59]

THE "EXPERIENCE BASE" AND
"ACCEPTABLE" EROSION

One more anomalous incident occurred before the end of 1983. This incident is a significant turning point in the history of decision making because for the first time the "experience base" became part of the technical rationale used to assess risk acceptability and thus part of the underlying structure of informal rules the work group evolved that normalized technical deviation. STS-6 was launched April 4, 1983. In postflight analysis, Thiokol engineers found that heat had reached (but not eroded) the primary O-rings in both the left and right SRB nozzle joints.[60] This was the first time that more than one ring was touched by hot motor gases and the first anomaly occurring on the nozzle joints. Nonetheless, the working engineers concluded again that the joint was an acceptable risk.

No erosion had occurred. Heat reaching the rings was clearly less serious than erosion. Anything less than the 0.053″ erosion seen on STS-2 was acceptable because it was within "the experience base," a phrase that reflected first-time reliance on past SRB joint flight experience as a standard for decision making. Also, this time anomalies occurred in the nozzle joint. Location mattered: the nozzle joint design differed significantly from the field joint design, making it less vulnerable to joint rotation and the effects of hot ignition gases. The nozzle joint was, in short, "less risky." Finally, blowholes in the putty again were identified as the cause. Research on the putty had been underway since the first erosion on STS-2. Now the working engineers knew more and made what they believed were appropriate corrections, accepting risk and recommending that the next flight proceed. This incident was not reported to Level II and Level I *because* it was within the experience base.[61] The Delta review concept governing FRR reporting dictated that only changes in test or flight experience that deviated from predictions, and thus required a changed technical rationale, be reported up the hierarchy.[62]

The remaining three flights of 1983 had no O-ring anomalies. Again, the success of the remedy affirmed the work group's diagnosis of the cause of the problem, further reinforcing the norms, beliefs, and procedures of the work group culture. Mulloy recalled, "We thought ol' Leon finally got the putty pattern right."[63]

But in 1984, 26 months after the first erosion, the tenth shuttle flight, STS 41-B, sustained O-ring damage that shook the work group's understanding. Primary O-rings in both the left SRB forward field joint and the right nozzle joint had eroded. O-ring erosion in two

joints was unprecedented. The putty changes that had been successful in eliminating the problem had not controlled erosion this time,
for the engineers again found blowholes in the thin layer of putty
that lined the booster segments.[64] Ironically, their own "fine-tuning"
was responsible.[65] A changed leak check procedure used at Kennedy
was adding to the blowholes created during stacking, further weakening the putty.[66] The leak check was a safety procedure. Since there
was no way to view the O-rings to check their position once the segments were joined, the original joint design included a leak check
port. Engineers could find out whether the rings were in proper position by injecting pressurized air into the space between the two rings
after booster assembly and checking for leaks by monitoring the pressure in that space.

Initially, the air pressure injected was 50 psi. However, for STS-8
and STS-9 it was increased to 100 psi on the field joints because 50
psi was believed insufficient to adequately test the sealing capability of the joint. No erosion occurred. For the 41-B mission, however,
the leak check pressure had again been increased, to 200 psi on the
field joints and 100 psi on the nozzle joints.[67] Because no erosion had
occurred since STS-2 and the recurrence on STS 41-B coincided with
the leak check change, the engineers concluded that the practice
they had initiated to help with the solution had now become a part
of the problem.[68] For the work group, this conclusion was based on
one incident and was therefore a working hypothesis at the time. A
bit of retrospective analysis done by investigators working for the
Presidential Commission allows us to see the very clear correlation
between the increase to a 200 psi leak check and the increase in erosion in the remaining flights before the final *Challenger* flight (see
appendix B, figs. B5.1–B5.3).[69]

What was most important in assessing the risk of erosion on the
two joints on STS 41-B was how much erosion had occurred on each
joint. The engineers had decided that the first incident of erosion on
STS-2 was an acceptable risk because postflight analysis and tests
convinced them that it was well within limits that assured the continued redundancy of the joint. Now, 26 months later, they again
needed to know the effect of erosion on redundancy: "If [the] primary O-ring allowed a hot gas jet to pass through, would the secondary
survive the impingement?"[70] In postflight analysis, they used the
procedures and norms they developed after STS-2. They measured,
using the safety margin and the experience base as guides. They
found that the amount of erosion was within the experience base: it
was less than the maximum found to date, 0.053" on STS-2.[71] It was

also within the 0.090" safety margin. They also knew that a ring sliced to simulate 0.095" erosion would seal at three times the pressure experienced at ignition.[72]

But with STS 41-B, a third factor affirmed their belief in redundancy: the notion of erosion as a "self-limiting phenomenon."[73] As flights continued, they constantly learned more about how the joint operated. From their ongoing tests and analysis, Marshall and Thiokol engineers had learned that the period of hot gas impingement on the rings was of short duration; therefore, only a limited amount of erosion could occur before the joint would seal. When pressure equalized in the joint, the gas flow stopped and impingement erosion stopped.[74] Thus, erosion was "self-limiting."[75] By virtue of increased knowledge about the operation of the joint, Thiokol engineers now had developed three standards—*the experience base, the safety margin, and the belief that erosion was a self-limiting phenomenon*—that they used to assess the risk of the erosion found on STS 41-B. They concluded that the anomalies they found were an acceptable risk.

However, from what they learned as a result of this flight, they predicted more anomalies in the future. Previously, O-ring anomalies occurred occasionally, the result of idiosyncratic causes that were corrected and did not recur. Now, they knew the leak check pressure might regularly produce erosion, so Thiokol engineers filed a problem report that was entered into MPAS so that the occurrence of erosion could be tracked. The March 1984 entry shows the work group's cultural construction of risk:

> Remedial action—none required; problem occurred during flight. The primary O-ring seal in the forward field joint exhibited a charred area approximately 1 inch long .03–.050 inches deep and .100 inches wide. This was discovered during post-flight segment disassembly at KSC. . . . *Possibility exists for some O-ring erosion on future flights*. Analysis indicates max. erosion possible is .090 inches according to Flight Readiness Review findings for STS-13 [i.e., STS 41-C]. Laboratory test shows sealing integrity at 3,000 psi using an O-ring with a simulated erosion depth of .095 inches. *Therefore, this is not a constraint to future launches.*[76] (Emphasis added)

As the emphasized sentences above indicate, after the 1984 flight of STS 41-B erosion became accepted and *expected* as a normal aspect of joint performance at NASA. After the *Challenger* tragedy, the revelation that NASA would use a procedure that created more blowholes when engineers knew that blowholes contributed to erosion seemed bizarre, to say the least. But that was not how the working engineers viewed the situation. Thiokol's Boisjoly testified that their decision was based on solid engineering logic and rigorous testing:

How that came about was that we were leak checking original-ly at 50 psi. We discovered just through many conversations and telecons that we have got to check the putty because the most important thing to determine in a leak check is whether or not the seal is in fact in position to seal and doesn't have any contamina-tion under it.

So we did a series of lab tests . . . and we determined that 200 psi under all circumstances of minimum tolerances would blow through the putty. So we instigated a double leak check, namely 200 psi to make sure the putty did not mask a leaking seal and then 50 psi subsequent to that test to actually test the seal. 50 psi is a very difficult test on an O-ring seal. So that you have the best of both worlds. First of all, the 200 psi ensures that the putty is not masking a leak, secondly, that the 50 psi proves the seal is indeed going to seal.[77]

The advantage of increasing the leak check pressure was that it assured that the rings would seal the joint. Erosion was defined as a harmless anomaly, traded for greater certainty that the O-rings were properly positioned to seal. Thiokol engineer Brian Russell indicated that Boisjoly's view represented a consensus position: "We thought it was of utmost importance to have a verified primary O-ring and so we increased leak check pressure to 200 psi to make sure that we would blow through the putty, realizing that blow holes are not desirable either, but yet it is more important to know that you have a good O-ring and have some putty blow through than otherwise."[78]

The distinction made in the work group culture between erosion and joint failure again becomes relevant here. The engineers did not consider erosion a threat to flight safety because erosion did not interfere with the performance of either the primary or the sec-ondary. Mulloy described how the distinction between erosion and joint failure figured into the way risk was constructed at the time: "Failure would be leakage of the joint . . . and there had been no leak-age of any joints. There had been observed erosion on an O-ring, but the O-ring sealed. . . . The erosion was considered an undesirable sit-uation that reduces margin for sealing, which we wanted to get rid of, and that's what we were working on. See, we were looking at hardware as we were getting it back and what we were seeing was hot gas impingement on O-rings. The problem was defined as hot gas impingement on O-rings, how can we eliminate? And that's what we were working on."[79]

Erosion, which after STS-2 was defined as an idiosyncratic incident, had happened again. Because the work group had made a change—the leak check pressure—that was likely to systematically produce erosion, they reported it in FRRs for the next flight. Mulloy gave a complete briefing of the erosion problem in accordance with NASA's Delta review concept for FRR reporting, which required that any change be discussed at all levels of the FRR hierarchy.[80] The words "acceptable erosion" were associated for the first time with erosion in FRR documents.[81] Also for the first time, data on O-ring erosion on both STS-2 and STS-6 were discussed at the two upper-level Marshall FRRs, the Shuttle Projects Office Board and the Center Board.[82] The probable cause of erosion on these three flights was identified as blowholes in the putty.[83] Mulloy presented the three-factor technical rationale developed by the working engineers at Marshall and Thiokol to all levels.[84] Based on Thiokol's analysis and conclusions, Mulloy's final chart recommended to Level I NASA administrators that NASA *"fly STS-41-C, accepting possibility of some O-ring erosion due to hot gas impingement."*[85]

At this point in the history of decision making on the SRB joints, all levels of the NASA hierarchy had been informed about the erosion problem and the technical rationale for accepting risk. Further, all levels were informed that erosion was expected in the future. For all, erosion had become a predicted—and thus normal—aspect of joint performance. This is not to suggest there was an absence of concern. John Q. Miller, Marshall's Chief Engineer, SRM Branch, wrote a memo to George Hardy identifying putty composition and the effect of the leak check pressure on it as an "urgent concern which requires expedition of previously identified full scale tests."[86] NASA Deputy Administrator Hans Mark issued an Action Item requesting Marshall's Larry Mulloy to conduct a full review of the SRB joint problems for NASA headquarters. The Action Item resulted in a letter from Marshall to Thiokol asking contractor personnel to "identify the cause of the erosion, determine whether it was acceptable, define necessary changes, and reevaluate the putty then in use."[87] Although Thiokol quickly produced a program schedule of ongoing and planned research, this extensive review was not complete until August 1985, some 15 months later.

The delay was not the result of indolence, scarcity, or obstacles to problem resolution created by top administrators. Analyzing the technology was difficult because of the liability of newness. The engineers had no precedent; each new condition required crafting engineering standards that could be used to assess risk. Tests, testing equipment, and other analytic tools had to be developed. This all took time, as did the analysis. Much research went into the putty problem. As with everything else, the work group had no precedent to guide them. The use of the putty itself was controversial; some, including an outside consultant, thought it was not needed at all.[88] Thiokol's Arnie Thompson, Supervisor, SRM Cases, took the position that "if we got the putty out of there, that would really solve the erosion problem."[89] Roger Boisjoly recalled: "We did have almost a year of discussions in the 1984 timeframe about removing the putty from the joints. We had discovered that the blow holes, especially a single blow hole in the putty was providing the source of jet impingement on the O-ring seals and eroding them. If we could remove that source of jet impingement by either substituting another material or putting many interruptions in the putty purposefully, we could take the sting of the heat away and the erosion would be minimized."[90]

Meanwhile, they were convinced by their three-factor technical rationale that it was safe to fly while the experiments went on. Erosion had occurred, but it was defined as an acceptable risk. Their analysis affirmed redundancy. Flight 41-C was launched April 6, 1984. It was not a surprise, given what they believed would be the effect of the increased leak check pressure, when anomalies occurred in two field joints. The conclusions of Thiokol's postflight assessment were that all primary seals had performed well. The expected erosion had materialized and was well within the parameters established for acceptable risk. Despite a recurrence of erosion, the flight of STS 41-C validated the construction of risk and the rationale and decisions that had preceded it.

E arly in 1984, yet another major turning point in the work group's normalization of the joint's technical deviation occurred. The decision pattern was repeated. We can chronicle once again how they expanded the boundaries of acceptable risk, maintaining and perpetuating their construction of risk as anomalies continued.

Signals of potential danger. STS 41-B, launched in February 1984, experienced O-ring erosion on primary rings in two SRB joints. Not only was this the first in-flight erosion since STS-2 in 1981, but also erosion on two rings was unprecedented. The evidence discovered at disassembly challenged the work group's definition of the situation.

Their solution to the erosion problem after STS-2 had eliminated erosion for six flights. What had happened?

Official act acknowledging escalated risk. Thiokol engineers filed a problem report with Level III Marshall management, who entered it into the MPAS. Officially, the SRB joints were acknowledged as a problem for the first time.

Review of the evidence. Working engineers from Marshall and Thiokol again identified the cause of erosion as blowholes in the putty between booster segments. Many believed that an increase in the leak check pressure before the 41-B launch had further weakened the putty at that particular joint, allowing hot gases to impinge on the primary. They quantified the erosion on both joints: it was within both the 0.053″ experience base established by STS-2 and the safety margin of 0.090″, which indicated the maximum amount of erosion a ring could sustain before it would not seal. In addition, ongoing lab tests about the dynamics of joint performance had convinced them that the erosion was a "self-limiting phenomenon" because the joint would seal and erosion would stop before the condition could become a threat to flight safety. Again they determined the joint was an acceptable risk, backing their recommendation to proceed with spaceflight with a technical rationale that now consisted of three factors: the safety margin, the experience base, and the self-limiting dynamics of the joint.

Official act indicating the normalization of deviance: accepting risk. The SRB joint O-ring damage found on STS 41-B, along with that from STS-2 and STS-6, was presented and the engineering analysis was discussed at each level of FRR for STS 41-C. At each level, the engineering charts presented erosion as "acceptable erosion": the most recent anomalies were, in their view, not threatening to joint redundancy. The charts conveyed the information that erosion would be expected in future flights because it was understood to be a by-product of the leak check increase, which would continue at that psi level.

Shuttle launch. The next flight, STS 41-C, returned from space with erosion. Because of the known effects of the leak check, erosion was expected. The amount was also as predicted: within the safety margin and the experience base. The postflight analysis confirmed the official construction of risk that the SRB joints were safe for flight. It affirmed the belief in redundancy. Moreover, the work group's belief in the norms and procedures for assessing risk—the safety margin, the experience base, and the concept of self-limiting behavior—was reinforced. The joint behaved exactly as predicted.

THE SOCIAL CONSTRUCTION OF RISK, 1984

Before 1984 drew to a close, three more shuttles were launched. The August 30, 1984, launch of STS 41-D has a significant place in this chronology for two reasons. First, it was yet another point at which the limits of acceptable performance were expanded to include something new and unexpected. It was in the postflight analysis of this mission that Thiokol engineers found the first occurrence of "blowby": a small amount of soot behind the primary O-ring of the nozzle joint. In addition, they found erosion on two primary O-rings. Second, the FRR documents for the next flight, STS 41-G, record the technical rationale that supported acceptable risk through the end of 1984, so we can see how deviation was normalized at that point. The boundaries of acceptable joint performance had again gradually expanded. Both erosion and blow-by had become part of the experience base against which future flight performance would be assessed. The documents show that the work group's definition of the situation was conveyed to the top of the launch decision chain, informing all key technical decision makers about the history of the O-ring problem and creating an official organizational construction of risk.

Blow-by was a sign of deeper gas penetration in the putty channel: hot motor gases had gotten through to the space between the primary and secondary O-rings of the nozzle joint *before* the primary sealed. The chief concern was redundancy: how had the O-rings performed in sealing the joint? Blow-by was new and alarming because of its newness. In postflight analysis, the engineers found the amount of soot blow-by was small and had not impinged on the secondary O-ring. In fact, the amount proved that the period during which hot gases passed the primary was short, verifying calculations that penetration by hot gases was a self-limiting phenomenon. Engineers had predicted that the primary would seal before any hot gases could impinge on the secondary, and the joint had performed as expected. In addition, the erosion, when measured, was less than on previous flights; it was thus within both the safety margin and the experience base. Therefore, Thiokol engineering concluded that the damage found on STS 41-D was an acceptable risk.

The problem, the engineering analysis, and the risk assessment were presented at all levels of FRR because the Delta review concept requires reporting changes in performance beyond that previously experienced, changes in the existing technical rationale or corrective actions, or changes in the composition of the component. The charts show us the construction of risk that prevailed at the end of 1984. Although we do not have records of the oral presentations and discussion, the charts show the typical format and how problems were reported up the hierarchy.

Figure 6.1 is the Thiokol chart about O-ring erosion, pulled from the large package of charts that they presented at both Level IV and Level III. Note that Thiokol engineers presented their postflight analysis and technical rationale (left-hand column) and their "resolution" to the O-ring damage found on the previous flight: "accept risk." Thiokol gave the technical rationale for accepting risk by quantifying the damage on STS 41-D (blow-by is indicated by the phrase "heat affect length of 6 inches") and comparing it with the safety margin and the experience base. The first bulleted entry refers to the safety margin, the second to the experience base. Because the full technical rationale had been discussed in the previous FRR and no changes were made in it, a repeat FRR discussion was not necessary. Therefore, the "to be discussed" column was marked "no."

At the Marshall Shuttle Projects Office Board FRR held a week later (fig. 6.2), Mulloy relayed the engineering conclusion that the O-ring erosion on STS 41-D was an "acceptable risk," along with the technical rationale for risk acceptance. Note that Mulloy presented five additional booster anomalies that occurred on that same launch. And in contrast to the other problems, which the right-hand column shows were still being "worked," the erosion and blow-by on STS 41-D were described as resolved at lower levels. The next day, at the Marshall Center Board FRR (fig. 6.3), Mulloy's recommendation was to "use-as-is," meaning no constraint to flight existed requiring redesign or a further fix, and his chart referred to the amount of erosion that was "allowable" (i.e., the safety margin). All of the key information—the amount of erosion and blow-by on STS 41-D, the technical rationale for accepting risk—was carried forward in charts for Level II and Level I FRR at NASA Headquarters, which also described erosion as "allowable" (fig. 6.4).

The last two flights of 1984 experienced neither erosion nor blow-by, confirming all preceding decisions. Interviews with work group members confirm the belief in redundancy to which these documents attest. After the *Challenger* tragedy, Thiokol engineer Brian Russell reflected that the blow-by discovered on STS 41-D was the first "big red flag."[91] But at the time, the work group's definition of

STS 41-G FLIGHT READINESS REVIEW CHARTS

STS 41-G FLIGHT READINESS REVIEWS

SRM Preboard (September 12, 1984)

PROBLEM SUMMARY (CONT)
STS-41D PRELIMINARY SRM POSTFLIGHT HARDWARE DAMAGE ASSESSMENT
- FIELD JOINTS

PROBLEM	CONCERN	RESOLUTION	TO BE DISCUSSED
RIGHT HAND FORWARD FIELD JOINT PRIMARY O-RING DAMAGED AT 277° WITH MAXIMUM O-RING EROSION DEPTH = 0.026 IN., PERIMETER AFFECTED 100° WITH HEAT AFFECT LENGTH OF 6 INCHES	INTEGRITY OF SRM FIELD JOINTS	ACCEPT RISK	NO
O LABORATORY TEST CONFIRMED THAT O-RING COULD SUSTAIN 0.095 IN. EROSION AND MAINTAIN PRESSURE SEAL AT 3000 PSIA			
O LESS SEVERE O-RING EROSION THAN SEEN ON STS-2 (0.053 IN.) AND STS-11 (0.040 IN.)			

Fig. 6.1

Shuttle Projects Office Board (September 19, 1984)

PROBLEM SUMMARY

PROBLEM	CONCERN	RESOLUTION FOR STS 41-G	TO BE DISCUSSED
– APU FUEL PUMP SEAL CAVITY DRAIN LINE ERODED ON 41–D	FIRST TIME OBSERVED CAUSE NOT APPARENT	EROSION PRECLUDES REUSE ONLY. DOES NOT AFFECT MISSION PERFORMANCE	YES – SPECIAL TOPIC
– RSS ANTENNA DAMAGE	FIRST TIME OBSERVED CAUSED BY DESCENT DEBRIS	NO SRB RSS FUNCTION POST SEPARATION – DAMAGE AFFECTS REUSE ONLY	YES – SPECIAL TOPIC
– REUSE OF SINGLE MISSION QUALIFIED PARTS (41–G)	EXCESSIVE IEA VIBRATION RESPONSE	WAIVER APPROVED; PARTS REFLOWN ON STS 41–D	YES – WAIVER
– CRACK IN I.D. OF BOLT HOLE ON 41–D RIGHT HAND MOTOR STIFFENER STUB	CAPABILITY TO REUSE SEGMENT	(NO PRE-EXISTING CRACKS IN STS 41–D STIFFENER STUBS)	YES – SPECIAL TOPIC
– O RING EROSION OF 0.26 X 6 INCHES ON 41–D RIGHT HAND MOTOR FORWARD FIELD JOINT	INTEGRITY OF JOINT SEAL	– ACCEPTABLE RISK – EROSION LESS THAN STS–2 AND STS–11 – TEST SHOWS MAXIMUM EROSION POSSIBLE LESS THAN EROSION ALLOW–ABLE (0.95") REF. STS–11 & 13 FRR	NO

Fig. 6.2

STS 41-G FLIGHT READINESS REVIEW CHARTS
(Continued)

Marshall Center Board (September 20, 1984)

PROBLEM SUMMARY
STS 41-D POST FLIGHT RETREIVAL INSPECTION
SIGNIFICANT OBSERVATIONS

OBSERVATION	CONCERN	RESOLUTION FOR STS 41-G	TO BE DISCUSSED
● ONE OF TWO APU FUEL PUMP SEAL DRAIN LINES ERODED ON L. H. SRB	CATEGORY 3 PREVIOUSLY OBSERVED	USE–AS–IS. MISSION PERFORMANCE NOT AFFECTED.	YES – SPECIAL TOPIC
● O–RING EROSION OF .026 X 6.0 INCHES ON R. H. MOTOR FORWARD SEGMENT FIELD JOINT	CATEGORY 1 INTEGRITY OF JOINT SEAL	USE–AS–IS. MISSION PERFORMANCE NOT AFFECTED. EROSION LESS THAN STS–2 (0.053) AND STS–11 (0.040) AND MUCH LESS THAN ALLOWABLE (0.095)	NO
● IO EROSION POCKETS ON L. H. SRM NOZZLE	CATEGORY 1 CONTINUED EXPOSURE TO NEGATIVE MARGIN OF SAFETY AT END OF BURN ON NOZZLE INSULATION THICKNESS	USE–AS–IS. STS 41–D RESULTS CONSISTENT WITH EXPECTATIONS UNTIL PLY ANGLE CHANGE MADE.	YES – SPECIAL TOPIC

CATEGORY 1   – MISSION PERFORMANCE

CATEGORY 2   – PRELAUNCH PROCESSING DELAY

CATEGORY 3   – REUSE/TURNAROUND IMPACT

Fig. 6.3

Level I (September 26, 1984)

PROBLEM SUMMARY
STS 41-D POST FLIGHT RETREIVAL INSPECTION
SIGNIFICANT OBSERVATIONS

OBSERVATION	CONCERN	RESOLUTION FOR STS 41-G	TECHNICAL ISSUE
O-RING EROSION OF .026 X 6.0 INCHES ON R.H. MOTOR FORWARD SEGMENT FIELD JOINT	CATEGORY 1 INTEGRITY OF JOINT SEAL	USE-AS-IS. MISSION PERFORMANCE NOT AFFECTED. EROSION LESS THAN STS-2 (0.053) AND STS-11 (0.040) AND MUCH LESS THAN ALLOWABLE (0.095)	NO

CATEGORY 1 – MISSION PERFORMANCE; CATEGORY 2 – PRELAUNCH PROCESSING DELAY
CATEGORY 3 – RESUSE/TURNAROUND IMPACT

Fig. 6.4

the situation was different. In the postflight analysis, the uncertainty created by the initial discovery was converted to certainty about risk acceptability, for their analysis led them to believe that they understood what caused it and that it was not a threat to redundancy. Russell described the repeated cycling of uncertainty and certainty as the deviant performance of the SRB joint was continually normalized by the work group in official decisions prior to 1985:

> We didn't like it. We—when it happened, I believe there was quite a bit of flurry of attention, and then a preparation for the next Flight Readiness Review that justified why we could continue with some analyses and tests, and then we did run some tests and in dealing with NASA we did make some changes in more control in the [putty] lay-up, for example, and these types of small changes to try and control the problem.
>
> And, as it kept recurring, we would—I guess I'd have to honestly say, until something new like soot blow-by occurred, or some greater erosion would occur, greater than we had seen before, then there would be another flurry of activity to try and understand the problem and some more tests done, and maybe some more little changes to the engineering to try and solve the problem.[92]

Spaceflight continued because each time an anomaly occurred, they believed they understood the cause and the limits of the phenomena they were seeing. Roger Boisjoly described Thiokol engineering's view on the booster joints at the end of 1984:

> Up until that time we had heat indications on O-rings, we had O-ring erosions, but we had never, ever compromised a primary seal, never. And we felt, you know, well, there's hot gas getting there, and then back in '84 there was a lot of subscale testing done that was characterized very well, the analytics were understood. They had a pretty good handle on the impingement characteristics. Everything was in a mode where—you don't like to see erosion on a seal, but if you can characterize where it is coming from, how it's getting there, and characterize its magnitude so that you can determine how safe or lacking safety you are, then you feel reasonably comfortable that while you are trying to fix the problem, everything is okay. And that's what happened, we were trying to fix the problem, were starting to look at fixing the problem on that basis, and whether or not to take the putty out of the joint, things like that, whether or not to put larger seals in.[93]

At the end of 1984, Boisjoly and the other working engineers at Marshall and Thiokol assigned to the SRB joint concurred that the joint was an acceptable risk and that correction—not redesign—was an acceptable remedy. Marshall's Mulloy affirmed that consensus existed in the work group at the end of 1984:

> What we were seeing was erosion of the primary O-ring. That mechanism was understood. How it happened was understood. How bad it could be was understood. And a good analytical model that correlated well with test data was done to make damn sure we understood. And we were able to predict pretty well how much erosion and we had large [safety] margins, so it's true then and it's still true today [April 2, 1986] that just hot gas impingement, erosion itself, you are tolerant of, and the risk—it was reported as a risk and it was accepted as a risk.
>
> We never perceived that we had to make a radical design change to the joint based on the observations that we were making from the return hardware. Everyone seemed to believe that a modification within the gas dynamics aspect of it as opposed to a physical modification would solve that problem. Find a way to stop hot gas jet impingement on the primary O-ring in the field joint and find a way to minimize the time that hot gas can blow by the primary O-ring and get to the secondary O-ring on the nozzle.[94]

NORMS, RULES, AND DEVIANCE IN THE WORK GROUP

Between 1981 and 1984, 14 shuttles were launched. Four times, an incident that first was seen as a deviant event was reinterpreted as nondeviant as a result of the redefinition of the bounds of acceptable performance for the SRBs. The construction of risk received repeated validation, stabilizing the norms, beliefs, and procedures that the work group had developed when deliberating about the joint in the early developmental period. The persistence of the work group culture seems remarkable, given that O-ring erosion recurred and altered in significant ways: both erosion and blow-by were considered acceptable risks at the end of this period. Since cultures are invented and, hence, susceptible to change, why did this culture persist? The answer, to a great extent, lies in the effects of environmental and organizational contingencies on decision making, to be discussed in later chapters. However, this chapter sensitizes us to the importance of another microlevel factor: the effect of the chronological sequence of events and patterns of information on the engineers' interpretive work as missions accumulated.

David Collingridge, in his book *The Social Control of Technology*, notes that many decisions about risky technology are most accurately described as "decision making under ignorance" because from a technical standpoint all conditions can never be known.[95] With no design precedent and no ability to test the vehicle under full environmental forces, data from in-flight performance was paramount in the normalization of technical deviation. Turner's distinction between ill-structured and well-structured problems applies.[96] After the accident, analysts wrote about the history of O-ring problems, retrospectively collecting all the evidence and presenting the public with the full collection of anomalies. For us, knowing the fatal outcome, the O-ring problem was a well-structured problem. In contrast, the work group saw the same information incrementally, as the problem unfolded. They dealt with one flight—and thus one incident—at a time.

As missions continued, in-flight anomalies had no predictable pattern. Erosion occurred on some flights but not others, and few joints were affected. Each shuttle had 16 rings, 8 on each SRB. In the first eight shuttle flights, 128 O-rings were used.[97] Engineers found erosion on one ring on STS-2 and heat effects on two rings on STS-6. They concluded that these effects did not reflect a design problem because, if they did, all joints would be affected. The infrequent occurrence and the irregular pattern created a temporal sequence that was extremely influential in shaping the construction of meaning in the work group: an incident would occur, followed by flights with no erosion, causing the group to conclude that they had correctly identified and corrected the problem. The effectiveness of the remedy affirmed their diagnosis. STS-2 appeared to be an "isolated case."[98]

Erosion did not occur again until the first flight of 1984. The incident was not viewed as a continuation or recurrence of what happened on STS-2 because it had a different cause: the leak check pressure had been increased. Because of the increased pressure, they expected erosion on future flights and were not surprised when it occurred. They continued to test and experiment with putty composition and lay-up. When no erosion occurred on the last two flights of 1984, they felt that they had again solved the problem. Again, the remedy confirmed the diagnosis. Once again, the engineers thought they had fixed the joint. At the end of 1984, consensus existed in the work group: erosion and blow-by were accepted and expected in future flights by the engineers and managers who worked most closely on it. Moreover, analyses and tests added legitimacy to the procedures that the work group began in response to the problem in the early developmental period: flying under circumstances that deviated from the performance predicted by the design and making corrections to improve the performance of the joint rather than redesigning it. The beliefs, norms, and procedures of the work group had become institutionalized. Thus was created a cultural predisposition for the work group to respond in similar ways to signals of potential danger the next time they occurred.

Tracing the work group's normalization of technical deviation during this period also allows us see the discrepancy between post-tragedy interpretations of key incidents and the meanings they held for work group participants at the time they occurred. Several actions suggested—after the disaster—that NASA managers were engaged in deviance and rule violations in these years of shuttle flight, thus establishing an apparent precedent of managerial wrongdoing that prompted the conventional explanation of the launch decision. Receiving the most attention in posttragedy accounts were these controversial decisions: continuing to fly with the existing design after the first erosion on STS-2; publicly declaring the shuttle opera-tional despite the flawed design of the SRB joints; waiving the newly imposed C 1 status one week before the launch of STS-6; and Level III Marshall management's inconsistent pattern of reporting informa-tion about the SRBs to upper-level NASA administrators in FRRs. For insiders, however, these decisions were acceptable and nonde-viant at the time they were made.

Continuing missions after the first incident of erosion on STS-2 clearly was calculated risk taking, as was every launch recommen-dation. But we must to pay attention to the origins of this decision within the organization. To attribute this decision to individual managers ignores the fact that risk assessments and launch recom-mendations were the product of a bottom-up decision FRR process that began with working engineers. The recommendation to fly after STS-2 was made first by Thiokol and Marshall working engineers and then conveyed to their managers in the work group. Consensus existed that erosion did not threaten the ability of either the prima-ry or secondary O-ring to seal the joint. Conforming to the guidelines of NASA's Acceptable Risk Process, engineers tested and did calcu-lations to learn the cause of the problem and the capabilities and lim-its of the SRB joint design, producing numerical boundaries for "deviant" (unacceptable) and "normal" (acceptable) performance. Because the erosion on STS-2 fell within these mathematical limits, it was redefined as an acceptable risk.

The decision to declare the shuttle operational after the July 1982 flight of STS-4 was defined as deviant by posttragedy analysts for three reasons: (1) the SRB joint design was known to be flawed, (2) simultaneously, NASA cut several safety procedures that were in place during the developmental period, and (3) President Ronald Reagan had made the speech publicly declaring the shuttle operational, and this symbolic act was widely interpreted as placing pressure on NASA to proceed with the operational phase. Consequently, it was alleged that NASA had "pushed ahead" with operational spaceflight, compromising safety in response to production pressures early in the decision-making history. It is important to note, again, where in the organization this decision originated—and when. This was not a decision made by Marshall managers assigned to the SRB Project. It was made in the early planning stages of the Space Shuttle Program by other NASA administrators who, when charting and coordinating the work flow for the program, established "program milestones": the deadlines for achievement of program goals. Operational spaceflight was a milestone contingent on successful completion of testing and flight prerequisites that assayed the readiness of the technology to meet the requirements of an operational system. The plan to reduce safety procedures after the fourth successful flight was also made at this early stage.

The question then becomes, Why in 1982 did top NASA administrators follow through with the initial plan after STS-4, when the booster design was known to be flawed? In the NASA organizational culture, "flying with known risks" was not deviant, but normative. The document describing the Acceptable Risk Process acknowledged that after all hazards that could be eliminated had been eliminated, certain residual risks would remain, leaving the shuttle with some overall aggregate risk that had to be assessed prior to each flight. So yes, they were flying with known risks. This was official policy, determined before shuttle flight began. The issue for decision makers at all levels was not whether NASA was flying with known risks, but how large those risks were. For this information upper-level administrators relied on a decentralized process that gave responsibility to work groups. And the SRB work group had approved the design as an acceptable risk and recommended that it was safe to fly.

The remaining two controversies involved alleged rule violations during these years by Level III Marshall managers. Mulloy's controversial C 1 waiver was, however, neither a rule violation nor deviant within the NASA culture. Waiving the redundancy requirement for C 1 items was a frequently used, formalized, recorded NASA procedure. The waiver system provided (1) a mechanism (C 1 status) that mandated extra review for a particular component and then (2) a mechanism for bypassing the redundancy requirement after engineering analysis indicated the component was an acceptable risk. If a C 1 item met engineering requirements for risk acceptability, flight could proceed by virtue of the precautionary measures of the waiver system. The method in which this waiver was processed was also not deviant in the NASA culture. Lunney's decision to rush the waiver through without convening his board conformed to an internal norm condoning a "walk-through" when the problem was routine and non-controversial and the launch schedule made speed of the essence. Schedule was always a concern. However, this action was not an instance of schedule taking precedence over safety. The documentary record makes clear that both Marshall and Thiokol working engineers agreed on risk acceptability. Because there was consensus, the waiver did not require the special attention of a full board meeting.

The second alleged rule violation during this period was what the Presidential Commission identified as Marshall Level III failure to report SRB joint problems in Level II and Level I FRRs. In my review of FRR documents, I examined reporting practices involving the SRB joints, comparing them with formalized reporting procedures and with reporting practices of other work groups. Although Mulloy's reporting practices appeared after the disaster to be rule violations, Level III personnel conformed to the NASA Delta review concept for flight-readiness reporting. At Levels II and I, the rule was, no change, no report.[99] And that is what Level III management did. This rule was designed to keep upper-level administrators informed about current developments and to keep problems previously resolved from making the FRR process more cumbersome than it already was at the top, when the whole system was reporting.

As we view the controversial NASA actions during 1981–84 within the social context of decision making at NASA, we uncover a previously unremarked aspect of the history of decision making on the SRB joints: a pattern of Level III management conformity to NASA launch decision rules, rather than a pattern of managerial wrongdoing.

Chapter Five

THE NORMALIZATION OF DEVIANCE, 1985

In 1985, nine shuttles were launched—almost twice as many as in the previous year. Seven exhibited erosion and/or blow-by. Moreover, two had damage beyond anything yet experienced. On STS 51-C, the first launch of 1985, field joint blow-by reached a secondary O-ring. This signal, ominous at the time, became even more so in retrospect, for STS 51-C was launched after standing on the launch pad during three consecutive nights of record low Florida temperatures, raising the question of the effect of cold temperature on the O-rings. Then, on the April 1985 launch of STS 51-B, a nozzle joint primary O-ring burned completely through. For the first time, hot gases eroded a secondary. After each of these flights, concern escalated at Marshall and Thiokol about the O-ring problem. Memos were flying back and forth between the two organizations, a full presentation about the problem was made to senior administrators at NASA Headquarters, and O-ring Seal Task Forces were formed at both sites to try to resolve the problem. Yet flight continued.

After the *Challenger* tragedy, the Presidential Commission and other analysts were incredulous that NASA proceeded with flight after these 1985 incidents—especially after internal memos were discovered in which some Thiokol engineers had expressed deep concern about possible catastrophe after STS 51-B. The worsening erosion, viewed within the context of the increased flight rate, led to speculation that managers were experiencing greater pressure to launch than before. Three additional findings of the Commission added to the general incredulity, reinforcing views about managerial wrongdoing. First, Marshall SRB Manager Mulloy imposed a Launch Constraint on the shuttle after the 51-B primary ring burn-through, supposedly halting flight until the problem was resolved. But he waived the constraint for the next shuttle flight—and for *each* shuttle flight remaining in 1985, proceeding with launches with apparent disregard for the reasons behind the constraint. Second, Mulloy did not report the Launch Constraint or its waivers to Levels II and I in FRRs. Third, in December, the O-ring problem was officially closed out at Marshall. The official closure six weeks before the *Challenger* launch appeared to be even more egregious than the other two actions. Erosion was still occurring. Why had the problem officially been closed? To many posttragedy analysts, it appeared that NASA managers were so intent on meeting the schedule that they violated rules that would impede spaceflight, then violated rules requiring that information be relayed up the hierarchy in order to conceal the deviant performance of the SRB joints from their superiors.

In this chapter, we follow the production of culture in the SRB work group in the year preceding the *Challenger* launch. In a year punctuated by dramatic signals of potential danger, we again witness the incremental expansion of the bounds of acceptable risk. Ethnographic thick description shows how (1) the work group continued to normalize the deviation of the joint from performance predictions and (2) insiders defined as acceptable and nondeviant actions that, after the tragedy, were defined as deviant by outsiders. First, we trace the interpretive work of the technical experts at the bottom of the launch decision chain, following the engineering analysis and logic that undergirded the normalization of technical deviation on STS 51-C and 51-B. Then, we examine escalating concern in the SRB work group about the O-rings, reconstructing the documentary record and revealing a shift in the cultural construction of risk. Finally, we find that, despite increased alarm in the work group, the norms, values, and beliefs of these technical experts continued to operate: the belief in redundancy prevailed. And contradicting conventional understandings, we find that (1) in every FRR, Thiokol engineers brought forward recommendations to accept risk and fly and (2) rather than amoral calculation and misconduct, it was a preoccupation with rules, norms, and conformity that governed all facets of controversial managerial decisions at Marshall during this period.

STS 51-C AND THE COLD

In January 1985, the Florida temperature at the Cape Canaveral launch site suddenly plummeted to record lows. Overnight temperatures reached 18°F–22°F for several days prior to launch of STS 51-C.[1] Scheduled for January 23, it was delayed by the Mission Management Team at the Cape because the overnight temperature had

dropped to 20°F and was still below freezing at launch time,[2] in violation of the Launch Commit Criterion requiring an ambient temperature range of 31°F–99°F for launch.[3] No concerns were raised by
Marshall or Thiokol personnel about the effect of the cold on the
booster joints. That night the ambient temperature dropped again,
then climbed steadily, reaching 66°F by launch time on January 24.[4]
When the boosters were examined after this mission, Thiokol engineers found erosion and blow-by on two field joints.[5] It was the blow-
by that alarmed them. For the first time, hot motor gases had reached
a secondary O-ring.[6] The secondary was not eroded, but the blow-by
was "jet black" and intermixed with the grease between the O-rings,
indicating two destructive forces at work: impingement erosion *and*
blow-by erosion. Marshall's Leon Ray, at the Cape for booster disassembly, believed the blackened grease indicated that the grease had
gotten hot. He concluded that there had been more opportunity to
sustain secondary O-ring damage than yet witnessed.[7] Thiokol's
Boisjoly described what they saw:

> SRM 15 (STS 51-C) actually increased (our) concern because that
> was the first time we had actually penetrated a primary O-ring on
> a field joint with hot gas, and we had a witness of that event
> because the grease between the O-rings was blackened just like
> coal . . . and that was so much more significant than had ever been
> seen before on any blow-by on any joint . . . the fact was that now
> you introduced another phenomenon. You have impingement ero-
> sion and bypass erosion, and the O-ring material gets removed
> from the cross section of the O-ring much, much faster when you
> have bypass erosion or blow-by.[8]

Encased deep in the booster structure, O-ring temperature always
lagged behind ambient temperature changes. Boisjoly believed that
the vehicle's standing on the launch pad for several days while
overnight low temperatures ranged from 18°F to 20°F had impaired
O-ring resiliency, the ability of the rubberlike rings to expand to fill
the gap during joint rotation. Reduced resiliency would slow sealing,
allowing hot ignition gases to bypass the ring until the joint sealed.
On calculating the O-ring temperature after the damage was found,
Marshall and Thiokol engineers learned that although ambient temperature had reached 66°F at launch time, O-ring temperature was
53°F. Three days of record low temperatures had cooled the rings.

Unprecedented damage and concerns about the cold notwithstanding, when Boisjoly and the other Thiokol engineers presented
their postflight analysis to Marshall SRM Manager Wear at the Level
IV FRR for the next scheduled flight, STS 51-E, their final chart stat-

ed, "STS-51E could exhibit the same behavior. Condition is not desirable but is acceptable."[9] They recommended proceeding with the next launch. Following the Level IV FRR, Mulloy sent a "Certified Urgent" message to Wear, requesting Thiokol prepare a detailed briefing on the O-ring problem for the Level III FRR on February 8 (see fig. 7).

In this major review, Thiokol engineers explained the engineering logic behind their finding of acceptable risk. They identified the cause of the blow-by as hot gases moving through blowholes in the putty. This was no surprise. When in 1984 they decided to increase the leak check pressure to 200 psi on the field joints to assure that the rings would seal, they knew that the probability of erosion would be increased. Consequently, the erosion was accepted and expected. The question was about the amount of erosion. The three-factor technical rationale was their basis for assessment. They found, first, that even though two different kinds of erosion forces had operated on the 51-C field joint primary ring, the amount of primary erosion (0.038") was well within the *experience base* (the 0.053" erosion found on STS-2). Second, the amount of erosion that occurred was within the 0.090" *safety margin*. Third, the primary ring sealed before the hot gases could erode the secondary, reinforcing the existing belief that the effect of hot gas flow on the rings (both impingement erosion and blow-by) was a *self-limiting phenomenon*.[10]

In short, the rings had operated exactly as Thiokol engineers had

MARSHALL REQUEST FOR BRIEFING AFTER STS 51-C

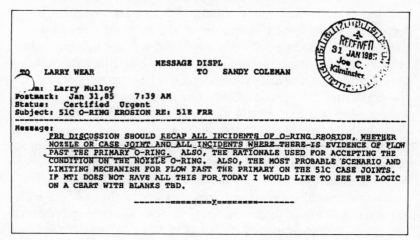

Fig. 7

predicted. The belief in redundancy was sustained. The primary had sealed; the secondary's ability to back it up had not been compromised. Despite blow-by erosion on the primary and the first-ever evidence of heat reaching the secondary, Thiokol engineers decided to accept risk for the next launch. Figures 8.1–8.3 show the last three charts from the 26-chart full review Mulloy requested, which show the temperature effects and the technical rationale that led the engineers to the decision to accept risk. Important at the time, they are extremely important in retrospect.

In figure 8.2, the "Rationale for Acceptance," we see the use of the three in-house standards that the engineers had developed for assessing joint performance. They justified continuing flight on the basis of the experience base (first and second bulleted items), the safety margin (third and fifth bulleted items), the self-limiting nature of the phenomenon (the reference to "momentary gas passage" in the second bulleted item; fourth and sixth bulleted items), and redundancy (seventh bulleted item).

But what about the effect of the cold? Thiokol's Level III FRR presentation included the first mention of the effect of temperature on the O-rings.[11] At this point they had no systematic test data on the relationship between temperature variation and O-ring performance, although no one doubted the obvious: rubber gets hard when it gets cold. Figure 8.1 on "Temperature Effects" stated the extent of their knowledge about cold weather at the time: "Putty becomes stiffer and less tacky. O-ring becomes harder." Thiokol's Boisjoly was present at the Level III review. He stated that his visual inspection of the joint at the Cape convinced him that "low temperature was indeed responsible for such a large witness of hot gas blow-by."[12] Boisjoly posited a correlation between cold temperature, resiliency, and the damage, on the basis of his observation of the blackened grease. He believed that the vehicle's exposure to record low temperatures for three days resulted in the postflight calculation of a 53°F O-ring temperature—the lowest calculated joint temperature for a launch. Marshall's Mulloy, taking his usual adversarial FRR role, questioned Boisjoly's conclusion that there was a causal relationship between cold temperature and erosion. Thiokol had already identified many different factors that caused O-ring erosion in the past: the putty pattern, a piece of lint in the putty, an O-ring in improper position, putty composition, unidentified imperfection in the O-ring composition, and so forth. If cold had caused the damage by reducing O-ring resiliency, why did damage occur on only some of the rings of the boosters? What engineering data did they have that showed it was

STS 51-E FLIGHT READINESS REVIEW CHARTS
(Analysis of 51-C Cold Weather Launch)

TECHNICAL ISSUES

* TEMPERATURE EFFECTS

  * THERMAL ANALYSIS SHOWS THAT THE SRM FOLLOWS DAILY TEMPERATURES TO A DEPTH OF FIVE INCHES

  * PUTTY BECOMES STIFFER AND LESS TACKY

  * O-RING BECOMES HARDER

    * A VITON O-RING WITH 70 DUROMETER AT 70°F CAN INCREASE TO 85 DUROMETER AT 20°F

    * TEMPERATURE PRIOR TO FLIGHT DROPPED TO 17°F

    * TEMPERATURE SQUEEZE REDUCED BY ~1% DUE TO A 4°F DELTA

Fig. 8.1

FLIGHT READINESS ASSESSMENT FOR STS-51E (CONT)

RATIONALE FOR ACCEPTANCE

* O-RING EROSION ON STS-51-C WAS WITHIN EXPERIENCE DATA BASE

* MOMENTARY GAS PASSAGE BY THE PRIMARY SEAL WAS SEEN ON THE STS-14A NOZZLE JOINT

* SECONDARY SEAL HEAT EFFECTS WERE WELL BELOW ANALYTICAL WORST CASE PREDICTIONS

* GAS JET PENETRATES THE PRIMARY SEAL PRIOR TO ACTUATION AND SEALING

* TESTS SHOW THAT O-RINGS WILL SEAL AT 3,000 PSI WITHIN 0.095 INCH OF MISSING MATERIAL (WHICH IS GREATER THAN THE WORST CASE PREDICTION AND ALMOST TWICE THE EROSION SEEN ON ANY SRM MOTORS).

* PRIMARY O-RING EROSION OBHSERVED TO DATE IS ACCEPTABLE AND WILL ALWAYS BE MORE THAN EROSION ON SECONDARY O-RING IF IT OCCURS.

  * PRIMARY O-RING LEAK CHECK PUSHES O-RING IN WRONG DIRECTION -- SECONDARY O-RING IS SEALED BY LEAK CHECK

  * GAS VOLUME IN FRONT OF PRIMARY O-RING IS 50% GREATER THAN FREE VOLUME BETWEEN O-RINGS

  * GAS WILL COOL AS IT PASSES PRIMARY O-RING AND DIFFUSES CIRCUMFERENTIALLY

* SECONDARY O-RING IS A REDUNDANT SEAL USING ACTUAL HARDWARE DIMENSIONS

Fig. 8.2

the cold as opposed to another factor or combination of factors that explained what they saw?

Boisjoly, commenting on how this issue was resolved, pointed out that they did not have the data to show a causal relationship: "And since Al (McDonald) and I didn't have any concrete evidence to prove our conclusion about the effect of low temperature, we agreed to the

FLIGHT READINESS ASSESSMENT FOR STS-51E (CONT)

* EVALUATION SUMMARY

    * STA-51-C PRIMARY O-RING EROSION ON TWO FIELD JOINTS

    * STS-51-C SOOT BETWEEN PRIMARY AND SECONDARY O-RINGS ON BOTH FIELD JOINTS -- FIRST TIME OBSERVED ON FIELD JOINT

    * EVIDENCE OF HEAT AFFECT ON SECONDARRY O-RING OF A68 (RIGHT HAND) CENTER FIELD JOINT BUT NO EROSION

* CONCLUSION

    * STS-51-C CONSISTENT WITH EROSION DATA BASE

        * LOW TEMPERATURE ENHANCED PROBABILITY -- STS-51-C EXPERIENCED WORST CASE TEMPERATURE CHANGE IN FLORIDA HISTORY

        * EROSION IN TWO JOINTS OBSERVED BEFORE -- STS-11 AND -14

    * STS-51-E COULD EXHIBIT SAME BEHAVIOR

    * CONDITION IS ACCEPTABLE

* STS-51-E FIELD JOINTS ARE ACCEPTABLE FOR FLIGHT

Fig. 8.3

following wording [for the chart presenting the technical rationale in subsequent levels of FRR], which at least referred to low temperature in a probability way, and these are the words that we used on the charts: 'Low temperature enhanced the probability. STS 51-C experienced worst case temperature change in Florida history.'"[13] In NASA culture, observation alone did not qualify as "evidence," "hard data," or "real data" in FRR presentations; it was viewed as "not scientific"—a critical point that will be elaborated extensively in the next chapter. Thiokol's SRM Project Manager Al McDonald said:

> We didn't have any real data, but the presentation says that we know that durometer of the O-ring gets harder with temperature. That is just a known fact. We didn't have actual data, but we know that elastomers all behave that way. That was part of the presentation. And we said that it may well have been that that was a contributor in some way, that it got colder than we had seen before, therefore the O-ring got harder, and maybe that made it more difficult to seat, and therefore allowed a little bit of gas to go by it before it did. That was our rationale at the time. We had no hard data.[14]

So Thiokol's final "recommendations" chart (fig. 8.3) stated: "STS 51-C consistent with erosion data base. Low temperature enhanced probability of blow-by. STS 51-C experienced worst case temperature change in Florida history. STS 51-E could exhibit the same behavior. Condition is acceptable."[15]

Thiokol's conclusion to accept risk was also influenced by the rarity of such low temperatures. Although three days of overnight temperature drops to the teens and twenties was the "worst case temperature change in Florida history," *its uniqueness lessened the risk: no one expected a recurrence.* After the *Challenger* accident, the phrase "worse case temperature change in Florida history" on this FRR chart looked like a dark warning. However, at the time the phrase held a different meaning for the Thiokol engineers who created the FRR charts. They included the phrase not only to identify temperature as a possible factor in the damage but because, paradoxically, this run of extreme cold weather *also* figured into their determination of risk acceptability. Three days of record lows was a "worst case condition that would not occur again."[16] Thiokol's McDonald recalled: "And we went to the Flight Readiness Review for the next flight, which we were required to every time to explain any anomaly, what we understood about it and why it is safe to fly the next flight. The conclusion we drew at that time . . . was that that particular flight was preceded by the three coldest days in Florida history, and it was cold. Now, we didn't expect to see the three coldest days in Florida history again, and therefore, the next flight should be acceptable because it shouldn't see this type of behavior."[17]

The issue of temperature effects was not dropped, however. Boisjoly returned to Utah and met with Arnie Thompson, Supervisor, SRM Cases, to discuss the blow-by scenario and the effects of cold temperature on O-ring resiliency—the ability of the ring to restore itself to its round shape after the "squeeze" on the seal was removed during joint rotation. Engineers had been preoccupied with the question of redundancy. Now, Thiokol engineers began testing the resiliency of the O-rings in varying temperatures.[18] Here we are again reminded of the ad hoc learning by doing that characterized their work, even as late as 1985. Cold temperature was never defined as a major concern for the shuttle when it was designed. To the contrary, the concern was about how the structure would withstand heat; for example, the temperature in the combustion chamber of the Main Engine at ignition would be 6,000°F, higher than the boiling point of iron.[19] Similarly, design and development of the boosters focused on heat extremes that they would have to withstand at ignition. Therefore, the rings were designed and tested extensively for heat tolerance. Some cold tests had been done to account for variation in the weather, but on the basis of historic temperature ranges at the Florida launch site, cold was not defined as a potential problem.

Prior to STS 51-C, no tests of O-ring resiliency had been performed by the SRB work group. Even after the three days of record cold, Boisjoly recalled, there was "no scramble to get temperature data."[20] The working engineers at Thiokol and Marshall did not expect a repeat of the run of 18°F–20°F Florida overnight temperatures. Boisjoly said: "Nobody had any idea that you would have like a 100 year storm two years in a row, and that's what that amounts to, tantamount to a 100 year storm. That statistically is so improbable as that it won't exist. . . . It was nobody's expectation we would ever experience any cold weather to that degree before we had a chance to fix it again, so that basically is why it [temperature data] wasn't pursued any further than that from my personal standpoint."[21] The engineers viewed the cold spell as an isolated, idiosyncratic incident. Their intuition suggested that the cold may have contributed to the erosion on STS 51-C, but they had no "real data," and the cold was not likely to be a factor in the future. So although they began their research on O-ring resiliency, it was not driven by a sense of urgency; and because of the definition of the situation, few engineering resources were devoted to it.

As the STS 51-E FRR continued at Marshall, Mulloy gave an 18-chart presentation, conforming to the Delta reporting concept that any new problem or new manifestation of an old problem be reviewed at all Marshall FRRs. The last chart was a duplicate of the recommendation chart Thiokol engineers had presented to Mulloy (fig. 8.3). Both the Marshall Shuttle Projects Office Board and the Marshall Center Board accepted the Level III recommendation to proceed with STS 51-E and certified it as flight-ready. At the Level II and Level I FRRs, Mulloy presented only two charts. NASA sometimes videotaped FRR presentations to instruct participants on how to make a proper presentation and review boards on how the engineering analysis was to be probed for flaws. Extremely significant in retrospect, this particular Level I presentation was videotaped, showing that (1) despite the reduction in charts from the Marshall FRRs, Mulloy's oral presentation was a full briefing that conformed to NASA requirements[22] and (2) all levels of the launch decision chain were fully informed about both erosion and the technical rationale for accepting risk at this 1985 review.

Mulloy identified the cause of the O-ring problem as gas paths through the putty. He presented the hazard analysis, explaining the three-factor rationale for accepting the conditions found on STS 51-C and proceeding with the launch of STS 51-E: the experience base, the safety margin, and self-limiting nature of the phenomenon. He did not mention the Thiokol engineers' concerns about temperature—a decision with which Al McDonald agreed—because no systematic data were yet available that proved the association between the cold and the damage found on STS 51-C. Only "solid engineering data" were admissible in FRR presentation. Recall Boisjoly's comment that the visual evidence of the black grease at disassembly was not considered "concrete evidence" and McDonald's comment about "no hard data." Mulloy concluded with the work group's construction of risk ("this represents an acceptable risk")[23] and recommendation to proceed with the next launch. Level I NASA administrators accepted the conclusion and recommendation, affirmed the aggregate risk of all shuttle components, and certified the readiness of STS 51-E.

STS 51-E (later renamed 51-D) dramatically displayed the discrepant responses to the technology at the top and bottom of the NASA organization in 1985. Aboard was Senator Jake Garn, Republican from Utah (home of Thiokol-Wasatch) and chair of a Senate subcommittee that oversaw NASA's budget.[24] Publicized as the shuttle's first "congressional observer," he was on board in his capacity as chair of the subcommittee. Listed as a payload specialist, Garn assisted in some in-flight experiments. Garn's inclusion on the flight in 1985 demonstrates the active efforts by NASA top administrators to manipulate the space agency's political and economic environment to assure access to scarce resources. But in addition to Garn's personal connection to key NASA resources, his flight crew membership tells us something important about the construction of risk at NASA. Despite the developmental nature of shuttle technology so clearly demonstrated by the continued learning and changes implemented in the SRB work group, top NASA administrators had lost sight of the concept of "aggregate residual risk" that undergirded the Acceptable Risk Process. Their inclusion of Garn suggests that, having publicly defined the Space Shuttle as operational, they bought their own definition: top NASA policymakers defined shuttle flight as sufficiently routine and safe that an important actor in NASA's political and economic environment could be taken along for a ride.

Ironically, STS 51-D sustained the most extensive erosion found on a primary O-ring to date. Despite moderate weather that put joint temperature at 67°F prior to ignition, Thiokol engineers found 0.068" erosion, surpassing the 0.053" found on STS-2.[25] STS 51-D's first-ever "outside the experience base" erosion did not alarm the work group, however. This damage was never given extra attention in FRRs, nor did it generate any administrative response that officially acknowledged it as a signal of potential danger.[26] The damage was on a nozzle joint. Because of the nozzle location on the boosters, the

design differed from field joint design. The position of the secondary
O-ring made the nozzle less vulnerable in case of a primary O-ring
failure. But more important, nozzle joint primary erosion was expected on this flight. Before launching, working engineers had decided for
the first time to increase the leak check pressure on the nozzle joints
to 200 psi. They knew that this strategy would increase erosion. So
before changing the procedure, Thiokol engineers calculated a safety
margin, the maximum amount of erosion that could safely occur,
and tested for the WOW conditions that would cause the nozzle joint
to fail. When they measured the erosion on STS 51-D, it was within
the quantified boundaries establishing risk acceptability. Rather than
challenging the prevailing construction of risk, the engineering
analysis affirmed it: the nozzle joint had acted exactly as predicted.

STS 51-B, BURN-THROUGH, AND THE
LAUNCH CONSTRAINT

Just 10 days after the return of Garn's mission, STS 51-B was
launched on April 29.[27] When the working engineers examined the
disassembled boosters at Kennedy, they found no field joint erosion.
Time elapsed while the nozzle joints were shipped by rail to Thiokol
for disassembly. When they arrived, Thiokol's refurbishment facility
was full of segments waiting to be analyzed from the test of a developmental motor.[28] Another shuttle was launched before STS 51-B's
postflight analysis was complete. The last week in June, Thiokol
engineers inspected STS 51-B's disassembled SRB nozzle joints at the
Utah plant. The engineers discovered an alarming new signal of
potential danger. Thiokol's Al McDonald stated: "We didn't get that
assembly back (to the Utah plant) and get the nozzle off it until the
end of June, so the flight had occurred the end of April, but we didn't
recognize we had a problem until the end of June, and we saw this O-
ring and were horrified because the primary O-ring seal had burned
all the way through. The secondary seal had some severe damage."[29]
    STS 51-B suffered the worst damage prior to the *Challenger*. The
primary O-ring of the left nozzle joint had eroded so extensively that
it failed to seal. For the first time, a primary O-ring had been violated, or "burned through," allowing hot gases *to erode a secondary O-
ring*. The amount of erosion on the primary was 0.171", exceeding
both the experience base and the safety margin. Marshall's Mulloy
stated:

> This erosion of a secondary O-ring was a new and significant event
> . . . that we certainly did not understand. Everything up to that

point had been the primary O-ring, even though it had experienced some erosion, does seal. What we had evidence of here was a case where the primary O-ring was violated and the secondary O-ring was eroded, and that was considered to be a more serious observation than previously observed. . . . [From] what we saw [in STS 51-B], it was evident that the primary ring never sealed at all, and we saw erosion all the way around that O-ring, and that is where the .171" came from, and that was not in the model that predicted a maximum of .090", the maximum of .090" is the maximum erosion that can occur if the primary O-ring seals. But in this case, the primary O-ring did not seal; therefore, you had another volume to fill, and the flow was longer and it was blow-by and you got more erosion.[30]

The nozzle primary had failed. The secondary, also eroded, had done its job as a redundant backup. But if, in the future, secondary erosion were sufficiently extensive, it would not be available as a backup, resulting in a catastrophic failure. Erosion and redundancy, which until this point had been defined as unrelated, were now defined as related technical issues.

In July 1985, soon after he was informed of the discovery, Marshall's Mulloy imposed a Launch Constraint on the nozzle joint for all flights after STS 51-B.[31] Imposing Launch Constraints was a two-step procedure at NASA. Contractor management decided which problems would be launch-constraining problems and identified them to NASA at the Level IV review. In response to the contractor's recommendation, NASA Project Managers "imposed" a Launch Constraint by submitting an entry to MPAS, the computer problem-tracking system. A Launch Constraint was an administrative act indicating that a problem was serious.[32] But a Launch Constraint did not necessarily halt flight. Like a C 1 designation, it was a procedural requirement designed to assure that serious problems received extra attention; they had to be addressed in each FRR and entered in the computer tracking system. Flight could proceed under two circumstances: (1) the problem was resolved, or (2) engineering analysis showed it was an acceptable risk.[33] Then the Launch Constraint was "lifted" by the Level III Project Manager.[34]

When Thiokol engineers analyzed the 51-B burn-through, they again found an idiosyncratic cause. For 0.171" of the primary to be eroded, blow-by had to have occurred in the first milliseconds of ignition.[35] This meant that the nozzle joint's primary O-ring *had never been in proper position to seal.* Wiley Bunn, Marshall Director of Reliability and Quality Assurance, said: "We had six joints on that vehicle. If the design is that darn bad, all six of them should have leaked. We only had one leak. Therefore, if we only had one leak, it had to be a Quality escape. And so we just renewed our vigor to find that Quality escape."[36] The Thiokol engineers ascertained that the "quality escape" was some flaw in the primary ring at installation that prevented it from sealing—a hair or a piece of lint could do it—and that the 100 psi nozzle leak check had not detected that the ring was not in proper sealing position. Although they had used 200 psi for the nozzle joints on the previous mission for the first time, for STS 51-B they had changed the putty lay-up and returned to 100 psi for the nozzles.[37] They concluded that, even with the putty change, a 100-psi leak check was not enough to adequately test the joint.[38] After this discovery, they decided to return to the 200-psi leak check pressure on the nozzle joint for the next launch.

Having identified an improperly positioned ring as the cause, the engineers assessed the risk of secondary erosion as they had primary erosion: quantify the conditions under which the secondary O-ring would both seal and fail. Every calculation and test affirmed redundancy. They used a mathematical model just completed by Thiokol scientist Mark Salita.[39] Salita had begun developing this model after the field joint erosion on the January cold temperature launch. Now he extended it to predict the behavior of nozzle joint O-rings.[40] Salita calculated the maximum nozzle joint secondary erosion that could occur under WOW conditions,[41] concluding, "number one, that it was improbable that jet impingement erosion would burn through a primary O-ring and number two, if you had blow-by of a primary O-ring . . . you would not expect to see jet impingement erosion compromise that [secondary O-ring] seal."[42]

Thiokol also ran cold-gas tests on an O-ring with pieces sliced out of it simulating erosion and conducted newly developed hot-gas subscale tests, believed to provide a more finely tuned, accurate assessment than previous methods. Thiokol's McDonald reported:

> We found that in the evolution of that test series that we could erode up to .125" [before a failure occurred], and we had a lot of tests there and below that we never saw a seal ever fail. We went up to .150" of the material removed and it still sealed. However, we had two tests, one at .145" and one at .160", that did fail. Now, both of those failed the primary seal, but eroded less than .010" off of the secondary seal. . . . If you took our worst measured [flight] erosion on the O-ring relative to what it took to really fail it [in tests], it was nearly a factor [i.e., safety margin] of three to one. . . . We did not feel it was that bad at three to one, and as long as we could retain the secondary seal during a good portion of the erosion time period, we felt good.[43]

As a result of postflight analysis, calculations, and the leak check corrective action, Thiokol engineers concluded the design was an acceptable risk. The norms, beliefs, and procedures of the past guided the working engineers' response to the 51-B nozzle joint erosion. If the most secondary erosion that could occur under worst-case conditions was 0.075" and the hot-gas test showed that the secondary always sealed with an erosion depth of 0.125", then the 0.032" of erosion discovered on the 51-B secondary was well within these worst-case conditions. By increasing the leak check pressure on the nozzle joints to 200 psi, Thiokol engineers expected nozzle joint performance to be similar to field joint performance: the primary O-ring would seal; the secondary would not erode. Salita's analytic model predicting the worst-case scenario and the hot-gas tests were fundamental to the engineers' decision to accept risk. If, under WOW conditions, the primary should fail initially, the secondary would still serve as an effective backup.

Using 10 Thiokol charts, Mulloy gave the 51-B analysis in a full briefing at the June 27 Marshall Center Board FRR.[44] All the other Marshall FRRs for the upcoming flight, STS 51-F, had occurred before the nozzle joint analysis was complete. As a consequence, the engineering work had been scrutinized by fewer people than was normal. To compensate for the missed reviews, a special review combining all Marshall boards was convened on July 1. Thiokol engineers presented 33 charts analyzing the entire O-ring erosion problem history.[45] The final chart reported the corrective action taken after the 51-B burn-through and its predicted result: increasing the leak check pressure to 200 psi would result in primary O-ring erosion "within experience base and no secondary O-ring erosion."[46] Thiokol engineers' flight-readiness recommendation was that "STS 51-F (SRM-19) nozzle joints are acceptable for flight."[47]

On July 2, Mulloy presented the work group's construction of risk to Levels II and I. The Thiokol engineering analysis was condensed to four charts. The "Problem Summary" chart (shown in fig. 9) reported the engineering logic behind the work group's conclusion that the 51-B damage was an acceptable risk. Following the standard format for Level I presentation, the anomaly is described in the left-hand column. The "Resolution" summarizes the engineering analysis. Read from the top down, the summary notes the earlier occurrence of the anomaly (first item), the cause (second item), the corrective action (third item), and the technical rationale for accepting risk (fourth item—redundancy; fifth item—self-limiting phenomenon; sixth item—safety margin). On the basis of the 200-psi corrective measure

STS 51-F FLIGHT READINESS REVIEW CHART
(Analysis of 51-B Secondary Erosion)

Level I (July 2, 1985)

PROBLEM SUMMARY

PROBLEM	CONCERN	RESOLUTION	STATUS
UNUSUAL EROSION OBSERVED ON PRIMARY AND SECONDARY O-RINGS OF STS 51-B NOZZLE TO CASE JOINT (SRM 16 A) — PRIMARY O-RING APPARENTLY NEVER SEATED RESULTING IN WORST EROSION YET OBSERVED — SECONDARY O-RING HAD 32 MILS EROSION (1ST TIME OBSERVATION)	FLIGHT SAFETY	• EVIDENCE OF HOT GAS PAST PRIMARY O-RING IS NOT UNPRECEDENTED • LEAK CHECK USED ON STS 51-B DID NOT VERIFY CAPABILITY OF PRIMARY O-RING TO SEAL • LEAK CHECK USING 200 PSIG STABILIZATION PRESSURE ON STS 51-F AND SUBS PROVIDES CONFIDENCE THAT PRIMARY O-RINGS HAVE CAPABILITY TO SEAL • LEAK CHECK ASSURES SECONDARY O-RING WILL SEAL AGAINST MOTOR PRESSURE • MAXIMUM EROSION THAT CAN OCCUR ON SECONDARY O-RING IN LIMITED TIME THAT FLOW EXISTS ON AND PAST PRIMARY O-RING IS 75 MILS (CONSERVATIVE ANALYSIS) • —2 SUBSCALE TESTS VERIFY THAT A PROPERLY SEATED O-RING CAN SUSTAIN A MINIMUM OF 125 MILS EROSION BEFORE SEAL IS COMPROMISED	CLOSED

Fig.9

and the three-factor technical rationale, the problem was presented as "closed"—resolved for that launch. Level I certified the next launch of STS 51-F as flight-ready. Then Mulloy, conforming to the requirement that all paperwork be completed prior to a launch, lifted the Launch Constraint prior to the flight of STS 51-F, permitting it to fly. This action was recorded in MPAS on July 15, 1985.[48]

The Presidential Commission viewed Mulloy's action as deviant, noting the "strange sequence of six consecutive launch constraint waivers prior to 51-L, permitting it to fly without any record of a waiver or even of an explicit constraint."[49] However, imposing Launch Constraints then lifting them was not deviant in the NASA culture. A Launch Constraint was a monitoring device. Marshall managers testified that a Launch Constraint was a designation intended to flag open problems so that they would be sure to receive attention in FRRs.[50] They asserted that a Launch Constraint was a *process*. What was singularly distinctive about this process was that the designation *mandated* that the problem be reviewed in FRR and tracked in MPAS.[51] Recognizing Commissioners' misunderstanding of the meaning of a NASA Launch Constraint, NASA managers acknowledged in testimony that the term "Launch Constraint" was a misnomer for describing a "process intended to give visibility to problems."[52]

NASA's "Constraining Problems List" for the External Tank, Main Engine, SRB, and SRM from the first shuttle flight in 1981 through the fatal *Challenger* flight extended for 14 pages and enumerated 131 constraints on the shuttle system.[53] Of these, only 66 were closed after one flight. The rest were lifted from two to twelve times. The average number of times was three. During the FRRs for the 1986 *Challenger* launch, for example, 13 Launch Constraints were lifted: nine on the shuttle's Main Engine, one on the External Tank, and three on the SRBs.[54] Mulloy testified: "By the way, this (lifting a launch constraint) is not at all unusual. During that period, you know, that the (51-B) launch constraint was being lifted, under this same procedure launch constraints were being lifted on other elements of the shuttle system."[55]

Here again, we see how the organizational culture was a barrier to outsider understanding. The Presidential Commission tried to understand the Launch Constraint; Mulloy tried to clarify:

> CHAIRMAN ROGERS: To you, what does constraint mean, then?
>
> MR. MULLOY: A launch constraint means that we have to address the observations, see if we have seen anything on the previous flight that changes our previous rationale, and address that at the Flight Readiness Review.
>
> CHAIRMAN ROGERS: When you say "address it," I always get confused by that word. Do you mean think about it? Is that what you mean?
>
> [In the reply, Mulloy describes one aspect of the Acceptable Risk Process.]
>
> MR. MULLOY: No, sir. I mean present the data as to whether or not what we have seen in our most recent observation, which may not be the last flight, it may be the flight before that, is within our experience base and whether or not the previous analysis and tests that previously concluded that was an acceptable situation is still valid, based upon later observations.[56]

Mulloy responded to further questions by reading from the July 1985 entry in MPAS officially recording his action.[57] This entry contained the engineering analysis that legitimated lifting the Launch Constraint.

> MR. MULLOY: What was done was analysis and test to substantiate that rationale. It is not a matter of nothing was done. . . . We were sure on the 51-F flight because of the 200 psi leak check that we had a good primary O-ring was a substantial part of the ratio-

nale. The second then was if the primary O-ring was violated, the maximum erosion that could occur on the secondary O-ring was only .075", which tests had shown could sustain .125". Now rightly or wrongly, that was the rationale.

COMMISSIONER WALKER: So you then were relying on the secondary O-ring in that case.

MR. MULLOY: We were relying on the redundancy, yes. We showed that we had redundancy that, should the primary O-ring fail, the secondary would function.[58]

Thiokol engineer Brian Russell, testifying before the Commission, concurred: "My understanding of a launch constraint is that the launch cannot proceed without adequately—without everyone's agreement that the problem is under control . . . and in this particular case on the 51-B nozzle O-ring erosion problem there had been some corrective action taken [the 200-psi leak check], and that was included in the presentation made as a special addendum to the next Flight Readiness Review, and at the time we did agree to continue to launch, which apparently had lifted the launch constraint, would be my understanding."[59]

Mulloy also conformed to all NASA requirements for reporting Launch Constraints up the hierarchy.[60] The Presidential Commission stated in its report that Mulloy failed to report the Launch Constraint, which was "contrary to the requirement, contained in the NASA Problem and Corrective Action Requirements System, that launch constraints were to be taken to Level II."[61] However, the reporting procedure was further specified in a program directive, in effect from 1983 through the launch of STS 51-L.[62] Level III Project Managers were to report (among other things) "all problems, open items, and constraints *remaining to be resolved* before the mission" to Levels II and I. For "significant resolved problems," the presentation should include *a brief status summary* with appropriate supporting detail and conclude with a readiness assessment (emphasis added).[63]

As a result of Thiokol's analysis and corrective action, the 51-B burn-through was a "significant resolved problem." Therefore (in NASA-speak) Mulloy "statused" it. He reported the *problem* up the hierarchy as required, but he did not call it a *Launch Constraint* because the problem that resulted in the Launch Constraint had been resolved at Levels IV and III, so that the Launch Constraint was no longer in effect for that launch when FRR moved to the upper levels at Marshall and Levels II and I.[64] Wiley Bunn differentiates between reporting the problem and calling it a Launch Constraint: "There is no need to say it's a launch constraint because in Flight Readiness Review the Center Director, the Program Manager, the Chief Engineer, everybody will get a discourse on why, even though we've had this anomaly, it's all right to fly . . . the people who are responsible, who have the expertise, have looked at this anomaly and have said: Yea, verily, we recognize that that happened, here's why it happened, here is why it's all right to fly. And that rationale is presented. It closes it from a launch constraint, but it doesn't close the problem because it still doesn't meet the engineering [design requirements]."[65]

STS 51-F was launched on July 29, 1985.[66] When it returned to earth, Thiokol engineers found a hot-gas path through the putty of one of the joints, but *no erosion*. The postflight analysis confirmed the work group's construction of risk and, as a consequence, both the technical and procedural decisions made in response to the damage found on STS 51-B: increasing the nozzle joint leak check pressure, the technical rationale for continuing flight, lifting the Launch Constraint, and continuing to fly.

Accepting burn-through was another major turning point in the work group's normalization of technical deviation. They expanded the boundaries of acceptable risk another time, now including the first-ever damage to a secondary O-ring. The five-step decision sequence shows how they maintained the cultural construction of risk—and the belief in redundancy that was its core—in the face of the most severe damage in the history of the work group's decision making.

Signals of potential danger. Postflight analysis of the April 29, 1985, launch of STS 51-B revealed the primary O-ring of the nozzle joint was burned through and did not seal, allowing hot gases to erode the secondary O-ring for the first time. This unprecedented damage was another deviation from predictions, challenging the prevailing cultural construction of risk. Erosion had threatened the secondary's ability to act as a backup. Erosion and redundancy, heretofore seen as two separate problems, for the first time were related.

Official act acknowledging escalated risk. Marshall's Mulloy imposed a Launch Constraint—an administrative response signifying a serious problem—on the booster joints.

Review of the evidence. Working engineers' postflight analysis indicated that the nozzle joint primary had failed at initial pressurization because it was not in proper sealing position. The norms, beliefs, and procedures of the work group culture continued to operate, normalizing the deviant performance of the SRB joint. The 200-psi corrective action, the safety margin, and the concept of erosion as a self-limiting phenomenon provided the basis for the work group's construction of risk. Salita's analytic mathematical model and the new hot-gas test results increased the quantified boundaries defining acceptable (and thus normal) joint performance, accommodating the new development on STS 51-B. The engineering conclusion was that the damage found on STS 51-B was an acceptable risk. The engineers' recommendation was to proceed with the next launch.

Official act indicating the normalization of deviance: accepting risk. The working engineers presented their analysis of the anomalies found on STS 51-B to Marshall. Mulloy subsequently conveyed the construction of risk developed by the work group to NASA administrators at Levels II and I. The joint's certification as flight-ready at all levels and Mulloy's lifting of the Launch Constraint were official acts indicating the normalization of deviant joint performance.

Shuttle launch. STS 51-F, launched July 29, 1985, experienced no O-ring erosion. Thus, the postflight analysis validated the norms, beliefs, and procedures of the work group culture. The belief that the SRB joint was an acceptable risk had survived its most serious challenge.

ESCALATING CONCERN: THE SHIFTING CULTURAL CONSTRUCTION OF RISK

Although the discovery of secondary erosion on STS 51-B did not overturn the belief in redundancy, it precipitated a marked shift in the cultural construction of risk. From 1978 through 1984, the voice of an engineer concerned about the SRB joints occasionally emerges from the pages of the recorded history. But in 1985, concern escalated. Memos warned of catastrophe. Task forces were formed to try to resolve the O-ring problems. In an August 19 briefing to senior NASA administrators at NASA Headquarters, Thiokol made a full presentation about the O-ring problem, emphasizing its seriousness. After the *Challenger* tragedy, analysts viewed these expressions of concern as themselves signals of potential danger sufficiently strong that NASA administrators should have halted flight while a new design was implemented. The Presidential Commission noted in its report, "The O-ring erosion history presented to Level I at NASA Headquarters in August 1985 was sufficiently detailed to require corrective action prior to the next flight."[67] The impression was that despite numerous efforts by engineers to draw its attention to the problem, NASA had done little. In view of these efforts, the December 1985 close-out of the O-ring problem at Marshall seemed a particularly willful disregard of flight safety.

However, when we examine the events after STS 51-B in chronological detail, resituating controversial actions in the historical sequence, we find that the situation was, again, much more complex than the conventional portrayal of this period. In particular, two findings emerge that challenge conventional interpretations: (1) The problem was not ignored; STS 51-B was an attention-getting signal at both Marshall and Thiokol, as indicated by the increased activity and resources devoted to the problem. (2) In FRRs, the working engineers at Marshall and Thiokol continued to find the joints an acceptable risk and to recommend launching through the remaining months of 1985. Despite the obvious escalation of concern and discontent with the status quo, the official construction of risk—developed in the engineering ranks—remained the same. The following sequence of key incidents shows that *escalating concern and a belief in risk acceptability coexisted* during this period.

Quotes taken from engineers' memos sent after STS 51-B were important to posttragedy interpretations that focused on managerial wrongdoing. Some of the most important memos are included in appendix B (figs. B6–B11). Below, I have quoted from or summarized them, extracting the same quotes from the whole document that other posttragedy accounts repeatedly used. However, I have also included other information as social context in order to more clearly convey the meaning the memos had for insiders at the time. Because reinserting the quotes in the full text also enhances/alters the meaning, readers would benefit from referring to these original documents.

June 1985

• When the 51-B burn-through was discovered in Utah, an engineer from NASA Headquarters was present. He wrote a memo to Michael Weeks, Deputy Associate Administrator, NASA Headquarters, discussing the 51-B secondary erosion. He reminded Weeks that Thiokol had not responded to NASA's April 5, 1984, Action Item requesting a full formal review at NASA Headquarters of the SRB field and nozzle joints.[68] At the time, Thiokol engineer Brian Russell had responded with a proposal for a full inquiry into the problem. But more than a year later, no full review had occurred. Weeks began looking into the matter.

July 1985

• Thiokol scientist Mark Salita demonstrated his model in an oral presentation at Marshall, laying out the engineering analysis for accepting risk to about 20 key people. Affirming the belief in redundancy, approximately 75 copies of his report were circulated at Marshall and Thiokol.[69] The same afternoon at Marshall, Thiokol's Arnie Thompson presented the results of Thiokol's resiliency tests.[70] These tests examined the time it took the O-rings to fill the gap created at joint rotation when temperature was varied.[71] The data indicated that as temperature decreases, the rubber gets hard, taking longer to seal the joint. These data did not threaten the belief in redundancy, however; they were preliminary, including a narrow range of temperature predictions. Also, a repeat of the January 1985 cold snap was viewed as improbable.

• The capture feature received verbal approval by NASA engineers for use on Thiokol's SRM steel segment cases. This decision was not a response to the most recent SRB joint anomaly; rather, it was part of a continuing effort to improve the margin of safety on the joint that had begun in the early development period. By now, the capture feature had gone through an extensive laboratory testing program, begun in October 1984 and ending in the spring of 1985. The results convinced Marshall S&E engineers that the capture feature was worthwhile.[72] Thiokol had begun phasing in the capture feature for the SRM steel cases early in 1985.[73] With the July approval by NASA engineers, Thiokol could order the steel segments that incorporated the design modification.

• In response to the secondary erosion on STS 51-B, Irving Davids, SRB Program Manager at NASA Headquarters, went to Marshall to review the O-ring erosion problem. He wrote a memo to Jesse Moore, Associate Administrator for Space Flight, giving an overview of both the nozzle joint and field joint situations, including all sides of the putty controversy (appendix B, fig. B6). Davids's memo concluded: "The present consensus is that if the primary O-ring seats during ignition, then subsequently fails [the WOW condition], the unseated secondary O-ring will not serve its intended purpose as a redundant seal. However, redundancy does exist during the ignition cycle, which is the most critical time." Davids recommended that Marshall give a NASA Headquarters briefing on the SRM O-rings.[74]

• Also in response to STS 51-B, Mulloy created an unofficial engineering O-ring seal task force at Marshall to get "more dedicated [i.e., full-time] manpower working on the problem."[75] At Marshall's insistence, Thiokol also formed an unofficial seal task force to solve the O-ring erosion problem. The Thiokol team consisted of Don Ketner (Chair), Roger Boisjoly, Scott Stein, Brian Russell, and Mark Salita; six others were assigned part time.[76] The Thiokol and Marshall teams were to work together on both short-term solutions (which could be implemented immediately) and long-term solutions (which required extended time to implement).[77] Marshall's John McCarty, Structures and Propulsion Lab, said: "The primary focus was on erosion, what caused erosion, what design changes could you make in order to preclude what caused erosion, and whether you could take putty out, bend it, put some other substance in, change the O-ring or the O-ring configuration, change in the leak check or the leak check configuration."[78] The Thiokol team was to try some putty and ring changes on the new lightweight motor with the capture feature, which was being test-fired in January. The January test-firing created a tight deadline.[79]

• With the January deadline in mind, Thiokol's Boisjoly wrote in his July 22 weekly activity report that the company needed to assign more people full time to the O-ring task force, removing them from other job responsibilities. He warned: "This problem has escalated so badly in the eyes of everyone, especially our customer, NASA, that NASA has gone to our competitors on a propriety basis and solicited their experiences on their joint configurations. . . . [If we do not] secure a timely solution, we stand in danger of having one of our competitors solve our problem via an unsolicited proposal. This thought is almost as horrifying as having a flight failure before a solution is implemented to prevent O-ring erosion."[80]

• On July 23, Richard Cook, a Resource Analyst at NASA's Comptroller's Office, wrote a memo to his supervisor about the SRB seals. Cook was a new employee. One of his first assignments was to do some background research on the SRBs in order to see whether any engineering questions were going to require additional funding that should be incorporated into NASA's next budget.[81] After consulting with engineers in the Office of Space Flight in Washington, he summarized the O-ring problem in his memo, concluding that the cause was unknown. He stated: "There is little question, however, that flight safety has been and is still being compromised by potential failure of the seals, and it is acknowledged that failure during launch would certainly be catastrophic. There is also indication that staff personnel knew of this problem sometime in advance of management's becoming apprised of what was going on."[82]

Cook's supervisor checked out the memo with the engineers in the Office of Space Flight, concluding that action was being taken to resolve the problem and that it was "in the proper technical channel."[83]

• On July 31, Boisjoly sent a memo to Robert Lund, Thiokol's Vice President of Engineering, urging that Thiokol's unofficial task force be given official status, with a dedicated team of members freed from their usual responsibilities in order to devote full time to the problem (appendix B, figs. B7.1 and B7.2). He concluded, "It is my honest and very real fear that if we do not take immediate action to dedicate a team to solve the problem with the field joint having the number one priority, then we stand in jeopardy of losing a flight along with all the launch pad facilities."[84]

August 1985

• Thiokol engineer Brian Russell wrote a letter on August 9 responding to two questions about the field joint that Marshall raised at the monthly Problem Review Board held in July.[85] Responding to the first question about how long it took the field joint secondary to seal, Russell cited the results of the Thiokol O-ring resiliency tests: "At 100°F. the (secondary) O-ring maintained contact. At 75°F. the O-ring lost contact for 2.4 seconds. At 50°F. the O-ring did not re-establish contact in ten minutes at which time the test was terminated. The conclusion is that secondary sealing capability in the SRM field joint cannot be guaranteed."[86]

These data were not alarming at the time, however, even to the engineers, because of the improbability of cold temperatures. The more significant message conveyed by this memo—and what continued to convince the engineers that the design was an acceptable risk—was in Russell's response to Marshall's second question, "If the primary O-ring does not seal, will the secondary seal seat in sufficient time to prevent joint leakage [i.e., failure]?" In response, Russell stated the engineering analysis that undergirded the belief in acceptable risk—redundancy, except under the worst-case condition that the primary failed late in the ignition transient: "MTI [Morton Thiokol, Inc.] has no reason to suspect that the primary seal would ever fail after pressure equilibrium is reached, i.e., after the ignition transient. If the primary O-ring were to fail from 0 to 170 milliseconds, there is a very high probability that the secondary O-ring would hold pressure since the case has not expanded appreciably at this point. If the primary seal were to fail from 170 to 330 milliseconds, the probability of the secondary seal holding is reduced. From 330 to

600 milliseconds, the chance of the secondary seal holding is small."[87] Of the first 170 milliseconds of ignition transient, Boisjoly said, "That was my comfort basis for continuing to fly."[88] The working engineers were convinced that the primary would seal the joint in that first stage of the ignition transient. Failures at later times were WOW conditions, unlikely events.

August 19, 1985

• Thiokol's Al McDonald, Thiokol senior administrators Mason, Wiggins, and Kilminster, Marshall's Mulloy, and three others gave a briefing on the SRB joints to Level I personnel at NASA Headquarters. Harry Quong, NASA's Director of Safety, Reliability, and Quality Assurance attended.[89] The O-ring problem was not the primary reason the meeting was called; it was an afterthought—one of several items added to the main agenda.[90] These Thiokol representatives were present to discuss another matter and so were asked ("while they were there") by NASA's Michael Weeks to make a presentation about the findings on the nozzle joint erosion that occurred on STS 51-B. Thiokol engineers took the opportunity to put together a full problem review.

McDonald presented 54 charts that reviewed the full erosion history and the Thiokol task force recommendations.[91] The charts stated that the erosion problem was a critical issue, illuminating the field joint as the more critical of the two joint types because of the different designs of the field and nozzle joints. Among the "Primary Concerns" Thiokol listed about the field joint were "joint deflection and secondary O-ring resiliency." Thiokol's resiliency data were presented.[92] Ten charts identified the full range of solutions Thiokol was working on to improve joint performance: "near term" (3–12 months to implement) and "long term" (12–27 months to implement)[93]. Full implementation of the capture feature on the SRB steel segment cases, approved by NASA engineers in July, was estimated at 27 months. McDonald reported that the work of the task force would be complete at the end of July 1986, when they would submit their formal report.

The coexistence of escalating concern and the belief in acceptable risk is documented in these charts. At the same time as they drew attention to the seriousness of the problem, they stated that the engineering analysis supported acceptable risk. McDonald included Salita's charts in the presentation, identifying the three-factor rationale—the experience base, the safety margin, and the self-limiting phenomenon—that led to the Thiokol engineering conclusion of acceptable risk.[94] Thiokol's last chart of "Recommendations" (see fig. 10) urged (among other things) speed in resolving the problem (seventh item) while stating explicitly the conditions that made it safe to fly (sixth item).

That evening, NASA's Michael Weeks, who chaired the meeting in the absence of Jesse Moore, briefed Moore on the presentation, conveying both messages from the Thiokol charts: it was "safe to continue flying existing design as long as all joints are leak checked

AUGUST 19, 1985, NASA HEADQUARTERS BRIEFING CHART

General Conclusions

- All O-ring erosion has occurred where gas paths in the vacuum putty are formed

- Gas paths in the vacuum putty can occur during assembly, leak check, or during motor pressurization

- Improved filler materials or layup configurations which still allow a valid leak check of the primary O-rings may reduce frequency of O-ring erosion but will probably not eliminate it or reduce the severity of erosion

- Elimination of vacuum putty in a tighter joint area will eliminate O-ring erosion if circumferential flow is not present - if it is present, some baffle arrangement may be required

- Erosion in the nozzle joint is more severe due to eccentricity; however, the secondary seal in the nozzle will seal and will not erode through

- The primary O-ring in the field joint should not erode through but if it leaks due to erosion or lack of sealing the secondary seal may not seal the motor

- The igniter Gask-O-Seal design is adequate providing proper quality inspections are made to eliminate overfill conditions

---

Recommendations

- The lack of a good secondary seal in the field joint is most critical and ways to reduce joint rotation should be incorporated as soon as possible to reduce criticality

- The flow conditions in the joint areas during ignition and motor operation need to be established through cold flow modeling to eliminate O-ring erosion

- QM-5 static test should be used to qualify a second source of the only flight certified joint filler material (asbestos-filled vacuum putty) to protect the flight program schedule

- VLS-1 should use the only flight certified joint filler material (Randolph asbestos-filled vacuum putty) in all joints

- Additional hot and cold subscale tests need to be conducted to improve analytical modeling of O-ring erosion problem and for establishing margins of safety for eroded O-rings

- Analysis of existing data indicates that it is safe to continue flying existing design as long as all joints are leak checked with a 200 psig stabilization pressure, are free of contamination in the seal areas and meet O-ring squeeze requirements

- Efforts needs to continue at an accelerated pace to eliminate SRM seal erosion

Fig. 10

with a 200 psi stabilization pressure, are free of contamination in the seal areas, and meet O-ring squeeze requirements" and efforts were underway to "accelerate the pace."[95]

After the *Challenger* tragedy, the minimal emphasis in the August 19 briefing on the relationship between temperature data and O-ring resiliency was perceived by some as a deliberate attempt to mislead senior NASA administrators by Marshall managers who made changes in Thiokol engineers' original charts.[96] An entry on an early version of the August 19 briefing charts stating that "lower temperatures aggravate the O-ring problem" was deleted. However, McDonald, who negotiated the changes with Mulloy, described them as minor.[97] The relationship between temperature and resiliency was not emphasized for three reasons:

1. The dramatic secondary erosion on STS 51-B nozzle was the chief concern of all and the problem for which the briefing was requested.

2. The 53°F calculated O-ring temperature on the January 1985 66°F launch was the result of three days of 18°F–20°F overnight low temperatures, unparalleled in Florida history. Because such weather conditions were not expected to recur, the temperature/resiliency data did not appear as important to engineers at the time as it does in retrospect. As Marshall's Ben Powers describes: "The temperature concern was not placed in the front row seat. The data was there, it was presented. I would say that, and certainly this is in hindsight, it should probably have been emphasized more, but here again, I want to emphasize that that is speaking today. . . . Extremely cold temperatures are certainly outside of the norm, and we don't expect to see that very often at KSC [Kennedy Space Center], and it is not something that you are working with often. . . . We were more concerned (with) the very heavy erosion that we were seeing on the nozzle joint, and that was a very serious concern of ours at that time."[98]

3. The engineers did not have sufficient data to make a "solid engineering argument" about temperature. McDonald, who made the presentation, stated that although one of the charts singled out resiliency as a "Primary Concern," the connection between temperature and resiliency "may have gotten lost" because "we hadn't run a very long range of temperatures when we got that data."[99] Another factor was very significant: STS 51-B, the focus of meeting attention because a nozzle joint had sustained the greatest erosion to date, had a calculated O-ring temperature of 75°F, contradicting Thiokol's position after STS 51-C that cold temperature was correlated with the most serious damage.

August 20, 1985

• Both Thiokol and Marshall seal task forces were given official status.[100]

August 22, 1985

• Arnold Thompson wrote a memo to Thiokol engineers stating: "The O-ring seal problem has lately become acute. Solutions, both long and short term are being sought, in the meantime, flights are continuing" (appendix B, fig. B8). Thompson recommended two "near-term" solutions be taken as soon as possible in order to increase initial compression on the O-rings: thicker shims and larger-diameter O-rings. Thompson concluded: "Several long-term solutions look good, but several years are required to incorporate some of them. The simple short term measures should be taken to reduce flight risks."[101]

His suggestions were followed. Said McDonald: "There were things done. In fact, the recommendation Mr. Thompson made was implemented as soon as we could implement it. The larger O-rings were put into the QM-5 test, which was the test we had going into the test bay at that very time to find out if they were going to work and improve the situation, as Mr. Thompson thought they would. We acted on that immediately."[102]

August 1985

• In response to the formal approval of NASA engineers in July, Thiokol ordered forgings for the capture feature.[103]

September 1985

• James Kingsbury, Marshall Director of S&E, sent a memorandum to Mulloy stating that he wanted to be briefed on Thiokol's plans to improve the O-ring seals. Kingsbury noted: "I have been apprised of general ongoing activities but these do not appear to carry the priority which I attach to this situation. I consider the O-ring seal problem on the SRM to require priority attention of both Morton Thiokol/Wasatch and MSFC [Marshall Space Flight Center]."[104] In response, Thiokol gave an interim task force report to Marshall on September 10.

October 1985

• Thiokol engineers complained that the O-ring task force was not getting sufficient support. In his October 1 weekly activity report to Al McDonald, Robert Ebeling, Manager of Thiokol's SRM Ignition System and Final Assembly and task force member, wrote:

HELP! The seal Task Force is constantly being delayed by every possible means. People are quoting policy and systems without work-around. MSFC is correct in stating that we do not know how to run a development program. . . . The allegiance to the O-ring Task Force is very limited to a group of engineers numbering 8–10. Our assigned people in manufacturing and quality have the desire, but are encumbered with other significant work. Others in manufacturing, quality, procurement who are not involved directly, but whose help we need, are generating plenty of resistance. We are creating more instructional paper than engineering data. We wish we could get action by verbal request but such is not the case. This is a red flag.[105]

After the disaster, this memo (appendix B, fig. B9) suggested that the task force difficulties arose because managers ignored the problem, even stonewalling engineers' attempts to get resources. Instead of concerted recalcitrance by managers, however, interviews indicate the task force problems were structural in origin. McDonald attributed the personnel difficulty to the matrix system. People with engineering specialties were usually assigned to several projects at once. Project Managers had to put in special requests for engineers to be assigned full time to projects. The best people were requested for full-time assignment by many Project Managers. McDonald said: "It wasn't easy to find available people without taking them off some other activities, because we wanted good people. We didn't want whoever just happened to be not working."[106] Thiokol's Bill Macbeth, Manager of Case Projects, described the equipment difficulties as typical of production organizations:

We are primarily a production facility, and so there's a lot of paperwork involved, and even a special task force with top priority has trouble getting things as quickly as they'd like to have them. It's easy to sit back and say, hey, I want such and such a test motor cartridge and I want it yesterday, but the fact is I have to wait until there's a batch of propellant available to cast the thing from, and then it takes four to seven days to get it cast and cured and ready to even put in the test fixture, assuming I had a test fixture ready for the cartridge. To the best of my knowledge, all of those test motors were available before they were ready to test them.[107]

In a memo, Scott Stein, a Thiokol task force member, blamed bureaucratic procedures and paperwork (appendix B, fig. B10).[108] Boisjoly identified Thiokol Vice President Joe Kilminster as a major obstacle (appendix B, figs. B11.1 and B11.2).[109] However, in later testimony before the House Committee on Science and Technology, Boisjoly clarified that it was not Kilminster per se, but the bureau-pathological government procurement system, that was the source of his frustration:

> MR. BOISJOLY: I was frustrated in procurement. We had a piece of equipment that we needed that was on the shelf at a company in San Diego that we could have simply gone down and picked it up and brought it back and used it and we had arranged to do something similar to that, while the procurement process and the rules that govern government contracts are such that it is just not as straightforward as that. . . . We just couldn't get the purchase orders written and go down and get them. Part of this was due to the rules and regulations of going out and getting single sources. That was the basis and source of those [Boisjoly and Ebeling] memos. . . . We, Morton Thiokol, are not at liberty to go out to single companies and purchase an item without going through a bid process.
>
> MR. ROE: . . . this is not a Thiokol process *per se* as a company, it is a governmental process, is that what you are saying?
>
> MR. BOISJOLY: That is correct.
>
> MR. ROE: All right, so the agency you deal with is NASA, so it is a governmental process that was creating this frustration, not that you couldn't get the material through the Thiokol leadership, is that correct?
>
> MR. BOISJOLY: That is correct.[110]

In testimony, Kilminster stated that the bureaucratic rules impeding the task force were necessary to protect against other possible risks:

> . . . Some of the Task Force members were looking to circumvent some of our established systems. In some cases that was acceptable; in other cases it was not. For example, some of the work that they had recommended to be done was involved with full scale hardware: putting some of these joints together with various putty layup configurations, then taking them apart and finding out what we could from that inspection process.
>
> DR. SUTTER: Why not do it?
>
> KILMINSTER: Well, we were doing it. But the question was, can we circumvent the system, the paper system that requires, for instance, the handling constraints on those flight hardware items? And I said, no, we can't do that. We have to maintain our handling system so that we don't stand the possibility of injuring or damaging a piece of flight hardware.[111]

• Thiokol task force members Boisjoly and Ebeling received NASA support to go to a San Diego seminar attended by "probably 600 of this country's finest seal people."[112] They presented their data on the problem and asked whether anyone else in the industry was seeing the same thing. Ebeling said it was an open plea for help: "Is there a neat way of fixing this, maybe quicker than what we are looking at? You know, some sort of interim thing. We've already tried the easy things, the putty layups; we tried a variety of different easy things. Putting grease in there, no grease. We tried all the easy things, and to no avail actually. . . . Anyway, it was so funny . . . no one heard that message . . . here you had the Boeings, the McDonnell-Douglases, the Dynamics and the whole slew of people, and they all had their people there. Not a one came forth."[113]

December 1985

• Bob Ebeling told fellow Thiokol task force members that he believed Thiokol should not ship any more SRMs to NASA until the problem was fixed.[114]

• The controversial O-ring problem "close-out" occurred. The episode began when Marshall's Larry Wear sent a memorandum to Thiokol's Joe Kilminster: "During a recent review of the SRB Problem Review Board open problem list,[115] I found that we have 20 open problems, 11 opened during the past 6 months, 13 open over 6 months, 1 three years old, 2 two years old, and 1 closed during the past six months. As you can see our closure record is very poor. You are requested to initiate the required effort to assure more timely closures."[116]

Word of the request was passed down to Thiokol engineers—the people responsible for originating problem close-outs. The erosion problem was one of the several they submitted for closure. On December 6, Thiokol engineer Brian Russell wrote Al McDonald requesting "closure of the Solid Rocket Motor O-ring erosion critical problems."[117] On December 10, McDonald relayed this request back to Marshall's Wear.[118] On December 18, 1985—just six weeks before the *Challenger* launch—the O-ring erosion problems were listed as closed in MPAS.

Because concern about the joints had escalated, the December 1985 close-out baffled the Presidential Commission. Adding to the confusion, it was Thiokol engineers who recommended closure—the very engineers who, in the January 27 teleconference, argued against launching the *Challenger* on the grounds that the O-rings were a threat to flight safety.[119]

> CHAIRMAN ROGERS: So you're trying to figure out how to fix it, right? And you're doing some things to try to help you figure out how to fix it. Now, why at that point would you close it out?

MR. BOISJOLY: I think if I may—

CHAIRMAN ROGERS: Just let me finish. Why would you write a letter saying let's close it out now?

MR. RUSSELL: Because I was asked to do it.

CHAIRMAN ROGERS: I see. Well, that explains it.[120]

But this action that appeared deviant to investigators in the wake of the *Challenger* tragedy was not defined as deviant by the Thiokol engineers at the time the decision was made. Wear's request referred to problems that were listed on the Marshall "SRB Problem Review Board open problem list," which was generated from the computerized MPAS. The purpose of MPAS was to give visibility to problems. The immediate effect of closure was that the problem would no longer be brought up before the monthly Problem Review Board meetings at Marshall. Certainly it mattered that, as Russell candidly testified, they were "asked to do it." Marshall had made similar requests in the past, but Russell noted that the previous requests had not been as strong as the December request.[121] In an effort to comply, Thiokol engineers discussed the problems that had been opened six months or longer. The SRB joint problem fell into that category.[122] But the engineers also understood that Marshall wanted problems closed out "with adequate justification."[123] However, they would not have completed the work necessary to close out the SRB problem legitimately before July 1986, when they would have completed tests and submitted task force recommendations to Marshall.[124]

In his December 6 letter recommending closure, Russell stated the rationale behind Thiokol's recommendation:[125] (1) It would not affect the ongoing work of the seal task force to resolve the SRB joint problem. To make that point, Russell listed 17 items describing ongoing and planned research indicating that work on the SRB problem was continuing. (2) It would not reduce the visibility of the problem. It was being monitored closely through other mechanisms: daily technical reviews at Marshall and Thiokol, FRRs, and two task forces. Thiokol engineers, including Boisjoly and Ebeling, thought the closure made good sense.[126] First, they believed that closure would have *no negative consequences for safety* because *they were not closing the problem out*—they were eliminating it from one of several tracking systems. Second, they stated that the close-out had *positive consequences for workload and information flows*. It was Thiokol engineers who had to report the information entered about each problem and its regular status updates on MPAS.[127]

Expressing exasperation with burdensome bureaucratic require-

ments and the negative effects of information overload, McDonald
defended his letter requesting closure:

> I wrote the letter, and I would like everyone to read the letter to
> see what it says. It says, "The subject critical problems are ongoing
> problems which will not be resolved for some time." So, right
> away, I don't think that tells anybody that we are going to forget
> about this and take it off of anything. It also says that, we "...
> request that subject critical problems be closed and removed from
> the next PRB [Problem Review Board] agenda list," and the reason
> being that we spend more time each month going through reading
> all of that same thing that we have been reading every month for
> two years, because somebody colors a square in down at Marshall,
> please keep track of this each month. . . .
> Marshall was very upset with us because we've got problems
> that have been on there for several years, and we haven't gotten
> them off of this list because it keeps getting thicker and thicker.
> And, if you ever want to get something where nobody will read it,
> you get it so thick that they finally pay no attention to it and that
> is exactly the thing the Problem Review Board was doing. . . . And
> so, we said, okay, if you want to get them off the list, then just take
> them off and we will handle them through this other mechanism
> that we're addressing with everybody that really knows about this
> problem.[128]

Harry Quong, Director of Safety, Reliability, and Quality Assurance
at NASA Headquarters, agreed, noting that the monthly MPAS
report he received was typically three-inches thick. It carried approximately 20 open problems on the SRM, compared to over 300 on the
Main Engine. Because of the comparatively few problems on the
SRM, he asserted that they was "easy to track." He "saw nothing
wrong" with the Thiokol engineers' action because it did not reduce
attention to the O-ring erosion problem, which was in the midst of a
"full blown test program."[129]

But both Marshall's Mulloy and Wear rejected the Thiokol close-
out recommendation.[130] They testified that this particular close-out
was an error: the problem was not resolved and so should not have
been removed from the tracking system. The error occurred because
McDonald's letter was "still in the review cycle" at Marshall when
the Marshall Problem Assessment Center received the Thiokol engineers' close-out entry to MPAS and acted on the recommendation.[131]
But even Marshall personnel thought that this system created reporting redundancy that was not useful. Wear, who conformed to the rules
by rejecting the request,[132] nonetheless agreed with the logic behind
Thiokol's action: "The Marshall Problem Assessment Center to me

was another fifth wheel that you had to contend with, but it didn't add anything. It didn't help you resolve anything; it didn't even help you remember. If you forgot it, they forgot it. But it was created in good intent, the paperwork all reads nice, but they never fixed a problem, they never found a problem that you didn't already know about."[133]

In the MPAS entry closing out the problem, the first paragraph gives the logic of acceptable risk at the end of 1985. The second records the Thiokol engineers' rationale for the close-out:

> Primary O-ring erosion is expected to continue since no correc-
> tive action has been established that will prevent hot gases from
> reaching the primary O-ring cavity. Steps have been taken to assure
> that the secondary O-ring will be seated and analytical analyses
> have indicated that under a worst case situation, erosion of the sec-
> ondary O-ring will not be severe enough to allow a leak path past
> the secondary O-ring (due to) an increase in the leak check pressure
> from 100 psi to 200 psi. . . . Analytical studies based on both
> impingement erosion and blow-by erosion show that this phenom-
> enon has an acceptable ceiling since implementing the above
> changes. Recent experience has been within the program data base.
> The seal improvement program plan will continue until the
> problem has been isolated and damage eliminated to the SRM
> seals. Status will continue to be provided in the Flight Readiness
> Review and in formal technical reviews at MTI and MSFC. At the
> conclusion of the program, a comprehensive report will be written
> to consolidate the results, conclusions, and recommendations.[134]

THE SOCIAL CONSTRUCTION OF RISK, 1985

As we have seen, concern about the SRB joints escalated in 1985. Many people at both Marshall and Thiokol paid more attention to O-ring problems after the April 51-B burn-through. The creation of official seal task forces and other formal organizational responses that increased the resources devoted to the problem suggest that the con-struction of risk had altered as a consequence. But the official con-struction of risk, emanating from the SRB work group and conveyed up the FRR hierarchy remained the same: accept risk. When the two task forces gave periodic interim progress reports, never did they rec-ommend halting flight until the O-ring problem could be perma-nently resolved.[135] For the six missions after STS 51-B, the Thiokol engineers brought to Marshall's Level IV and III management recom-mendations to launch based on the three-factor rationale.

After the disaster, the Presidential Commission struggled to rec-oncile the Thiokol engineers' FRR recommendations to accept risk with the 1985 O-ring damage and their memos warning of "catastrophe," stating that the problem was "acute," and asking for "HELP!" To continue launching under the circumstances was, in the Commission's view, clearly deviant. They pushed SRB work group engineers for answers. Consistently, the engineers' testimony affirmed that the three-factor technical rationale and the tests and calculations on which it was based, as documented in FRR charts, convinced them that the joints were an acceptable flight risk. They stated that the correctness of their corrective actions and risk assessments was affirmed by flight performance. After STS 51-B, two of the six missions had no erosion; the other four sustained some primary O-ring erosion that was within the experience base. No secondary erosion occurred. In FRRs for the rest of 1985, erosion was reported as "within experience base" and "within previously accepted experience."[136] Again we see the importance of patterns of information to microlevel understandings. Each postflight analysis convinced the work group that increasing the leak check pressure to 200 psi had controlled erosion. From the vantage point of post-*Challenger* investigations, the continuing erosion looked like a problem out of control; to the engineers assigned to the problem, analyzing it one launch at a time, the joints were operating exactly as engineering calculations had predicted.

Questioned by the Commission, Thiokol task force member Brian Russell defends engineering decisions after STS 51-B:

> There was talk of that [delaying the program while the SRB problem was fixed] among the working level people at Thiokol. Still, we as a company hung our hat on the test data, the flight problem data, namely, the amount of erosion and blow-by and so forth, and how that equated with some rather extensive analysis that we had done on the problem. I personally felt that we were still able to operate. In essence, I agreed with the ultimate recommendation to continue with flight, realizing that we still would see some blow-by and some jet impingement, damage to the O-ring, and yet none of us liked what we saw, but we thought that it was still acceptable, is the word I would choose.[137]

A Russell memo written three weeks before the disaster validated the above testimony. Russell wrote that the increased leak check pressure assured that the secondary O-ring would seal and articulated the three-factor rationale supporting the belief in redundancy.[138] Presidential Commissioner Sutter questioned Russell about his memo; Boisjoly interjected to explain memos he and Ebeling wrote:

COMMISSIONER SUTTER: It seems to me that this [memo] conveys an attitude then that, okay, we've got troubles with this joint but it is okay to keep flying with it even though we want to fix it, that you're willing to keep flying with the joint even though you know you've got problems? And I'm just wondering, what was the attitude.

MR. RUSSELL: The attitude was that it was an undesirable condition but still acceptable for flight, based upon the history that we had seen up until that point.

MR. SUTTER: The history of the erosion and the blow-by and the changing the leak checks and all of that did not really build into your people that should have been responsible, that this was a critical item?

MR. RUSSELL: I don't think that's true. I think we considered it to be a critical problem, which is the reason that we had the Task Force created to fix it. But, critical to stopping the program, it obviously wasn't.

MR. SUTTER: Well, you were going to run, say another 20 or 30 flights without doing anything?

MR. BOISJOLY: No. Those who were intimately familiar with it—that is why we were writing these memos, to try to get a little bit of flame turned up underneath it, to be able to get something done, because we were concerned, we were extremely concerned. I think that is what those memos reflect. . . . We were attempting in that team to short-circuit company procedures. We were attempting to get things done that are not normally done, so that we could try things that we could tweak the system and get as many pieces of information in as short a period of time as possible, so that we could effect a change and get it implemented as quick as possible, and that was a big source of frustration because we weren't getting that support.[139]

Concerns notwithstanding, Boisjoly, too, believed the joint was an acceptable risk prior to the *Challenger* teleconference. In testimony, he reiterated the technical rationale supporting redundancy that appeared in 1981–85 FRR documents:

We were treating the seal as though it were 1R [redundant]. We were treating the seal as though it were 1R because we were saying that if you should bypass a primary O-ring during an ignition transient, you still have a secondary seal to pick up the slack . . . and the rationale, that actually could be seen in the paperwork we always presented at the Flight Readiness Reviews, is because [of] the amount of squeeze that was still mechanically there on the secondary O-ring when you pressurized. . . . So just by the very fact that we were presenting that [compression data] in the FRR, it defaults to a 1R type presentation.[140]

These exchanges throw into stark relief the contrast between the interpretations after the accident that to continue flight under the circumstances was deviant and the meaning to work group participants at the end of 1985. The following exchange epitomizes that contrast, showing how cultural meanings shaped engineers' definition of the situation. In an interview conducted by Emily Trapnell, an FAA attorney who headed the team of investigators assisting the Presidential Commission, Thiokol engineer Boyd Brinton tries to explain the engineering position. In sequence, he refers to (1) the engineering maxim "Change is bad," (2) the basing of flight decisions on risk assessment outcomes, and (3) the routine use of the word "catastrophe" (which the investigator interpreted as a serious warning in one memo) by insiders to mean a failure consequence of any component designated C 1 or C 1R—it was used interchangeably with those symbols at NASA.

> MR. BRINTON: Making a change on a working system is a very serious step.
> MS. TRAPNELL: When you say working system, do you mean a system that works or do you mean a system that is required to function to meet the schedule?
> MR. BRINTON: What I was trying to say is in the colloquialism, "If it ain't broke, don't fix it."
> MS. TRAPNELL: Did you consider that system to be not broken?
> MR. BRINTON: It was certainly working well. Our analyses indicated that it was a self-limiting system. It was performing very satisfactorily.
> MS. TRAPNALL: Had you ever heard an engineering opinion that there was a potential for catastrophic failure based on the O-ring erosion and blow-by problem?
> MR. BRINTON: Oh, yes. I think all the engineers know that that would be the result of a leak through both O-rings. I, again, was aware of the analyses that were done that indicated that that was self-limiting.
> MS. TRAPNALL: Well, then, I guess I don't understand. You say, on the one hand, that it was a self-limiting situation.
> MR. BRINTON: Yes.
> MS. TRAPNALL: But on the other hand, you say that the engineers were aware of the potential catastrophic result of burn-through.
> MR. BRINTON: Well, let me put it this way. There are a number of things on any rocket motor, including the Space Shuttle, that can be catastrophic—a hole through the side, a lack of insulation. There are a number of things. One of those things is a leak through the O-rings. We had evaluated the damage that we had seen to the O-rings, and had ascertained to ours and I believe NASA's satisfac-

tion that the damage that we had seen was from a phenomenon that was self-limiting and would not lead to a catastrophic failure.

MS. TRAPNALL: Are you aware of the memo written by Roger Boisjoly I think during the summer of '85 where he expressed concern that the failure to put enough effort into fixing the O-ring situation could lead to a catastrophic failure? Does it surprise you to hear that one of Thiokol's own engineers believed that this O-ring situation could lead to a catastrophic failure and loss of life?

MR. BRINTON: I am perfectly aware that the front tire going out on my car going down the road can lead to that. I'm willing to take that risk. I didn't think that the risk here was any stronger than that one.[141]

Asked whether concern ever escalated to the degree that anyone at Thiokol recommended or considered recommending that launches be delayed until a solution was found, Thiokol engineer Bill Macbeth answered:

Mr. Boisjoly did not so recommend in the July 31st memo. He said, if such and such occurs it may lead to a catastrophe. A little bit of cage-rattling there to try to get somebody to react and do something that he wanted done. And he didn't know at the time that it was being done and already had been under way for approximately two weeks time. Mr. Thompson referenced a memo he wrote in August, and that's in the testimony. And I don't believe that that memo says we ought to stop flying. I don't think either memo really says we ought to stop flying. They say we have got a potential problem, and we already knew that.[142]

At Marshall, too, concern coexisted with the belief in acceptable risk. Marshall engineers reflected on their understandings at the end of 1985. Leon Ray, S&E engineer:

I thought that it was safe to fly. As long as it wasn't cold, it was safe to fly.[143]

Ben Powers, S&E engineer:

The memos we wrote and the presentations we made did not specifically say stop the program. They said that the design is not adequate. That's what they said, and that it needs fixing. And we were. Meanwhile, back at the ranch, this field joint thing was being a good boy and wasn't doing anything wrong, and it was working.[144]

Keith Coates, Chief Engineer in Marshall's Special Projects Office:

How did I perceive it? I wanted it fixed to where it would not reoccur. It was a risk. Evidently not, in my mind, an unacceptable risk because I didn't lay down in front of any trains.[145]

Jim Smith, Marshall's Chief Engineer, SRB Program:

> It would be my opinion that the engineers that were working on it were concerned, no doubt about it, but I have to say I believe that people honestly felt we understood it well enough that we were not going to have a shuttle accident. I don't know that the same degree of concern was shared by people: some were very concerned, some maybe not as concerned. We needed to keep our hands on the pulse. We needed to watch the situation. But we had analysis that led us to believe we had captured the problem . . . if you go take these steps, control the hardware, control how you build it, operate it this way, keep these engineering controls there . . . we think it's safe to go fly that vehicle, even though, had we our druthers, we'd go redesign that joint, as we were doing. We thought we had it captured, engineering-wise. If we stay within these constraints, we will be safe to fly. Nobody ever said, in any Flight Readiness Review I was involved in, that it's not safe to fly.[146]

NORMS, RULES, AND DEVIANCE IN THE WORK GROUP

The damage that occurred during 1985 was dramatic and attention-getting, both to work group participants at the time and to post-*Challenger* analysts. The joints of STS 51-C revealed that hot gases had reached a secondary O-ring for the first time. Then, on the April launch of STS 51-B, hot gases burned through a nozzle joint primary, eroding a secondary O-ring. Relocating these incidents in the original stream of actions shows us how the work group normalized even these signals of potential danger in official risk assessments.

The damage occurring on STS 51-C and 51-B was on field and nozzle joints, respectively, and had two different causes as well. Post-flight analysis of January's STS 51-C convinced engineers of acceptable risk, because the data conformed to predictions about the experience base, the safety margin, and the self-limiting aspects of joint performance. Three days of record cold Florida temperatures may have slowed STS 51-C's primary sealing, which accounted for the hot gases reaching the secondary, but the primary had sealed; erosion was as predicted. Ironically, the cold temperature itself also figured importantly in the normalization of technical deviation: the string of three overnight temperatures in the 18°F–20°F range in Florida was considered a deviant event, unlikely to recur. In April, when nozzle joint primary burn-through and secondary ring erosion occurred on STS 51-B, alarmed engineers identified the cause as an inadequate nozzle joint leak check. Again they were guided by the engineering beliefs, norms, and procedures of the past: chief among them, a safety margin for the secondary. They implemented a corrective action previously tried and found true—increasing the leak check pressure. Because the work group thought they had properly understood the cause and corrected it, and because the analytic model for calculating worst-case scenarios predicted that the most erosion that could possibly occur would not fail the joint, they concluded that the SRB joints were an acceptable risk. Therefore, flight could proceed.

At the same time, work on the capture feature, begun years earlier, was moving into the test phase. The work group appeared to be creating and living with contradictory official constructions: acceptable risk, therefore fly; unacceptable risk, therefore modify the design. The persistence of the dominant paradigm alongside the reinvention of the cultural construction of risk represents a major cultural contradiction in the work group. How did the belief in acceptable risk prevail under these circumstances? In the next two chapters we turn to the question of cultural persistence, paying attention to the structural influences on work group decision making. Before this discussion, however, one more microlevel factor needs emphasis: "learning by doing" was another facet of incrementalism that mattered in the work group's definition of the situation.[147] Levitt and March point out how organizational experience accumulates in the structure of routines. Similarly, routines and beliefs change in response to direct experience. Trial-and-error experimentation will increase the likelihood that a routine will be used when it is associated with success and that it will not be used when associated with failure. This general organizational tendency took on even greater significance at NASA because trial-and-error learning was the only way to learn about their unprecedented technology. Moreover, scientific methods and apparatus themselves were valued highly, lending credibility to results. As the working engineers gained experience with the SRB joints, three sources of new technical information— tests and analytic calculations, experimentation with corrective measures, and postflight analyses—reinforced both the routines and the belief that the design was an acceptable risk.

First, engineers learned more about the kinds of physical tests and models that would be useful as they went along. Tests and analytic models became more refined, leading to new numbers (the safety margins) that quantified the boundaries between deviant (unacceptable) and normal (acceptable) joint performance. In 1985, the newly developed Salita model for calculating worst-case scenarios played a definitive role in the persistence of the belief in SRB joint risk acceptability. Second, the working engineers also were developing new corrective measures for improving joint performance that grew out of their expanded understanding of joint dynamics. In 1983 and 1984, for example, they learned that increasing the leak check pressure from 50 to 100 to 200 psi had effectively controlled field joint performance. So, in 1985, they increased the leak check pressure to 200 psi on the nozzle joint with similar positive results. Finally, the work group participants were reassured by flight performance.[148] No field joint erosion occurred in 1985 after January's STS 51-C. When the leak check pressure on the nozzle joints was increased after STS 51-B, performance conformed to predictions. For the rest of 1985, as in the past, the effectiveness of the remedy affirmed the diagnosis. There were no more signals of potential danger. Instead, the new information indicated that the field and nozzle joints were operating as expected. Once again, the technical deviation that on first inspection was interpreted as a signal of potential danger had been normalized in subsequent engineering analysis. At the end of the year, the work group believed they had, in the words of Marshall's Jim Smith, "captured the problem."

What about NASA rule violations in 1985? Again, this ethnographic history shows the contrast between insider and outsider definitions. In the aftermath of the *Challenger* disaster, the Presidential Commission questioned Marshall and Thiokol employees intensely concerning three controversial actions that following the discovery of 51-B damage appeared to violate safety rules: Marshall Level III Project Manager Mulloy's "waiver" of the Launch Constraint, Mulloy's failure to report the Launch Constraint or its subsequent "waiver" to Levels II and I, and the official closure of the problem at the end of 1985. For the Commission and other posttragedy analysts, these actions suggested that production pressures led Marshall managers to violate requirements constraining flight, in order to meet the schedule, as well as to violate reporting and tracking requirements, in order to conceal the extensiveness of the O-ring problem from upper-level NASA administrators. In fact, Marshall managers conformed to rules in each of these controversial Level III decisions. Equally important, we see writ large the deleterious effects of a rule-minded bureaucratic system on the working engineers, who in 1985 were struggling to get out from under rules that were impeding their work. These rule-related incidents sensitize us to a major cultural conflict at NASA that will engage us in the remaining chapters of this book: the difficulty of doing innovative R&D engineering in an organization that simultaneously compels its members to adhere to all the proscriptions and prescriptions of a large, aging, bureaucratic structure.

Marshall managers conformed to rules about Launch Constraints. Members of the Presidential Commission interpreted a Launch Constraint as a NASA designation intended for a rare shuttle problem so serious that flight should not continue (i.e., launches should be "constrained") until the problem was fixed.[149] The House Committee on Science and Technology report stated that a Launch Constraint "would seem to indicate that the shuttle should not be launched until the problem giving rise to the constraint was solved."[150] In NASA culture, a Launch Constraint is the most serious designation that could be given to a problem,[151] as the Presidential Commission surmised, but it did not necessarily halt shuttle launches. A Launch Constraint was a designation frequently used in the NASA system to assure that problems so designated received special attention in reviews. If contractor engineers recommended accepting risk and their engineering analysis withstood the challenges of FRR, the Launch Constraint was lifted—automatically, as a result of the contractor engineering analysis and recommendation to proceed.[152] If they did not recommend proceeding, or if the analysis backing a recommendation was challenged, launch was delayed pending further tests, analysis, and/or appropriate corrective action. Each time Mulloy lifted the Launch Constraint, he was responding to the risk assessment of the Thiokol engineers.

Also criticized for not reporting the Launch Constraint up the FRR hierarchy, Mulloy had again followed the rules to the letter. The Presidential Commission, earlier confused about waivers and their significance in the NASA culture, in this instance apparently conflated two very different procedures. As Mulloy remarked, "You don't waive launch constraints; you lift them."[153] Mulloy "addressed" the SRB joint problem at all levels for all subsequent flights, as required. He did not mention it as a *Launch Constraint* in FRRs beyond Level III because that was not how NASA did it. The *problem* resulting in the Launch Constraint received a full review at all levels the first time it occurred. For each flight thereafter, it was fully reviewed at Marshall, and a "brief status summary" was given at Levels II and I. Finally, Mulloy officially recorded the Launch Constraint in MPAS on July 10, 1985, then notified MPAS when the constraint was lifted for each subsequent 1985 launch, recording the actions as required.[154]

The December 1985 close-out illustrates simultaneously Marshall management's preoccupation with conforming to bureaucratic procedures and Thiokol engineers' difficulty doing technical work in such a system. After the *Challenger* accident, the Marshall close-out appeared to be another managerial attempt to suppress information about the O-ring problems from higher-ups. It was not. Although Marshall management requested that problems open longer than six months be closed, it was Thiokol working engineers who selected the O-ring problem as one to be closed. They were faced with a conflict: conform to a Marshall request to reduce the number of problems carried in MPAS or conform to the Marshall requirement that open problems be carried in MPAS. Unable to simultaneously conform to both, they opted to conform to the Marshall request by violating the Marshall requirement. People violate laws, rules, and regulations in the workplace for many reasons. If they perceive a rule to be irrelevant, that perception may influence their willingness to abide by it.[155] The rule may be thought irrelevant (1) to a particular task or (2) to the larger purposes of the organization. Apparently, Thiokol engineers violated the Marshall reporting requirement for both these reasons.

The engineers did not define the close-out request as a deviant action, according to the logic of the memos sent to Marshall. Neither was it a surreptitious attempt to hide the problem. The existence of the seal task forces was known at Marshall, Thiokol, and NASA Headquarters, and full briefings had been given to all FRR levels. The Thiokol engineers viewed MPAS as a redundant and ineffectual back-up to these oral systems of communication. Therefore, closing out the problem was a "paper change" that did not affect erosion problem visibility, engineering efforts to resolve it, or the risk of shuttle flight. In their view, MPAS was an example of bureaucracy run amuck: it had so many entries that problems got buried. Moreover, Thiokol engineers were the ones who created the entries; therefore, tracking the problem in this system made additional work for them. For all these reasons, the rule that open problems be carried in this system seemed irrelevant.

We have reached the end of this reconstruction of the history of decision making, begun in order to understand repeated NASA decisions to fly with a flawed design in the years before the *Challenger* tragedy. Ethnographic thick description showed the process of culture creation. We saw how the Level III and IV managers and engineers assigned to the boosters made their risk assessments, made sense of technical deviation, and converted uncertainty into certainty prior to each launch. We witnessed how they incrementally developed a work group culture in which the belief in redundancy supported findings of acceptable risk, and how they passed that defini-tion of the situation up the FRR hierarchy, creating a collective con-struction of risk at NASA. If we can distill a single important insight from this complex sequence of interactions, perhaps it is about the potential of small precedents established early to have dispropor-tionately larger consequences later as culture, once created, shapes subsequent choices.

The work group's official construction of risk at the end of 1985 is significant because the belief in redundancy was an important com-ponent of the worldview that many key participants brought to the *Challenger* teleconference on January 27, 1986. This preexisting def-inition of the situation was the all-important frame of reference within which the discussion that night took place. But before we revisit the eve of the launch, there is more to this story. We have seen how the work group culture developed; now we want to know why it persisted in the face of abundant evidence that challenged the group's worldview. Because other work groups at NASA and even the SRB work group had initiated delays in other instances, we know that engineers *were* capable of halting the inexorable progression in which decisions led to similar decisions, as incidents that defied engineering expectations were repeatedly normalized. Why not in the case of the SRBs?

The work group culture was a negotiated order wrought from engi-neering disagreements.[156] It persisted because of key environmental and organizational contingencies. The process of culture creation occurs within larger structures that predate a particular interaction, shaping local situations. Therefore, the production of culture in the work group must be viewed within this wider social context. The goal of the next two chapters is to lay out in full a theory of the nor-malization of technical deviation in the work group in the years pre-ceding the *Challenger* launch decision. To the production of culture we add two structural factors—the culture of production and struc-tural secrecy—which complete the framework readers need in order to understand what happened on the eve of the launch.

Chapter Six

THE CULTURE OF PRODUCTION

The chilling question is, Why did worried engineers repeatedly recommend launching in the face of continued and worsening signals of potential danger? Why was the first set of responses repeated, creating a pattern where none had existed before, *becoming* culture? Why did the work group's definition of the problem and method of responding to it remain unaltered as information about the O-rings altered? This paradox—a worldview that survives despite evidence that repeatedly challenges its basic assumptions—is well known among researchers interested in the sociology of science and technology, who often have noted the "obduracy" of established perspectives or paradigms in the face of anomalous information.[1] The best-known of this genre of study is, of course, Thomas Kuhn's classic *The Structure of Scientific Revolutions.*[2] In the Kuhnian sense, a paradigm is a fundamental component of scientific culture. It is a worldview based on accepted scientific achievement, which embodies procedures for inquiring about the world, categories into which these observations are fitted, and a technology that includes beliefs about cause-effect relationships and standards of practice and behavior.

Science and its process are deeply and inextricably rooted in the condition of everyday laboratory life. To uncover how practitioners develop and elaborate their paradigms requires that the "black box" of science—the unrevealed nitty-gritty day-to-day work that produces scientific knowledge—be opened up to reveal how, in laboratories and testing facilities, scientific work gets done.[3] In the previous three chapters, we opened up the "black box" of the history of decision making about the SRB joints by working engineers, showing how this small scientific community created a worldview about a technical artifact. The belief in redundancy was a product of the slow accretion of history, ideas, and interactions. It rested on organized

196

knowledge having the characteristics of a Kuhnian scientific paradigm.[4] The outcome was the normalization of SRB joint deviation throughout the NASA organization from 1977 to 1985 or—to put it differently—a scientific paradigm that persisted despite repeated challenges.

To trace microlevel decision making as we have done is to study ethnocognition: the way members of a particular group produce a definitive mental model as they construct, negotiate, and assemble knowledge to give meaning to an object. But cultural influences from the environment affect local situations. Engineering calculations, lab experiments, and testing go on within a wider social order that figures importantly in the discovery and interpretation of scientific facts.[5] Institutionalized belief systems—the framework of rules, roles, and authority relations—affect choice by generating cultural scripts according to which certain social relationships and actions are taken-for-granted.[6] They stand as common understandings, deeply embedded, seldom explicitly articulated. The patterns found in much of social life derive from the taken-for-granted quality of these institutionalized cultural scripts, which are then reproduced in other structures.[7] We can think of scripts as "rules of appropriateness" that constrain choice by shaping the menu of possible options people consider, making some choices viable and precluding others.[8]

In *The Functions of the Executive*, Chester Barnard stated that management sets the premises of decisions in organizations. However, as Richard Scott points out, overall agreement about organizational policies and the areas to which they apply is a result of institutional, not organizational, processes.[9] Each organization—and the individuals in it—knows about the production process and division of labor from participating in the same institutional environment. So powerful are widely shared cultural beliefs and taken-for-granted categories and procedures that they need not be formally expressed in organizational processes and structures.[10] Nonetheless, they are. The result is that societal rules for interaction combine with rules specific to the organization (standard operating procedures, formal rules, the organization chart, resource allocation), creating a bias for "what has come before."[11] This is meant, not to deny the decision maker's experience of free will and rational choice when making a decision, but to point to the subtle prerational dynamic by which institutional and organizational arrangements determine the range of choices that people will see as rational in a given situation.[12]

This chapter frames the production of culture—the engineering decisions described in the previous three chapters—within the structure of the culture of production.[13] It makes vivid the macro-micro connection, demonstrating how the institutionalized beliefs of the culture of production were reproduced in the work group.[14] We see how the premises of decision making about the SRB joints existed before any specific managerial actions. The salient structural condition was NASA's institutional history: politics, competition, and scarcity. The culture of production contributed to the persistence of the work group's definition of the situation because their actions were normative and acceptable within this larger cultural context. The culture of production comprised the institutionalized belief systems of the aerospace industry, the engineering profession, the NASA organization, and Marshall Space Flight Center. Each cultural layer was in many ways distinctive; nonetheless, aspects of one carried over and interpenetrated the next. The greater the degree of institutionalization, the greater the resistance of cultural understandings to change.[15] Deeply embedded in this framework of institutions, the work group's cultural construction of risk was less vulnerable to challenges and intervention. Nested culture—the overlapping and mutually reinforcing characteristics of these systems of meaning—contributed to and stabilized the group's definition of the situation.[16]

This chapter, too, contradicts aspects of historically accepted understandings. First, it shows the culture of NASA and, in particular, Marshall, to be more complex than generally portrayed in post-tragedy accounts. The milieu of scarcity, competition, and political bargains in which the Space Shuttle Program was situated did create the production pressure so well acknowledged after the *Challenger* disaster. But other historic changes altered the structure and emphasis of the space agency, making bureaucratic activities and concerns a major cultural preoccupation.[17] The result was a problematic culture dominated by three imperatives: those of the original technical culture, bureaucratic accountability, and political accountability. Second, this chapter contradicts images of managerial wrongdoing and amoral calculation in the years preceding the *Challenger* launch by demonstrating how production concerns were prerational and taken-for-granted at all levels of the hierarchy, thus affecting all facets of organizational life.

Becker reminds us that culture is "apparently" shared.[18] Given what we know about variation in culture, we cannot assume because people are grouped together by a shared characteristic (e.g., membership in the engineering profession) or belong to the same organization that they share a culture.[19] Drawing conclusions about the extent to which culture is shared ultimately rests on microlevel data. It is therefore significant that the Level IV and III managers and engineers in the SRB work group agreed on launch recommendations through the end of 1985. We would expect some significant differences in worldview to exist between managers and engineers because of the distinctive responsibilities associated with their respective positions.[20] However, from the history of decision making, we have evidence that the culture of production informed their sensemaking in common directions prior to the *Challenger* launch. Managers at NASA are trained engineers, so "managers" and "engineers" are, to a great extent, similarly positioned in larger institutional structures. Shared cultural meaning systems give otherwise diverse groups an understanding of the requirements of each other's roles, enabling them to negotiate accommodations during conflicts that grow out of role necessities. Managers' and engineers' agreement on official launch recommendations during these years was scripted into their common social mission.

Clifford Geertz acknowledges the difficulty of locating general cultural meaning systems in local social contexts, noting that many attempts fall into the category of "perfected impressionism": they tend to be evocative, resting on suggestion and insinuation, with the result that "much has been touched but little grasped."[21] But in close examination of the microcosmic, he argues, "small facts speak to larger issues."[22] To avoid the pitfall of perfected impressionism in this chapter, I first describe relevant aspects of the culture of production found in the engineering profession, the NASA organization, and Marshall Space Flight Center, showing their overlapping, nested character. In doing so, I simplify these complex cultures in order to isolate the principles that cut across them.[23] Then, to identify the macro-micro link, I depend on those "small facts." I show how these cultural belief systems were expressed in the behavior of work group members, showing the components of worldview that reinforced the persistence of the work group's cultural construction of risk.[24] I make no sweeping analysis of American society and the aerospace industry, drawing inferences at the end only on the basis of what the work group actions make palpable. If the reader finds some redundancy in my tracing of the connection between the macrolevel factors and microlevel work group behavior, it accurately reflects the layered belief systems of the culture of production. They were, in the language of engineers, redundant systems: made stronger and fail-safe by multiple backups.

ENGINEERING CULTURE

In *Craft and Consciousness*, Bensman and Lilienfeld show the relationship between worldview and the occupational technique and methodology of many occupations and professions.[25] They convincingly argue that craft-based procedures and assumptions to handle specialized materials and symbols creates habits of mind that give each occupation its distinctive character. The social administrative arrangements of occupations and professions further affect worldview. We examine engineering as a craft and engineering as a bureaucratic profession in order to later demonstrate how these two aspects of engineering culture contributed to the persistence of the SRB work group's cultural construction of risk.

ENGINEERING WORK: UNRULY TECHNOLOGY

The technique and methodology of engineering create common understandings that are part of the occupational worldview. The public is deceived by a myth that the production of scientific and technical knowledge is precise, objective, and rule-following.[26] Engineers, in this view, are "the practitioners of science and purveyors of rationality."[27] This myth grows, in part, from the success of producers of technology in presenting an image of both process and product as efficient and infallible and, in part, because the formal language and specialized skills associated with technological products obscure the process behind the product from public view. When technical systems fail, however, outside investigators consistently find an engineering world characterized by ambiguity, disagreement, deviation from design specifications and operating standards, and ad hoc rule making.[28] This messy situation, when revealed to the public, automatically becomes an explanation for the failure, for after all, the engineers and managers did not follow the rules. But the myth of precise science-based technology is further perpetuated, says Charles Perrow, by the inherent bias of all accident investigation findings. The engineering process behind a "nonaccident" is never publicly examined.[29] If nonaccidents were investigated, the public would discover that the messy interior of engineering practice, which after an accident investigation looks like "an accident waiting to happen," is nothing more or less than "normal technology."[30]

Normal technology, according to Brian Wynne, is unruly.[31] By "unruly," he refers not to postaccident perceptions that rules were not followed, but to *the absence of appropriate rules to guide fundamental engineering decisions.* Wynne points out that investigations of accidents and ethnographic research in nonaccident technical settings repeatedly show that engineering behavior *is* rule-following. The rules engineers follow, however, are "practical rules": operating standards consisting of numerous ad hoc judgments and assumptions that are grounded in evolving engineering practice. Rules are experience-driven. Wynne states: "Beneath a public image of rule-following behavior and the associated belief that accidents are due to deviation from those clear rules, experts are operating with far greater levels of ambiguity, needing to make uncertain judgements in less than clearly structured situations. The key point is that their judgements are not normally of the kind—how do we design, operate and maintain the system according to 'the' rules? Practices do not follow rules; rather, rules follow evolving practices."[32]

Two sets of rules coexist: a formal, preestablished system of rules that sets forth general principles, and the informally evolved arrangements that engineers create as the technology develops in ways not covered by the general principles. In the implementation and operation of complex technological systems, new rules and relationships are continually being invented and negotiated. These ad hoc systems of rules have official status, either by representation in documents recording the logic of engineering decisions that are then stored, forming the organizational memory, or by incorporation in updated statements of formal rules. Thus, an important but unacknowledged aspect of engineering work consists of generating rules that are tailored to a specific technical object. Why is this so? The formal rules, established prior to experience with the technical system, are sufficiently broad to apply to diverse circumstances. However, they are not helpful in guiding decision making in specific situations. Their universal quality does not meet the needs of a system in which tasks are nonroutine and lack precedent and the knowledge base is in continual flux.[33]

In normal technology, the production of engineering knowledge is a craft. The task is to design and implement a technical object for use in a specific technical system and environment-of-use. Niche specificity requires the development of "local knowledge."[34] Local knowledge develops from a process of learning that has at its base tacit understandings about how to go about the work and about the product itself that are difficult to convey. Tacit understandings are knowledge that has not been formulated explicitly and, therefore, cannot be stored, copied, or transferred effectively by impersonal means, such as written documents or computer files.[35] Yet tacit understandings must be organized into visible, discrete, and measurable units to make communication and decision making possible;[36] to do so, engineering work relies on the methods of scientific positivism to quantify, rank, classify, standardize, and simplify. At the same time, the production of technical knowledge also depends on more ephemeral, intuitive skills that enhance discovery, interpretation, and learning from experience. These two aspects of engineering are joined as engineers craft not only the object but also the rules that make decision making possible so that they can get on with the job.

Fundamental to evolving rules from practice is learning by experience. In all organizations, learning is "routine-based, history dependent, and target-oriented."[37] Landau and Stout state that "in an intelligently managed organization, the information generated by anomaly, by discrepancy between expected and actual outcomes, becomes the means by which fallible rule sets are corrected and moved toward solution sets."[38] This general tendency is profoundly realized in engineering work. Learning proceeds through iteration.[39] Petroski, in *To Engineer Is Human*, states that a design is a hypothesis to be tested.[40] Because the test of an engineering hypothesis is a comparison of its predictions with performance, experience becomes the quintessential learning device. Petroski points out: "The very newness of an engineering creation makes the question of its soundness problematical. What appears to work so well on paper may do so only because the designer has not imagined that the structure will be subjected to unanticipated traumas or because he has overlooked a detail that is indeed the structure's weakest link."[41] Absolute certainty can never be attained for many reasons, one of them being that even without limits on time and other resources, engineers can never be sure they have foreseen all possible contingencies, asked and answered every question, played out every scenario.[42] Departures from traditional designs are expected to be quirky—the aircraft industry is illustrative.[43] The implementation of new technologies generally reveals them to be more complex and sensitive than anticipated. Bottlenecks, surprises, and glitches are the rule, not the exception. Designs that worked on paper have to be "debugged through use"; "corrective actions" to strengthen a weak link are the norm.[44] This "fine-tuning" confounds interpretation of performance, muddying the causal issues.[45]

The ambiguity of the engineering craft is complicated by "interpretive flexibility."[46] Not only do various tests of the same phenomenon produce differing results, but also the findings of a single test are open to more than one interpretation. In addition, many testing instruments are relatively insensitive, making both use and interpretation of results "more art than science."[47] Many technologies—especially large technical systems like the Space Shuttle—cannot be tested in lab conditions. Tests are conducted on models, which can only approximate the complex systemic forces of nature and technical environment. This situation creates risk: the world outside the laboratory becomes the setting for experiments.[48] Weingart observes that "technical systems turn into models for themselves: the observation of their functioning, and especially their malfunctioning, on a real scale is required as a basis for further technical development and also for increasing their safety."[49]

For all these reasons, judgments are always made under conditions of imperfect knowledge. Marcus notes, somewhat ironically, that decisions about all the scientific and engineering problems that affect society lie someplace between the extremes of perfect knowledge and perfect ignorance.[50] It is not surprising, under these circumstances, that engineering work is guided by a system of flexible rules tailored and retailored to suit an evolving knowledge base. The essence of engineering as a craft is to convert uncertainty to certainty, figuring probabilities and predictions for technologies that seldom stay the same, because of the very corrections that "debug" them. In the workplace, engineers formulate the rules as they go along, attempting to capture the unruly technology with numbers, experienced-based theories, and practical rules. They attempt to close technical debates by amassing results from tests and experience. The debates necessarily are resolved, for many reasons, including social circumstances that require their closure.[51] Closure is more often an outgrowth of what lawyers refer to as "a preponderance of the evidence"—a number of tests with the same result, cumulative experience, or a correlation of findings between tests and experience—than of a single breakthrough experiment constituting definitive scientific proof.

Although the results may convince a wider audience, the "core set"—the working engineers and scientists most closely associated with the technology—understand the precariousness of this closure, for they are most intimately aware of the test result that does not conform to the others, the limitations of design, the ambiguity surrounding the various engineering interpretations that are embedded in day-to-day engineering work.[52] Even in closure, there is ambiguity. Closure is always temporary, for the design hypothesis is tested again and again in use. No matter how often experience affirms the hypothesis, it is never the case that the hypothesis is "proved." It may only be that the situation that would disprove it has not yet occurred. The developmental character of technical knowledge production and the crucial role of learning by experience generate commitment to a particular design. To change is to start over, swapping the evils known for the evils unknown—thus the axiom "Change is bad" that permeates engineering work. With design change, the process of knowledge production must begin again, initiating the development of a new system of ad hoc practical rules.

THE SOCIAL LOCATION OF ENGINEERING: THE BUREAUCRATIC PROFESSION

The social administrative arrangements of engineering also contain cultural scripts that are integral to occupational worldview. Engineering is a "bureaucratic profession."[53] Rather than being retained as consultants, engineers are typically hired as staff in technical production systems, situated in a workplace organized by the principles of capitalism and bureaucratic hierarchy. Engineers are prepared for this existence in technical schools and universities whose programs are underwritten by corporations and government projects that effectively monopolize technical intelligence. Educational systems craft engineers to become units in the industrial system, training them to meet changing industrial specifications. The goal of engineering activities is the application of technology in production. Specialization cements ties to the interests of industry, limiting job alternatives and mobility.[54]

Their "place" in the hierarchical system is clear. The daily existence of engineers in production organizations exemplifies what Braverman identified as the "separation of conception from execution" that began with Frederick Taylor's introduction of "scientific management" into the workplace in the early twentieth century.[55] Workers' control over their craft was altered when planning responsibilities were taken from the individual craft worker and shifted to managers, leaving the worker to follow orders, implementing plans without access to the full picture. In the modern workplace, technologists generally have little responsibility beyond the development, testing, calculations, and paperwork related to the applied work they do. Their creative work is bracketed by program decisions made outside (and above) the lab. Stinchcombe writes, "The product of the engineer is a proposed decision by the client."[56] To a great extent, the job is reactive and consists of providing clients, both in-house and external, with designs that fit prescribed plans and information for making technical decisions. Assigned to projects that typically have a long duration, engineers become further specialized, adding to already limited mobility.[57] Possible career advancement consists mainly of movement into management positions in the same organization.

These institutional arrangements have led many scholars to conclude that engineers are "servants of power": carriers of a belief system that caters to dominant industrial and government interests.[58] The argument goes as follows: Located in and dependent on organizations whose survival is linked to the economies of technological production and to rationalized administrative procedures, the engineering worldview includes a preoccupation with (1) cost and efficiency, (2) conformity to rules and hierarchical authority, and (3) production goals. These cultural meaning systems are reinforced by educational and professional arrangements. Diversity of training and specialization fragment the profession, limiting its power to direct the habits of mind of engineers away from production-system–induced interests to the human consequences of their technical activities.[59] Engineering loyalty, job satisfaction, and identity come from the relationship with the employer, not from the profession. Absent the professional's traditional autonomy, engineers adopt the belief system of the organizations that employ them. Engineers do not resist the organizational goals of their employers; they use their technical skills in the interest of those goals.

Recent events in engineering history show that engineers do not unquestioningly adhere to business values, however.[60] The profession is internally divided between the elite in the engineering community, for whom business professionalism is a powerful influence, and rank-and-file engineers, who struggle against business professionalism in a new climate of critical thinking about technology.[61] Moreover, engineers are not powerless in the workplace.[62] They are central to the mission of the organization and important to its competitive advantage. Technical production systems depend on engineers for product design, development, and analysis for decision making. Although working engineers do not have control over schedule, resources, their product, or its disposition in the market, they have autonomy to choose how they go about their own jobs because they possess technical skills and information that the technical elite do not.[63] Zussman notes, "If a supervisor interferes with or overrules this competence, he acts not in the interest of industrial authority but against it by disrupting the smooth flow of work within the division of labor."[64] In their day-to-day technical work and decision making about the technology, engineers are constrained more by their own elaborate technical routines and mechanical methods than by direct managerial interference.

Despite dissent expressed in professional associations and autonomy in technical decision making, the principles of capitalism and bureaucratic organization that characterize technical production systems remain integral to the worldview that engineers bring to the resolution of technical issues. Cost and efficiency are ineluctable core concerns that, to a great extent, define engineering. Two important points must be made about the role of economic considerations in engineering work, however. First, these concerns are central to the engineering worldview.[65] Zussman, in his important study *Mechanics of the Middle Class*, asked engineers to assess the effect of profit maximization on their problem solving. He found that engineering work was inherently economic, with cost itself a criterion in technical efficiency. He concluded that engineers "do not experience a tension between the logic of efficiency and the logic of profit maximization because their very conception of efficiency is shaped by considerations of profitability."[66] Cost and efficiency in engineering work are among the taken-for-granted goals of the profession.

The second point about cost and efficiency is that they are not the *only* central concerns in an engineering worldview. Dorf describes engineering as "problem-solving under constraints":[67] "Engineering is concerned with the proper use of money, a capital resource, as well as all other resources. Engineering requires the economical use of energy and materials so that there is a minimum of waste and a maximum of efficiency. It is also the safe application of the forces and materials of nature. Safety is an important quality and must be incorporated in all objectives."[68]

Solutions are always a compromise among a number of standard requirements: safety; reliability; long-term economy; minimum of labor; practicality; ease of manufacturing and installation, maintenance, and operation; and aesthetics.[69] These, too, are integrated into the engineering worldview during training, and reinforced in the workplace. In technical decision making, "satisficing," not maximizing, is the norm.[70] Petroski quotes designer David Pye on the unavoidability of compromise in design:

> The requirements for design conflict and cannot be reconciled. All designs for devices are in some degree failures, either because they flout one or another of the requirements or because they are compromises, and compromise implies a degree of failure.
> Failure is inherent in all useful design not only because all requirements of economy derive from insatiable wishes, but more immediately because certain quite specific conflicts are inevitable

once requirements for economy are admitted; and conflicts even among the requirements of use are not unknown.

It follows that all designs for use are arbitrary. The designer or his client has to choose in what degree and where there shall be failure. Thus the shape of all designed things is the product of arbitrary choice. If you vary the terms of your compromise—say, more speed, more heat, less safety, more discomfort, lower first cost— then you vary the shape of the thing designed. It is quite impossible for any design to be "the logical outcome of the requirements" simply because, the requirements being in conflict, their logical outcome is an impossibility.[71]

In the satisficing that necessarily follows, conflict between cost and safety is an endemic struggle. Safety, too, is integral to the engineering worldview. Engineers design to avoid failure. The dilemma they face is that failure is avoided through sound, strong structures— yet excessive strength is uneconomical. Thus, the challenge of engineering design is to build safe structures more economically.[72] Therefore, engineers routinely deal with yet another ambiguity: how strong is strong enough?[73] This particular ambiguity is worked out, as other engineering ambiguities are, through numbers. Calculating factors of safety and margins of safety is how engineers deal with all the uncertainties of design, construction and use. These calculations "provide a margin of error that allows for a considerable number of corollaries to Murphy's Law to compound without threatening the success of an engineering endeavor."[74] In determining how strong is strong enough, engineers walk the line between a margin of safety so high that the structure is uneconomical and thus "overdesigned" and a margin so low that there is no room for the unexpected. In the engineering profession, many design-related factors are juggled in production systems, cost and safety among them.

Bureaucratic rules and lines of administrative authority in the complex organizations for which engineers work are also taken-for-granted in the engineering worldview. Discipline is fundamental to the success of bureaucracies, which require unusual conformity to prescribed actions in order to achieve reliable performance; it is even more essential in technical production systems, which must assemble parts to produce objects on a schedule. The organization must instill in each member an awareness of the responsibilities and routines of the job as well as limits to individual competence and authority. By promoting career interests through the devices of salary increments, retirement funds, and promotion, which reward conformity to official regulations and orders from above and the successful achievement of organizational goals, the bureaucratic organization induces discipline in its workers. Rule following and conformity are reinforced by their centrality to both engineering as a craft and the bureaucratic setting of the profession. The engineer's precision-laden calculation, methodical technical routines, systematic comparison, prediction, and testing, and the positivistic principles on which they are based instill the importance of mechanical, meticulous rule following during professional education.

Of bureaucracy Merton has written, "The chief merit of bureaucracy is its technical efficiency, with a premium placed on precision, speed, expert control, continuity, discretion, and optimal returns on input."[75] In that sentence, one might readily substitute "engineering as a profession" for "bureaucracy." In addition, both involve reliance on categorization: the classification of individual problems on the basis of established criteria, the treatment flowing from the classification. Both involve proceeding by fixed routines. Both require extensive paperwork, which is stored in files as memory and precedent. Both are guided by universal rules intended to give order, predictability, and certainty.[76] Both cultivate respect for the chain of command and a sense of limited responsibility by virtue of functional specialization.[77] And both the engineering profession and bureaucratic production systems must, by definition, be concerned about cost and efficiency. To succeed as an engineer is to conform both to bureaucratic procedural mandates, chain of command, and production goals and to the rules for technical decision making learned while training for the engineering craft.

Research has established the engineering profession's acceptance and endorsement of bureaucratic authority relations and capitalistic concerns about the costs and efficiencies of production systems. Engineers experience administrative decisions about deadlines, project assignments, and resource constraints in the workplace as inevitable and legitimate.[78] They are used to and expect working conditions created by the upper echelon that include production pressure, cost cutting, limited resources, and compromises.[79] Although they criticize individual supervisors, they do not object to supervision in principle, endorsing hierarchy both as a structure of command and a structure of opportunity.[80] Their support for the hierarchical arrangements of bureaucracy is borne out by their own aspirations toward upward mobility via the management track.[81] Long-term career interests tie them to the organization, creating "a stake in the maintenance of the industrial order and in the rules of the game, on which the expectation of advancement is based."[82]

NASA CULTURE

As the centerpiece of the American aerospace industry, NASA is a major employer of engineers, so we would expect the characteristics of engineering culture to be reproduced at NASA. Attesting to the nested quality of culture, NASA's entire Shuttle Program exhibited the "unruly technology" that characterizes the engineering craft when complex technical systems are involved: interpretive flexibility, absence of appropriate guidelines, unexpected glitches as commonplace, "debugging through use," extensive systemwide problems with technical components, practical rules based on experience that supplemented and took precedence in technical decision making over formal, universal rules, and cost/safety compromises as taken-for-granted. Moreover, the Space Shuttle Program existed in a NASA organization that exemplified the principles of capitalism and bureaucratic organization characteristic of the technical production systems in which engineers often work.

But at NASA, the cultural belief systems of engineering as a craft and as a bureaucratic profession were also elaborated on in distinctive ways. During the Apollo era, according to space analyst Howard McCurdy, NASA had a "pure" technical culture appropriate for the engineering of unruly technical systems.[83] It consisted of "a commitment to research, testing and verification; to in-house technical capability; to hands-on activity; to the acceptance of risk and failure; to open communications; to a belief that NASA was staffed with exceptional people; to attention to detail; and to a 'frontiers of flight' mentality."[84]

This pure technical culture was the original source of the NASA self-image that after the *Challenger* tragedy was referred to as "can do." "Can do" was shorthand for "Give us a challenge and we can accomplish it." Organizational culture includes the organization's self-image as well as the norms, rules, and technical know-how that organize beliefs and actions in the light of this image.[85] Originally, the space agency's self-image had two sources: the ethos and practices of the technical culture *and* the history of triumphant success that was its outcome—the imaginative, cutting-edge science and engineering; the successful missions that reached to the moon and beyond; the daring of the silver-suited astronauts; and the sophistication of the high-performance products. Both sources were fundamental to the self-image, for the successes depended on engineering skills and discipline.

The original technical culture and the "can do" self-image coexisted in a NASA organization that the Apollo Program itself made more structurally complex and thus "more bureaucratic." At the outset, NASA was not set up for a project as large as Apollo. To accomplish the goal of reaching the moon, NASA received an initial and immediate 61 percent increase in budget allotment from Congress, triggering a period of expansion that altered organizational structure.[86] NASA established the Manned Spaceflight Center in Houston (later known as Johnson Space Center), launch facilities at Cape Canaveral (later renamed Kennedy Space Center), the Vehicle Assembly Building at the Cape (then "the largest enclosed space in the world"), and the Mississippi Test Facility.[87] In addition, NASA increased its reliance on "contracting out": hiring contractors from the aerospace industry for the technical and manufacturing expertise necessary to the Apollo Program. Still taking pride in the in-house technical know-how and hands-on engineering that was the core of the technical culture, NASA leaders viewed this as a temporary measure. To implement contracting out, further structural expansion was necessary, including the addition of an intermediate layer between top NASA administration and project work groups to coordinate activities and structures linking NASA and contractors.[88]

Despite the bureaucratic expansion, McCurdy's research indicates that the NASA of the 1960s was able to maintain the strong technical culture that had preexisted Apollo.[89] In the 1970s, however, as NASA's environment dramatically shifted from one of munificence to one of scarcity, powerful elites took actions that again altered NASA's mandate, strategies, and tactics. As a result, NASA was transformed into the epitome of the bureaucratic technical production system. Already a bureaucracy, it became "bureaupathological." A business ideology emerged, infusing the culture with the agenda of capitalism, with repeating production cycles, deadlines, and cost and efficiency as primary, as if NASA were a corporate profit seeker. McCurdy's research shows these elite actions eroded the strong research-oriented technical culture of Apollo: it still existed, but in a changed NASA that made the practices associated with the original culture more difficult to carry out and maintain.[90] These elite actions fall into two categories: (1) the actions of Congress and the executive branch, which jointly developed the policy that the Space Shuttle "should, in a reliable fashion and at an internationally competitive cost, provide for most of the Free World's space launch needs"[91]—but then failed to provide the resources commensurate with achieving that goal, and (2) the actions of NASA top administrators who sought to live up to this policy despite the budget constraints. We now explore how NASA's institutional environment, identified in chapter 1 as the structural source of disaster, affected the culture of decision in the workplace, adding bureaucratic accountability and political accountability to the original technical culture.

BUREAUPATHOLOGY AND BUREAUCRATIC ACCOUNTABILITY

The preoccupation with bureaucratic proceduralism and hierarchical authority relations in NASA culture during the Shuttle Program was triggered by the election of Ronald Reagan, whose administration increased the interface between government and business. All agencies were subjected to increased pressure to do business with business. Contracting out, begun as a temporary measure, became institutionalized, making business enterprise a permanent aspect of NASA structure.[92] Because contracts between organizations specify the authority relations that govern an exchange,[93] NASA's expanded use of contractors resulted in additional administrative structures and procedures to coordinate and control NASA-contractor relations. During this same period, all aspects of government were contending with increased oversight, including central clearance for procurement, budget formulation and execution, auditing, accounting, and personnel administration. To comply, NASA increased the size of its nontechnical staff, expanding its administrative structure still further. At one point, NASA headquarters staff was growing three times faster than the agency as a whole in order to handle the more complicated requirements.

As a consequence of these developments, the space agency's pure technical culture began to be compromised by changes in the premises of accountability at NASA.[94] According to the research of Romzek and Dubnick, during Apollo the space agency relied on professional accountability: control over organizational activity rested with the employees with technical expertise. It was a system built on trust of and deference to the skills of those at the bottom of the organization. The separate space centers had autonomy, with centralization used only when necessary to implement new programs. In the 1970s, however, professional accountability struggled to survive as the agency adopted the trappings of bureaucratic accountability. Control at the top, superior-subordinate relationships, orders, close supervision, rules and regulations, and hierarchical reporting relations began to dominate NASA's technical culture.[95] Many engineers, instead of doing hands-on engineering work, were shifted to supervisory oversight of contractors' work. Less time was spent in the labs and more spent at the desk and/or traveling. Project Managers and engineers alike were burdened with the procedural and paperwork demands of central clearance as they responded to the organization's burgeoning nontechnical administrative apparatus.

BUSINESS IDEOLOGY AND POLITICAL ACCOUNTABILITY

The business ideology typical of technical production systems became a part of NASA culture when the space agency lost the institutional consensus for its mission that gave it abundant resources. Political accountability became central to the space agency's technical mission.[96] Top space agency administrators shifted their attention from technical barriers to space exploration to external barriers to organizational survival. Deal making was necessary to forge support for the program. NASA's administrator from 1971 to 1977 was James Fletcher, a physics Ph.D. from the California Institute of Technology, a technically accomplished leader in the aerospace industry, and former president of the University of Utah. Fletcher negotiated a consensus in an extremely competitive environment, but in the process, scientific goals that spawned the technical culture were joined with administration and Department of Defense military aspirations. Fletcher's hard-won "victory" allowed a compromised Space Shuttle Program to begin, but only with the promise of cost-effective space transportation. The original technical culture, developed in an R&D agency engaged in programs with a defined end point, was further subverted by a commercial, pay-as-you-go, continuing project.

Scarcity and the need for political alliances created institutional pressures for efficiency and a requirement to fulfill the promises that had induced Congress to fund the program. Top NASA administrators were absorbed with "myth managing":[97] attaining legitimacy (and thus resources) by projecting and living up to a cultural image of routine, economical spaceflight.[98] In the beginning, this image was promoted in the initial budget proposal, in the Mathematica report that the shuttle would "pay its own way," in the hoop-la after the fourth shuttle flight when the program was declared operational, and in annual budget reports to Congress. It was a legitimating myth that was far removed from both the realities of the normal, unruly, technology characteristics of technical systems and NASA's rising Shuttle Program costs. The gap between the image and the reality grew as NASA-contractor relations in the Fletcher years and after were plagued by huge overcharges, misplaced or stolen equipment, excessive waste, and blatant fraud that worsened NASA's economic situation.[99]

The gap was not acknowledged by top NASA administrators, however, who failed to bring production goals in line with system capabilities. Instead, they pushed a production schedule that perpetuated the myth of routine, economical, operational spaceflight.[100] Apparently, they bought their own myth. No actions at the top acknowledged the acceptance of risk and failure as possibilities—a premise of NASA's original technical culture. We might legitimately infer—as did Astronaut John Young—that from the beginning of the Space Shuttle Program, the actions of NASA top administrators were influenced by NASA's legendary "can do" attitude.[101] The initial conceptualization of a "space bus" with a regular schedule was most surely a vision in keeping with the space challenges the agency had accepted and met in the past. And administrators went ahead with the program despite budget constraints. However, originally spawned from both the ethos and engineering practices of the technical culture *and* the successes they produced, the "can do" image for NASA elites was apparently based on past achievements alone, not the engineering practices responsible for those achievements, for the new elites were removed from hands-on technology.

Ranson, Hinings, and Greenwood note that "frames of meaning embody a conception of the organization and therefore a view of the appropriate allocation of scarce resources."[102] Top administrators reduced budget allocations for safety personnel, altered safety procedures, and reduced testing requirements.[103] They arranged for payload specialists like Christa McAuliffe and Senator Jake Garn to fly, without asking whether the official tasks and public relations use of these crew additions justified the risking of their lives.[104] They engaged in and publicized mission adventurism while ignoring various external reviews, including the 1984 annual report of the space agency's own Aerospace Safety Advisory Panel, which warned top administrators that this was no operational system.[105] Jack Macidull, FAA investigator working for the Presidential Commission, wrote, "Valid safety awareness was screened by first misdefining, and then not reevaluating the capability of the system based on new data, which led to dangerous operational assumptions and procedures."[106]

The agency culture was defined by its conflicting cultural imperatives. The introduction of business ideology and political accountability into the space agency contributed further to the erosion of NASA's technical culture, effecting the final step that converted NASA to the bureaucratic and capitalistic mode typical of technical production systems. As early as 1981, the year of the first shuttle flight, top administrators emphasized the need to reduce costs and turnaround time between flights.[107] But the emphasis on cost and efficiency hardly needed articulation to a workforce of engineers aware of the massive aerospace industry layoffs in the 1970s, subsequent congressional budget cuts, the struggle to get funding for the Shuttle Program, and NASA spending reductions that affected safety personnel and programs. Moreover, the policy established at the top was brought home resoundingly to the workforce through the production schedule itself. Jack Buchanan, Thiokol Operations Manager at Kennedy Space Center, said: "We were all used to working—well, I've worked as many as 23 hours in a day and many, many, many times 15, and Saturdays and Sundays, and called in 2:00 in the morning, get in your car and go to work. I have a son that works here also. My son was working seven days a week and overtime on top of that. You've got to feel that if that is going on habitually that somebody is worried. Nobody tells you, but you're in a working environment which you believe that schedule is important."[108] The launch schedule mandated working on many launches at various stages of production simultaneously, creating a deadline-driven organization that focused attention on procedural conformity and the prompt completion of technical procedures and paperwork.

Competition is integral to technical production systems. NASA's original technical culture was grounded in an ideology of cooperation that was fruitful for R&D between the separate space centers, although competition also existed between them.[109] However, when scarcity forced NASA to compete for funding and payloads, the dilemma of the parent organization was reproduced in the subunits. Scarcity, bureaucratic accountability, and political accountability altered the relationships between the space centers. Intercenter rivalries were keen during the Shuttle Program, breaking down the cooperative ethos that characterized NASA culture of the Apollo era.[110] Each center was accountable to top NASA administrators for its contribution to the production schedule and to the political agenda established to insure agency survival. They competed for scarce resources and also to meet program goals.

But cooperation was still necessary. Each center had separate responsibilities essential to the program: Johnson for systems engineering and integration, operations integration, management integration, and the Orbiter; Marshall for the entire propulsion system; and Kennedy for launch operations. And, obviously, the mistakes of one center, one office, one laboratory, could result in failure for all.[111] The production schedule and launch safety were necessarily interrelated. Avoiding harmful technical incidents and disasters is a significant precept in the cultural meaning systems of engineering work, in engineering as a bureaucratic profession, and in the NASA organization. Indeed, we could argue, citing the history of decision making in the Shuttle Program, that delaying the production schedule for safety reasons was institutionalized at NASA. Only 8 of 25 launches had no delays in the period between the Level I FRR and completion of countdown; the rest averaged two delays each.[112] Although the American press castigated NASA for incompetence every time a launch was delayed, the reality was that NASA's launch delays were based on decisions made by work groups concerned about safety and unwilling to allow a launch to proceed unless certain conditions were met.

The triumvirate of cultural imperatives—the original technical culture, bureaucratic accountability, and political accountability—permeated the organization. The consequences of Fletcher's deal forging, in the form of congressional and administration expectations for the shuttle, fell on Center Directors, who in turn relied on Project Managers to move the hardware elements toward launch. It was the Project Managers who bore the burden of the three cultural imperatives. They were the connecting links between the bottom-up engineering risk assessments and the actions of top administrators. They were responsible for upholding the tenets of the original technical culture in risk assessments that assured the safety of the components; the bureaucratic accountability and proceduralism in which risk assessment was embedded; and the cost, efficiency, and production concerns on which the agency's political accountability rested. And at the bottom of the launch decision chain, working engineers were making risk assessments in a transformed NASA culture in which bureaucratic rules and procedures were a major preoccupation and cost, efficiency, and production deadlines were a taken-for-granted aspect of work.

MARSHALL SPACE FLIGHT CENTER CULTURE

The tensions between the cultural imperatives of the NASA organization may have been greater at Marshall than at the other centers. NASA's much-respected original technical culture was born at Marshall, out of a military heritage that made discipline a core cultural element. The center grew out of the army's Redstone Arsenal (named for the battlefield missile developed there, the Redstone rocket). After World War II, the Defense Department established the Army Ballistic Missile Agency (ABMA) in Huntsville, which designed and tested rockets for military use. The ABMA was operated at the Red-stone Arsenal, run by a rocket team of 120 German engineers that had escaped to the United States after the war. There, under the leadership of Wernher Von Braun, German rocketry wizard and aerospace legend, they re-created their strong precision/verification German research culture.[113]

When NASA was established in 1958, the Von Braun team left the army for the space agency. Marshall was officially opened in Huntsville in 1960, with Von Braun as Director. The propulsion knowledge of the German rocket team fit perfectly with the needs of Project Apollo, establishing Marshall as the technical locus of America's race with the Soviets to get to the moon. Von Braun and his associates inculcated the technical exactitudes, superior knowledge and expertise, mandated hands-on strategies, awareness of risk and failure, and open communication that gave Marshall its original technical culture. He instituted and required the "dirty hands approach" to engineering: staying in close touch with the technology; doing, rather than seeing that it gets done.[114] This prescription applied to management as well as working engineers.

As work on Project Apollo progressed, Von Braun had to alter the organization structure to suit Marshall's role in the project and the demands of increased contracting out. The center developed into a matrix organization, an innovative development at the time viewed as appropriate to the management of several separate rocket projects, adding a new and separate division to direct and monitor contractor operations.[115] Another layer was added to the burgeoning bureaucracy. Yet, under Von Braun's leadership, the original technical culture persisted. It was Marshall's strong research culture, together with the pioneering rocketry successes that resulted, that was the origin of NASA's legendary "can do" attitude identified with the Apollo Program. The expertise and strict engineering discipline of Marshall personnel commanded respect from and often intimidated representatives from the other centers and contractors. Although publicity (and thus glory) went to Cape Canaveral, the astronauts, and Mission Control in Houston, Marshall was responsible for the technical breakthroughs behind Apollo's success.

Concerns with cost and schedule nonetheless existed in the technical culture at Marshall during the Von Braun era. Phillip Tompkins, organizational communication expert conducting research there at that time, observed that "only the easy decisions at Marshall were made by scientific evidence or demonstration. The difficult decisions created a rhetorical problem because the solution could not be demonstrated scientifically. At MSFC, the three master *topoi*, Aristotle's term for possible lines of argument in a rhetorical situation were reliability, time, and cost. . . . It is important to grasp that the difficult aerospace decisions involved trade-offs among the three main *topoi* of reliability (or safety), time (or schedule), and cost."[116] However, the point to be made is that in the Von Braun era, cost, schedule, and safety satisficing went on in a project environment very different from that of the Shuttle Program: no production cycling, plenty of money.

But that changed. As the Nixon administration began, budget cuts compelled NASA to cancel the two final Apollo missions. *Apollo 17*, the sixth and last lunar landing, was launched in December 1972. William Lucas became Center Director in 1974, three years after Fletcher took control of the space agency.[117] According to Malcolm McConnell, author of *Challenger: A Major Malfunction*, Lucas had been a young staff member under Von Braun at the Redstone Arsenal, well versed in both the tenets of the original technical culture and military discipline.[118] The organizational structure, permanency of contracting out, and repeating production cycles of the Space Shuttle Program made Marshall a very different place than during Apollo. Despite the changes, the center had a strong technical culture, continuing in the Von Braun tradition. The people at Marshall were confident and proud of their skills and achievements; they "saw themselves as heirs to Von Braun's visionary engineering excellence."[119] Marshall's engineering conservatism (the source of complaints from Thiokol engineers) was an acknowledged part of a rigorous Marshall technical culture concerned about safety. SRB Chief Engineer Jim Smith, who began in 1958 at the Redstone Arsenal and transferred to Marshall in 1960, said: "I would categorize the organization (MSFC) since day one as a very conservative type organization from an engineering standpoint. I guess we overdesign. Our approach to almost any matter is that, whatever the design requirements are, that we would tend to go overboard to be sure that we have the right design margins for the situation."[120]

Marshall's Robert Schwinghamer, who ran the Materials Processing Lab in S&E, stated: "You know, hell, they call us the dinosaurs. We're always being beat on for being conservatives. We are from the Paleolithic age. We don't, you know, why don't you hurry up and do this or let's—no, let's test it first, you know. . . . We have always had that reputation, and the people in headquarters have always poked fun at us because of it and so have the people at Houston."[121]

But maintaining the practices associated with the original technical culture was more difficult at Marshall than at the other centers because the influences of bureaucratic structure and capitalism were particularly keen. Marshall was responsible for the shuttle's propulsion system. Thus, Marshall people managed three of the four major shuttle hardware elements. Consequently, the center was dominated by mission deadlines and preoccupied with completing engineering work and qualification during the many phases of overlapping FRRs. To coordinate work on the three projects and the many contractors they required, Marshall had extra administrative apparatus (e.g., the Level "2 1/2" FRR—the layer of review between Level III and the Center Board) and additional procedural controls (such as MPAS—the problem tracking system discussed earlier) that other centers did not.

The political accountability created by the myth managing at the top trickled down to Marshall's Lucas. Competition had existed before the Shuttle Program, especially between Marshall and Johnson. Now, however, it increased. Johnson was given overall leadership of the new Space Shuttle Program, and Marshall had to take direction from Johnson as well as NASA Headquarters in Washington.[122] Preoccupied with maintaining engineering excellence and Marshall's superiority in the competition between centers, Lucas "wanted the center to do the very best job at all times."[123] He was reputed to be very strong technically, asking penetrating technical questions. He emphasized engineering precision. Further, he believed in doing things exactly by the book. According to McConnell, Lucas was a "master bureaucrat" who "created an atmosphere of rigid, often fearful, conformity among Marshall managers."[124]

Saddled with a vastly more complex Marshall organizational structure and apparently lacking Von Braun's charisma and skill at maintaining personal contact with people at different levels in the Marshall hierarchy, Lucas relied on hierarchy and formal mechanisms to transfer information.[125] He insisted on bureaucratic accountability for monitoring and controlling internal operations.[126] Lucas's management style, combined with the production pressure the center was experiencing, not only exacerbated the intercenter rivalry but resulted in competition between the three Marshall projects. Each Project Manager vied with the others to conform to the cultural imperatives of the original technical culture, the bureaucratic mandates, and the business ideology of production. They competed to meet deadlines, be on top of every technical detail, solve their technical problems, conform to rules and requirements, be cost-efficient, and, of course, contribute to safe, successful space-flight. Not to do so would imply that they were not doing their jobs and that, therefore, Marshall was not doing its job. No Project Manager wanted his hardware or people to be responsible for delaying launch; no Project Manager wanted his hardware or people to be responsible for a technical failure. To describe the pressure at Marshall simply as production pressure is to underestimate it. It was, in fact, performance pressure—pressure to meet the demands of all three cultural imperatives—that permeated the workplace culture.

The Marshall Center Board FRR was the quintessential embodiment of Marshall culture. Although Marshall's Level IV and III FRRs were adversarial and rigorous, they paled in comparison to the Lucas-embellished culture of the more formal, large-audience Center Board review. The Center Board was the final in-house review before Marshall Level III Project Managers made their assessments of flight readiness at Level II and Level I before Johnson and NASA top administrators, respectively. Lucas presided. Here we see the distinctive Marshall performance pressure to obey *all three cultural imperatives* clearly manifested. SRM Program Manager Larry Wear gave a compelling description:

> The Center Board would be held in a humongous conference room that looks like an auditorium. It's an open meeting. There might be one hundred—one hundred and fifty people there. Be a whole raft of people, ninety percent of whom weren't going to ask you any questions. . . . It's great drama. Sometimes people give very informative, very interesting presentations. That's drama in a way. And it's an adversarial process. I think there are some people who have, what's the word, there is a word for when you enjoy somebody else's punishment . . . masochistic, they are masochistic. You know, come in and watch Larry Wear or Larry Mulloy or Thiokol take a whipping from the Board. There are people who I think actually come to watch that element.
>
> It's serious work. There are reputations at stake, not just individuals, but the Center itself. . . . You don't leave the Center to give a significant briefing unless the Center senior management is aware of what you're doing, and that Board was a means of doing it. He (Lucas) was looking after the institution's image and his own. One reason is because he becomes a member of this thing, as it gets on up higher. He doesn't want to go to the high-level board and sit there and then be embarrassed by what his people are saying.
>
> There are standards to be upheld. He challenges the technical information. We have this saying, "only one-chart deep." You can't go to the Board knowing only what's on your [engineering] charts. Lucas and the Board ask very hard questions, going into details

much deeper than what's presented to them in the charts. You've got to be able to answer any question. And he challenges the style of the presentation. He [Lucas] requires you to present the things up to the standards of Marshall Space Flight Center, and I heard him say several times, he'd stop something and he'd turn to the man's boss, and he'd say, "Jim, I just don't believe this represents the standards of this Center, do you?" Of course, Jim would say, "No, sir, I certainly don't." And that meant the briefing is not crisp and clear.

It is intense information. It's important information. It's important from the standpoint of whether the vehicle is going to fly and its danger and the national assets. This is a billion dollar program, and national assets are involved. In a lot of ways, you always thought the future of the Center hinges on whether we do right or don't do right. You know, it is not an insignificant affair. For one thing, human life is involved. For that reason alone. But also the image of the Center is at stake. The continuance of the program is at stake. If you lose this program, you know, we lose 8,000 jobs. It is a high, important, dramatic situation. There's big money at stake. There's your job at stake. There's national prestige, prestige of the Center, all those things are at stake. So it's a lot of pressure.[127]

Marshall's concerns about looking good among other centers translated into competition between the SRB, Main Engine, and External Tank Projects at the Center Board FRR. Impression management was the name of the game. Performance pressure resulted in managing impressions by *leaving no stone unturned*. This thoroughness created intense preoccupation with procedural conformity and "going by the book" for all three cultural imperatives: (1) bureaucratic accountability—following rules and procedures for decision making and relaying information up the hierarchy, (2) political accountability—meeting the schedule by getting the necessary flight qualification work done in time for the Center Board, and (3) original technical culture—being able to support work group recommendations and conclusions with data, engineering analysis, and technical rationale that were responsive to every conceivable question.

On bureaucratic accountability, we know from the history of decision making that Marshall Level III and IV managers did conform to rules and procedures for reporting problems up the hierarchy in FRR. Wear said:

The last thing that you wanted to do was to be up on your feet [in FRR] and to tell Dr. Lucas that you had everything resolved and were all ready to fly, and then to have some fellow from the 14th row stand up and [say], "Dr. Lucas, they forgot to tell you about 'x.'" That's the last thing in the world you want to have happen to you. Whatever people remember about Bill Lucas, one thing he would not tolerate was that you attempted to slip something by him without full disclosure. That was death, and so on a few occasions, I have seen that happen. I have experienced that myself on occasion, which kind of experience one did not wish to have happen more than once or twice in his career. So the inclination was to make sure you had reported every problem thoroughly.

Both bureaucratic and political accountability were reflected in preoccupation with conforming to requirements for preflight qualification of hardware to meet FRR deadlines. SRB Project Manager Larry Mulloy remarked:

> We were absolutely relentless and Machiavellian about following through on all the required procedures at Level III. It was "Don't make me go tell Bill Lucas (chair of Center Board) that we haven't completed an igniter qualification, and we've got an igniter sitting down there ready to launch," or "Don't make me go tell anybody that we've got a piece of paper which is a certificate of qualification which has been in the system for three months and is now holding up launch because we haven't completed the paperwork." That stuff got done overnight. That was the beauty of the [FRR] process. It just kept things moving along nicely.[128]

To meet the demands of the original technical culture, Marshall work groups experienced pressure to achieve the most rigorous data analysis possible to back engineering recommendations. Leslie Adams, Deputy SRB Project Manager, stated: "Any member of that [12-member] Board can ask questions, and they're not discreet about asking them. They ask any questions that they can come up with. The full Board has to be satisfied that there are no safety of flight issues . . . have any problems understood and a full hard basis for flying, it is not a safety of flight issue, be able to have the rationale for flight, or otherwise you will not get through those boards. We've been there many times where we've been sent back with Action Items to go get more data, to develop a further understanding for the Board."[129] The emphasis was on science-based technology. But science, in FRR presentations, required numbers. Data analysis that met the strictest standards of scientific positivism was required. Observational data, backed by an intuitive argument, were unacceptable in NASA's science-based, positivistic, rule-bound system. Arguments that could not be supported by data did not meet engineering standards and would not pass the adversarial challenges of the FRR process. At Marshall, solid technical arguments and extensive testing and analysis covering every aspect of a technical issue were expected in every FRR presentation after Level IV. Explaining why Mulloy did not include his temperature concerns in the FRR presentation following the 1985 cold temperature launch of STS 51-C, Thiokol's Boisjoly affirmed that the requirement for quantitative, data-based, "engineering-supported" positions was normative:

> It was our feeling at the time that nothing gets presented to Dr. Lucas unless the people that are doing the presenting are absolutely sure that all bases have been covered technically and that they have all the answers to all the questions that potentially could be asked, and this was a case where clearly there were no answers available because it was just a question of observation as to what we were presenting. I have been personally chastised in FRR at Marshall for using the words "I feel" or "I think," and I have been crucified to the effect that that is not a proper presentation because "I feel" and "I suspect" are not engineering supported statements, but they are just judgmental. And so when people go in front of Dr. Lucas, they know full well that if they use words like that or if they use engineering judgment to try to explain a position, that they will be shot down in flames. And for that reason, nobody goes to him without a complete, fully documented, verifiable set of data.[130]

Having traced the culture of production from its origins in the engineering profession through its manifestation at NASA, then at Marshall, we now follow it to the SRB work group. These nested cultural systems affected work group decision making, contributing to and stabilizing the group's definition of the situation: continuing to recommend launch in the circumstances they had made sense within the culture of production. Completing the macro-micro link, we see how the original technical culture, political accountability, and bureaucratic accountability shaped their decisions about risk in the years preceding the *Challenger* launch.

THE SRB WORK GROUP: ALIGNING ACTIONS AND THE NORMALIZATION OF DEVIANCE

People are both the carriers and creators of culture. Cultural persistence is enacted in everyday life, as we try to make sense of events by drawing on largely unconscious knowledge for our responses. We tend to make the problematic nonproblematic by formulating a definition of the situation that makes sense of it in cultural terms. We achieve consistency by aligning our actions with the past, either by "interpreting our acts in cultural terms or by taking account of culture in the framing of action."[131] As a result, people can see their own conduct as culturally approved and conforming, even when the behavior is objectively deviant. Aligning actions with culture legitimizes questionable conduct, affirms the rightness of norms, and maintains the course of joint action.[132] In this way, we reproduce cultural meaning systems in both small and large choices, relegitimating the past, providing a medium for the present, and setting the stage for the future.[133]

In all the practices related to technical decision making, the work group behavior aligned with, or conformed to, the cultural meaning systems of the engineering craft. These cultural definitions contributed to the group's definition of the situation as normal technology, feeding into the persistence of the belief that the design was an acceptable risk. SRB technology and the work group's response to it was not deviant, but aligned with behavior typical in what Wynne identified as the unruly technology of technical systems. Deviation from specifications was common in an innovative design; to proceed with anomalies was normative for engineers when certain conditions were met. Although erosion itself had not been predicted, its occurrence conformed to engineering expectations about large-scale technical systems. At NASA, problems were the norm.

The word "anomaly" was part of everyday talk; all postflight reports and FRR documents listed "Discrepancies" from design expectations and contained "Unsatisfactory Condition Reports" with many entries. The whole shuttle system operated on the assumption that deviation could be controlled but not eliminated. This knowledge led to an FRR process in which each of the million shuttle parts were assessed before and after each flight, deviations corrected, and parts refurbished for reuse. The work group's decisions to correct and fly (so controversial after the *Challenger* disaster) were consistent with the engineering practice of "debugging through use." As Roger Boisjoly said, "Problems are expected in the aerospace industry because of limited ability to test. If you identify a problem and don't correct it, that's ludicrous."[134] The Rationale for Retention that accompanied every CIL entry indicated the institutionalized acceptability of deviation from universal guidelines, the formation of practical rules to guide decision making, and the correct-and-fly belief. Within this social context, what the engineers were dealing with was normal, natural trouble.[135]

Interpretive flexibility existed. Ambiguity, arbitrariness, and disagreement evident in SRB work group deliberations also were normative. Controversies centered on what tests were appropriate, how much leakage was permissible, what gap size should be, or whether the joints needed putty. Despite interpretive flexibility, the group's job was to measure risk and make a judgment about safety. The Acceptable Risk Process that they followed in learning about the hardware was aligned with universally accepted engineering practice in risk assessment and management, the term "acceptable risk" itself was not unique to NASA or the Shuttle Program but well known to engineers working on a variety of hazards.[136] The work group's definition of flight "success" was not determined simply by whether a flight returned (as appeared to be the case for top NASA policymakers), but by assessing each part and comparing performance with predictions. The frequent use of the term "experience base" in documents and testimony conforms to the common engineering belief in performance as the ultimate design test and the importance of learning by doing in the craft of engineering.

Finding no precedent for their design and receiving industry advice to "test to make sure the thing would work" and develop in-house standards for determining risk acceptability, the working engineers proceeded to develop ad hoc rules, also accepted practice when making engineering decisions about unruly technology. Constraints on launch that could be lifted and rules that could be waived are examples of how NASA created a formal system to mediate the gap between its set of universal rules and the flexible system of rules required to accommodate organizational learning about particular components. As Larry Mulloy said about the C 1 waiver, "The waiver was merely to put the configuration requirements in consonance with the actual performance of the joint."[137] In an occupation where deviation from design expectations was routine, where risk was a matter of degree rather than an all-or-none proposition, where rules were crafted and then recrafted as more was learned, and where ambiguity was the norm, labeling the performance of a joint that was redundant except under a rare coincidence of worst-case conditions as a C 1 item was the cautious way to proceed, not a risky action. Its unruliness evident from the start, the performance of the joint fit the formal standards for neither the C 1 nor the C 1R category, so the work group shifted it to the category that would give it more review attention.

Following the imperatives of their craft, they relied on numbers, using the scientific principles and positivistic methods that were the common tools of their trade: predictions, analytic models, bench tests, field tests, and comparison of predictions with performance. For each SRB part, success was a variable, worked out in terms of learned experience about performance capability. To this end, the group tested to the failure point, determining safety margins, as is customary in engineering work. As experience with the SRB joint accumulated, the first rules were expanded and new rules were added as new information dictated. Cobbled together as experience accumulated, the rules became standard operating procedures, some of which became formalized, all of which embodied organizational learning.[138] These standard procedures were themselves a technology that incorporated beliefs about cause-effect relations and standards of practice and behavior, undergirding the work group's scientific paradigm.

The SRB work group also aligned its actions with the cultural meanings typifying engineering as a bureaucratic profession located in a technical production system: (1) the legitimacy of bureaucratic authority relations and conformity to rules, and (2) the taken-for-granted nature of compromise between cost, schedule, and safety. Consider, first, the group's response to procedural rules and hierarchical authority relations. Though engineers are often cited as an example of Merton's ritualistic bureaucratic personality, little data are available about the alleged ritualistic rule-following tendencies of engineers. Deviance from accepted practice is often essential to the ongoing enterprise, even when it appears to conflict with official goals.[139] We might expect engineers to both conform and deviate in the workplace, as most employees do. Although limited in the work group actions that we can assess by the methodology of historical ethnography, the data do show some examples of conformity by both managers and engineers in the SRB work group to procedural rules and hierarchical reporting relations.

In the years before the *Challenger* launch, there were complaints about the design among the working engineers, but at no time did anyone act in any way that defied or resisted the hierarchical authority structure. Even written engineering complaints about the joint followed the authorized chain of command; thus, for example, Marshall engineer Leon Ray's memos of the late 1970s were written in the name of his superior, John Miller, to the next in command, Robert Eudy, and Roger Boisjoly's July 1985 memo asking for additional resources for the seal task force was cosigned by his immediate superior, then directed to the next in line, Robert K. Lund, Vice President of Engineering at Thiokol. FRR reporting requirements were followed to the letter, both by the working engineers, who followed the standardized formats established for postflight reports and flight-readiness presentations, and by Level IV and III SRB Project Managers, who reported between levels exactly as Marshall requirements dictated. In the two instances when work group participants deviated from formal rules, their behavior conformed to other system requirements and the requests of higher authorities. Even these instances of deviance were interpreted by participants as acceptable and nondeviant.

The first was the 1982 Level II management failure to convene the appropriate board in order to consider a waiver of the SRB joint's C 1 status so that flight could proceed. Instead of a formal meeting, the documents were circulated for approval and signature to make sure the paperwork was completed by the prelaunch deadline, an informal path frequently taken when no controversial issue was at stake. The second was the Thiokol engineers' December 1985 closure of the joint problem from MPAS. The contractor engineers closed this issue, in violation of the problem reporting requirement, in order to respond to a request from Marshall to reduce the number of problems carried in MPAS. Because the SRB joint problem was being tracked by several methods, and because the Thiokol engineers felt that MPAS was overloaded and ineffective, they removed the problem to get rid of what they viewed as an unnecessary, "make-work" requirement.[140] In neither case did those responsible define their actions as contrary to safety goals. In neither case were the actions surreptitious. Both actions were recorded, as required. This point is significant because it indicates that within the participants' definition of the situation these actions were acceptable and nondeviant.

Consider, second, the work group's actions vis-à-vis the cost, schedule, and safety compromises characteristic of engineering in technical production systems. The task of risk management in the aerospace industry is defined as "the ability to make the programmatic or engineering changes necessary to enhance the safety and performance of flight systems while controlling costs and schedule."[141] Satisficing where cost, schedule, and safety were concerned was institutionalized in the criteria for NASA's own change approval process: "NASA compares the cost and schedule impacts of the proposed change against the performance improvement that is anticipated. Of particular concern are the safety aspects related to the change (e.g., What analyses and tests must be conducted to insure that the change does not directly or indirectly have a negative impact on the system's safety or reliability?)."[142]

The question we want to answer is, In the years preceding the *Challenger* tragedy, what was the effect of cultural expectations about cost and schedule (political accountability), procedural rules and hierarchical authority relations (bureaucratic accountability), and safety (the original technical culture) on risk assessments? McCurdy argued that the incursion of bureaupathology and production goals into the NASA culture did not eliminate the priority of safety concerns that typified the original technical culture, but eroded it by making the practices associated with it more difficult to carry out.[143] The work group participants, struggling to conform to all three, affirm that this was the case. Their words vividly demonstrate how structures of power can alter the environment of decision in the workplace.

THE ORIGINAL TECHNICAL CULTURE, BUREAUCRATIC ACCOUNTABILITY, AND POLITICAL ACCOUNTABILITY

Tensions between the concerns and methods of the original technical culture, bureaucratic accountability, and production ideology affected the work group on a daily basis. They were well aware of cost and schedule pressures from the inception of the Shuttle Program. Marshall's Larry Wear remarked:

> You know, you're faced with a dilemma. Once you start the flight program, once the program progresses to that point, you know, to shut everything down and go fix these problems before you fly again is a big deal. I mean, that is a big traumatic event, to ground the whole program. There are 50,000 people around the country working on it, and you suddenly say, "We are going to take a one-year hiatus and fix this one problem." That's a big deal. And, I'm sure if you ask everybody involved, they would say, "If I thought that was warranted, I would have certainly done it." And I'm sure that everybody would say that the fact that there is this many jobs and people and organizations et cetera in place would not have bothered me a bit. That's easy to say after the fact. In reality, in real time, it's a pressure on anyone. Can we continue? Can we do what we're doing safely?[144]

Work group cost and schedule concerns had to be balanced with safety, not just because safety was one of the prescriptions of engineering as craft and engineering as bureaucratic profession, but because failure—and the responsibility for it—was something that neither Marshall nor the work group members wanted. Political accountability and occupational consequences were not the only reasons; as Jim Smith pointed out, "We know the astronauts."[145] Further, the SRB work group members at Marshall were, almost without exception, socialized in Marshall's original technical culture. Indeed, work group members (and many at Marshall not in the SRB work group but who had some responsibility for the SRBs—including Lucas) had been at Marshall for 20 to 25 years prior to 1986. Although this information, in itself, cannot be taken as proof that the tenets of the original technical culture still were actively pursued in the work group, other evidence indicates that it was.

Collins points out that the "core set"—the people most closely associated with complex technical systems—are fully aware of the ambiguity inherent in their unruly technology.[146] The SRB work group did not appear to be at all deluded by the public definition of the system as operational. They had not lost sight of risk and the possibility of failure. Recall Jim Smith's description of each launch as a "gut check," Leon Ray's comment "There are no small mistakes in our business. They are all big," and Larry Wear's observation that "high performance equipment is, by nature, unforgiving." But actions speak louder than words. Evidence of the presence of the original technical culture in the SRB work group comes from their vigilance: the teleconferences between Thiokol and Marshall engineers held several times each day; the many meetings and "data dumps"; the extensive detail in Thiokol postflight reports to Marshall and in the Level IV and Level III FRR packets; the paper trail of memoranda and letters pushing for answers to the technical problem; the rigor with which engineering analyses were challenged in Level III FRRs; the trips to professional meetings and manufacturing firms in the aerospace industry for consultation; and the running engineering disagreements between Marshall and Thiokol engineers and the many technical issues raised with Thiokol by Marshall's S&E engineers Ben Powers and Leon Ray, responsible for overseeing the contractor engineers. The Thiokol task force attempted to get resources, trying to manipulate a bureaupathological system to obtain help and materials. Engineers said the Marshall managers in the work group were attentive to safety. Ray noted: "Larry Mulloy was sensitive to the problem. He would come down to the Cape for post-flight assembly. He would fly himself down on Saturday and Sunday. He was a concerned, responsible manager."[147] These activities by the work group represent striving, caution, vigilance, and exploration consistent with the original technical culture, *not* inattention, complacency, and resting on the laurels of NASA's history of success.

But perhaps the best indicator that the work group had not forsaken Marshall's original technical culture in the interest of cost and schedule concerns was their willingness to delay launch. Ben Powers was asked by FBI investigators about unwritten rules or pressure on engineers to stay on schedule so that Marshall did not become responsible for a no-launch recommendation or launch delay. Powers vigorously denied this was so, citing an incident two months prior to

Challenger as proof. (Also salient in the following incident is the professional accountability—deference to the technical expertise of working engineers—that was so essential to the technical culture during Apollo.) The left forward SRM case segment that was to be used on the *Challenger* (51-L) was damaged at the Cape in November 1985. Thiokol was responsible for the damage. Powers recalled:

> I raised hell about it. Thiokol wanted to fly it, and it was the same thing, a field joint . . . these same players, Boisjoly and Thompson, specifically Thompson, were in charge of evaluating whether or not they could use that damaged cylinder. And I flatly refused to use that thing. My logic was we did not qualify a damaged cylinder. . . . And the pressure there [Thiokol] was it costs too much money to scrap it. It has propellant in it, and gosh, what do you want us to do, ship that thing all the way back to Thiokol and measure it and all that stuff? I said, hey, I don't think you can use that thing. So I threw up a barrier to them and got it stopped. But these were the same guys [that objected to the *Challenger* launch]. They sent those guys down on a plane to try to persuade me to turn around on it. There was a discrepancy report written on it and I disapproved the thing. So they said, hey, you haven't looked at it and you don't understand it. So they sent half a dozen of their engineers down here to talk to me about it, to tell me, hey, you know, we've got this fix for it. . . . I listened to their arguments and told them they had not convinced me. It was unacceptable.
> Our management then backed me up on that. And they took that cylinder and set it to the side. And I don't know how many millions it's worth, but it's worth a lot. . . . I was asked to pass judgment on it and I did and was supported up the line. My division chief sat in the meeting, listened to it, didn't comment one way or the other, supported me, recommended up the line, and that thing was shoved over to the side. And Marshall in that case let cost be damned and went with my recommendation. So I can't say the management won't follow what I recommend. That thing could have easily turned into a launch delay, that damaged cylinder, because they had to find another one and get it down there. Okay? Well, as it turned out, they could take one that was already there and swap out with it. But that was luck. Most of the time they would have had to produce one and send it down here, so that could have turned into a launch delay for 51-L.[148]

Marshall S&E engineers all concurred that there was no policy at Marshall, spoken or unspoken, that Marshall was not to be responsible for a no-launch situation or a delay.[149] Safety did not take a back seat to schedule, they said. Instead, Marshall had a reputation for being very conservative in engineering decisions. Keith Coates recalled Marshall's November 1983 decision to destack STS-9 and the month-long delay that resulted. The SRB work group delayed the launch because of a suspect exhaust nozzle in the right SRB that was discovered only after the shuttle was on the launch pad. In an operation called a "destack," the shuttle was taken back to the gigantic Vehicle Assembly Building and the Orbiter was separated from the External Tank and the SRBs so that the suspect nozzle could be replaced.[150] Coates said, "I feel that if Marshall personnel felt that there was an unacceptable risk, they did not hesitate to take the steps necessary to correct that, even to the extent of delaying launches and changing hardware."[151] Marshall's Leon Ray said:

> Any manager is driven by costs and schedule. He's got a budget, and he's constrained by it. The technical people say, "We want it done right," and the manager says, "We can't afford it. How important is this to mission safety?" Then they have to make a management risk decision. He says, "Is there some way I can do it? Will we lose part of the data? Will we lose the mission?" They won't delay unless they can justify it, because they are going to have to justify it to their superiors. You can't deny the fact that it causes you to do a lot of thinking. If a guy is too conservative, he's "shootin' down launches"; if he's too conservative, you will find someone else in that position. But do they say, "To heck with the risk"? No, they just don't do it, they just don't do it.[152]

Marshall's John P. McCarty, Structures and Propulsion Lab said:

> I think there is always an interest in observing schedules, because you can't run any kind of program without a schedule. Typically, scheduling is not a discussion topic that we have. Generally, we address more of the technical issues and the technical problems and a resolution of those problems whatever the schedule requirement. Now, occasionally we get asked a question of, you know, it's going to take us three days longer to do this. Do you really think you still need to do it? And if we feel we need to do it, we do it. . . . I can remember hardware changes that required schedule delays. I can remember postponing the launch 48 hours instead of 24 hours in order to effect safety checks, checks that we felt were appropriate, so I don't think—schedule is important, but it's not the deciding factor, and anytime a technical issue is to be decided, cost is even less [important] than schedule.[153]

Marshall's Powers remarked:

> When you start talking about launching 10, 12, 14 vehicles a year and going through the process of evaluating the hardware and going through the Flight Readiness Review, you can't help but worry about how in the world you are going to get all that work done plus the other work you have to do, because that's a big work load. And

yes, there is an influence of pressure there to a degree because you know you have to get that amount of work done. But from the standpoint of anybody telling you, hey, you are going to go launch so many this year come hell or high water, no, I don't think anybody really believes that, especially when they stop a launch because of the weather down at Timbuktu or something. Boy, that doesn't make sense to me at all. When somebody says "I've got to have beautiful weather in all these different stations downrange before I can launch," that just doesn't match with "we're going to have so many a year" to me . . . you would never get the weather for it.[154]

Marshall managers affirmed what their working engineers said. They asserted that the major impact of production pressures in the operational period was to increase the workload, but not at the expense of safety. Leslie Adams said:

I think the emphasis is always on safety, number one. And, in fact, the flight review process, which is, at the rate we're launching, almost the long pull in launch to launch [i.e., time consumer], has continued and all we do is work longer hours to get through that process. It works to understand every anomaly that occurred on a previous flight, any change that's been made from the last flight, is there a rationale for it, is there a basis that that will not affect the performance of the vehicle. . . . If there was schedule pressure you would have had cut out some of the Flight Readiness Reviews. But we don't cut them out. We just go through more. I see absolutely no indication of trade-off between schedule and safety of flight. That's just something that's not allowed.[155]

Here we see the intersection of political accountability, bureaucratic accountability, and the original technical culture. As the number of launches per year increased, so did the workload. In this way, the production schedule made the practices associated with the original technical culture more difficult to carry out, as McCurdy's research indicated. In both the waiver "walk-through" and the close-out decision, we glimpsed first managers then engineers struggling to do their jobs within a bureaupathological system. Thiokol's Boisjoly and Stein attributed task force resource problems to government procedures that delayed response to contractor requests. And the increased launch schedule increased the workload because there was less time to complete the paperwork associated with flight-readiness qualification. Marshall's Wear commented on the workload, describing how production pressure became institutionalized and taken-for-granted:

The program is set up, there is a plan, there is a plan to launch, twelve this year and fourteen next year and sixteen the next year. That is in the cards. In order to do that, you have to do certain things, on Monday, something else on Tuesday, and something else on Wednesday, just to make the machine work. So, you can call it pressure: I call it work. That says I've got to get ready to fly the next flight quicker than I did the one before. And I have to get ready to fly that one quicker than I did the one before that, because you can't go from six a year to fourteen a year without either expanding the work day or working faster. It's as simple as that. I don't know whether that is pressure. To me, I regard it as just more work to do. It comes with the territory.

Larry Mulloy makes palpable the work group's effort to conform to all three cultural mandates:

> The problem was the increasing launch rate. We were just getting buried under all this stuff. We had trouble keeping the paperwork straight, and were accelerating things and working overtime to get things done that were required to be done in order to fly the next flight. And they [FRRs] kept coming. The system was about to come down under its own weight, just because of the necessity of having to do all these procedural things in an ever accelerating fashion. It didn't mean you took unwarranted risks, it just meant you had to work longer and harder and faster and smarter. But no, that didn't mean that you short-circuited the [hardware] qualification, it just meant that you were working weekends, in three shifts, to get it qualified.[156]

From 1981 through 1985, the original technical culture, bureaucratic accountability, and production ideology all played a part in work group decisions to continue to fly rather than delay launch in order to alter the design. The production schedule was important. In order to halt the production schedule while a new design was implemented, a design change had to be necessary. Change was considered necessary if (1) the problem was a threat to the mission or (2) living with the problem was expensive in the short run. The work group continued to come forward with launch recommendations, despite continuing O-ring damage, because neither of these criteria were met. First, the SRB joint problems were not defined as a threat to mission safety, as the three chapters on the history of decision making attest. In 1985, even as concern escalated at Thiokol and Marshall, the engineering analysis indicated that the joint was an acceptable risk. Roger Boisjoly stated, in a personal interview:

The memo I wrote in July 1985 marked the end of my tolerance. [In this memo to Lund, Vice President of Engineering at Thiokol, Boisjoly had warned of the possibility of losing a flight unless the task force received more resources to work on resolving the problem quickly.][157]

ME: If your tolerance ended then, why did you and other Thiokol engineers continue to make recommendations in FRR that it was still OK to continue flying with the design for the rest of 1985 and into 1986?

BOISJOLY: We corrected the problem that occurred on the April [1985] flight. Because all erosion and hot gas blow-by after [that] were within predictions and the experience base, we thought it was an acceptable risk.

ME: Did the political aspects of the Shuttle Program and production pressure affect your FRR engineering analysis or conclusions?

BOISJOLY: There was never one thing we flew where I or any of my colleagues were forced to sign off on anything we didn't agree with or create a phoney number. None of my colleagues would have stood for that, and you know I would not. We always had autonomy in engineering decisions. That is why the night of the [Challenger] launch was such a shock. We lost our autonomy.[158]

The normalization of the technical deviation of the booster joints and belief in acceptable risk was reinforced by the work group's adherence to the practices of the original technical culture. The working engineers were convinced of risk acceptability by their belief in the positivistic methods and principles of science and what they produced: (1) correlation between findings of tests and analytic models, (2) no failures in any of the test motors that were fired (the most recent in May 1985), and (3) flight performance after correction, which was, as Boisjoly stated above, significant data. Presidential Commissioner Richard Feynman examined the analytic model that the work group used to predict performance, arguing it contained an error that led to flawed predictions.[159] This is another example of the luxury of hindsight: the tragedy led to reexamination, reinterpretation, skepticism, and criticism of all the group's engineering methods and the data they produced. Prior to the disaster, however, that information had a different meaning to the SRB work group. Not only was the Salita mathematical model well thought of, but the predictions it yielded were affirmed by data from other sources, also a tenet of positivism. Mulloy stated, in response to Feynman's questions: "We did not just use the math model. What we did was build a math model that was correlated to test. There was a test fixture that was built to empirically determine the maximum erosion that could occur . . . then Dr. Salita's math model was shown to correlate very well with that."[160]

But it was not just the data alone that led the work group to believe that the joint was an acceptable risk. The "can do" attitude affected the persistence of the work group's construction of risk. For the Level IV and III managers and engineers responsible for technical decisions, the "can do" attitude had origins that were distinctly different from the history of successful launches that was its basis for NASA's top administrators. In contrast to posttragedy imagery of "can do" at NASA as macho-techno risk taking, for the work group its source was in *rule following:* rigorous adherence to the engineering methods, routines, and lessons of the original technical culture and to the bureaucratic proceduralism of the organization. Recall SRB Chief Engineer Smith's comment: "We had analysis that led us to believe we had captured the problem . . . if you go take these steps, control the hardware, control how you build it, operate it this way, keep these engineering controls there . . . we think it's safe to go fly that vehicle."[161] In interviews, I asked what the "can do" attitude, widely reported in the media after the *Challenger* tragedy, was; did it exist in the SRB Project, and if so, did it have an effect on SRB joint decisions. The work group members I talked to agreed that it did exist and that it had affected decision making by reinforcing their belief in their technical analysis. Describing it, several work group members stated, "We believed in our people and our procedures." They were assured in their decisions because they had "long-term personnel with a history of hands-on hardware design that lead to experience and first-hand knowledge"; because they were "flying to design controls"; because "we followed every procedure"; because "the FRR process is aggressive and adversarial, examining every little nit"; because "we went by the book"; and because "we did everything we were supposed to." Aligning actions with the procedural rigor that was a hallmark of both the original technical culture and bureaucratic accountability at Marshall affirmed their belief that their diagnosis was correct.

Nor did the joint problems meet the second criterion for halting production: living with the O-ring problem was not expensive in the short run. Prior to the night of the *Challenger* launch, of all the problems Mulloy faced, the leading economic problem in his view was the parachute problem, not the SRB joints. He stated: "I was spending far more time each day dealing with parachute problems. We kept tearing the parachutes on the SRBs. This was a serious problem because it had economic consequences. If the parachutes didn't hold, the SRBs were not recoverable, and this was expensive. They sank to the bottom of the sea. On the joints, we were just eroding rubber O-rings. That didn't have serious economic consequences. The parachute problem was of far greater economic consequence, and that was a much more serious problem."[162]

When weighing the short-run economic consequences of living with the O-ring problem, Mulloy was not including the economic consequences of a mission loss due to SRB joint failure because the joint problems were not defined as a threat to mission safety. Nor was he including the economic consequences of redesign. Money for hardware redesign was not an obstacle, according to Mulloy, who was responsible for the budget for his project:

> We never felt strapped for money. The budget basis was the cost of hardware refurbishing of the SRBs for a certain number of launches. It was based on over-optimistic launch rates. In 1985, we originally were budgeted to launch fifteen. After cuts, they gave us a budget to launch twelve. We actually launched nine. The inflated mission model assured us of plenty of money. I had plenty of money to run the Project. I spent it on the pressure points, the #1 flight constraints, like O-rings and joints and nozzles—the things that would take us out of the sky. That was priority. Then you worked on the less essential things. The last thing you worked on was building up a stock of reusable hardware to support a fictitious mission schedule for launching more vehicles than they would ever be able to.[163]

In order to interrupt the production schedule for a design change, the engineers had to demonstrate that the proposed change would provide more than a marginal improvement in performance over the design currently in use because of the unknown safety hazards associated with change. This took time, for making a change required new local knowledge and a new set of "practical rules" to reduce the introduction of new hazards and assure the improvement was more than marginal. This, too, was a credo of engineering as a craft and of NASA's original technical culture. The gradual integration of the design change while flight continued was satisfactory to the work group because (1) it did not interfere with the production schedule, (2) the cost of living with the problem was only the cost of replacing rubber O-rings, and (3) they believed the present design was an acceptable risk. They spent the money, but they also took the time to conduct a 27-month research program testing the capture feature and additional corrective actions, with a report expected in mid-1986.

In continuing to recommend launch through 1985, the SRB work group was conforming to the nested cultural scripts of their unique culture of production, those "rules of appropriateness" that isolate the range of choices as rational in a given situation. We can see the cultural scripts from the engineering profession, NASA, and Marshall reproduced in SRB work group actions. They aligned with (or conformed to) cultural meanings associated with engineering as a craft and thus were not deviant in that context. At the same time, they demonstrated their allegiance to organizational rules and procedures, which, in combination with the practices of the original technical culture, contributed to the "can do" attitude that reinforced the technical choices they made. Their choices were constrained by cost, schedule, and safety concerns, exemplifying the satisficing characteristic of engineering in technical production systems. Moreover, they aligned their actions with the tenets of bureaucratic authority relations and business ideology characteristic of engineering in technical production systems.

As a result, the decisions from 1977 through 1985 that analysts and the public defined as deviant after the *Challenger* tragedy were, to those in the work group making the technical decisions, normal within the cultural belief systems in which their actions were embedded. Continuing to recommend launch in FRR despite problems with the joint was not deviant; in their view, their conduct was culturally approved and conforming. Zucker notes that "institutionalization is rooted in conformity—not conformity engendered by sanctions (whether positive or negative), nor conformity resulting from a 'black box' internalization process, but conformity rooted in the taken-for-granted aspects of everyday life. . . . Institutionalization operates to produce common understandings about what is appropriate and fundamentally meaningful behavior."[164]

The alignment between the SRB work group activities and these cultural meaning systems shows that they were part of the worldview, or frame of reference, that the work group brought to their decision making. So salient were the triumvirate of cultural imperatives in their culture that they affected decision making in a prerational way, not requiring conscious calculation. The cultural scripts in which work group actions were embedded were powerful because they overlapped, occurring in the training of professional engineers for their craft and social location in technical production systems, reproduced in the NASA organization and its subunit, Marshall Space Flight Center. Moreover, certain of the cultural meaning systems central to this story are embedded in the wider cultural belief systems of American society: capitalism and competition are "the" economic way; concerns with cost, production goals, and efficiency dominate industries; bureaucracy and hierarchical authority relations are the most frequently occurring form of organization; and technology and technological achievement are prized. In making the choices that they did, the work group was not only responding to these overarching and overlapping cultural beliefs but reproducing, affirming, and reinforcing them. They conformed, aligning their actions with the institutionalized cultural meaning systems and expectations that constituted their worldview. In this manner, the culture of production contributed to the normalization of deviance in the work group.

Chapter Seven

STRUCTURAL SECRECY

We now need to go a step further. The culture of production served as a context for decision making in other parts of the Shuttle Program. Presumably, all work groups were affected, to a greater or lesser extent. Yet prior to the *Challenger* teleconference, NASA's production schedule was halted and delayed many times. Despite the influences of the culture of production, in other circumstances signals of potential danger were not normalized. Even the SRB work group initiated delays, the extraordinary performance pressures at Marshall notwithstanding. If they delayed in some cases and not for the SRB joint, we must ask, What was different that caused the belief in acceptable risk to persist?

The work group was assessing risk on the basis of a material object. The behavior of the technical component and the information generated about it figured importantly in the normalization of deviance.[1] Although after the disaster, it appeared that Marshall managers intentionally concealed information, thus preventing top NASA administrators from intervening to resolve the O-ring problem, it was structural secrecy, not individual secrecy, that was important to the definition of the situation prior to 1986. By structural secrecy, I refer to the way that patterns of information, organizational structure, processes, and transactions, and the structure of regulatory relations systematically undermine the attempt to know and interpret situations in all organizations. At NASA, structural secrecy concealed the seriousness of the O-ring problem, contributing to the persistence of the scientific paradigm on which the belief in acceptable risk was based.

This chapter clarifies the role of information and knowledge at NASA in the years prior to the *Challenger* launch. It puts in place the final piece of the puzzle of why NASA continued launching despite anomalies, thus completing an explanation of the normalization of deviance during that period that includes the production of culture in the work group, the culture of production, and structural secrecy. To lay out the case for structural secrecy, we first examine the evidence for individual secrecy that became the conventional wisdom in post-tragedy accounts. Then we look at the sources of structural secrecy, tracing their deleterious effects on information and its interpretation. Finally, we consider the subtle but significant link between a person's position in an organization, access to information, ability to interpret it, worldview, and, ultimately, action.

INDIVIDUAL SECRECY

The Presidential Commission reported that in the years preceding the *Challenger* launch, NASA Levels II and I were not aware of the "history of problems concerning the O-ring and the joint,"[2] the "degree of concern of Thiokol and Marshall about the erosion of the joint seals in prior Shuttle flights, notably 51-C and 51-B,"[3] or "the launch constraint, the reason for it, or the six consecutive waivers prior to 51-L."[4] Noting that these "failures in communication" were based on "a NASA management structure that permitted internal flight safety problems to bypass key Shuttle managers,"[5] the Commission concluded that there appeared to be "a propensity of management at Marshall to contain potentially serious problems and to attempt to resolve them internally rather than communicate them forward."[6]

The implication was that if the Level II and I officials in the launch decision chain had only known, they would have intervened years before, averting the tragedy. These findings were repeated in the press, leading many to the inexorable conclusion that communication was the problem: top NASA administrators did not act aggressively to resolve the O-ring problems prior to the *Challenger* launch because Level III Marshall managers did not pass information upward, in violation of reporting rules. This impression was reinforced by other posttragedy reports describing the distinctive culture at Marshall under Center Director Bill Lucas as a high-pressure tank in which Lucas's concern with the Marshall's image, rules, and performance quelled subordinates' ability to convey bad news to the top of the organization. The Presidential Commission based its conclusion about communication on (1) what it mistakenly identified as failures to follow reporting procedures in the FRR process and (2) what it correctly identified as "procedural inadequacies" in computerized problem reporting practices.[7]

The House Committee on Science and Technology, whose report was published later and received less publicity, found the opposite: communication was open. It concluded that the system did not inhibit communication and that it was not difficult to reveal problems.[8] Although agreeing with the Commission about procedural inadequacies in computer tracking systems, the House Committee pointed out the far greater importance of the highly developed oral system for communicating about technical issues at NASA. Listing a full page of oral mechanisms (in addition to FRRs), the House Committee concluded: "There are many regularly scheduled meetings and teleconferences at all levels of management throughout the Shuttle Program. In addition, 'special' meetings and telecons are routine. No evidence was found to support a conclusion that the system inhibited communication or that it was difficult to surface problems."[9] More striking is the House Committee's conclusions about FRR: "The procedure appears to be exceptionally thorough and the scope of the issues that are addressed at the FRRs is sufficient to surface any problems that the contractors of NASA management deem appropriate to surface."[10] Finally, the Committee found no evidence that Level III Marshall managers "suppressed information that they themselves deemed to be significant,"[11] asserting instead that "the underlying problem which led to the *Challenger* accident was not poor communication or inadequate procedures as implied by the Rogers Commission conclusion. Information on the flaws in the joint design and on the problems encountered in missions prior to 51-L was widely available and had been presented to all levels of Shuttle management."[12]

The House Committee conclusion is supported by my analysis showing that FRR reporting conformed to reporting requirements (documented in chaps. 3, 4, and 5), and by the interview transcripts in the National Archives. Both contradict individual secrecy as the reason why senior NASA administrators did not intervene to resolve the O-ring problems. Without exception, all of the engineers, technical assistants, managers, and safety personnel assigned to the SRB Project agreed that communication was open at Marshall.[13] Access from bottom to top was easy, they asserted; problem reporting was encouraged, and no reprisals accompanied the bearing of bad tidings. This climate is represented in two statements from people whose jobs required them to report bad news—Wiley Bunn, Director, Marshall Reliability and Quality Assurance:

> In the [Presidential Commission] testimony, I've heard there appears to be some thread of thought that says that people—that Dr. Lucas doesn't want to be bothered with problems, or maybe

people can't carry their problems to Dr. Lucas. I don't have that problem. I don't have any problem speaking my mind to people that I need to speak to, Jim Kingsbury, George Hardy, Bob Lindstrom, Stan Reinartz, all those people. If I have a technical issue that doesn't get resolved I feel perfectly free to go to Dr. Lucas with it. . . . If there is any perception on the part of anybody at Marshall Space Flight Center that he doesn't want to know about a problem, that's wrong.[14]

Marshall's Leon Ray:

I feel at ease in picking up the phone and talking to anybody. It doesn't make any difference who it is, and I have, many times. . . . I think the communication channels are wide open as far as you want to use them. . . . I have never been chastised by any of my superiors for not going through the chain of command. I've never done that. It is quite different from the military service. You know, you do that one time in the military service and you are hung.[15]

In FRRs, where information was presented exactly when momentum for an impending launch was reaching its zenith, participants said that both the FRR structure and process made secret keeping impossible. The Lucas emphasis on standards and rigor, the big audience representing various interests and specializations, the adversarial style, and competition worked against willful concealment in FRR. George Hardy, Marshall SRB Project Manager prior to Mulloy, believed that "it is inconceivable, impossible, that any one individual or group of individuals would decide to keep a problem from the normal [FRR] process. Too many people are involved. It would take a conspiracy of mammoth proportions. Those O-rings are looked at by many, many people: government people, contractor people."[16] Larry Wear, Marshall's SRM Manager, said:

I believe there was truth in flight readiness reporting. It was open, the whole environment was entirely open. What keeps everybody honest is, Lord knows, a room of 150 people. You just couldn't keep problems secret. Even if you'd wanted to, you couldn't. If I had a problem and I wanted to just ignore it and hope it goes away, I wouldn't get away with it for ten minutes. When I stood up and said, I have no problems, Ben Powers or Jim Smith, or one of the guys in my own office, or a guy from Thiokol, or anybody sitting in the room would have said, "Wait a minute, Larry. Didn't you forget about the umpty-ump?" The impression of a lot of that [post-*Challenger*] literature is that Marshall could sweep things under the carpet and no one would ever know about it. Well, they weren't in the same meetings I was in. You were in a fishbowl to start with.

One of those guys would, and rightly so, with good candor, be the first to say, "The Project hasn't told you about the scorched O-rings?" Also, that's the role that Science and Engineering is supposed to play. They not only solve problems, they keep you honest. And Quality [Assurance] is another guy able to stand up and say, "They forgot to tell you about that one." I don't recall a single one of those [FRRs] that did not have a representative from Houston and from headquarters in attendance, and they had their opportunity to either say, "Bill Lucas, I know something that they did not talk about," or, when Larry Mulloy went to Level II to do his presentation, to say, "Mr. Aldrich, at Marshall they had five problems and here with you they've only mentioned three."[17]

With no dissent, managers and engineers admitted experiencing pressure not to be the ones responsible for holding up a launch. However, they stated that rather than concealing technical issues, they presented them hoping that someone else would have a bigger problem. According to Marshall engineer Keith Coates: "For years, there's been the expression 'get under that umbrella.' If Johnson has a problem with the orbiter tiles, and you are pushing hard to make a schedule on something and it looks like it may be nip and tuck, and lo and behold, they are slow getting their tiles put on, you are thankful for them because you can 'get under that umbrella.'"[18] Project Managers had their own term for it—Marshall's Mulloy: "You always went to those reviews hoping that somebody else would have a worse problem than you did so that it wouldn't be your problem that held up a launch. We Project Managers joked among ourselves about it. We called it 'being the long pole,' the lightning rod, the one that absorbed all the attention and electricity, so to speak, by having the problem that delayed launch."[19]

The notion that individuals at Marshall intentionally suppressed information to keep top administrators in the dark about the O-ring problem in the years before the *Challenger* launch is wrong. The question is not whether people knew about the SRB joint problems prior to the *Challenger* tragedy. Everybody in the launch decision chain who was supposed to know knew. Rather, the question is, What was their construction of risk? Unfortunately, the focus after the accident on intentional secret keeping by managers deflected attention from the way that obstacles to understanding are systematically structured into information exchange in organizations.

Structural secrecy at NASA had three sources that worked together to reinforce the normalization of deviance: (1) patterns of information affected the work group's definition of the situation; (2)

NASA organizational structure, processes, and transactions affected information conveyed from the work group to the top of the launch decision chain, creating a collective construction of the joints as an acceptable risk; and (3) the structure of safety regulation inhibited the ability of safety regulators to alter the scientific paradigm on which the belief in acceptable risk was based. Thus, the construction of risk persisted.

THE WORK GROUP: THE EMBEDDEDNESS OF SIGNALS

Patterns of information affected the work group's definition of the situation. Emerson notes the influence of "holistic effects" on decision making. Workers in organizations tend to process cases (or, at NASA, "technical issues") by taking into account the implications of other cases for the present one, and vice versa.[20] He suggests that the "stream of cases" (the overall caseload, the sequence of cases) affects decisions: workers tend to evaluate a problem in relation to some larger problem set. This tendency is especially keen in situations of uncertainty and when the task is to assess the seriousness and priority of the problem in order to treat or control it. We saw that numerous other problem sets were an important aspect of the SRB decision context at NASA. Many other technical problems existed, both on the SRBs and on other shuttle parts. The joint anomalies were occurring in an environment in which problems were normal and expected—and occurring—on every element of the Space Shuttle.

Perhaps more important for our story, work group decision making also took place in a context shaped by other decisions on the *same* problem. A sequence of decisions made about a given case, or issue, creates a decision stream.[21] At NASA, the production schedule required a risk assessment for each technical component prior to each launch, thus creating a decision sequence. When each anomaly occurred, the working engineers responded to it as if it were a signal of potential danger, as the Acceptable Risk Process required. A signal of potential danger is information that deviates from expectations, contradicting the existing worldview. By definition, it is a strong signal. The work group evaluated each anomalous incident on its own merits *and* within the decision stream for the SRB joint problems. The result was that information initially interpreted as a signal of potential danger was mediated for working engineers because each signal was embedded in patterns of information that mattered in the interpretation and production of technical knowledge. Patterns of information have the capacity to confirm worldview, to challenge and alter it, or, as was the case with the booster joints, to allow it to persist unchallenged and unchanged. Signals of potential danger lost their salience as a result of the risk assessment process. Accumulating incrementally, information about O-ring anomalies looked very different to the work group than it did to outsiders, who viewed it knowing the disastrous outcome. Signals were *mixed:* information indicating trouble was interspersed with and/or followed by information signaling that all was well. Signals were *weak:* information was informal and/or ambiguous, so that the threat to flight safety was not clear. And, as 1985 drew to a close, signals were repeated, becoming *routine* as the frequency and predictability of erosion institutionalized the construction of risk.

First Decisions: Signals, Salience, and Precedent

Emerson observed that the first decision establishes a precedent for subsequent decisions in the same stream. This observation held true for NASA.[22] The first challenge to the belief in redundancy was the 1977 discovery of joint rotation. The salience of this early signal of potential danger was reduced through the research process, which affirmed redundancy. Both Marshall and Thiokol working engineers concluded on the basis of the data that the design was an acceptable risk, despite disagreement about the details of joint dynamics. This belief was reinforced by industry experience: joint rotation was common, gaps were common, and the industry norm for innovative technology was to test in order to determine risk acceptability, establishing in-house standards for hazard assessment. The second challenge was erosion on STS-2. Erosion was a new and different problem—a different signal, not more of the same. For the working engineers, the salience of this first incident of erosion also was mitigated in post-flight analysis. They ran tests and did calculations to determine the safety margin. Having determined that the maximum amount of erosion possible still would not fail the joint, they again concluded that the design was an acceptable risk. An unpredicted but localized anomaly had occurred. They believed they understood what caused it, they fixed it, and they altered their ongoing research to incorporate new tests in order to improve joint performance.

These two decisions established precedents that would guide future decision making on both redundancy and erosion.[23] In both cases, certain engineering theories, practices, beliefs about cause and effect, and corrective procedures normalized the deviant performance of the joint, prompting the belief that the design was an acceptable risk. In both cases, the outcome was to proceed with launches using a design whose performance deviated from design expectations. The scientific paradigm begun in the developmental period provided the foundational actions and ideologies from which the work group culture would grow. Its persistence cannot be understood from these two decisions alone, however, but from what followed.

Mixed signals. Signals of potential danger were followed by signals that all was well, reinforcing the belief in acceptable risk. After the discovery of joint rotation in 1977 (a signal of potential danger), the engineers made changes to correct the problem. Subsequently, eight full-scale tests and the first shuttle flight, STS-1, showed no joint problems (signals that all was well). After the STS-2 erosion, further corrective actions were taken. Launches for the next three years showed no erosion, convincing the engineers that they had identified the cause of the problem and corrected it. When erosion began occurring in 1984, there were mixed signals of a different sort. Erosion appeared sometimes on an aft field joint, a forward field joint, a center field joint, or a nozzle joint. But most joints were unaffected. Therefore, the engineers concluded that the problem was not the design. In addition, they found that each incident had a different cause: localized putty defects, a brush hair or lint on a ring, cold temperature, or a ring pushed out of sealing position when segments were joined. Each time, the working engineers learned more about joint dynamics, relying and elaborating on their procedural precedents, developing more sophisticated tests, analytic models, and corrective measures. Each time an anomaly occurred, they believed they understood it and implemented what they considered to be an effective solution. Each time, their decision making was affirmed by postflight analysis that signaled that all was well.

Weak signals. The suspected correlation between cold temperature and erosion on the January 1985 launch of STS 51-C was the signal of potential danger viewed by posttragedy analysts as perhaps the most serious threat to flight safety that was ignored at NASA. But the working engineers defined cold temperature as a weak signal after the postflight analysis for STS 51-C. The information they had was based on the simultaneous occurrence of cold and the observation of blackened putty. They intuited a relationship between the cold temperature and the damage they saw. Many factors had contributed to erosion in the past. The engineers had no systematic data that showed a causal connection between the cold and the damage. The information they had was observational, and thus informal and ambiguous—a weak signal. Moreover, all formal, quantified data (a stronger signal than observational data at NASA) affirmed risk acceptability. The postflight analysis showed that the amount of erosion experienced was well within predictions for what the joint could safely tolerate. In addition to the weakness of the correlational evidence, two other factors contributed to the engineers' definition of the situation. They knew that when rubber gets cold, it gets hard, so that the rings would take longer to seal the joint, increasing the probability of erosion in cold temperature. However, they did not expect a repetition of the three coldest days in Florida history, so the salience of cold temperature as a signal was further weakened by the improbability of a recurrence. (Significantly for the *Challenger* teleconference, this perception stemmed any aggressive attempt to conduct tests and accumulate data on cold temperature and joint performance.) Finally, the decision stream played a role: the next flight sustained the greatest erosion in program history. With a calculated O-ring temperature of 75°F, STS 51-B contradicted the posited relationship between damage and cold temperature, further weakening its salience as a possible causal factor.

Routine signals. The frequent event, even when acknowledged to be inherently serious, loses some of its seriousness as similar events occur in sequence and methods of assessing and responding to them stabilize.[24] Knowledge and expectations about performance that are borne out time after time encourages decision makers to classify each similar event as a normal or "typical" case of X (in the SRB work group, a typical case of erosion).[25] The more frequent the similar events, the more routine the individual case. Seriousness itself becomes routine, a taken-for-granted characteristic of the case, reducing the experience of seriousness for the worker.[26]

After the tragedy, analysts viewed the increase in erosion during 1984 and 1985 as itself a signal of danger. At the time, however, working engineers defined most of these incidents as signals that the joint was performing as predicted. In February 1984, when they increased the leak check pressure to assure redundancy on the field joint, they did so knowing that the probability of erosion would increase as a consequence. When erosion then occurred, the working engineers were not surprised; further, the amount of erosion conformed to engineering predictions. They *were* surprised when, on the 1985 launches of STS 51-C and 51-B, erosion deviated from expected performance. But again they found idiosyncratic causes: cold weather on STS 51-C, a nozzle joint ring improperly positioned on STS 51-B. In April 1985, they increased the leak check pressure to correct the nozzle joint problem. Subsequently, nozzle joint erosion, too, was expected.

On each of the remaining flights in 1985, erosion had become a routine signal, occurring in a predictable manner. This is not to say that the work group ignored the problems or stopped worrying about them. Concern and activity to resolve the problem escalated during 1985. Nonetheless, repeated flight performance that conformed to engineering predictions assured the SRB work group that the design was an acceptable flight risk, as they increased their efforts to resolve the problems.

PRECEDENT, FEEDBACK, COMMITMENT, AND CULTURAL PERSISTENCE

The first decision in a decision stream does not survive as precedent unless that original decision is validated by its outcome and by the outcomes of subsequent decisions. So the decision stream feeds back into that first decision, determining its continuation. For the working engineers, *two* decision streams contributed to the persistence of the work group's cultural construction of risk, transforming that first collection of practices and beliefs into a paradigmatic precedent: (1) FRRs for individual launches and (2) cumulative decision making for all launches.

Individual FRR. The manifest function of FRR was to reduce ambiguity about residual risk. Open technical issues were closed as analysis, testing, and corrections necessary to qualify each component for flight were completed. But, in addition, the FRR process had ritualistic, ceremonial properties with latent consequences that also reduced ambiguity, affecting the perceptions of risk held by work group members. Negotiating in FRR, creating the documents, making the engineering analysis and conclusions public, and having them accepted in an adversarial review system contributed to the persistence of the cultural construction of risk. Arguments were scrutinized and challenged at each level. If controversial or incomplete, the working engineers did additional tests and analysis. Weak arguments were either strengthened or dropped, research continuing until the engineering analysis was itself acceptable. The engineering theories, methods, and corrective procedures underlying the general conclusions and recommendations became more coherent in the process, as the hypothetical character and interpretive flexibility of the related technical issues were converted into organizational facts that were binding.[27]

Any decision process tends to generate commitment to the group stance.[28] As these issues worked their way up the hierarchical levels, the managers and engineers in the work group became committed to the line of action they had chosen. The solidification of argument and the dropping of ambiguity that goes into negotiation and documentation affects not only the audience, but the creator.[29] Matza observes that "the very act of writing or reporting commits the author to a rendition of the world."[30] The documents themselves, apart from their contents, assert consensus through the matter-of-fact tone of the formal mode of discourse and affirms it both to the audience and to the documents' creators.[31] An additional factor that binds people to their actions is "going public."[32] When a person participates in and is identified publicly with a decision, that person will resolve inconsistencies to produce attitudes consistent with that choice.[33]

As the engineering logic behind the work group's decision to accept risk was approved at each succeeding FRR level in the hierarchy, the corrective actions, the ongoing research, and the engineering hypotheses, tests, and calculations that constituted their scientific paradigm became more public, receiving additional affirmation, becoming more objectively "real."[34] Challenged and surviving in what was defined by all concerned as a tough "getting raked over the coals" process, the engineering arguments gained in legitimacy. Legitimacy was formally conferred by the ceremonial signing of the Certification of Flight Readiness at each level.

Cumulative FRRs. The second important source of affirming feedback was the decision stream that included all launches over time. By design, launch decisions were interdependent. Because the shuttle parts were designed to be reused, their physical condition and performance from one launch was critical input in the FRR for the next. The effect was a continuous feedback loop, joining one launch decision and its outcome to the next. For the SRB joints, each succeeding postflight analysis fed back into the engineering analytic process, affirming the technical logic behind the belief in risk acceptability. Commitment to one behavior has implications for other behaviors.[35] For each launch, the work group publicly committed itself to its analytic methods, conclusions, and recommendations about risk acceptability. Moreover, the engineers committed to a specific line of action that flowed from their technical analysis: accept risk and fly.

This sequential, longitudinal, rolling commitment to multiple audiences up the FRR hierarchy for a single launch and the rolling affirmation from postflight technical assessment from launch to launch contributed to the institutionalization of the work groups' cultural construction of risk. Commitment to a line of action builds over time in decision sequences. For the work group to reverse itself at any point prior to the 1986 *Challenger* launch teleconference would have required a rejection of the scientific paradigm advanced in FRR for all previous launches. To do so was an occupational risk, jeopardizing the engineers' professional integrity, causing them to lose face in all those forums. Either they were wrong then or wrong now. Larry Wear said:

> Once you've accepted an anomaly or something less than perfect, you know, you've given up your virginity. You can't go back. You're at the point that it's very hard to draw the line. You know, next time they say it's the same problem, it's just eroded 5 mils. more. Once you accepted it, where do you draw the line? . . . Once you've done it, it's very difficult to go back now and get very hard-nosed and say I'm not going to accept that. I can imagine what the Program [Level II] would have said if we—we being me, Larry Mulloy, Thiokol, or anyone in between—if we had stood up one day and said, "We're not going to fly any more because we are seeing erosion on our seal." They would have all looked back in the book and said, "Wait a minute. You've seen that before and you told us that was OK. And you saw it before that, and you said that was OK. Now, what are you? Are you a wimp? Are you a liar? What are you?"[36]

In order to delay or halt the launch production schedule, as Marshall's SRB Project Chief Engineer Jim Smith pointed out, "you have to be able to show you've got a technical issue that is unsafe to fly."[37] The problem had to emerge from the Acceptable Risk Process with evidence supporting a no-fly recommendation. But the working engineers did not have such evidence. They had signals that lost their salience through engineering analysis, tests, and corrective actions; they had signals that were mixed, weak, and routine. Indeed, they had the opposite: a three-factor technical rationale justifying acceptable risk. And it was this that they passed up the hierarchy.

NASA organizational structure, processes, and transactions were the second source of structural secrecy. As information was passed up the launch decision chain, these factors affected the definition of the situation held by others, constraining them from altering the scientific paradigm. As a result, the work group's definition of the situation was transformed into a collective organizational construction of risk.

ORGANIZATIONAL STRUCTURE, TRANSACTIONS, AND PROCESSES

Secrecy is built into the very structure of organizations.[38] As organizations grow large, actions are, for the most part, not observable. The division of labor between subunits, hierarchy, and geographic dispersion segregate knowledge about tasks and goals. Distance—both physical and social—interferes with the efforts of those at the top to "know" the behavior of others in the organization—and vice versa. Specialized knowledge further inhibits knowing. People in one department or division lack the expertise to understand the work in another or, for that matter, the work of other specialists in their own unit. The language associated with a different task, even in the same organization, can be wondrously opaque. Changing technology also interferes with knowing, for assessing information requires keeping pace with these changes—a difficult prospect when it takes time away from one's primary job responsibilities. To circumvent these obstacles, organizations take steps to increase the flow of information—and, hypothetically, knowledge. They make rules designating when, what, and to whom information is to be conveyed. Information exchange grows more formal, complex, and impersonal, perhaps overwhelmingly so, as organizations institute computer transaction systems to record, monitor, process, and transmit information from one part of the organization to another.

Ironically, efforts to communicate more can result in knowing less. Rules that guarantee wide distribution of information can increase the amount to the point that a lot is not read. Transaction systems, designed to promote knowing, can also be a source of structural secrecy, concealing even as they reveal. Information is relayed on standardized forms, in standardized formats, then coded into categories, based on special organizational language. Fine distinctions tend to disappear in the interest of computerized processing, storing, and retrieving masses of information. If a category is not established, that information is not readily available. The whole enterprise is undermined by the ease of producing information with computer technology. Masses of information, even though in categories, is not useful, except in a symbolic sense. The ability to produce, accumulate, store, and exchange information using sophisticated transaction systems has become a symbol of legitimacy for many organizations.[39] Obfuscation parades as clarity: they produce too much, obscuring rather than enlightening.

Systematic censorship is a natural response to this abundance. By systematic censorship, I mean that information is reduced in patterned ways that are intrinsic to organizations, even when every effort is made to send that information forward. Information is lost, sometimes to deleterious effect. Three mechanisms systematically censor the information available to top decision makers. First, *official organizational practices* can be designed intentionally to restrict the amount and form of information going to a particular position to reduce information overload and assure that certain information receives adequate attention. Second, *specialization* remains an obstacle to interpretation. One's role-related specialized knowledge and experience can interfere with understanding matters outside one's own job description. Of the information received, some is absorbed, but not all, leaving a residue of uncertainty. Uncertainty results in a third form of systematic censorship. When unable to discriminate, top decision makers tend to *rely on signals*.[40] Relying on signals is a shortcut, a way of isolating bits of "telling" information from what is available. Decision makers cannot know each individual case thoroughly because confirming or disconfirming each and every fact takes time and energy, often prohibitively costly. Instead, to assess a situation they rely on familiar cues and signals—readily observable characteristics—so that they can turn their energies to other matters.[41] This strategy helps them decide what to do next by simplifying the manner in which the present situation is interpreted. But this reliance on signals, though efficient, can also interfere with knowing, for some information, though transmitted, is again reduced and thus remains in essence secret.

The Space Shuttle Program merged private enterprise with NASA's own massive decentralized structure. Together, these various organizations combined in a loosely coupled system characterized by the unique technology, expertise, language, and geographic location of the separate parts. In an effort to circumvent the structural secrecy built into its sprawling unwieldiness, the system was loaded with mechanisms for monitoring, recording, and transmitting information necessary to assure safety. Redundancy was built in via computer systems for tracking and reporting problems. However, while they promoted exchange of information, these transaction systems were also a source of obfuscation. The production cycle, the unruly technology, and the consequent learning by experience created a constant flow of new information about the technology. Between the exchange of information about the technology and the procedural bureaupathology, the overall effect was a "blizzard of paperwork":[42] multiple transactions between units, on a daily basis, in highly technical language. The House Committee on Science and Technology noted that NASA was a place where "large amounts of information are disseminated on a routine basis, often with little or no indication of its importance to all of the recipients."[43] The information control revolution, which James Beniger so ably portrayed at a societal level, was acted out daily at the space agency.[44]

Language was fundamental to structural secrecy at NASA. Talk about risk, in NASA culture, was by nature technical, impersonal, and bureaucratic—full of what to the uninitiated are meaningless acronyms, engineering terms, and NASA procedural references ("Action Item," "FMEA," "CIL," "waivers"). Routinely used and taken-for-granted, the language did not lend itself to sending signals of potential danger. For example, the fatal outcome of a mission loss in event of SRB joint failure was described on engineering risk assessment documents impersonally and tersely, in standardized format and language. "Failure effect summary: Actual loss—Loss of mission, vehicle, and crew due to metal erosion, burnthrough, and probable case burst resulting in fire and deflagration."[45]

Although after the *Challenger* tragedy, this entry appeared to outsiders to be a clear and obvious signal of danger, in the NASA culture this statement was standard in every CIL entry for each C 1 and C 1R item. Words like "anomaly," "hazard," "discrepancy," "acceptable risk," and "Rationale for Retention" were part of the common vocabulary, routine in meaning because they reflected routine aspects of daily engineering work. Even the word "catastrophe," which after the tragedy seemed to be a signal of danger when it appeared in memos and documents about the SRB joints, did not convey urgency because it was a term routinely associated with C 1 and C 1R failures. Everyone knew these failures would be "catastrophic." The word designated the consequences of a failure in which crew, vehicle, and mission were lost. A failure could result in mission loss only, but to qualify as a catastrophe, all three losses had to occur. "Catastrophe" was stripped of anything but technical significance in official language. This is not to assert that the people working on the shuttle were no longer sensitive to risk, danger, and the traumatic consequences of failure, but that the symbols available in the official language were used so frequently that they were not a useful means of conveying risk and danger. In response to this massive technocratic failure, the indigenous language of engineers showed the development of special risk talk that enabled them to make the distinctions that the official language did not. They distinguished between a "safety hazard" and a "real safety hazard." Instead of using "catastrophe," the deeper significance of an action an engineer believed might result in loss of crew, vehicle, and mission was reflected in the vernacular by euphemisms that, as euphemisms do, emphasized the perceived seriousness of the consequences by understating them. The engineers signaled the inadvisability of an action by predicting that as a consequence they would all have "a long day," or "a bad day." Grappling with this problem formally, the space agency designed a system of categories and symbols to identify the most risky shuttle components. Intended to single out serious problems for extra review, these categories and designations accomplished that goal. Yet so many problems fell into each category that the designations themselves were ineffective as indicators of serious problems. Take, for example, Launch Constraints. All open problems were considered Launch Constraints, and many open problems remained even as FRR began. In addition, many formal Launch Constraints were imposed to assure that particular technical issues received extra attention in review. The same was true of the C 1 designation. Recall that the O-ring problem had, in 1982, been designated a C 1 item, but that a similar designation was assigned to 748 parts of the shuttle, including 114 on the SRBs, at the time of the *Challenger* launch.[46]

What sort of signals effectively communicated potential danger from SRB work group engineers to top administrators in such a technocratic system?

Verbal Complaints and Memos as Weak Signals

After the *Challenger* tragedy, the verbal complaints and memos sent by working engineers at Marshall and Thiokol appeared to be signals of potential danger that managers ignored. However, as Starbuck and Milliken pointed out, retrospective analysis of bad organizational outcomes tends to focus attention selectively on the road not taken that might have altered the outcome.[47] For insiders at the time they occurred, these verbal complaints and memos were not signals of potential danger. Their ability to challenge the definition of the O-ring problem was undermined by their informal nature, the social context, and the fact that in all collective exchanges the engineers who originated them consistently took the position that the joints were an acceptable risk.

Apparently, toward the end of 1985, some Thiokol engineers expressed the view that shuttle flight should stop while the problems were fixed. Thiokol engineer Bob Ebeling, Manager of SRM Ignition System and Final Assembly, was the only person who testified to holding that view, however. He said that in a December 1985 O-ring task force meeting, he voiced his opinion that "we shouldn't ship any more rocket motors until we got it fixed." But this was a weak signal because, as he remarked, he did not voice this concern "to the right people."[48] Thiokol's Al McDonald, who worked very closely with Ebeling and the task force, testified that he was unaware that any of his engineers felt that the program should be halted while the joint design was fixed.[49] The Thiokol seal task force was in close touch with the seal task force at Marshall, so the working engineers were talking to their equals. Also, Thiokol's group made several formal progress reports to Marshall. At no time did they suggest delaying launch, either in task force meetings or in formal exchanges with Marshall, although engineers themselves testified that complaining directly to managers was acceptable practice and something they routinely did.[50] Keith Coates testified: "I don't know of any hesitancy on the part of engineers in the laboratories to pick up the phone and call their counterparts and even sometimes your counterparts' bosses if they have particular concerns about something, or for engineers to call even the Chief Engineer's Office and even the personnel in the Projects Office to express concerns, or vice versa, for the Thiokol personnel to call us."[51]

Larry Wear, who, at Level IV, was in touch with Thiokol engineers many times a week in informal and formal situations, said:

I was convinced entirely that the Thiokol engineering organization had the same view that I did. . . . There was no discussion and no revelation on anybody's part that what we're doing here is flying with something that is in an absolutely unsafe condition and you ought to stand down until you get it fixed. There was never any statement made by anyone to that effect, including the people I have heard testify otherwise since them. If they believed that, they never told me. . . . The conclusion from all those analyses had been that we know how much heat is possible there. We have seen erosion. You analyze that. For right or wrong, we have applied statistical approaches to it, as to how much erosion you think that could be. And then we run tests on the side to say all right, let's assume it is worse than that, how bad can it be and the seal fails? The conclusion is that it's three times what you've seen. None of these [Thiokol] people that I talked to regularly about it have ever said that data lies, that data is wrong. . . . I was in a meeting at Thiokol last fall with the Task Team. We reviewed the progress they were making and the things they had going. No one there said you guys from NASA ought to be aware that we have a split vote out here. Some of us think what you are doing is irrational and irresponsible. Nobody said that.[52]

But managers at both Marshall and Thiokol did get memos from engineers expressing concern about the O-ring problems.[53] In most large organizations, memos as a form of communication are a weak signal because they are informal, usually between individuals, and subject to discretionary decision making that does not guarantee that they will bring action or that they will reach the upper echelon of the organization. Memos may have been even less effective at NASA than in many organizations. At NASA, formal procedures were the way to get things done. Many people acknowledged the weakness of memos as a signal in such a system, stating that "memos get you nowhere." Further, the content and social context of these particular memos made them weak signals to insiders at the time. Take, for example, the much-publicized 1978–79 memos of Marshall's Leon Ray. As an S&E engineer, Ray's job was to write memos and reports expressing dissatisfaction with contractors. Also, S&E memos were always written in language that was "hard" and "blunt."[54] Keith Coates, former S&E engineer, commented: "When memos are submitted, they are usually not to pat someone on the back. You could develop a very negative outlook on life from the [S&E] laboratory because our job is to look at contractor designs and find the things wrong. And you squawk the things that are wrong."[55] Finally, Ray subsequently participated in engineering work that led him to concur with Thiokol engineering analysis and conclusions to accept risk and fly, and he reported this in FRRs.[56] Seen within the context of Leon Ray's position in S&E and his agreement with work group decisions to accept risk and fly, the memos were weak signals, less salient when they were written than to outsiders after the disaster. Reflecting on the impact of both his verbal complaints and his memos, Ray himself said of management afterward, "They thought I was 'crying wolf.'"[57] Wiley Bunn observed that contemporaneous tests also neutralized their significance:

> Sitting here today, after 51-L, you can't come to any other conclusion than the harbingers of doom on the joint [were correct], and there were one or two or three people that felt very strongly about that joint design. Ben Powers had problems with it. Leon Ray had problems with the thing. Somebody being concerned about it, in my business, draws some concern out of me, too, and you watch with bated breath at the tests, because even though he's a voice in the wilderness, sometimes those people are right. But we got through the test program in pretty good shape. . . . After we ran through those and we ran demonstration tests and then the Qual [qualification] tests, and we got through the first few flights, the

joint rotation ceased to be a primary concern to me because the thing obviously worked and it was something about it that Leon Ray or Ben Powers didn't fully understand.[58]

Thiokol engineers sent memos stating concerns; none suggested flight be stopped.[59] Arnie Thompson, Thiokol Supervisor, Rocket Motor Cases, and Roger Boisjoly's superior, wrote a memo immediately after the August 19, 1985, briefing stating that "the O-ring seal problem has lately become acute." He did not recommend flight be halted, instead recommending short-term measures be taken to reduce flight risks.[60] Roger Boisjoly's July 31, 1985, memo to Thiokol Vice President Robert Lund warned about the possibility of a "catastrophe" if a task team with personnel assigned full-time was not put in place. At the time he wrote the memo, however, Boisjoly did not believe flight should be stopped, testifying that he wrote his memo in order to "turn the flame up a little bit" to get the job done more quickly.[61] In addition, the authors of these memos restricted distribution, lessening potential impact by narrowing the audience. Ray did not send copies of his memos to Thiokol; Boisjoly limited circulation to Thiokol personnel by typing "Company Private" on his memo;[62] Thompson limited circulation to a few engineers at Thiokol. No memos went to Marshall that suggested that Thiokol engineers believed it was not safe to fly.

Unsurprising in an organization preoccupied with bureaucratic procedures, a formal mechanism was available at both Marshall and Thiokol that was a stronger signal than memos. Any engineer could submit an ECR. It assured that engineer requests for changes were carried up the hierarchy and received a mandatory review and written response.[63] Leon Ray confirmed that ECRs were attention-getting signals that any engineer could send, describing an ECR as a "Pearl Harbor" letter distributed to all the powerful people that says, "I don't like the design and here's why."[64] Engineers at Marshall and Thiokol said they did not submit ECRs because they did not have the data necessary to do so. Ray said he did not write an ECR because engineers only submit them when the threat to safety is sufficiently "cut and dried" that there is consensus among the working engineers. It was convincing data that generated consensus. Explaining his decision, Ray made the distinction between a "safety hazard" and a "real safety hazard":

> Some of us thought it was a safety hazard. We had lots of opportunity to voice our opinion. We could never get consensus on it. Normally you get consensus first. If you don't have it, it [an ECR] won't force anybody to do anything. The ECR will be disapproved. Unless

I thought it was a real safety hazard, I wouldn't submit an ECR. It's a last resort. It has to be a real safety hazard. I would never blow smoke to get results. I won't give a false answer, or exaggerate. You lose credibility in a hurry. We thought it was OK to fly until the cold weather came up again, then we didn't want to fly it.[65]

Although the Thiokol engineers' memos reflected their frustration with the lack of resources allocated to the task force and their desire to get the O-ring problems resolved as quickly as possible, there is no indication that, had those resources been made available more quickly, the task force would have recommended delaying launch until the problem was resolved. Despite the verbal complaints and memos of 1985, the task force's own engineering analysis convinced them it was safe to fly. Engineers were empowered or disempowered to take formal action by their data. The original technical culture mandated that engineering recommendations be backed by "solid technical arguments." The subjective, the intuitive, the concern not affirmed by data analysis were not grounds for formal action at Marshall. Although engineers were concerned, no one took formal action because the data continued to indicate that the joint was an acceptable risk. The social construction of risk determined the procedural response. The data did not indicate the joints were a "real safety hazard." Therefore, every display of discontent by individual engineers was by informal means—weak signals in the NASA system.

The most important conduit of information about risk was FRR. It was the engineering analysis and conclusions presented in this formal setting, not verbal complaints and memos, that were the salient signals for launch decision making. NASA's top technical decision makers at Level II and Level I depended on the work group for technical analysis and launch recommendations. In every FRR prior to the *Challenger* launch, the message conveyed from the work group through the hierarchy was to accept risk and fly.

FLIGHT READINESS REVIEW AND THE SYSTEMATIC CENSORSHIP OF INFORMATION

FRR was created specifically *to circumvent structural secrecy* at NASA by mandating that engineering risk assessments go through the hierarchy to top NASA administrations. It was designed to pull together all parts of the organization. Despite this goal, FRR actually mirrored the problems of information exchange in the parent organization, duplicating the structural secrecy inherent in it, the consequences of which were played out in the microcosm. FRR was hierarchical; the participants reflected the specialization of the position they occupied. Members of the review boards were subjected to a "blizzard of paperwork" as they prepared and received FRR packets of technical analyses prior to the scheduled meetings. The language was scientific, technical, and coded into organizational categories. NASA top technical decision makers at Levels II and I depended on work groups for data analysis and conclusions. As the construction of risk developed by working engineers was relayed up the launch decision chain, systematic censorship processes common to all organizations—official organizational practices, specialization, and reliance on signals—affected understanding at the top levels. The belief in acceptable risk was perpetuated throughout the hierarchy, resulting in a collective organizational construction of risk.

Official Organizational Practices

Typically, the packet of charts and materials for the Level IV FRR at Thiokol was about half an inch thick. By the time of the Level I FRR, the original packet was reduced to 10 to 15 pages. Three official organizational practices systematically censored information as it went from bottom to top in FRR, shrinking the amount of information presented.

Problem resolution. By the time of the Level II and I FRRs, there were fewer pages in the FRR packet because many problems had been resolved.[66] Thiokol's Roger Boisjoly remarked: "Documents got reduced because of a boiling-down process. We were not hiding things. That's not so. Things got closed out, resolved, at lower levels. They were not brought up, or maybe just a quick overview—sometimes just a single 'bullet'—at the top levels."[67]

At NASA, resolution of problems for each shuttle element was the responsibility of all Level III Project Managers. Although the Presidential Commission criticized the "propensity of management at Marshall to contain potentially serious problems and to attempt to resolve them internally rather than communicate them forward,"[68] that was the job of Level III. Levels II and I were for certification of flight readiness, not problem resolution. When FRR began about two weeks prior to launch, almost all work was completed. Because (1) the major shuttle elements were reusable and had to be refurbished and requalified each time, (2) the entire technical system was "unruly," and (3) because of the operational launch schedule, any changes/corrections necessary from the last flight were, by definition, last minute changes. Thus, problems were still being resolved during FRR. But by Levels II and I, most of problems had been resolved—hence, the smaller packet.

Reporting practices. Information was also reduced by rules put in place to avoid information overload at the top and to assure that serious problems got attention at Levels II and I. The rules specified not only what kinds of problems were reported and under what circumstances but also the amount of information presented. "All problems, open items, and [launch] constraints *remaining to be resolved* before the mission" were reported. For "significant resolved problems" (i.e., lifted Launch Constraints), the presentation included a brief status summary with some supporting detail and a readiness assessment (in NASA language, they were "statused").[69] NASA's Delta review concept automatically excluded many technical issues from discussion in order to focus attention on others. This was NASA's policy of "management by exception" that governed Level II and I reporting: any change or data from the previous flight that fell outside the expected performance for the hardware had to be reported.[70] If data from the previous flight was within expectations, the technical rationale on which risk acceptability was based remained the same and thus was not repeated.

As Marshall's Mulloy put it, "So the rule was, if you haven't seen anything new in O-ring erosion, you don't talk about O-ring erosion at Level I."[71] Boisjoly confirmed the practice. When asked why temperature concerns from the January 1985 launch were presented in the 51-E FRR but not in subsequent reviews, he described the Delta review practice: "Because the way they work it is once they have addressed an issue, they don't have to readdress that issue in subsequent FRRs. And that particular one was addressed specifically [in the 51-E FRR] because it had occurred in the preceding flight on 51-C, and therefore it automatically becomes a problem, an open problem, and an issue in the next Flight Readiness Review [51-E] because the way the FRRs are structured is that all previous flight inspection data has to be presented and closed out prior to flying the next vehicle."[72]

Obviously, not every anomaly could be reviewed by top administrators as all parts of the system came together in the final stages of the launch decision process. Although the advantage of "management by exception" was that it narrowed the scope of responsibility at Levels II and I so that the most serious problems received the attention they deserved, the disadvantage was that many anomalies were not regularly reviewed by top administrators.

Time constraints. The final official practice that reduced information at the top was a time limit imposed for the Level I FRR. It was always scheduled for (but not restricted to) four hours, in contrast to the Level IV FRR, which could last a full day or more.[73] The whole shuttle system was coming together. The agenda included nine flight-readiness reporting areas, requiring 12 presentations.[74] In-depth briefings on the remaining open items and "significant resolved problems" by 12 presenters would have required a "three-day marathon."[75] This four-hour limit was imposed for the express purpose of reducing presentations to succinct summaries that condensed masses of information. The time limit prevented competing Project Managers from trying to convince top administrators of their project's superiority by presenting every detail of a problem and its resolution. Instead, it created a competition to communicate the important information as quickly as possible (thus shedding additional light on the pressure by Marshall's Lucas to make sure presentations at Levels II and I were "concise and clear" and up to "Marshall standards"). Presenters conveyed the essentials according to a prescribed format: a statement of the problem, its cause, its problem, and whether it was a constraint to flight.[76] The videotape of Mulloy's full briefing in April 1985 at the Level I 51-E FRR illustrates how these points were covered.

The brevity of presentations at the top was accepted practice for two reasons: (1) Behind these summaries were pages and pages of information representing thousands of hours of engineering work that had been subjected to criticism and review at lower FRR levels. (2) Level I administrators had structured opportunities for a deeper look. Additional information in response to Action Items assigned by the Level I Board was discussed in follow-up reviews (Delta reviews; see chap. 3) and in the L-1 review.[77] L-1 review was the final, formal certification of flight readiness, by which time all work was to be completed. If any problems were still unresolved, the launch was slipped; once these problems were resolved, the L-1 review was rescheduled, the launch date was reset, and FRR certification was completed.

Specialization

Specialization also was a source of systematic censorship. It interfered with the ability of top administrators to interpret the information that they did receive; thus further information was "lost." Although these top Level II and Level I administrators were trained engineers, they had broad responsibilities, more administrative than technical. For example, Jesse Moore, Associate Administrator for Space Flight at Level I, had management responsibilities for Space-lab, expendable launch vehicles, and other programs in addition to the Space Shuttle Program. He was responsible for the manpower, the facilities, and the overall resources necessary to keep the space centers operating as institutions.[78] Astronauts Robert Crippen and John W. Young, who participated at Level I as Director of Flight Crew Operations and Chief of the Astronaut Office, respectively, also were responsible for training and evaluating 91 astronauts as well as myriad tasks related to design, development, and operation of space vehicles and stations.[79] The position and role of all Level II and I participants kept them away from hands-on engineering work, limiting their ability to understand, challenge, or reject the technical information that they received from the work groups. As Marshall S&E engineer Ben Powers remarked, the people at Levels II and I "were not sitting in the seat of understanding."[80]

FRR was designed to take this obstacle to understanding into account. The main responsibility for technical work and assessment was delegated to work groups, with the expectation that Levels II and I would oversee and certify flight readiness, not resolve problems. Interviews indicate that Level II and I reviews were adversarial and rigorous and that questions were asked that required Action Items and further research. But the deep, tense probing of the Marshall Center Board was absent here. These top officials may have asked probing questions, but their questions were of a different sort than those asked at other lower levels. Their questions were based on the principles of their engineering training: general rocketry principles about compression sealing, vacuums, combustion, thrust, safety loads and safety factors, heat effects on various materials, and so forth. They used these general principles to examine *the engineering logic* of the technical analysis, test results, performance standards, conclusions, and recommendations. But their understanding was limited because they were distant, in physical location, daily experience, and responsibility, from the "core set"—the Level III and IV managers and engineers in the work group who knew every nuance, every hiccup of joint performance.

Notice, for example, the number of times Associate Administrator Moore said "I guess" and "I think" when he discussed the FRR following the April 1985 mission with a Commission investigator:

> And then we had a Flight Readiness Review, I guess, in July, getting ready for a mid-July, a late July flight, and the action had come back from the project office. I guess the Level III had reported to the

Level II FRR, and then they reported up to me that—they reported the two erosions on the primary [O-ring] and some 10 or 12 percent erosion on the secondary on that flight in April, and the corrective actions, I guess, that had been put in place was to increase the test pressure, I think, from 50 psi to 200 psi or 100 psi—I guess it was 200 psi is the number—and they felt that they had run a bunch of laboratory tests and analyses that showed that by increasing the pressure up to 200 psi, this would minimize or eliminate erosion, and that there would be a fairly good degree of safety factor margin on the erosion as a result of increasing this pressure and ensuring that the secondary seal had been seated.[81]

He got it right, but this is not the statement of a person intimately familiar with the booster joint engineering. This interpretive handicap at top levels was obvious when I reviewed the videotapes of the televised hearings of the Presidential Commission; although nearly everyone who testified brought notes, the managers and engineers at Levels III and IV in the SRB work group relied on those notes only for dates, not for explaining the dynamics of the SRB joint behavior and the engineering rationale behind work group decision making. Level II and I personnel relied on their notes, and even with them (and, presumably, study and briefings before they testified), their uncertainty about the technical analysis was visible.

Reliance on Signals

Specialization had Level II and I administrators making judgments under conditions of uncertainty, which commonly results in reliance on signals as a means of making assessments. Despite the reduction of information in each packet, packets came from nine reporting areas. Time to read them, much less pursue independent inquiry, was limited by other job responsibilities. Some Level II and I FRR participants apparently did not read the materials ahead of time, relying totally on oral presentations. Others skimmed the materials, looking for key charts, but concentrating on the engineering recommendations and conclusions. Because the costs of independent inquiry were high, Levels II and I depended on the people in work groups and the signals they received in the FRR sessions themselves.

Weak signals. Working engineers' concerns about the effect of cold temperature on O-ring performance were carried forward as a weak signal to the Marshall Center Board and omitted at Level II and I FRRs. Cold temperature received minimal treatment at the August

19 NASA Headquarters briefing as well.[82] The engineers did not have the "fully documented, verifiable set of data" that constituted appropriate "engineering-supported" evidence admissible in FRR presentations.[83] Observational data, backed by an intuitive argument, constituted a weak signal in NASA's quantification-oriented, science-based system. Arguments that could not be supported by data did not refute arguments that could be. Those that did not meet the engineering standards of the original technical culture would not pass the adversarial challenges of the FRR process. If a hypothesis was not supported by "solid" evidence at Levels IV and III, it either was not carried forward, or was carried forward in ways that strictly reflected the available data.[84] In FRR, the cultural emphasis on scientific positivism and quantitative arguments systematically excluded nascent engineering theories that challenged institutionalized ones that met these standards, creating a predisposition to continue launching.

Mixed signals. The documents and arguments presented in FRR always contained mixed signals. Since the purpose of the flight-readiness process was to identify problems *and* resolve them, FRR presentations contained both signals of potential danger (any anomalies or changes since the previous flight) and signals that all was well (data analysis and corrective action, followed by a recommendation to accept risk and proceed with flight). Thus, each time the SRB work group made a presentation, they were simultaneously announcing a problem and relaying the message that it was all right to fly. Each time, the former signal was rendered insignificant by the latter.

The famous NASA Headquarters briefing of August 19, 1985, occurred outside of the FRR process. It contained the most extensive information presented from the SRB work group to top administrators prior to the *Challenger* tragedy. About this meeting the Presidential Commission concluded, "The O-ring erosion history presented to Level I at NASA Headquarters in August 1985 was sufficiently detailed to require corrective action prior to the next flight."[85] But it also contained mixed signals that shaped the interpretations of top officials. Although the presentation overall conveyed the signal that the O-ring problem had a long history and was a major concern, once again the people at the top were advised that *as long as certain conditions were met, it was safe to fly* (see chap. 5, fig. 10). The House Committee, acknowledging the effect of these mixed signals on top administrators, contradicted the Presidential Commission:

In hindsight, the August 19th briefing clearly identified a serious problem. Perhaps the Thiokol engineers understood the seriousness of the problem; however, Thiokol's own summary and recommendation at the conclusion of the August 19th briefing stated:

"Analysis of existing data indicates that it is safe to continue flying existing design as long as all joints are leak checked with a 200 psig [pressure converted to "absolute" units] stabilization pressure, are free of contamination in the seal areas and meet O-ring squeeze requirements."

This conclusion was accepted by all who heard the briefing, and this was the information that was transmitted throughout NASA. The evidence does not support a conclusion that the top decision makers would have arrived at a different conclusion from the managers at Marshall.[86]

Routine Signals. The FRR engineering analysis, conclusions, and recommendations about each upcoming launch were the salient signals for Levels I and II. Consistently, the SRB work group conveyed their three-factor technical rationale and recommendation to accept risk and fly to top administrators. In judgments about seriousness, the more frequently a particular kind of case occurs, the more familiar and routinized it becomes to those who deal with it.[87] For those at the top as for those at the bottom, erosion was a regular, expected, and taken-for-granted aspect of joint performance at the end of 1985.

THE STRUCTURE OF SAFETY REGULATION

One way to overturn an entrenched scientific paradigm is with contradictory information that is an attention-getting signal, too strong to explain away, refute, or deny. Since the SRB joint deviated from design expectations from the beginning, in theory safety regulators had years to intervene, sending signals that contradicted the paradigmatic worldview of the SRB work group and top administrators, thereby averting the tragedy. But this did not happen. The Presidential Commission found that NASA's safety system failed at monitoring shuttle operations to such an extent that the report referred to it as the "Silent Safety Program."[88] The ability to regulate safety had been undercut by a third source of structural secrecy at NASA. The structure of regulatory relations created obstacles to social control, limiting information and knowledge about the O-ring problems. It inhibited the ability of safety regulators to alter the scientific paradigm on which the belief in acceptable risk was based. Regulatory effectiveness was blocked by *autonomy,* or the fact that regulators and the organizations they regulate exist as separate, independent organizations, and *interdependence*, or the fact that regulators and regulatees are linked so that outcomes for each are, in part, determined by the activities of the other.[89] Autonomy and interdependence were not peculiar to NASA, but are fundamental to all regulatory relationships. Elsewhere, I have written in detail about their general implications and, in particular, how they affected safety regulation at NASA.[90] Here I condense those earlier discussions in order to extend the argument in a new direction: the consequences of regulatory relations for decision making at NASA.

AUTONOMY, INTERDEPENDENCE, AND NASA's SAFETY SYSTEM

Regulatory organizations can be physically located external or internal to the organizations they regulate. Each strategy has unique advantages, but each also has disadvantages that routinely undermine regulatory effectiveness.[91] External regulatory bodies are autonomous entities, separate physically from those they regulate. The advantage of autonomy is objectivity, a fresh viewpoint, and an adversarial relationship that guides the discovery, monitoring, and investigation necessary for social control. The disadvantage is that external agents are not consistently present in the regulated organization, have limited access to information as a result, and often have difficulty understanding specialized language and keeping up with technological developments. Thus, external regulators frequently become dependent on the regulated organization for information and its interpretation. As a result, regulators' definitions of what is a problem and the relative seriousness of problems often are shaped by definitions of the situation created by the regulatee.

Further, external regulators can become interdependent with the regulated organization. When this happens, true adversarialism gets lost, resulting in bargaining and compromise that does not always result in the most effective resolution. One kind of interdependence is when regulator and regulatee share a goal or interests: when harm or good fortune befalls one, the well-being of the other is similarly affected. In this situation, regulatory activities and outcomes can be softened. Also, resource exchange can mediate the adversarial advantage external regulators have. In contrast, internal regulators have definite advantages in monitoring, discovering, and investigating problems. They have access to information, know the language, and can keep up with changes. However, interdependence is a problem. The internal regulator depends on its parent for resources, and shared interests and goals are intrinsic to the relationship. Both lessen the likelihood that threats and sanctions will be aggressively invoked.

NASA designed a safety system that combined internal and external regulation in order to offset the disadvantages of either strategy alone and provide the balanced surveillance essential for preventing accidents.[92] At first, NASA had internal safety units only: the Safety, Reliability, and Quality Assurance Program (SR&QA) and the Space Shuttle Crew Safety Panel (SSCSP).[93] Both were created with the expectation that NASA's own personnel, informed about its technology, management systems, culture, and language and with access to day-to-day activities, could be most effective at the close monitoring essential to safety. Interdependence was viewed as an advantage. Because of a presumed shared interest in avoiding accidents, they were expected to bring to their assessments the objectivity necessary for independent review. Shortly after the 1967 Apollo launch pad fire that killed three astronauts, Congress added an external regulatory body composed of aerospace experts: the Aerospace Safety Advisory Panel (ASAP). This legislative action was guided by the notion that expert outsiders, with technical experience and reputation throughout the aerospace industry, would have both the objectivity, the knowledge, and the influence to balance NASA's self-regulating system. All three regulation bodies used a compliance strategy—the best strategy for accident prevention.[94] The goal of a compliance strategy is early detection and intervention, negotiating to correct problems before they can cause any harm. Punishments are also part of a compliance strategy, but they are used as leverage in the negotiations to assure that safety standards are met.[95] To enforce a punishment is a sign that the regulatory system has failed.

Despite this well-conceived design, autonomy and interdependence interfered with regulatory ability to investigate, discover, and monitor safety hazards.

Internal regulators: SR&QA and SSCSP. SR&QA bore the major responsibility for safety oversight at all space centers. Its responsibilities extended to NASA contractors, which it monitored from offices located at contractor facilities. Each component of Safety, Reliability, and Quality Assurance had distinctive functions. *Safety* engineers were responsible for (1) in-plant safety and (2) flight safety. For in-plant safety, they prepared and executed plans for accident prevention and were responsible for industrial safety requirements. In addition, they were to assure that hardware was not damaged in handling. For flight safety, they participated in postflight analysis and FRR and reviewed and monitored potential hazards and risk assessments. *Reliability* engineers determined that contractors' components and systems could be counted on to work as predicted. Their job was conceptual and analytical: they reviewed contractor data and project predictions. *Quality Assurance* engineers were responsible for procedural controls: (1) creating, assessing, and monitoring hardware inspection programs and (2) identifying and reporting problems by monitoring computer problem-tracking systems.

The Presidential Commission's report concluded that SR&QA did not make problems visible with sufficient accuracy and emphasis, failing to (1) identify discrepancies in internal documents on the CIL that sometimes listed the booster joint as C 1 and sometimes as C 1R, (2) compile and circulate trend data about the in-flight O-ring erosion, and (3) establish, maintain, and adequately monitor requirements for reporting shuttle problems up the NASA hierarchy in computer tracking systems. The Commission attributed these failings to interdependence. SR&QA depended on NASA for resources and legitimacy; both had been cut.[96] Between 1970 and the *Challenger* tragedy, NASA trimmed 71 percent of its SR&QA staff.[97] At Marshall, SR&QA staff size was cut from about 130 to 84.[98] In 1982, when the shuttle was officially declared operational, NASA either reorganized SR&QA offices or continued them with reduced personnel. These staff reductions made it understandably difficult to monitor adequately a system as complex as NASA's.

The effects of interdependence on the second internal regulator, the SSCSP, were even more extreme. The panel was composed of 20 people from Johnson, Kennedy, and Marshall, Dryden (the NASA facility at Edwards Air Force Base, Calif.), and the Air Force.[99] Panel members were selected from S&E, Project Management, and the Astronaut Office, which trains astronauts and provides flight crews for space vehicles.[100] The panel's job was to (1) identify possible hazards to crews and (2) advise shuttle management about these hazards and their resolution. It prioritized safety issues for NASA's attention, submitting a formal report to the Level II Program Manager after every meeting. From 1974 through 1979, it met 26 times, addressing such issues as mission-abort contingencies, crew escape systems, and equipment acceptability.[101] Although the O-ring problems began in 1977, my review of meeting summaries shows that they were never discussed by this panel. The SSCSP was designed to function only during the shuttle's developmental period.[102] When the Space Shuttle officially became operational, the SSCSP first was combined with another panel, then eventually discontinued.[103]

NASA's control over internal resources and their exchange mattered in these outcomes. A logical assumption is that organizations producing high-risk technological products will play an active role in their own regulation, if only because they alone possess sufficient technical knowledge to do so. Susan Shapiro notes that self-regulating organizations are "fundamentally suspect."[104] But in addition, self-regulation of risky technical enterprise may, by definition, be accompanied by dependencies that interfere with regulatory effectiveness. The disadvantages of regulatory interdependence and its contribution to NASA's "Silent Safety Program" are writ large in this history. NASA top officials created a regulatory structure that was consistent with their definition of the situation in 1982: the safety redundancies of a developmental system could be dropped because they were "not productive for the operational phase of the shuttle."[105] Based on her analysis of engineering projects and safety, Carol Heimer notes that organizations tend to pay more attention to the costs of the main activities and neglect the costs of activities that do not contribute significantly to meeting deadlines: "When rewards are tied to specific measures of performance, people tend to do what is required to produce those measures and to neglect other things."[106] Following preestablished criteria for an operational system and ignoring the developmental nature of the shuttle's "unruly technology," top NASA officials purposely weakened the internal safety system.

The external regulator: the ASAP. Nine outside experts—aerospace industry leaders—were appointed to the ASAP by the NASA administrator for six-year terms.[107] Their job was to assess the safety of technical components and their management. They conducted fact-finding sessions at NASA centers and contractor sites. Technical issues came to the attention of panel members in several ways. The NASA administrator and the ASAP staff director (a NASA employee) requested technical investigations for the panel's on-site visits. Occasionally, congressional committees requested specific information from the panel. Sometimes, ASAP members discovered problems at an investigative site. They tried to negotiate safety issues at the lowest possible level, then if an issue was not resolved, took it up the hierarchy, as necessary. ASAP members gave personal feedback to the people with whom they had contact, also sending findings and recommendations to the centers, contractors, and Congress. The centers and contractors had to respond in writing to each panel recommendation.

Although the ASAP's annual reports indicate active exploration of problems and risks, the Presidential Commission found no indication that the booster joint design or in-flight anomalies were assessed. The Commission attributed this failure to the panel's "breadth of activities," citing its broad scope of oversight responsibilities.[108] The ASAP confronted the traditional barriers of autonomy. Working on average 30 days per year, the panel had trouble covering all aspects of safety in the vast NASA enterprise and therefore could "not be expected to uncover all of the potential problems nor can it be charged with failure" because the "ability to function effectively depends on a focused scope of responsibilities."[109] Further, we find evidence that interdependence was a barrier to effective regulation. The panel's original mandate required it to investigate issues raised by the NASA administrator, the staff director, and Congress. Because many of its activities were defined by others, the panel's capacity for independent discovery and review was curtailed.

INFORMATION DEPENDENCIES

Despite these pernicious effects of autonomy and interdependence, all three safety regulators did discover other problems, force resolution of other technical issues, and rectify other hazardous situations. Why not the SRB joint problems? To varying extents, all three safety units depended on the engineers in the SRB work group for information. These information dependencies affected the regulators' definition of the situation. What constituted a problem and how serious the problem was judged to be were shaped by the information that was given to them by the engineers in the SRB work group.

Neither the SSCSP nor the ASAP identified the O-ring problems. For both, other responsibilities curtailed the amount of time that could be dedicated to proactive investigation, so they may not have made contact with engineers in the work group. Their dependence on the work group for information suggests that the problem either was not mentioned to them at all or, if mentioned, may have been reported as a "resolved problem"—which was how engineers defined it after each postflight hazard assessment. Thus, the ASAP and SSCSP (the latter went out of business at the start of the operational phase, when erosion was considered resolved) would not have defined O-ring performance as a problem requiring attention from them.

SR&QA engineers *did identify the problem*, however. They did not overlook it or miss it. Because they depended on the work group to originate and interpret data, they assessed and defined the data similarly. Specialization in organizations means that one internal department never exactly duplicates the work of another. SR&QA engineers did not originate engineering data and analysis. They *reviewed* the determinations of the working engineers at Marshall and Thiokol. As Leon Ray said, "Safety people are not in a 'doing mode'; they are in a 'reviewing mode.'" The Safety engineers knew about each incident of erosion, the technical rationale, and the corrections that were made after each incident. They knew about the increased leak check pressure and its association with increased frequency of erosion. They not only witnessed but participated in, contributed to, and concurred with the work group's cultural construction of risk in the years prior to the *Challenger* launch.[110] Further, Safety representatives were present at every FRR level as the construction of risk was passed up the hierarchy.[111] Wiley Bunn, Marshall Director of Reliability and Quality Assurance, described the definition of the situation shared by all SR&QA representatives throughout the hierarchy prior to the January 1986 *Challenger* teleconference: "Laying down in front of the truck is maybe a good analogy. Why didn't we lay down in front of the truck? Because when you read the analysis and the rationale as to why you could fly like that, it made sense. There wasn't anything worse than that going to happen. You could expect the kind of events that had occurred to happen, but you need not expect anything worse to happen. And there was [safety] margin even if it did happen worse. And people really believed that. There wasn't any—the truck wasn't even coming."[112]

The Safety engineers were exposed to the same patterns of information as the SRB working engineers and concurred with their construction of risk. Consider, in particular, what this situation tells us about the consequences of the absence of trend data, data that the Presidential Commission said would have alerted NASA's Levels II and I to the "ominous" increase in the incidence of erosion that began in 1984.[113] The trend was, in fact, not "ominous" to the work group in the years preceding the *Challenger* launch. That a trend existed (defined as repeated incidents increasing in frequency) was well known. The working engineers believed the increased leak check was responsible and defined each incident as an acceptable risk, as an outcome of engineering analysis. Trend data, had it been created by SR&QA engineers, would have been a direct reflection of, rather than a cause for reflecting *on*, the working engineers' construction of risk. Since Safety engineers did not originate data or replicate tests (S&E did both), any trend data they created would have been developed from data provided by Marshall and Thiokol working engineers in the SRB work group; the interpretation of those data would have been consistent with the technical analysis and rationale that the work group formulated after each postflight analysis, with which SR&QA concurred.

In retrospect, we know that trend data that examined the relationship between temperature and O-ring anomalies would have been extremely important. The Presidential Commission stated: "A careful analysis of the flight history of O-ring performance would have revealed the correlation of O-ring damage and low temperature. Neither NASA nor Thiokol carried out such an analysis; consequently, they were unprepared to properly evaluate the risks of launching the 51-L mission in conditions more extreme than they had encountered before."[114] We cannot know, had SR&QA compiled trend data prior to STS 51-L, whether temperature would have been included, but it seems unlikely. No one—Thiokol engineers in particular—thought cold temperature was going to be a problem, even after the January 1985 launch of STS 51-C. The Thiokol task force generated no temperature trend data, nor did it request that NASA generate it. And it was a pencil-and-paper task that would not have been a drain on resources. Because of the shared construction of temperature as a nonproblem, no one had the idea.

SR&QA was harshly criticized after the disaster. The House Committee noted that SR&QA "had failed to exercise control over the problem tracking systems, had not critiqued the engineering analysis advanced as an explanation of the SRM seal problem, and did not provide the independent perspective required by Senior NASA managers at Flight Readiness Review."[115] Concurring, the Presidential Commission concluded that as a result, decision making was impaired: people throughout the organization were not well informed about the O-ring problem. The report stated, "A properly staffed, supported, and robust safety organization might well have avoided these faults and thus eliminated the communication failures."[116] The report asserted that if safety resources had not been cut, SR&QA might have intervened earlier, averting the tragedy.[117]

It is not clear, however, that more people assigned to safety would have altered the definition of the situation. Those who were present participated in work group decision making almost daily. The engineering analysis made sense to them. They reviewed it and agreed. There is no evidence of looking the other way, cooptation, hesitancy to speak, or willful concealment on their part. Apparently, they became encultured; thus, their definition of the situation was the same as the work group's. Similarly situated in the culture of production and exposed to the same incremental decision making, the same patterns of information, decision streams, and commitment processes that the FRR rituals generated, they also normalized deviance. From this, we can only speculate that more Safety engineers participating in the same way on a daily basis would not have altered the scientific paradigm that prevailed in the years preceding the *Challenger* disaster.

A COLLECTIVE CONSTRUCTION OF RISK: POSITION, INFORMATION, INTERPRETATION, AND ACTION

This chapter completes our examination of the history of decision making, 1977–85. We have an explanation of the normalization of deviance at NASA that includes the production of culture in the work group, the culture of production, and structural secrecy. In combination, they explain how an official collective construction of risk originated and persisted at the space agency, despite anomalies that increased in frequency and seriousness. Three sources of structural secrecy affected insider understandings prior to the *Challenger* launch: (1) As the problem unfolded, signals of potential danger were embedded in patterns of information that affected their meaning to the work group. After hazard assessment, signals initially perceived as a threat to flight safety were reinterpreted as weak, mixed, and eventually became routine. The work group's definition of the situation was affirmed by FRR decision streams and the ritualistic aspects of the decision process. (2) As information was relayed from the work group up the hierarchy to top administrators in the launch decision chain, organization structure, processes, and transactions perpetuated the work group's scientific paradigm. Systematic censorship reduced the information available to Levels II and I, resulting in a risk assessment at the top that mirrored the conclusions of the work group, sans the work group's awareness of the unruliness of the technology. (3) NASA's regulatory system also was a source of structural secrecy. Autonomy and interdependence systematically inhibited the ability of safety regulators to both generate and interpret information that might have altered the construction of risk throughout the hierarchy. In particular, information dependencies affected regulators' definition of the situation. Rather than being an effective source of signals that challenged the prevailing construction of SRB joint risk, NASA's safety system fed into it, contributing to the persistence of the collective construction of risk.

By clarifying the role of information and knowledge at NASA in the years prior to the *Challenger* launch, this chapter challenges conventional interpretations published after the accident. This analysis shows that locating responsibility for continuing to launch during this period with individual managers who intentionally suppressed information about O-ring problems is incorrect. What people knew and understood was a consequence of their position in the organizational structure. This chapter brings home the striking connection between social location, information, worldview, and response to events and activities.[118] In the previous chapter, we saw how people's positions in the culture of production created common understandings that shaped their sensemaking and action. Here, we saw that sensemaking and action were also affected by people's positions in the NASA-contractor system, as it determined access to and interpretation of information.

The shared official construction of risk is particularly interesting in light of the way understanding varied with position in the organizational structure. Clearly, at NASA, the "core set"—the engineers in the SRB work group doing the hands-on engineering analysis—had the most information and were fully sensitive to the complexities of the technology. Level II and I personnel received less information and "were not sitting in the seat of understanding." We must wonder about other possible interpretive differences arising from the different positions of managers and engineers. David Bella, engineer and organization analyst, writes about "systematic distortion." He argues that as systematic censorship reduces information, it is the information unfavorable to the organization's agenda—that does not support the ambitions and survival needs of the organization—that is routinely filtered out.[119] Unfavorable information is lost, not by malicious intent, purposeful concealment, or reluctance to say something superiors do not want to hear (all psychological in origin), but as a collective and systemic consequence of organizational structure and roles: people deliberately do not seek out unfavorable information. He notes, "The technological consequences of such distortions can be disastrous."[120]

Lee Clarke, author of *Acceptable Risk?* identifies another source of distortion that operates on the information that makes it to the top.[121] A "disqualification heuristic" prevents risk experts and decision makers from accurately perceiving the likelihood of a failure. According to Clarke, a disqualification heuristic is an "ideological mechanism or mind-set that leads experts and decision makers to neglect information that contradicts a conviction—in this case, a conviction that a sociotechnical system is safe."[122] It enables them to selectively focus on confirming information while relegating disconfirming information to secondary or even trivial status. In the

Exxon *Valdez* case that Clarke studied, the failure to take into account information that disconfirmed beliefs about system safety led to inadequate preparation for major oil spills.[123]

Both systematic distortion and the disqualification heuristic operated in global estimates of risk at NASA. By global estimates, I mean personal perceptions of the possibility of failure of the shuttle as a whole. Gathering his own data in seat-of-the-pants interviewing with NASA insiders, Presidential Commissioner Richard Feynman observed: "It appears that there are enormous differences of opinion as to the probability of failure with loss of vehicle and of human life. The estimates range from 1 in 100 to 1 in 100,000. The higher figures come from working engineers, and the very low figures from management."[124]

The variation in these global estimates is a powerful demonstration of how position in an organization affects access to and interpretation of information, worldview, and consequently, action. The systematic distortion Bella refers to occurs because the different positions of managers and engineers put them in touch with different data from which to make their estimates. Working engineers had access to a variety of tests, calculations, and other observations that those at the top of the decision chain did not have. Although Feynman does not give us all the details, at least one engineering estimate he included came from a range safety officer who witnessed the firing of developmental and qualifying motors. His failure estimate for the shuttle system was 1 in 25, an estimate based on the data available to him: tests readily observable in his position.[125] Apparently, top managers, distant from the testing and other close observations of the technology, were basing their estimates on the successful flight rate, which, prior to the *Challenger* was 24 launches with no failures.

The disqualification heuristic further enlightens us about the difference that Feynman found between the global estimates of managers and those of engineers. We know that one's position in an organization brings with it certain expectations, experiences, and assumptions about the way the world works that are brought to bear on the interpretation of information.[126] Managers and engineers occupy distinctly different cultural niches in organizations. While both managers and engineers pursue production goals and safety goals (along with many others), expectations and rewards for engineers are primarily for technical achievements, whereas for NASA's top technical decision makers (also trained engineers), production and cost/efficiency concerns are a major part of the "expectation-performance-evaluation-reward" package. These additional responsibilities give managers a frame of reference distinctly different from that of working engineers.

This managerial worldview allows the disqualification heuristic to operate in global estimates of vehicle failure, aiding and abetting a subtly selective focus on information that affirms the imperatives of that worldview and minimizes the relevance of contradictory information. This situation was exacerbated in the case of NASA senior administrators in Washington who were *not* in the launch decision chain, who had no technical decision-making responsibilities and therefore no contact with daily engineering analysis. It is easy to see how a "can do" attitude could develop at the top, and how this frame of reference readily could lead to the kind of actions administrators took: mission adventurism and distortions of program capability that persuaded the public, Congress, and payload customers that the shuttle was a routine, operational system. Unlike the range safety officer, they were not positioned so that disconfirming evidence was routinely brought to their attention, nor, apparently, did they seek it out.

However, we want to know the extent to which these biasing factors may have operated in FRR decision making, since FRR was the key mechanism for sending information about risk from bottom to top. When we shift from individual global estimates of vehicle success or failure to the calculated risk assessments for specific technical components that occurred in FRR, a different picture emerges. The influence of systematic distortion and the disqualification heuristic can be reduced when rules are in place to counteract these tendencies. In fact, Bella argues that one way to combat the filtering of unfavorable information is to create checks and balances that systematically force it to surface. This strategy would also be an antidote to the ideological mind set that drives Clarke's disqualification heuristic.

In FRR, NASA had such a system. Unlike the unregulated global estimates described above, launch decisions were regulated by a system of engineering rules and bureaucratic checks and balances that would interfere with both the selective dropping of unfavorable information and the unconscious operation of the disqualification heuristic. Clarke and Short remind us of the importance of power in framing the terms of debates about risk.[127] FRR debates were governed by rules, procedures, and norms deliberately created to seek unfavorable information that challenged the decisions to accept risk and fly that were being presented: a matrix system that brought in a variety of specialists; the "fishbowl" atmosphere; proactive inquiry via "probes," "challenges," and Action Items; competition between projects; adversarialism; and the certification at each level that implicated all levels in the outcome. Both structure and process were designed to bring conflicting viewpoints and contradictory data to the fore in FRR discussions.

These precautions notwithstanding, Bella's type of systematic distortion did result, but in a perhaps more subtle way than he imagined. As FRR progressed, ambiguity was eliminated from technical arguments. Open problems were closed. This must have contributed to different global perceptions of risk, depending on position. The people at the top saw tight engineering analysis and resolved problems; the uncertainties well known to people in work groups were less visible. In addition, those at the top of the launch decision chain learned of fewer technical issues and so would tend to see the shuttle system as one with fewer and less complex problems than working engineers would. In the engineering risk assessments for the booster presented in FRR, however, there is no evidence that at any management level the disqualification heuristic operated in FRR in the years preceding the *Challenger* tragedy. The message received at the top of the launch decision chain was consistent with what the working engineers and safety regulators believed. Although the information was condensed, ambiguity lost, and analysis strengthened as it went up the FRR hierarchy, the essentials of the engineering opinion were retained. The weak, mixed, and routine signals that top decision makers dealt with accurately reflected the determinations of risk acceptability developed by technical experts at the bottom of the launch decision chain.

Acknowledging the role of information dependencies, the House Committee declared that the information that reached top decision makers was "filtered": it was interpreted for Levels II and I by those beneath them in the hierarchy, whose position and expertise enabled them to interpret data and pass it on.[128] Noting that FRR operated "as well as its design permitted,"[129] the Committee report stated that "the fundamental problem was poor technical decision making over a period of several years by top NASA and contractor personnel, who failed to act decisively to solve the increasingly serious anomalies in the Solid Rocket Booster joints."[130] The Committee asserted that "poor technical decision making" by top administrators occurred because the administrators only knew what work groups told them: "Flight Readiness Reviews are not intended to replace engineering analysis, and therefore, they cannot be expected to prevent a flight because of a design flaw that [Project] management had already determined represented an acceptable risk. . . . Had the engineering analysis led Marshall to a different conclusion about the severity of the SRM seal erosion problem, the system would have reacted to these concerns long before the 51-L Flight Readiness Review."[131]

In the years preceding the *Challenger* tragedy, the engineering analysis led the work group to conclude that the joints were an acceptable risk. SR&QA and Levels II and I concurred. With the clear vision hindsight provides, both the Presidential Commission and the House Committee fully agreed on one point: "Neither NASA nor Thiokol fully understood the operation of the joint prior to the accident."[132] Commissioner Feynman observed, "The origin and consequences of the erosion and blow-by were not understood . . . officials behaved as if they understood it, giving apparently logical arguments to each other often depending on the 'success' of previous flights."[133]

Only in the wake of the tragedy was it clear they had not understood. At the end of 1985, they believed they did.

Chapter Eight

THE EVE OF THE LAUNCH REVISITED

We now return to the eve of the launch. Accounts emphasizing valiant attempts by Thiokol engineers to stop the launch, actions of a few powerful managers who overruled a unanimous engineering position, and managerial failure to pass information about the teleconference to senior NASA administrators, coupled with news of economic strain and production pressure at NASA, led many to suspect that NASA managers had acted as amoral calculators, knowingly violating rules and taking extraordinary risk with human lives in order to keep the shuttle on schedule. However, like the history of decision making, I found that events on the eve of the launch were vastly more complex than the published accounts and media representations of it. From the profusion of information available after the accident, some actions, comments, and actors were brought repeatedly to public attention, finding their way into recorded history. Others, receiving less attention or none, were omitted. The omissions became, for me, details of social context essential for explanation.

By restoring social context in three ways, this chapter challenges conventional interpretations of the eve of the launch. First, the organization of the book places the *Challenger* launch decision in its proper position as one decision in a decision stream begun many years before. From the early development period of the Space Shuttle through the end of 1985, the SRB work group had consistently defined the SRB joints as an acceptable risk. Behind this determination was a scientific paradigm that established the redundancy of the joint. The belief in redundancy and the scientific paradigm behind it were institutionalized prior to 1986. They were crucial components of the worldview that many decision makers brought to the teleconference on the eve of the *Challenger* launch. Understanding the launch decision rests on knowing this important fact.

Second, the events of that night are resituated in a more detailed chronology. The chapter 1 version was abbreviated—a stereotype I constructed from aspects of the event that were reproduced time and again in posttragedy accounts. Here, the chapter 1 version is repeated in boldface type, juxtaposed against another version that restores voices, actions, and details omitted from nearly all other accounts. Reconstructed in ethnographic thick description, this restoration of the confusion, diverse viewpoints, complexity of the technical issue and engineering arguments, and little-known aspects of interaction is, in itself, stereotype-shattering.

Third, readers are now in a position to restore context. When the tragedy occurred, the public had little knowledge of the complex culture in which the launch decision was made. Understanding the culture, as I discovered in the first year or so of my research, is absolutely essential to understanding what went on. In chapters 3 through 7, readers learned the culture, acquiring new information that now is a part of the frame of reference, or worldview, you bring to the interpretation of the information presented here. You are aware of the culture of production and structural secrecy and how they affected decision making about the SRB joints. You know something about how booster joints work, the organizational structure, NASA's technical and bureaucratic language, FRR, the Acceptable Risk Process, reporting rules, residual risk, management risk decisions, data dumps, teleconferences, and other intricacies of NASA culture that were taken-for-granted aspects of launch decision making for insiders. Understanding the cultural context gives meaning to activities and circumstances, allowing a glimpse of the "native view," so readers now bring to this reading a different sensibility than they did before.

The reconstruction that follows draws principally on the over 9,000 pages of interview transcripts at the National Archives. For the history of decision making, an extensive paper trail of documents created at the time corroborates the interview data gathered after the accident. For the eve of the launch, only the charts from the Thiokol engineering analysis and some handwritten notes made by some of the participants exist against which we can evaluate what they say. Because the interviews were gathered in the aftermath of the tragedy, we need to assess its impact on the accounts that constitute this chapter. We know that testimony to the Presidential Commission and the House Committee on Science and Technology was given under oath, and that the National Archives interviews were legal depositions collected in March and April 1986 by experienced investigators. Nonetheless, recollections of the teleconference are distorted in a number of predictable ways. First, in the interview transcripts many people voluntarily mentioned that they had difficulty remembering all that was said that night. Marshall's George Hardy testified:

> It is not easy just within your own heart and mind to try to separate yourself from what you know has happened. I don't mean by that the [responsibility for] cause or failure. I am talking about the tragic incident itself. And I am not suggesting that anyone, and in any testimony before this committee or this Commission has knowingly in any way presented untruth. But I have talked to some of my colleagues, and I have found that they found it very difficult to remember precisely not only everything said and everything done, but even more than that, some of the motivations or some of the thoughts that took place at that time.[1]

Although many took notes during the teleconference, as they customarily did in engineering discussions, their notes consisted of technical points and calculations made as they considered the technical issues. They were not clairvoyant. They did not foresee the disaster; they did not realize that what they said to each other during the teleconference would be subject to microscopic analysis by millions. As Marshall engineer Keith Coates remarked, "I just wasn't registering it for posterity."[2] This obliviousness to possible future public scrutiny is patently obvious in the candidness of Mulloy's "When do you want me to launch, next April?" and Hardy's "I'm appalled" comments. Many participants commented on the impact of the media blitz on their ability to recall. The statement by Thiokol's Jack Buchanan is representative: "It's hard to keep clear in your mind what you heard on Monday night [the eve of the launch] and what you've heard on all the following testimony and the hundreds of newspaper articles I've read and all the TV reporting. It sort of blends after a while."[3]

Moreover, distortions occurred because people felt their jobs were at risk. As Marshall's Jim Smith recounted: "A lot of people became scarce when the accident happened. It was very obvious they tried to divorce themselves from much knowledge of any facts and I guess they felt their job was going to be in jeopardy too. It was obvious people were concerned about whether they literally would lose their jobs, or be totally removed from their position and put someplace else, in a corner, or whether there would be a possibility of some legal action."[4] Teleconference participants at both Thiokol and Marshall reported being advised by both company legal counsel and supervisors to tell the truth to official investigators. In particular, the working engineers at both places asserted to investigators that they did not receive any warnings or urging that they follow a particular "company line." Thiokol's Brian Russell reflected statements made by others, "I was told by every manager that I dealt with to tell the truth as I saw it and as I felt it, and so there has been no influence by anyone, NASA, Thiokol, the outside, anywhere, to have me come up with a structured statement that may not be my own feelings, and that's true even today."[5] Roger Boisjoly, in a 1987 paper, also affirmed that they were encouraged to express their opinions, but added that Thiokol employees had been advised by company attorneys "to answer all questions with only yes or no and not to volunteer anything freely" in the Presidential Commission hearings.[6] Regardless of the legal advice they said they received, it would be naive to assume that the participants' testimony was given without thought to work-related consequences.

Finally, we can only surmise the effect of being personally involved in the deaths of the astronauts and in a event that was immediately acknowledged as a calamity of historic significance. It is very difficult, if not impossible, to assess the effect of this trauma on postdisaster accounts. How many were affected and to what extent is not known. From my personal interviews and from the interview transcripts at the National Archives, it is clear that the tragedy was traumatic for all associated with it. At the very least, we can assume that the teleconference participants responded to the tragedy in the way that people normally do when their everyday lives are suddenly and dramatically altered. Peter Berger notes that the events that constitute our lives are subject to many interpretations, not just by outsiders, but by ourselves.[7] When an unexpected event occurs, we need to explain it not only to others, but to ourselves. So we imbue it with meaning in order to make sense of it. We correct history, reconstructing the past so that it will be consistent with the present, reaffirming our sense of self and place in the world. We reconstruct history every day, not to fool others but to fool ourselves, because it is integral to the process of going on.[8] So we would expect that, in addition to the initial failure to register the teleconference fully, the effects of forgetting, the media effect on personal recollections, and the intentional self-protection in response to occupational risk, accounts of that evening would also be affected by the unconscious editing that goes on as people attempt to rescue order from disorder—perhaps driven in this case by a need to be guiltless, a need to have been "correct."

Given these sources of distortion, to what extent is the following a credible representation? The teleconference itself had 34 participants, who by their presence acted as controls on each others' accounts. Although teleconference participants were divided about the launch decision and many expressed strong feelings afterward, no one charged any of the other participants with intentionally falsifying or omitting information about the teleconference. Further, I found consensus about the general sequence of events. Most people at both Thiokol and Marshall held common recollections of chronology. I think this consensus existed because the subject was an engineering controversy that had a sequence to it, and engineering was their business. They remembered the engineering arguments—many took notes that recorded aspects of the data and technical discussion—which served as anchor points for the myriad comments and actions that took place. Some people recalled things that others did not mention, but for many of these points, there were several corroborating accounts. Rarely did only one person mention an activity or comment, and then it was usually some act or thought of his own. In such cases, I included those individual recollections, citing the source. From all of the pieces, there was an overall logical consistency of parts to the whole chronology.

However, the meaning that the teleconference participants ascribed to the actions and comments was varied. This variation is not surprising, for the nature of social reality is that there is no single "truth" that accounts for it, but any action or incident has many truths, one for each participant. This variation does not imply that these truths are random. Rather, as we saw in the last chapter, meaning is scripted into one's position in a structure. Something happens: social location determines one's access to information about the event, ability to interpret it, and frame of reference in which it is observed. As historical ethnography, this book is intended to create a social history that reveals how participants interpreted actions and events. Because of the deep historic record of the eve of the launch, this chapter, more than others, is better able to show the diversity of participants' views and the effects of distinctive social locations on understandings. This diversity reveals, not untruths, but multiple meanings, and thus multiple truths. Karl Weick observes that whenever people act in a context of choice, irreversible action, and public awareness, their actions tend to become binding;[9] they become committed to their actions, then develop valid, socially acceptable justifications. Committed action, justifications, and meaning become linked. Actions "mean" whatever justifications become attached to them. The *Challenger* launch decision exemplifies committed action: a choice, irreversible action, and public awareness. Each person, both by what he did say and by what he did not, acted in ways that became binding. Their accounts are the interpretations they hold, and as such, constitute the native view of each participant.

N ASA's Space Shuttle *Challenger* originally was scheduled for launch January 22, 1986. A crew of seven was assigned. Commander Richard Scobee, Pilot Michael Smith, and Mission Specialists Ellison Onizuka, Judith Resnik, and Ronald McNair were astronauts. Gregory Jarvis, an aerospace engineer, and Christa McAuliffe, a teacher from New Hampshire, were payload specialists. McAuliffe's assignment—to teach elementary school students from space—gave the *Challenger* a special aura. Officially known as Space Transportation System (STS) mission 51-L, it became publicly identified as the "Teacher in Space" mission, despite the scientific and technical assignments of the other crew members.

The Level IV FRR for STS 51-L was held at the Thiokol plant in Utah on December 11, 1985. On the basis of their hazard analysis, the Thiokol working engineers reported to their management that the SRBs were an acceptable risk and recommended that STS 51-L be launched. No information about the past performance of the booster joints was presented in their briefing charts because NASA's Delta reporting concept required contractors to report only on new evidence found since the previous flight.[10] Since then, the engineers had uncovered no new information that altered the three-factor technical rationale behind the belief in redundancy. They would have no further information until the postflight analysis of *Columbia* (STS 61-C), which was still on the ground. As a consequence, the chart entitled "STS 61-C Performance," identified the SRB joints as an open item: "TBD (To Be Determined)."[11] Thiokol certified in writing at the Marshall Level IV FRR that the SRB hardware was ready to support STS 51-L, pending satisfactory completion of open items. On January 3, the SRBs were certified as flight ready by Mulloy's Level III Board at Marshall, also pending resolution of open items.

According to plan, *Challenger* would be the first launch of 1986. However, the launch date had to be "slipped" several times—first to January 23, then to January 25, then to January 26—because seven launch delays over a 25-day period postponed the December launch of *Columbia* (STS 61-C).

The *Columbia* launch had been delayed for safety reasons:

• December 18: Ground crews fell behind and were given an extra day to catch up.

• December 19: Countdown was stopped with 14 seconds to go because of an out-of-tolerance turbine reading on one of the SRBs.

• January 4: The astronauts, just back from the holidays, were given extra time to refresh their training on flight simulators. This was in addition to normal crew training requirements, which had been completed.

• January 6: Countdown was stopped with 31 seconds to go because of a computer-identified problem with the fuel fill-and-drain valve.

• January 7: Countdown was held while the launch team waited for weather at two Transatlantic Abort Landing (TAL) sites to clear. Weather at Kennedy also was marginal. Launch was postponed until January 9.

• January 8: Launch was rescheduled for January 10 when a computer revealed an obstruction in a main engine valve.

• January 10: Launch was scrubbed because of heavy rains and driving winds at Cape Canaveral.[12]

Setting a NASA record for false starts, STS 61-C was launched January 12.

On January 13, the Marshall Center Board FRR for the *Challenger* mission met.[13] The major portion of Mulloy's presentation was devoted to reviewing changes in the SRB parachute recovery system.[14] It was that, and not the SRB joints, that had been the most troublesome item for the SRB Project in the past year. Since April 1985, the SRB joints had been operating as predicted. Reporting on the flight readiness of the SRBs, Mulloy forwarded the SRB work group conclusions up the hierarchy: "No findings from continuing analyses that changes previously established rationale for flight."[15] Because STS 61-C had just been launched the day before, Mulloy had only ascent data: the chart stated that "all SRB systems functioned normally" and "no anomalies."[16] The Level II and I FRRs for *Challenger* were held on January 15. Mulloy made no mention of SRB joint anomalies on previous missions, conforming to the Delta review concept.[17] Since Mulloy still had only ascent data, he reported "no 61-C Flight Anomalies."[18] The SRBs were certified in writing as flight-ready by Levels II and I, pending closure of the few remaining open items—one of which was the booster performance on the still-orbiting STS 61-C.

On January 18, STS 61-C landed at Edwards Air Force Base. In the postflight inspection, Thiokol engineers found erosion on three joints.[19] Thiokol concluded that the SRB joints were an acceptable risk for STS 51-L. Erosion was an expected result of the 200-psi leak check pressure, and the amount found was within the experience base.[20] Therefore, the data conformed to predictions, flight performance again affirming the belief in redundancy. Conforming to the NASA requirement that items remaining open after Level I be reviewed by the Mission Management Team, Mulloy discussed the work group's analysis and conclusion: accept risk and fly.[21] In the FRR for *Challenger*, no deviation from required NASA procedure occurred.[22]

Efforts for the January 26 *Challenger* launch from Kennedy Space Center, Cape Canaveral, Florida, were coordinated by the top technical managers and administrators in NASA's four-tiered launch decision chain. Among them were Jesse Moore, Associate Administrator for Space Flight, NASA Headquarters, Washington (Level I); Arnold Aldrich, Program Manager, Johnson Space Center, Houston, Texas (Level II); William Lucas, Director, Marshall Space Flight Center, Huntsville, Alabama; Stanley Reinartz, Manager, Shuttle Projects Office, Marshall; Lawrence Mulloy, Manager, Solid Rocket Booster Project, Marshall (Level III); and Allan McDonald, Director, Solid Rocket Motor Project, Morton Thiokol, Utah (Level IV). Following *Columbia*'s precedent for delay, early countdown activities were terminated because the forecast indicated that weather at Kennedy would be unacceptable throughout the launch window. NASA rescheduled for January 27. That day, countdown was proceeding normally when microswitch indicators showed that the exterior hatch-locking mechanism had not closed properly. By the time it was fixed, the wind velocity exceeded the Launch Commit Criteria for allowable crosswinds at the Kennedy Space Center runway used in case of a return-to-launch-site abort. The launch was scrubbed at 12:36 P.M. and rescheduled for January 28 at 9:38 A.M. EST.

An afternoon launch window was also available for the January 28 launch, pending continued acceptable weather at Dakar, Senegal, a TAL site with lighting for a night landing. NASA calculates and announces in advance launch windows that give a range of dates over which a launch might occur. For each possible launch date, a primary and a secondary (backup) launch window are established.[23] The duration of each launch window is three hours because that is the maximum amount of time the crew is allowed to remain strapped in their seats during unplanned holds. For the *Challenger* launch, the primary window opened at 9:38 A.M. and closed at 12:38; the secondary window opened at 3:48 P.M. and closed at 6:48 EST.[24]

NASA personnel at the Cape first became concerned about cold temperature at approximately 1:00 P.M. on January 27. The forecast for the eve of the launch predicted clear and uncharacteristically cold weather for Florida, with temperatures expected to be in the low 20s during the early hours of January 28. Marshall's Larry Wear, Solid Rocket Motor Manager, asked Morton Thiokol-Wasatch, in Utah, manufacturer of the Solid Rocket Motor (SRM), to have its engineers review the possible effects of the cold on the performance of the SRM. This was not the first time concerns had been raised between Marshall and Thiokol about SRM performance. On several previous launches, hot combustion gases produced when the propellant ignited at liftoff had charred and sometimes even eroded the surface of the rubberlike Viton O-rings designed to seal the joints between the SRM case segments of the Solid Rocket Boosters (SRBs) that help get the shuttle off the launch pad and into the sky.

In response to Wear's inquiry, Thiokol's Robert Ebeling, Ignition System Manager, convened a meeting at the Utah plant. The Thiokol engineers expressed concern that the cold would affect O-ring resiliency: the rings would harden to such an extent that they would not be able to seal the joints against the hot gases created at ignition, increasing the amount of erosion and threatening mission safety.

After the January 27 scrub, the Marshall and contractor launch representatives for the Main Engine, External Tank, and the SRBs were advised of the predicted temperatures for January 28 and polled by Stan Reinartz, Manager, Marshall Shuttle Projects Office, as to feasibility of launch the next day.[25] All parties reported to Reinartz that there were no SRB constraints for a launch at 9:38 A.M., January 28. Although both Al McDonald and Jack Buchanan were at the Cape representing Thiokol, neither mentioned any temperature-related concerns for SRM items to Reinartz. However, Marshall's SRM Manager Larry Wear, in Huntsville, recalled the low temperatures preceding the January 1985 launch of STS 51-C and that Thiokol had made some remarks about the effect of low temperatures. He asked Boyd Brinton, Thiokol Space Booster Project Manager who was at Marshall for the launch, whether he recalled what Thiokol had said in 1985 about cold temperatures. Brinton did not. Wear asked Brinton to call Thiokol in Utah to find out if they had any temperature concerns about the *Challenger* launch. Meanwhile, Wear went to his files and looked up the 1985 Thiokol assessment on STS 51-C in FRR documents to see what they had said.[26]

Brinton's call reached Thiokol's Bob Ebeling in Utah at about 11:00 A.M. MST. Ebeling asked members of the O-ring task force and some others to assemble in his office at 12:30 P.M. because that was when all their lunch hours ended.[27] Among those attending were Ebeling, Don Ketner (Chair of the O-ring Seal Task Force), Roger Boisjoly, Arnie Thompson, Brian Russell, and Jerry Burn. Discussing the blow-by that occurred on the cold January 1985 launch, they feared that the predicted cold might impair the ability of the O-rings to seal the joint on the *Challenger.* They discussed their resiliency tests. Russell stated, "We had done tests at 100 degrees, 75 degrees, and 50 degrees and had seen that the O-ring became much more sluggish to respond to receding metal surfaces at 50 degrees and the fact that we were expecting temperatures much colder than that gave us concern."[28]

Ebeling informed Thiokol's McDonald at the Cape that the engineers in Utah wanted to discuss their temperature concerns with Marshall and asked McDonald to get the latest weather forecast.[29] McDonald called back relaying temperature predictions for a low of 22°F at approximately 6:00 A.M. EST, reaching about 26°F at the intended launch time of 9:38 A.M., and warming through the 40s and 50s in afternoon.[30] Engineering representatives then reviewed their concerns with Thiokol Vice Presidents Lund and Kilminster, who agreed that NASA should be contacted. McDonald called Cecil Houston, Marshall's Resident Manager at Kennedy, informing him that Thiokol needed to discuss with Marshall some concerns about the effect of the cold temperatures on the O-rings and that they did not know whether the launch should go or not. Houston conveyed that message to Marshall personnel at the Cape and in Huntsville.

A three-location telephone conference to discuss the situation was set for 5:45 P.M. EST. Participating were some of the managers and engineers associated with the SRB Project located at Thiokol, Marshall Space Flight Center, and Kennedy Space Center. During the teleconference, Thiokol expressed the opinion that the launch should be delayed until noon or after. A second teleconference was arranged for 8:15 P.M. EST so that more personnel at all three locations could be involved and Thiokol engineering data could be transmitted by fax to all parties.

The first teleconference lasted about 45 minutes. Thiokol actually took no official position during this teleconference. A brief unstructured data discussion occurred, in which Thiokol engineers expressed some of their concerns about cold temperature.[31] However, the telephone connection was bad and the people on non-NASA telephones could not hear—in particular, Marshall's George Hardy, who was at home, and Stan Reinartz, who was in his room at the Merritt Island Holiday Inn at the Cape, where all of the Marshall people were staying. Further, no one had any documents to lead or follow a discussion, and Marshall's Larry Mulloy, essential to any SRB-related decision, had not been located. So a second teleconference was arranged to include a full complement of personnel with specializations appropriate to discuss the issue. At that time, Thiokol would make a formal presentation with conclusions and launch recommendation, following FRR format.

Jud Lovingood, Deputy Manager, Shuttle Projects Office at Marshall, who was running that office while Reinartz was at the Cape, suggested that the teleconference be between Marshall and Thiokol personnel, with a plan to go to Level II if Thiokol recommended not launching.[32] Thiokol engineers requested that the second teleconference begin at 6:15 P.M. MST (8:15 EST), because they wanted an hour and a half to prepare their presentation.[33] This suggestion was agreeable to people at Marshall and Kennedy because they also needed time in order to notify personnel at all three locations and to assemble them at teleconference facilities. At Kennedy, the ground crews were coming in at 10:30 P.M. in preparation for fueling the External Tank, which was to begin around 1:00 A.M. Jim Smith, Chief Engineer, SRB Project at Marshall, noted: "There obviously was a time constraint to try to get a decision if there was a recommendation not to launch. It's quite a lot of effort to go in and tank, then go detank. They'll detank every time if they are not going to be successful and go that day. And its a big task to go through that operation."[34]

Between the first and second teleconferences, the following took place.

At Kennedy. The first teleconference ended at about 6:30 P.M. EST. Reinartz found Mulloy having dinner at the Holiday Inn and briefed him about Thiokol's concerns and the planned teleconference. Lovingood called Reinartz back, telling him that if Thiokol came forward with a no-launch recommendation, they should not launch, and suggesting that Level II—Arnold Aldrich—be advised that the teleconference was being held so that Aldrich would be prepared to initiate a Level I meeting in event of recommendation for delay. As Mulloy and Reinartz were leaving their motel to return to Kennedy, the two visited Lucas and Kingsbury in their rooms, informing them of the upcoming teleconference. Reinartz did not contact Aldrich.[35] Reinartz testified, "We did not have a full understanding of the situation as I understood it at that time, and felt that it was appropriate to do so before we involved the Level II into the system."[36]

At Marshall. The first teleconference ended in Huntsville at about 5:30 CST. Many people had already left for the day. As a result, Marshall's Jud Lovingood and Jim Smith had difficulty locating the appropriate people for the second teleconference. Following NASA's matrix system of assembling personnel with interdisciplinary skills to make technical decisions, they tried to pull together people of all specializations necessary for a discussion of O-ring performance and temperature.[37] The relevant personnel were from two laboratories in Marshall's S&E Directorate: (1) the Structures and Propulsion Lab had specialists on the dynamics of the booster joints and (2) the Materials and Processes Lab had specialists on the materials constituting the O-rings and putty. Supervisors were called and asked to bring their relevant "support staff." Most were located, but some key personnel were not found.

The Director of the Structures and Propulsion Lab, Alex McCool, was out of town. His deputy, John McCarty, was working out at the gym. When Smith located him, McCarty asked Smith to call his people for him. Bob Wegrich, who worked on thermal design, misunderstood and, thinking the meeting was at 7:15 A.M. the next morning, did not come. McCarty wanted Ben Powers, solid rocket design engineer in Structures and Propulsion, to support him. Jerry Peoples, also in Structures and Propulsion and who worked closely with the Thiokol O-ring task force, was not asked because Powers was able to represent that department. No one called Marshall's Leon Ray, the S&E engineer who had worked closely with Thiokol engineers on the SRB joint from the beginning and who had the reputation of knowing more about the joint than any one at Marshall. Powers mistakenly thought Ray was already in transit to the Cape for the launch.[38] Ray was at home.

The Director of the Materials and Processing Lab, Bob Schwinghamer, telephoned Bill Riehl.[39] Riehl was Chief, Nonmetallic Materials Division, where tests were being run on the putty and O-ring materials. Riehl told Schwinghamer that all of the Division's tests were at ambient temperature or above. No low-temperature testing had been done in their division because of the agencywide assumption that the materials would be called upon to withstand extremes of heat at ignition and warm Florida temperatures, not cold. Riehl phoned three people in the Polymers and Composites Branch to ask what the literature produced by O-ring and putty manufacturers stated about the capabilities of their products. None remembered any data about cold temperature. However, they suggested Riehl call John Schell, former Chief of Marshall's Polymers and Composites Lab, known as an outstanding rubber chemist, who retired in 1981.

Schell remembered data about the effect of cold on Viton, met Riehl at Schell's old office, and found the relevant literature in the files.[40] They met with Schwinghamer, reporting that as the temperature dropped Viton began to stiffen, and that there was a point at which it began to stiffen more rapidly. That point—the "knee in the curve"—was between 20°F and 25°F. Schell and Riehl concluded that the O-ring would be all right at 25°F because when the ignition pressure hit it and jammed it into the gap, it would seal. Schell said: "It would have been a little harder, there's no question about that. I mean, there's data all over the world to show that it would have been a little less resilient and a little harder, but at those pressures, it would have sealed."[41] Riehl and Schell conveyed their assessment to Schwinghamer: "If Thiokol comes in and says,'We can't go to launch in the 20 degree range,' we can't argue with them, because we're just going off the cliff in stiffness in those temperature ranges."[42] The three left for the Marshall teleconference room.

At Thiokol. The first teleconference ended in Utah at about 4:30 MST. Thiokol engineers began working on the engineering presentation for the second teleconference. Vice President of Engineering Bob Lund led the group discussion. Since the flight of STS 51-C on January 1985, no field joint erosion had occurred, and the engineers had felt that the field joint problems were resolved. It was the nozzle joint—a very different design than field joints—that had been having all the recent problems.[43] However, its design and location made the nozzle joint less vulnerable to rotation. The field joints were vulnerable and were thus the issue now. Thiokol's main concern was that cold would reduce O-ring resiliency, creating a potential for increased primary O-ring blow-by in the field joints during joint rotation. The question was, as it always had been, one of joint dynamics, ignition pressure, and timing. But now they had the uncertainty added by the predicted cold temperature: if the primary did not seal, would the secondary O-ring be energized by ignition pressure and form a seal before significant joint rotation occurred and the expected reduced resiliency became a more significant factor?

All the Thiokol people with relevant technical expertise were available because they had been notified during work hours prior to the first teleconference. However, not everyone was in the room the entire time; thus some missed parts of the discussion. Larry Sayer, Director, Engineering and Design, went to the white board and began writing down what they had discussed prior to the first teleconference as a way of working out their technical argument, deciding what charts they would need, and what information would be on the charts. Collectively, they decided they would need 13 charts. Lund assigned the responsibility for the charts, dividing it between people, based on their specializations. Engineers Roger Boisjoly, Jerry Burn, Don Ketner, Joel Maw, Arnie Thompson, and Brian Russell went to their separate offices to prepare them. Joel Maw, a thermal engineer from the Heat Transfer Section, sat at a computer terminal calculating O-ring temperatures for six SRMs that the full group had selected: four SRMs tested in cold temperatures and two launches that had experienced blow-by on field joints. Don Ketner sat beside him, making up the chart as Maw did the calculations.

Those who remained in the conference room while the engineers did their work began discussing the content of Thiokol's conclusions and recommendations charts. They relied on the preceding discussions, without the benefit of the charts containing temperature data that would be used in the presentation. At about 5:30 MST, Jerald Mason, Senior Vice President, Wasatch Operations, and Calvin Wiggins, Vice President and General Manager, Space Division, came into the conference room. Kilminster had telephoned them at Thiokol's Brigham City plant, where they were working for the day. Because a no-launch recommendation by Thiokol on the eve of a launch was unprecedented, they had returned to Wasatch for the teleconference.[44]

When they arrived, Lund, Sayer, and Bill MacBeth, Manager, Project Engineering, were at the white board roughing out the conclusions and recommendations. Those present decided that Thiokol would recommend no launch unless the O-ring temperature was equal to or greater than 53°F, which was the calculated O-ring temperature on the January 1985 SRM-15 (STS 51-C) and the lowest temperature in their flight experience base. No one at Thiokol objected to the recommendation, nor did anyone argue in favor of launch. However, as Bill Macbeth stated: "We discussed the fact that we didn't seem to have a very strong technical position either way, for or against launch. But as very conservative people, since cold might be an indicator, we decided that we ought to not launch until the motor was a little bit warmer, more like the SRM-15 [January 1985 STS 51-C], which was rather ironic and we admitted it. The SRM-15 at that temperature had had problems."[45]

As the engineering data charts were completed, they were circu-

lated among the assembled Thiokol personnel. At 6:00 MST, with
the teleconference 15 minutes away, some charts were still in preparation. Brian Russell gathered those that were finished in order to
send them by telefax to Marshall and Kennedy. The operator on duty
that night did not know how to work the fax machine, however. So
transmission was delayed while Russell called the operator's supervisor to learn how. Maw's chart of O-ring temperature calculations
was among the last to be completed. The full set of Thiokol charts is
displayed in figure 11.

The handwritten charts were prepared that afternoon; the printed
ones were taken from previous presentations. Note that throughout
Thiokol uses their own nomenclature of SRM numbers for identification rather than the STS numbering system NASA uses, which
only appears on figure 11.2. Consider Maw's chart, which was very
important to Thiokol's decision making because it was the only systematic temperature data on full-scale hardware performance (fig.
11.11). From ambient temperature (third column) and wind conditions (last column), he calculated O-ring temperatures (fourth column) for selected motors, listed in the first column. Reading down
the first column, note the data base. The first four items were all the
full-scale static tests of developmental motors and qualification
motors fired at ambient temperatures below 50°F; the next two items
were the only two instances of in-flight field joint blow-by. The last
item was Maw's calculation of predicted O-ring temperature for STS
51-L at launch time: 29°F.[46] Although the chart shows both 29°F and
27°F, the lower temperature was the worst-case prediction, based on
winds more extreme than those expected. Thiokol's engineering
analysis, as stated in their charts, was summarized by Jack Kapp,
Manager, Applied Mechanics Department:

If those [Maw's] numbers were near correct, and we felt good about
that, then we were really firing about 24 degrees Fahrenheit below
our lowest experience base. That was engineering's primary concern. We indicated anew that that [STS 51-C] was the lowest [O-
ring] temperature that we had seen and substantially the worst
blow-by. We indicated that the O-ring squeeze [on STS 51-L] would
be lower due to the lower temperature; that the hardness of the O-
ring would increase essentially 20 points on the shore hardness
scale; that the grease would be more viscous; that the O-ring actuation time—the time for the primary O-ring to come across the
groove and be extruded into the seal gap—would have to be
increased [i.e., greater]. If that time was increased, then the action
of the primary seal could be compromised and it was not at all sure
that late in the ignition transient [i.e., the 600-millisecond period

JANUARY 27, 1986, TELECONFERENCE CHARTS

Temperature Concern on

SRM Joints

27 Jan 1986

Fig. 11.1

HISTORY OF O-RING DAMAGE ON SRM FIELD JOINTS

	SRM No.	Cross Sectional View			Top View		Clocking Location (deg)
		Erosion Depth (in.)	Perimeter Affected (deg)	Nominal Dia. (in.)	Length Of Max Erosion (in.)	Total Heat Affected Length (in.)	
61A LH Center Field**							
61A LH CENTER FIELD**	22A	None	None	0.280	None	None	36°--66°
51C LH Forward Field**	22A	NONE	NONE	0.280	NONE	NONE	338°-18°
51C RH Center Field (prim)***	15A	0.010	154.0	0.280	4.25	5.25	163
51C RH Center Field (sec)***	15B	0.038	130.0	0.280	12.50	58.75	354
	15B	None	45.0	0.280	None	29.50	354
41D RH Forward Field	13B	0.028	110.0	0.280	3.00	None	275
41C LH Aft Field*	11A	None	None	0.280	None	None	--
41B LH Forward Field	10A	0.040	217.0	0.280	3.00	14.50	351
STS-2 RH Aft Field	2B	0.053	116.0	0.280	--	--	90

*Hot gas path detected in putty. Indication of heat on O-ring, but no damage.
**Soot behind primary O-ring.
***Soot behind primary O-ring, heat affected secondary O-ring.

Clocking location of leak check port - 0 deg.

OTHER SRM-15 FIELD JOINTS HAD NO BLOWHOLES IN PUTTY AND NO SOOT NEAR OR BEYOND THE PRIMARY O-RING.

SRM-22. FORWARD FIELD JOINT HAD PUTTY PATH TO PRIMARY O-RING, BUT NO O-RING EROSION AND NO SOOT BLOWBY. OTHER SRM-22 FIELD JOINTS HAD NO BLOWHOLES IN PUTTY.

Fig. 11.2

after ignition during which the joint was rotating], and I believe the number there was about 200 milliseconds, it would seal. Then there was some resiliency data that indicated that the secondary seal may not be in position [to seal the joint]. We indicated the fact that the secondary seal may or may not be in position, depending on when and if the primary sealed.[47]

Before the teleconference began, however, a few people in the room noticed something when comparing the charts that was not obvious when examining them individually. The charts contained data that conflicted with Thiokol's argument that cold weather was correlated

JANUARY 27, 1986, TELECONFERENCE CHARTS
(continued)

PRIMARY CONCERNS -

o   FIELD JOINT - HIGHEST CONCERN

    o   EROSION PENETRATION OF PRIMARY SEAL REQUIRES RELIABLE SECONDARY SEAL
       FOR PRESSURE INTEGRITY
       o   IGNITION TRANSIENT - (0-600 MS)
          o   (0-170 MS) HIGH PROBABILITY OF RELIABLE SECONDARY SEAL
          o   (170-330 MS) REDUCED PROBABILITY OF RELIABLE SECONDARY SEAL
          o   (330-600 MS) HIGH PROBABILITY OF NO SECONDARY SEAL CAPABILITY

    o   STEADY STATE - (600 MS - 2 MINUTES)
       o   IF EROSION PENETRATES PRIMARY O-RING SEAL - HIGH PROBABILITY OF
       NO SECONDARY SEAL CAPABILITY
          o   BENCH TESTING SHOWED O-RING NOT CAPABLE OF MAINTAINING CONTACT
          WITH METAL PARTS GAP OPENING RATE TO MEOP
          o   BENCH TESTING SHOWED CAPABILITY TO MAINTAIN O-RING CONTACT DURING
          INITIAL PHASE (0-170 MS) OF TRANSIENT

Fig. 11.3

FLD JOINT   PRIMARY CONCERNS   SRM 25

   o   A TEMPERATURE LOWER THAN CURRENT DATA BASE RESULTS
     IN CHANGING PRIMARY O-RING SEALING TIMING FUNCTION
   o   SRM 15A — 80° ARC BLACK GREASE BETWEEN O-RINGS
     SRM15B — 110° ARC BLACK GREASE BETWEEN O-RINGS

   o   LOWER O-RING SQUEEZE DUE TO LOWER TEMP

   o   HIGHER O-RING SHORE HARDNESS

   o   THICKER GREASE VISCOSITY

   o   HIGHER O-RING PRESSURE ACTUATION TIME

   o   IF ACTUATION TIME INCREASES, THRESHOLD OF SECONDARY
     SEAL PRESSURIZATION CAPABILITY IS APPROACHED

   o   IF THRESHOLD IS REACHED THEN SECONDARY SEAL MAY
     NOT BE CAPABLE OF BEING PRESSURIZED

Fig. 11.4

JANUARY 27, 1986, TELECONFERENCE CHARTS
(continued)

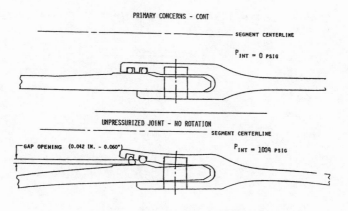

Fig. 11.5

Fig. 11.6

with O-ring damage. On Boisjoly's chart (fig. 11.6), he compared the only two instances of field joint blow-by: SRM-22 (STS 61-A, October 1985) and SRM-15 (STS 51-C, January 1985). His chart made the point that blow-by on the January cold-temperature launch was the worse of the two. However, his chart did not include temperature data. But on Maw's chart (fig. 11.11) SRM-22, which Boisjoly used in his comparison, had a calculated O-ring temperature of 75°F. In the only two cases of field joint blow-by, one occurred at an O-ring temperature of 53°F,

JANUARY 27, 1986, TELECONFERENCE CHARTS
(continued)

O-RING (VITON) SHORE HARDNESS VERSUS TEMPERATURE

°F	SHORE HARDNESS
70°	77
60°	81
50°	84
40°	88
30°	92
20°	94
10°	96

Fig. 11.7

SECONDARY O-RING RESILIENCY

DECOMPRESSION RATE
2"/MIN (FLIGHT ≈ 3.2"/MIN)

TEMP (°F)	TIME TO RECOVER (SEC)
50	600
75	2.4
100	*

* DID NOT SEPARATE

RIGHT DUROMETER (?)

Fig. 11.8

the other at 75°F. Here was a flaw in the engineering rationale behind their launch recommendation. Larry Sayer remarked: "We talked about that a little bit, because that certainly weakened the engineering position that we didn't want to launch below SRM 15 (STS 51-C), which was 53°. . . . The way we resolved it in our own minds was that Mr. Boisjoly, who had seen both of those, said that the blow-by on SRM 15 was much more severe than we saw on SRM 22."[48]

Don Ketner, speaking of the conflicting data, stated: "We had

JANUARY 27, 1986, TELECONFERENCE CHARTS
(continued)

BLOW - BY   TESTS   (PRELIMINARY)

ARGON

	TEMP.(°F)	RESULTS ($\frac{IN^3}{IN\ SEAL}$)
	75	No LEAKAGE
	30	No LEAKAGE

F-14

	75	No RESULTS YET
	30	No RESULTS YET

400   170 ms
330 ms

1000 psi in 0.6 sec
3 to 4 kts at each condition
.020 compression

Fig. 11.9

FIELD JOINT O-RING SQUEEZE (PRIMARY SEAL)

MOTOR	FWD	CTR	AFT
SRM 15 A	16.1 (.045)*	15.8 (.044)	14.7 (.041)
SRM 15 B	11.1 (.031)	14.0 (.039)**	16.1 (.045)
SRM 25 A	10.16 (.028)	13.22 (.037)	13.39 (.037)
SRM 25 B	13.91 (.039)	13.05 (.037)	14.25 (.040)

* 0.010" EROSION
** 0.038" EROSION

Fig. 11.10

blow-by on the hottest motor and on the coldest motor. So it was not conclusive to me just how bad the blow-by erosion would be on a cold motor. It was intuitively obvious that cold is worse and therefore, because it was so much colder than any previous flights, that we should not launch. And that was the same opinion that others in the room expressed."[49]

The teleconference call came at 6:15 MST (8:15 EST), as scheduled, but Thiokol was not yet prepared. Thiokol's Boisjoly noted: "We had

JANUARY 27, 1986, TELECONFERENCE CHARTS
(continued)

HISTORY OF O-RING TEMPERATURES
(DEGREES - F)

MOTOR	MGT	AMB	O-RING	WIND
DM-4	68	36	47	10 MPH
DM-2	76	45	52	10 MPH
QM-3	72.5	40	48	10 MPH
QM-4	76	48	51	10 MPH
SRM-15	52	64	53	10 MPH
SRM-22	77	78	75	10 MPH
SRM-25	55	26	29	10 MPH
			27	25 MPH

1-D THERMAL ANALYSIS

Fig. 11.11

CONCLUSIONS:

TEMPERATURE OF O-RING IS NOT ONLY PARAMETER CONTROLLING BLOW-BY

SRM 15 WITH BLOW-BY HAD AN O-RING TEMP AT 53°F
SRM 22 WITH BLOW-BY HAD AN O-RING TEMP AT 75°F
FOUR DEVELOPMENT MOTORS WITH NO BLOW-BY WERE TESTED AT O-RING TEMP OF 47° TO 52°F

DEVELOPMENT MOTORS HAD PUTTY PACKING WHICH RESULTED IN BETTER PERFORMANCE

AT ABOUT 50°F BLOW-BY COULD BE EXPERIENCED IN CASE JOINTS

TEMP FOR SRM 25 ON 1-28-86 LAUNCH WILL BE 29°F 9AM
38°F 2PM

HAVE NO DATA THAT WOULD INDICATE SRM 25 IS DIFFERENT THAN SRM 15 OTHER THAN TEMP

Fig. 11.12

very little time to prepare data for that telecon. When it was decided in a group in our conference room that we would take a systematic approach and assign certain individuals certain tasks to prepare, we all scurried to our individual locations and prepared that information in a timely manner. That is why the charts for the most part are hand printed or hand written, because we didn't have time to get them even typed."[50] Thiokol's Arnie Thompson had a different view: "I

JANUARY 27, 1986, TELECONFERENCE CHARTS
(continued)

RECOMMENDATIONS :

° O-RING TEMP MUST BE $\geq$ 53 °F AT LAUNCH

DEVELOPMENT MOTORS AT 47° To 52 °F WITH
PUTTY PACKING HAD NO BLOW-BY
SRM 15 (THE BEST SIMULATION) WORKED AT 53 °F

° PROJECT AMBIENT CONDITIONS (TEMP & WIND)
To DETERMINE LAUNCH TIME

Fig. 11.13

was very content with the pitch that we had put together. We had been working on it all day long, and I felt real good about it."[51]

Thiokol held the teleconference line open until 6:25, when Russell faxed the first charts to Marshall and Kennedy. At about 6:45 MST, Thiokol was sufficiently organized to begin.

Thirty-four engineers and managers from Marshall and Thiokol participated in the second teleconference (see appendix B, fig. B1). Thiokol engineers in Utah presented the charts they had faxed containing their technical analysis of the situation. They argued that O-ring ability to seal the booster joints at ignition would be slower at predicted temperatures than on the coldest launch to date: a January 1985 mission when the calculated O-ring temperature was 53°F. On that flight, hot propellant gases had penetrated a joint, eroding a primary O-ring to such an extent that it had not sealed at all, allowing the hot gases to "blow by" the primary toward the secondary O-ring. The secondary performed its intended function as a backup and successfully sealed the joint, but more extensive blow-by could jeopardize the secondary O-ring and thus mission safety.

The 34 participants did not divide cleanly into the categories "managers" and "engineers." According to Marshall's Larry Wear: "All of the people that participated in that meeting are engineers. So it's difficult to say that engineering is saying one thing and management is saying something else. They are all engineers and they are all engineering managers. There aren't any "pure management people" in the whole stack. Everyone who is there is a trained engineer, has worked as a working engineer. Even though they may not be assigned to engineering organization per se, they are, by trade, engineers and by thought process engineers."[52]

The recruitment of specialists for the teleconference brought Level III and IV managers and engineers in the SRB work group together with people who had not been part of the daily and weekly interactions about the SRB joints. This practice, established in order to use NASA's matrix system to best advantage when making launch decisions, was designed to bring together all relevant disciplines and fresh insights to flight-readiness deliberations. From each location, names of participants were announced, as was the custom. Although some could identify the speakers at other teleconference sites by voice, some could not identify all, and some could identify none. Some did not even recognize all the names. Thiokol Senior Vice President Mason and Vice President Wiggins were not acknowledged to the other locations as being among those at Thiokol, which, according to Marshall's Larry Wear, was "very unusual."[53]

Thiokol's Jack Buchanan said that teleconferences normally were not orderly affairs, and this one was no exception. There were the usual distractions that prevented people from hearing everything that was said. At Thiokol, Russell faxed additional charts as they were ready, leaving the conference room to do so (and missing parts of the discussion). The conclusions and recommendations charts were not finished. The Thiokol managers and engineers had agreed on what they wanted to say, but there were some problems putting the statements in logical order. Jerry Burn still was cutting and pasting together those last two charts. The charts arrived at Kennedy and Marshall in spurts and out of sequence. They had to be retrieved from fax machines that were in other rooms, put in sequence, copied, and distributed.

People would whisper comments or questions to the person next to them in response to a chart or to information that came over the teleconference net. Side conversations were frequent because people tended to sit next to the person they supported so that the people closest to the technical problem could give information to those immediately above them in the hierarchy. Marshall engineer Ben Powers remarked: "I saw that John McCarty, my deputy laboratory boss, was there, and since he would be the guy that I would support, he would be the voice. I went and sat right next to him so I could work with him. That's the way we work. Rather than sit at the other end of the room and try to avoid him, it's better just to sit right next to your boss so you can talk to him."[54] In Huntsville, Schell, Riehl, and Schwinghamer, late because they had been investigating the manufacturers' literature on cold temperature and O-ring materials, came into the Marshall teleconference room after the initial introductions. They, too, sat hierarchically: Schell (out of retirement for the evening), next to Riehl (Chief, Nonmetallic Materials Division), next to Schwinghamer (Director, Material and Processes Lab).[55] Soon after arriving, Schwinghamer got up from the table and showed the data on cold temperature and Viton to his superior, Wayne Littles, Associate Director of Engineering.

Russell began the Thiokol presentation with the first chart that focused on the history of damage on the SRM field joints (fig. 11.2). The audio hook-up was clear at all locations, but no television hook-up accompanied it. Russell had taken an old chart and added new information to it that pertained to the two field joints that had experienced blow-by, indicated by double and triple asterisks: the top two lines of data on STS 61-A (SRM-22A) and the last two lines on the page, giving more information about the two flights that would be central to their temperature argument. Note that since the field joint was Thiokol's major concern, nozzle joint performance was omitted. As the discussion got underway, Marshall's Resident Manager at Kennedy, Cecil Houston, realized that people were using two or three different temperature estimates for the morning of the launch. Houston went to his office, called Gene Sestile, Acting Launch Director, told him about the teleconference, and asked for predicted hourly ambient temperatures through noon on the day of the launch. The predicted temperatures matched the ones McDonald had given Thiokol earlier in the day: 30°F at midnight, 22°F at 6:00 A.M., rising to 25°F by 9:00, 26°F at launch time, and warming through the 40s and 50s in the afternoon.[56] Houston read them over the net.[57]

The Marshall people did not contest the Thiokol engineering analysis at this time, only asking questions for clarification as Thiokol went through the charts. Thiokol's Jack Buchanan, at Kennedy, observed: "It was nothing really earthshaking, but they were asking in detail about questions and explanations and give and take, which was normal, I think, of a technical professional meeting and with people who had worked together for many years back and forth. So about the O-ring, they said it was sooted but if you look at the chart carefully you'll see the O-ring was undamaged. So they asked, is it really undamaged? Well, we looked at it with a magnifying glass and we couldn't find damage. There's that give and take."[58]

Thiokol's Bill Macbeth, in Utah, recalled in testimony to the Presidential Commission being surprised when Russell presented the first chart:

> I recognized that it was not a strong technical position, but yes, I basically supported that position. I had become very concerned during the presentation, however, when one of the people seemed to indicate, at least I interpreted it that way, that he had forgotten or didn't know about one of the recent warm temperature firings that had also had a problem.
>
> MR. MOLESWORTH: Can you identify that person?
>
> MR. MACBETH: Roger Boisjoly. The statement that I understood him to make—and I was sitting close to him when Brian Russell put the first chart up with the history [fig. 11.2]—this was a formal chart he had presented before [but] he had added on this recent data of the SRM 22 [STS 61-A], which was roughly a 75° shot—I thought I understood Roger [Boisjoly] to say, "I didn't know that." And that then jogged my mind to start thinking more carefully back over some of the history that I could remember . . . and I felt that really weakened our position, because our whole position was based on the performance of SRM-15, and it had been successful, and yet it had blow-by and erosion. But there had been other motors that had blow-by and erosion that had been warm shots. In fact, I think everything else we fired from the Cape was above 60 degrees. And so it began, to my way of thinking, to really weaken our conclusions and recommendations. And I was already wishy-washy. And that one [chart] really hit me home when I began to think, gosh, you haven't really thought this out as thoroughly as you should have.[59]

Thiokol's Brian Russell, recalling that the full group at Thiokol had not discussed the implications of the 75°F O-ring temperature data, remarked: "It caught Bill, for one, by surprise that we had had soot blow-by on STS 61-A. And that tended to cloud the issue, especially in his mind. . . . I imagine that we tended to focus on the data that we thought were telling us something: namely, that we have a higher risk situation with the cold temperature. And others not necessarily as familiar wouldn't have realized that data at that time because we really hadn't mentioned it until we had to write it down and present it later. . . . It might have been mentioned, but I doubt it because, like I say, Bill for one was surprised to see that data."[60]

Following Russell's discussion of field joint history, Boisjoly, who had been at the Cape for the disassembly of SRM-15 in January 1985, took over with several charts—some old, some new. His first one had been presented at the August 19 NASA Headquarters briefing. It stated Thiokol's primary concerns at that time, giving the rationale for continuing to fly (fig. 11.3). Boisjoly recalled:

That was the chart that I pulled right out of the [August 19, 1985] Washington presentation without changing one word of it because it was still applicable, and it addresses the highest concern of the field joint in both the ignition transient condition and the steady state condition, and it really sets down the rationale for why we were continuing to fly. Basically, if erosion penetrates the primary O-ring seal, there is a higher probability of no secondary seal capability in the steady state condition. And I had two sub-bullets under that which stated bench testing showed O-ring not capable of maintaining contact with metal parts, gap, opening rate to maximum operating pressure.

I had another bullet which stated bench testing showed capability to maintain O-ring contact during initial phase [i.e., 0–170 milliseconds of ignition transient]. That was my comfort basis of continuing to fly under normal circumstances, normal being within the data base we had. I emphasized when I presented that chart about the changing of the timing function of the O-ring as it attempted to seal. I was concerned that we may go from that first beginning region into that intermediate region, from 0 to 170 being the first region, and 170 to 330 being the intermediate region, where we didn't have a high probability of sealing.[61]

Boisjoly then presented a chart giving an overview of the engineering concerns for the *Challenger* mission (fig. 11.4), followed by charts that backed up those points with details (figs. 11.5–11.7 and 11.10). Summarizing them, Boisjoly recalled:

Now that [the ring] would be harder. And what that really means basically is that the harder the material is, it would be likened to trying to shove a brick into a crack versus a sponge. That is a good analogy for purposes of this [Commission hearing] discussion. I also mentioned that thicker grease, as a result of lower temperatures, would have a higher viscosity. It wouldn't be as slick and slippery as it would be at room temperature. And so it would be a little bit more difficult to move across it. . . . We would have higher O-ring pressure actuation time [i.e., the rings would be slower to respond to ignition pressure]. That was my fear . . . the secondary seal may not be capable of being pressurized, and that was the bottom line of everything that had been presented up to that point.[62]

Then, for the first time, someone at one of the other teleconference locations saw the contradiction in the Thiokol data, pointing out on the net that the blow-by at 75°F (SRM-22; fig. 11.6) indicated that temperature was not a discriminator. Since O-ring damage was known to have had a number of causes in the past, Boisjoly was asked, what evidence did Thiokol have that damage on STS 51-C was the result of the cold? Boisjoly recalled:

I was asked, yes, at that point in time I was asked to quantify my concerns, and I said I couldn't. I couldn't quantify it. I had no data to quantify it, but I did say I knew it was away from goodness in the current data base. Someone on the net commented that we had soot blow-by on SRM-22, which was launched at [an O-ring temperature of] 75°. I don't remember who made the comment, but that is where the first comment came in about the disparity between my conclusion and the observed data because SRM-22 [STS 61-A, October 1985] had blow-by at essentially a room temperature launch. I then said that SRM-15 [STS 51-C, January 1985] had much more blow-by indication and that it was indeed telling us that lower temperature was a factor. This was supported by inspection of flown hardware by myself. I was asked again for data to support my claim, and I said I have none other than what is being presented, and I had been trying to get resilience data, Arnie and I both, since last October, and that statement was mentioned on the net.[63]

The people at Marshall and Kennedy continued to have what all participants described as a neutral engineering exchange about the data and what it meant. At some point during the discussion, Thiokol's Jack Buchanan, Resident Manager at Kennedy, leaned over to McDonald and whispered that when STS-9 was destacked at the Cape a year earlier, they found rainwater had accumulated in the joint. Neither McDonald nor Buchanan mentioned it to the others in the room or brought it up on the teleconference line.[64] After Boisjoly, Arnie Thompson presented his charts. The first gave resiliency data from lab tests showing that as temperature decreased, the O-ring would be slower to respond (fig. 11.8). Thompson had presented these same data to a group at Marshall in July 1985. His next chart, however, generated some discussion (fig. 11.9). It showed the results of some subscale tests, from which he had extracted the relationship between temperature and blow-by. It contained data again contradicting Thiokol's position: comparing static tests for blow-by at two temperature extremes, this chart showed no blow-by at either a 75°F or 30°F temperature. When someone over the net asked about this discrepancy, Thompson responded that those static tests were not very good tests because they did not adequately simulate joint operation under real ignition conditions. Russell faxed the conclusions and recommendations charts at about 9:20 P.M. EST. The last data chart presented was Joel Maw's temperature chart, which showed the contrasting 75°F and 53°F O-ring temperatures of the two flights with blow-by (fig. 11.11). No one on the net had any questions. As Maw finished, the conclusions and recommendations charts arrived at the space centers. **Thiokol Vice President of Engineering Robert Lund presented the**

Thiokol engineering conclusion to teleconference participants: O-ring temperature must be equal to or greater than 53°F at launch.

Marshall's Larry Mulloy asked Thiokol management for a recommendation. Thiokol Vice President Joe Kilminster responded that on the basis of the engineering conclusions, he could not recommend launch at any temperature below 53°F. Immediately, Marshall managers in Huntsville and at the Cape began challenging Thiokol engineers' interpretation of the data. Mulloy stated that since no Launch Commit Criteria had ever been set for booster joint temperature, what Thiokol was proposing to do was to create new Launch Commit Criteria on the eve of a launch. Mulloy then exclaimed, "My God, Thiokol, when do you want me to launch, next April?" Reinartz asked George Hardy, Marshall's Deputy Director of Science and Engineering who was in Huntsville, for his view. Hardy responded that he was "appalled" at the Thiokol recommendation.

Reinartz then asked Kilminster for comments. Kilminster requested a five-minute off-line caucus for Thiokol managers and engineers in Utah.

At NASA, teleconferences were the norm. As John W. Young, Chief of the Astronaut Office, remarked, "Those kinds of meetings go on at NASA before launch every hour of the day and night."[65] Engineering disagreements also were the norm. Typical of engineers who work on large-scale technical systems, the Marshall and Thiokol working engineers had disagreed many times since the inception of the Space Shuttle Program about how the SRB joint was performing and why it was behaving as it was. However, in all the years of association between Thiokol and Marshall, this was the first time the contractor had come forward with a no-launch recommendation.[66] Thiokol's Bob Ebeling stated: "This is the one and only time of all my 13 years on this project [that] anything like this ever took place. We had disagreements with NASA all the time, and that's normal, and their engineers, we disagreed with them quite often. They're superconservative people. Anyway, this is the one and only time that I've ever seen Thiokol ever ask for a delay. Usually we bend over backwards to meet the schedule. Now here we are requesting that it not be launched because we are a bit concerned about it."[67]

After Lund read the conclusions chart and the launch recommendation, "O-ring temp must be equal to or greater than 53°F at launch" (figs. 11.12 and 11.13), he added that NASA might want to consider delaying launch until the afternoon launch window or slipping it to a warmer day. Mulloy then asked if that was the end of the presentation. Lund said yes. Mulloy then asked Thiokol's Joe Kilmin-

ster, formally responsible for presenting the official company position to NASA, for his recommendation. Kilminster responded that
based on the engineering recommendation, he could not recommend
launch. At this point, Marshall managers began aggressively challenging Thiokol's presentation and no-launch recommendation.
Mulloy said that he had a dissenting opinion. He talked for about five
minutes. He stated that the data presented, when considered within
the context of all engineering data available, did not support
Thiokol's recommendation not to launch unless the temperature
were equal to or greater than 53°F. When all data were considered, (1)
temperature did not appear to be correlated with erosion and blow-
by and (2) even if the predicted cold slowed the primary, the data indicated redundancy: the secondary would seal the joint. He then went
through Thiokol's analysis, referring to notes he took during the presentation, pointing out data in Thiokol's charts that contradicted the
temperature correlation and data that supported redundancy:[68]

> Where I was coming from was we had been flying since STS-2
> with a known condition in the joints that was duly considered and
> accepted by Thiokol, it was accepted by me, and it was accepted by
> all levels of NASA management through the Flight Readiness
> Review process and through special presentations that we had put
> together and provided up here to the headquarters people. The
> [three-factor technical] rationale for accepting that condition had
> been previously documented.
> I stated, after that beginning—"This is a condition that we have
> had since STS-2 that has been accepted"—that blow-by of the O-
> rings cannot be correlated to the temperature by these data. STS
> 61-A had blow-by at 75 degrees. Soot blow-by has occurred on more
> than one occasion, independent of temperature. This is the nature
> of the challenges: think about this, think about your data.
> "Primary erosion occurs due to concentrated hot gas passed
> through the putty." I just wrote that down to say we know why we
> get erosion. We have done tests, we have done analyses, we under-
> stand the limits that the erosion can be, and we understand by tests
> how much we can withstand. The colder temperature may result
> in greater primary O-ring erosion and some heat effect on the sec-
> ondary because of increased hardness of the O-ring, resulting in
> slower seating . . . [but] it was stated in the Thiokol data that was
> presented that we had successfully shown that an O-ring with a
> durometer of 90 would extrude into the gap and seal. It wasn't too
> hard to extrude into the gap [Mulloy referred to 30°F temperature
> data in figs. 11.7 and 11.9]. . . . The other positive point is that
> "Squeeze at 20 degrees is positive," it is greater than 20 thou-
> sandths [the minimum manufacturer's requirement]. We were
> starting out with a squeeze of some 36 to 40 mils.

"The secondary seal is in a position to seat." It is in a position to seat and seal because of the 200 psi and 50 psi pressurization at leak check. The primary may not seal due to reduced resiliency and increased durometer—the worst case condition. If the worst thing happens, however, during the period of flow past the primary the secondary will be seated and seal before significant joint rotation occurs, which is less than 170 milliseconds. My conclusion was that that condition [redundancy] has been recognized at all levels of NASA management and is applicable to 51-L.[69]

All participants said that Thiokol had a weak engineering argument. Marshall's Ben Powers, who agreed with the Thiokol recommendation, made this observation:

I don't believe they did a real convincing job of presenting their data. . . . The Thiokol guys even had a chart in there that says temperature of the O-ring is not the only parameter controlling blow-by. In other words, they're not coming in with a real firm statement. They're saying there's other factors. They did have a lot of conflicting data in there. I can't emphasize that enough. Let me quote. On [one] chart of theirs [fig. 11.9], they had data that said they ran some sub-scale tests down at 30 degrees and they ran blow-by tests in a fixture at 30 degrees, and they saw no leakage. Now they're talking out of both sides of their mouth, see. They're saying, "Hey, temperature doesn't have any effect." Well, I'm sure that that was an extremely important piece of data to Mr. Hardy, and certainly to my boss, Mr. McCarty, and Mr. Smith.[70]

Thiokol's engineers agreed. Brian Russell:

The fact that our recommendations were questioned was not all that unusual. . . . In fact, it's more the rule. We are often times questioned on our rationale, which is the way I believe it should be. . . . We felt in our presentation of the data that we had to include all of the data that could possibly be relevant, even though not all of it tended to support our point of view. . . . So his [Mulloy's] argument was rational and logical in my mind. I didn't agree with it necessarily. Well, I did not, because I felt that the data of the soot blow-by which was extensive on 51-C and the resiliency tests were telling us something. Namely, that the colder the temperature, the higher the risk. And the other data, though they didn't support that, I tended not to give them as much value in my own mind.[71]

Thiokol's Larry Sayer:

I'm the first to admit that we had a very weak engineering position when we went into the telecon. . . . When Larry (Mulloy) came through and made his reply to our recommendation, you know, it was a very sound, rational appraisal of where he felt we were in error.[72]

Thiokol's Roger Boisjoly:

> My observation is that when we as engineers presented our data and were unable to quantify it, it left it somewhat open from the standpoint of we only had data down to 50 degrees.[73]

Thiokol's Jack Kapp:

> Most of the concerns that we had presented were qualitative in nature. At that particular time we had a very difficult time having enough engineering data to quantify the effects that we had been talking about. A lot of it was based on "engineering feel." ... Everything that Mr. Mulloy had brought up we had discussed earlier [as they prepared for the teleconference]. Engineering, I think, was generally aware that a great deal of our recommendation and our supporting data was subjective and kind of qualitative, but nonetheless it was there, and the engineering judgment was there.[74]

Thiokol's Al McDonald:

> NASA concluded that the temperature data we had presented was inconclusive and indeed a lot of the data was inconclusive because the next worst blow-by we had ever seen in a primary seal in a case-to-case field joint was about the highest temperature we had launched at, and that was true—the next worst blow-by.[75]

Thiokol's recommendation for 53°F as the baseline temperature for decision making was central to the controversy. The discussion became more heated as the participants began to argue about temperature requirements for launch and to question 53°F as a numerical cutoff for the launch decision. NASA had many temperature guidelines for the entire shuttle system as well as for its individual components under varying conditions, so many, in fact, that a NASA index of each component and the original document in which its temperature specifications could be found extended for 26 pages.[76] This proliferation notwithstanding, no Launch Commit Criteria had been established for O-ring temperatures. Although Thiokol's working engineers had been worried about cold temperature after STS 51-C in January 1985, they did not attempt to initiate and establish a temperature guideline with their own or NASA management.[77] As Boisjoly testified, "It was nobody's expectation we would ever experience any cold weather to that degree before we had a chance to fix it again, so that basically is why it wasn't pursued any further than that from my personal standpoint."[78]

In the absence of a formalized, test-derived rule about O-ring temperature that also took into account pressure and sealing time, uncertainty prevailed. Thiokol created a rule, using the experience base: do not launch unless O-ring temperature is 53°F or greater. But people at Marshall and Kennedy were surprised at the choice of this number. First, it was contradicted by data from tests done at 30°F presented in Thiokol's own charts, as the above comments confirm. Second, it contradicted other temperature guidelines. There were serious differences in opinion about which one applied. The manufacturers' guidelines for the Viton O-rings that Schell and Riehl had studied said Viton would retain resiliency down to 25°F. Although no one had tested O-ring behavior within Thiokol's design at a full range of temperatures, Schell and Riehl thought that the enormous ignition pressures would force them to seal down to 25°F. Marshall's Reinartz interjected that he thought the motor was qualified at a propellant mean bulk temperature ranging from 40°F to 90°F. Since Thiokol had fired the fully assembled motor at those temperatures in Utah, he thought 40°F was a legitimate data point. Thiokol's Jack Buchanan recalled that others thought Reinartz's point was valid, but they did not argue that the 40°F applied to the seals because "they saw the charts and the charts said that the theoreticians had calculated an O-ring temperature of 29° based on the weather forecast. The Marshall people did refer to a Johnson document that talked about 31 degrees."[79] That document established a Launch Commit Criteria of 31°F–99°F ambient temperature for the entire Space Shuttle.[80]

Third, the 53°F cutoff struck many as peculiar because it contradicted Thiokol's recent behavior. For 19 days in December and 14 so far in January, the ambient temperature at the Cape had been below the 53°F limit Thiokol was urging on the eve of the launch.[81] During this time, NASA had been having trouble getting STS 61-C off the launch pad and scrubbed launch four times for reasons unrelated to temperature. Thiokol had never raised the effect of cold temperature on O-rings as an issue for STS 61-C, although the predicted ambient temperature was less than 53°F for several of those scheduled launches, going as low as 41°F for one of them.[82] Moreover, on the morning of the January 27 teleconference, the ambient temperature was 37°F at 8:00 A.M. but Thiokol had not raised any concerns about the cold temperature. The *Challenger* launch scheduled for that morning was scrubbed, not because of temperature, but because of crosswinds at the launch site. John Schell, the retired Marshall rubber chemist participating in the teleconference, said, "It [the 53°F limit] struck me kind of funny because if I read the paper right, it was 40°F and they were going ahead and launch, and so it kind of surprised me that one day they were going to launch at 40° and the next they were saying 53°."[83]

Fourth, as Mulloy pointed out, the *Challenger* launch recommendation contradicted the position Thiokol engineers had taken after the very January 1985 53°F flight that was the crux of their no-launch argument. In the next FRR, Thiokol engineers recommended launch because their data analysis showed the safety margin was very much intact: the primary would still seal with three times the erosion STS 51-C had experienced. Mulloy recalled:

> The conclusions as presented by Morton Thiokol—and I certainly had no dissent at that time—was that 51-E, which was going to be launched in a much warmer temperature, may have exactly the same type of erosion and blow-by, that we might expect that; that the tests and analysis that had been done showing that did show that we were tolerant to that. And the conclusion was that 51-E was acceptable to launch with full expectation that we might see again exactly what we saw in 51-C.
>
> We tested that logic and rationale in great detail. I signed an Action Item in my Solid Rocket Booster FRR Board [for Thiokol] to provide additional data relative to that particular recommendation on 51-E . . . that Action Item was answered [by Thiokol]. The conclusion relative to temperature was that that temperature effect would still give us adequate squeeze and joint operation. Now that was the basis that I was coming from, which led me to conclude that this was a rather surprising recommendation.[84]

Finally, he argued that by suggesting a 53°F temperature cutoff Thiokol was, in effect, establishing a Launch Commit Criterion on the night before a launch. He said that they were basing very serious conclusions and recommendations on a 53°F limit that was not, in his view, supported by the data. With this as a criterion, it might be spring before they could launch another shuttle. Mulloy asked, "My God, Thiokol, when do you want me to launch, next April?"

Mulloy, responding to Presidential Commissioner Hotz, who inquired about the meaning of his controversial statement:

> It is certainly a statement that is out of context, and the way I read the quote, sir—and I have seen it many times, too many times—the quote I read was: "My God, Thiokol, when do you want me to launch, next April?" Mr. McDonald testified to another quote that says, "You guys are generating new Launch Commit Criteria." Now both of those I think kind of go together, and this is what I was saying.
>
> The total context, I think, in which those words may have been used is, there are currently no Launch Commit Criteria for joint temperature. What you are proposing to do is to generate a new Launch Commit Criteria on the eve of launch, after we have successfully flown with the existing Launch Commit Criteria 24 pre-

vious times. With this LCC, i.e., do not launch with a temperature greater than 53 degrees [*sic*], we may not be able to launch until next April. We need to consider this carefully before we jump to any conclusions. It is all in the context, again, with challenging your interpretation of the data, what does it mean and is it logical, is it truly logical that we really have a system that has to be 53 degrees to fly?[85]

Mulloy was not alone in his view.[86] Marshall's Larry Wear stated:

Whether his choice of words was as precise or as good or as candid as perhaps he would have liked for them to have been, I don't know. But it was certainly a good, valid point, because the vehicle was designed and intended to be launched year-round. There is nothing in the criteria that says this thing is limited to launching only on warm days. And that would be a serious change if you made it. It would have far reaching effects if that were the criteria, and it would have serious restriction on the whole shuttle program. So the question is certainly germane, you know, what are you trying to tell us, Thiokol? Are we limited to launching this thing only on selected days. . . . The briefing, per se, was addressing this one launch, 51-L. But if you accept that input for 51-L, it has ramifications and implications for the whole 200 mission model, and so the question was fair.[87]

Marshall's Bill Riehl said:

I was very much surprised. . . . The 53° took me. Not only that, but 53°, the implications of trying to live with 53° were incredible. And coming in the night before a launch and recommending something like that, on such a weak basis was just—I couldn't understand. If they had come in with 30 degrees, based on what I saw [on their charts], I would have understood it. But when they came in with 53 degrees, if I had been in the conference at the head table, I would have challenged it and said: Come here. Where'd you come in with the 53°? And they said because they had flown at 53° before, which is no reason to me. That's tradition rather than technology.[88]

After Mulloy's "April" comment, Reinartz asked Marshall's George Hardy for his view. Hardy stated that he believed Thiokol had qualified the motor at a temperature substantially lower than 53°F and that he was "appalled at the Thiokol recommendation" not to launch until the temperature reached 53°F. He went on to say that the information that was presented was not conclusive because the engineering data did not support a correlation between temperature and O-ring erosion; other factors had been responsible in the past, as Thiokol had noted in the first statement of their conclusions (fig. 11.12). Hardy said he agreed with Mulloy that the engineering data did not seem to change the basic rationale for flight to date—that if the primary should fail in a WOW condition, the secondary would provide redundancy. Then he concluded with another statement remembered by all participants: "I will not agree to launch against the contractor's recommendation."[89]

After Hardy spoke, there was dead silence from the Utah teleconference participants for a few seconds. Several Thiokol engineers felt that Hardy's use of the word "appalled" was an unusually strong response, which, joined with Mulloy's remarks, put pressure on them. Thiokol's Brian Russell: "And that was a strong comment which hit me pretty strongly. Now, I've not had that many personal dealings with Mr. Hardy. . . . I have to admit here again, my experience with him is pretty limited. But in my experience, it did catch me by surprise. He is very highly respected by his own people at Marshall Space Flight Center. So that was a very strong comment which tended to hit me hard. In my mind, I didn't think that the recommendation was out of line based on the data."[90]

Thiokol's Joel Maw: "I could see a few smiles in the room, you know, like maybe they expected it. I don't know. In talking to some people afterward, you know, I guess this is something that this fellow might say, but at the time I felt like he was intimidating us."[91]

Thiokol's Bob Ebeling: "[It] was a very strong, stern comment, and Mulloy and Hardy both used sternness in their vocabulary [so] that we knew they weren't real happy with us enlightening them of this problem."[92]

Thiokol's Jack Kapp: "I have utmost respect for George Hardy. I absolutely do. I distinctly remember at that particular time, speaking purely for myself, that that surprised me, that that [use of the word "appalled"] was a little bit out of character for Mr. Hardy. I'd have to say it was. And I do think that the very word itself had a significant impact on the people in the room. Everybody caught it. It indicated to me that Mr. Hardy felt very strongly that our arguments were not valid and that we ought to proceed with the firing."[93]

Other engineers, from both Thiokol and Marshall, interpreted Hardy's comments differently.[94] Thiokol's Jack Buchanan: "I think that was normal give and take. I think there was concern by the Marshall people that they just couldn't have a system that you could only launch in the summer. But of course, as you know, in Florida, really our cold spells are not that long. So I think they were exaggerating to make a point. They were making a point, driving a point home."[95]

Thiokol's Boyd Brinton testified that he did not think his colleagues in Utah were being pressured to change their recommendation, nor that there was anything unusual about Marshall's response to the presentation. He said, "It's kind of a way of life. I have certainly been in other meetings that were similar to this, and I don't think that there was anything that was extremely outstanding."[96]

Thiokol's Bill Macbeth agreed:

No, it certainly wasn't out of character for George Hardy. George Hardy and Larry Mulloy had difference in language, but basically the same comment coming back, [they] were indicating to us that they didn't agree with our technical assessment because we had slanted it and had not been open to all the available information. . . . I felt that what they were telling us is that they remembered some of the other behavior and presentations that we had made and they didn't feel that we had really considered it carefully, that we had slanted our presentation. And I felt embarrassed and uncomfortable by that coming from a customer. I felt that as a technical manager I should have been smart enough to think of that, and I hadn't.[97]

Marshall's Ben Powers and Keith Coates, who both agreed with Thiokol's no-launch recommendation, were at Marshall with Hardy. Powers said:

He had a pad out. He had his sleeves rolled up. You know, he was working and he was looking at what they were doing, and when he would ask them a question he would write it down. . . . My best assessment of what he was doing was trying to understand that data the very best he could, because I thought he was preparing himself to try to defend that position if he stopped the launch. . . . And the reason for that is, if he tells his boss he's going to stop the launch, then he has to understand why he's stopping the launch. You know, it's that kind of thing, because stopping a launch is not an easy thing. There's a lot of people committed to launching. You've got a lot of things going in that direction, and you don't just stop a launch just for fun. You have to have a good reason for it. And I think he was trying to nail down what the reason was so that he could then defend that position. . . . And I think that the Thiokol people may have been understanding him as like he was being hard in that he was asking a lot of questions rather than just accepting what they were saying, you know, straight out.

But that's not the way we do business. You can't just let some guy give you a pitch that he's writing out like this and he's doing it quickly without asking him a lot of questions. You know, it's our job to make that guy think and make sure that his rationale is correct because in the haste of this thing he may get tangled up in his underwear and really give you a story that in the light of the next day just doesn't look like it makes any sense at all after you've had time to sit down and think about it. I would not consider what he was doing as being too forceful. You know, I was right across from him, and I know he was really working hard and doing a good job of trying to fully understand that data . . . [then] he more or less lateralled the ball over to them, you know, saying okay, you guys, if you're really going to stick with this story, then we're not going to overrule you.[98]

And Keith Coates said:

There's another statement that George made that I think I ought to note. George was used to seeing O-rings eroded, and I'm sure that he recognized that it was not a perfect condition, and he did not feel that the 20° difference between 53° to 33°, or somewhere in that range, should make that much difference on seating an O-ring, since you've got 900 psi cramming that O-ring into that groove and it should seal whether it is at 53° or 33°. He did say, I do remember a statement something to this effect, "For God's sake, don't let me make a dumb mistake." And that was directed to the personnel in the room there at Marshall. I think he fully appreciated the seriousness of it, and so I think that that is on the other side of the coin that he was amazed or appalled at the recommendation.[99]

At Thiokol, no one spoke, then Jerry Mason turned to Joe Kilminster and said, "Joe, ask them for a five-minute caucus off-line." Kilminster requested the caucus. All participants who were asked why the caucus was called responded that it was because Thiokol's engineering analysis was weak.[100] Thiokol's Larry Sayer explained: "We had a very short time to work the problem, and there was not a lot of data available, and so my assessment is that we had a weak engineering position to present so it was open for discussion again at that time as far as I was concerned. . . . I guess anytime you have a customer that voices that strong an opposition, you have to reconsider what you've said, and it is maybe not an intimidation factor but it certainly is a leveling agent that takes you to a different position than where you were when you were making the initial recommendation."[101]

Thiokol's Bill Macbeth said: "Mr. Mulloy and Mr. Hardy had essentially challenged us to rethink our position and revisit our data, and that's motivation for a caucus. Now, whether one of them said, you better go caucus, or Joe Kilminster asked for a caucus, which is kind of my recollection, that doesn't surprise me. That's a fairly standard procedure in this type of a telecon. I think that quite a few of us in the room felt that we had been properly challenged because we hadn't done a good technical job, and we were glad to have the caucus and go back over the data and try to rethink our position and what we had seen."[102]

The last thing before they broke for the caucus, McDonald, at the Cape, spoke for the first time on the teleconference net. His comment turned out to have a significant effect. Contrary to McDonald's intention, which was to make an objective observation of facts that were simultaneously "good news and bad news," *all* participants interpreted what he said as an affirmation of redundancy, and therefore a statement in favor of launch.[103] Recalled McDonald: "To the best of my memory, I said that the cold temperatures are in the direction of badness for both O-rings, but it is worse for the primary O-ring than the secondary O-ring because the leak check forces the primary O-ring into the wrong side of the groove, while the secondary O-ring is place in the right side of the groove, and that we should consider this in our evaluation of what is the lowest temperature we can operate it."[104]

At Thiokol, the working engineers heard this as an argument for redundancy.[105] Roger Boisjoly said, "When Al first came on the line and explained the situation about the primary and secondary O-ring and how the secondary O-ring was in position, I had a flash of confusion in my mind as to what he was saying . . . there was a definite misunderstanding about Al's statement, there's no question in my mind, because I had it, and I was probably closer to that joint than most anybody else."[106]

At Kennedy, Larry Mulloy thought: "At that point he made the first comment that he had made during this entire teleconference. . . . He stated that he thought what George Hardy said was a very important consideration, and he asked Mr. Kilminster to be sure and consider the comment made by George Hardy during the course of the discussions, that the concerns expressed were for primary O-ring blow-by and that the secondary O-ring was in a position to seal during the time of blow-by and would do so before significant joint rotation had occurred. . . . Mr. McDonald was sitting here and it clearly was a supportive comment."[107]

Bob Schwinghamer, at Marshall, recalled: "One of the things that we always believed, and indeed, that Thiokol had told us, was never mind blow-by of the primary because if it does that, it will only seat the secondary from the driving gas pressure. And one of the last comments that Al McDonald made before we went off the loop was 'Don't forget what George said about the secondary, you know, there's always the secondary.' And what we, all of us, inferred was, 'Yeah, there's always the secondary you can count on and depend on.'"[108]

The caucus continued for about 30 minutes, during which all three sites were on mute. In Utah, after some discussion, Thiokol

Senior Vice President Jerry Mason said, "We have to make a management decision," thus excluding Thiokol engineers from the decision making. Included were four senior Thiokol managers, among them Robert Lund, who had supported the engineering position. Engineers Arnie Thompson and Roger Boisjoly again tried to explain the engineering position. Thompson sketched the joint and discussed the effect of the cold on the O-rings. Showing photographs of the rings on the two flights with blow-by, Boisjoly argued for the correlation between low temperature and hot-gas blow-by. Getting no response, Thompson and Boisjoly returned to their seats. The Thiokol managers continued their discussion and proceeded to vote. Three voted in favor of launch; Lund hesitated. Mason asked him to "take off his engineering hat and put on his management hat." Lund voted with the rest.

Mason took charge of the Thiokol caucus. He stated that they had been flying with the possibility of blow-by and erosion for a long time because they had confidence that these conditions were acceptable risks. Now they had to focus on the temperature issue, separating it from the blow-by and erosion risks that they had been comfortable taking.[109] He suggested they go back over their engineering argument and consider Mulloy's critique of it. Mason repeated Mulloy's points, which, since he had missed all of the preteleconference engineering discussion at Thiokol and participated only in the conclusions and recommendations session, Mason viewed as "new data" that Thiokol had not considered: (1) they had problems with warm motors as well as cold motors, so they did not really have a temperature correlation; (2) tests showed the joint would seal even if the primary had three times more erosion than experienced on the January 1985 launch; and (3) subscale tests showed no blow-by at 30°F. Mason reiterated the belief in redundancy: these data indicated that the primary would not be compromised, but if it were slow to seal and blow-by occurred, the secondary was in position and would seal. Thiokol's Jack Kapp reflected: "In my perception, Mr. Mason certainly was [speaking for launch]. It was obvious to me that based on what he had seen and heard, Jerry then, for whatever reason, wanted to fire."[110]

In response, Boisjoly and Thompson vigorously defended the engineering stance presented on the Thiokol charts. They said that they did not know what the secondary would do.[111] They repeated that STS 51-C was the coolest motor Thiokol had ever fired and had the most blow-by. They argued that if they launched at predicted temperatures, they were outside the experience base of 53°F by 24°–25°.

Even if they used 40°F as a temperature floor, the SRMs were still so far out of the data base in an area of known vulnerability that they simply should not fire.[112] Mason then stated that they all knew the seal was vulnerable to erosion, but they had run extensive tests to show that it could tolerate a lot of erosion. Boisjoly reiterated that he could not quantify it, but in launching below the data base they were moving away from goodness. Then Mason asked them to recall McDonald's comment that if the primary failed, the secondary would be in position and would seal. Boisjoly and Thompson stressed that the resiliency tests provided the most important data. Mason responded that those tests did not incorporate the pressure that acted on the O-ring during ignition. When tested with pressure, the O-ring acted as it was designed to operate, even in 30°F tests.

Boisjoly and Thompson dominated the discussion; most of the other engineers were quiet.[113] A few interjected comments of a factual nature but took no position. And then Mason said that if engineering could not produce any new information, it was time to make a management decision. Mason recalled: "We'd been getting the comments for maybe 15 to 20 minutes, and we, in my mind, had reached the state where everyone had said what they thought several times, and my comment, I think, was that 'Well, we're now starting to go over the same ground over and over again. We're really going to have to make up our minds.' And my message there was that if anybody had anything new to add we needed to get it out on the table, but there was not going to be anything gained [by repetition]."[114]

Thiokol's Joe Kilminster explained Mason's decision as typical when engineering disagreements could not be resolved by data that pulled all to a consensus: a management risk decision was necessary: "There was a perceived difference of opinion among the engineering people in the room, and when you have differences of opinion and you are asked for a single engineering opinion, then someone has to collect that information from both sides and made a judgment."[115] Bill Macbeth affirmed: "And when you get that kind of an impasse, that's the time management has to then make a decision. They've heard all of the evidence. There was no new evidence coming in, no new data being brought up, no new thinking, no new twists being put on it from our previous position, and we were just rehashing. And so Mr. Mason then said, 'Well, it's time to make a management decision. We're just spinning our wheels.'"[116]

None of the engineers responded to Mason's request for new information.[117] Mason, Wiggins, Kilminster, and Lund began to confer.

Alarmed that their exclusion from further discussion indicated a pending management reversal of the engineering position, Boisjoly and Thompson left their seats to try once more to restate their position. Thompson began to sketch the joint, describing again the engineering argument. Boisjoly placed on the table in front of them photographs of the blow-by on SRM-15 and SRM-22, which showed the difference in the amount of soot on the two launches. But when it was clear that their arguments were having no effect, they stopped trying and returned to their seats. Mason reiterated some of Mulloy's points and said that under the conditions they had discussed, he could not disagree with what Mulloy said. He would recommend launch. Mason then asked each of the executives, in turn, how they felt about it. Cal Wiggins said he recommended launch. Joe Kilminster said he recommended launch. Mason turned to Bob Lund, who sat hesitating, making a few observations about the data, and shaking his head.[118] After a few moments, Mason said, "It's time to take off your engineering hat and put on your management hat."

Thiokol's Larry Sayer observed: "I could feel he [Lund] was in a very tough position and one that he was very uncomfortable with. He turned around and looked at the people that support him that put the engineering data together, and no one said anything. I think he was really looking for help—is there anything else we can do from an engineering point of view that would make our original recommendation strong enough that we could stand behind it?"[119] Thiokol's Jack Kapp described it:

> I certainly understood that Bob Lund was under a lot of pressure. You know, I don't care what anybody says, I felt pressured that night. These things were apparent to me. NASA had heard our arguments in detail and essentially had rejected them out of hand. You know, Larry Mulloy made the statement "I have a contrary opinion" right after hearing all of our arguments, so that was a factor. Mr. Hardy's statement that he was appalled at the recommendation was a factor providing pressure. These are knowledgeable, brilliant men. I have nothing but respect for Mr. Hardy and Mr. Mulloy. I worked with them for years. That was a factor. The fact that our manager here, our general manager, had voted that we go ahead and fire and two principals had—that just left the weight of the world on Bob Lund's shoulders. He was effectively making a decision for the whole darn company. I remember thinking, boy that's an awful lot to put on one individual where you've essentially got what I would perceive as 10 or 12 participants in the program all saying we've heard, now we want to fire, and here's one individual. Now it all comes down to him.[120]

Mason, harshly criticized later for polling management and disenfranchising engineers, recalled that moment:

> And I said something to the effect, to Cal or Joe, "What do you think?" And that ended up being a "poll." I think Joe said, "Well, my feeling is it's okay." What do you think, Cal? And he said, "Well, I think it's all right." And then Bob—and then I said to Bob, and everyone remembers this, I said, "Bob, you've got to put on your management hat, not your engineering hat." And my message was intended to be that we had all been spending our time there as engineers, looking at numbers and calculations and so forth. We now had to take that information and do some management with it. And there have been a lot of people who have characterized that expression like I was saying, "Don't be an engineer, be a manager," and that isn't the case. I think there's engineering management that has to take place, where you take engineering data and made a management decision with it. So that's how I intended it, but how it was interpreted was variable depending on the people that were there.[121]

Lund recalled:

> Jerry says, well, it is obvious we are not going to get a unanimous decision here. He says, we are going to have to decide. And so he went back to the rationale, and he says, this is the way I think that the doggone thing is coming out, and he asked the rest of us whether we thought that was right. You know, when you are a manager, you have pressures to make decisions from all directions, and certainly that was pressure to make a decision. He wasn't asking me to do something dumb. You know, of all the things he didn't want to do was fly the *Challenger* and have a—you know, that would be dumb. There is no question about that. And so he was asking for a judgment, you know, he was asking for the judgment that I could best make, and I tried to provide that judgment to him.[122]

Lund voted with the rest. When Mason, Wiggins, Kilminster, and Lund were asked why they concurred in the original engineering opinion not to launch unless the O-ring temperature were 53°F or greater, then changed their minds, all said that they were influenced by facts not taken into account before their initial recommendation. These facts supported redundancy: thus, they believed that the secondary would seal the joint. All identified the same technical points as reasons they changed their votes, but gave them slightly different priority.[123] Lund articulated them:

> Number one, it was pointed out that the conclusion we had drawn that low temperature was an overwhelming factor was really not

true. It was not an overwhelming correlation that low temperature was causing blow-by. We have blow-by both at low temperature and high temperature. The second thing, we went back, and we had some new data, not new, but data that was brought out, you know, we have also run lots of tests where we have put a lot of erosion on [i.e., simulated erosion by cutting away parts of] O-rings, and we could experience a factor of three times more erosion on the O-ring and still have it work just fine, so we had a very large [safety] margin on erosion. The last one was the McDonald [point] that during that original pressurization, if it blows by there is only going to be so much blowing because that doggone thing is right there and seated. It is still going to seal, and you are not going to have a catastrophic set of circumstances occur.[124]

All four stated that McDonald's sole teleconference statement, which they, like everyone else, heard as supporting redundancy, was influential. Mason's comment was representative:

[McDonald said] "Even though it [the primary] moves it will move slower because it's cold. Be sure to evaluate the effect of the fact that the secondary O-ring is already in the sealing position." And that comment by McDonald was something we hadn't addressed in our earlier discussions [preparing for the teleconference at Thiokol]. We were spending all our time figuring out the probability of the primary seating. It was the first time that I recognized the importance of the secondary O-ring being in a sealing position. Because, what that said to us is that even if the primary was so slow that we had a massive blow-by and it never seated, that that would immediately pressurize the secondary O-ring and once it was pressurized, then we were okay. . . . The engineers, Boisjoly and Thompson, had expressed some question about how long it would take that [primary] O-ring to move, [had] accepted that as a possibility, not a probability, but it was possible. So, if their concern was a valid concern, what would happen? And the answer was, the secondary O-ring would seat.[125]

After Lund voted, no one spoke. Ebeling described the moment: "All of the paper shuffling and all of the other extraneous activity that was going on in that room with 14 people in it came to a screeching halt, and there was a big quiet."[126] Kilminster began writing notes from their discussion in preparation for reactivating the Thiokol connection to the teleconference network. The five-minute caucus had gone on for half an hour.

Because the teleconference connection was on mute while the Thiokol caucus was underway, participants at Marshall and Kennedy had not heard the conversation. At both places, the expectation was

that Thiokol would come back with arguments sustaining their recommendation to delay launch.

At Kennedy. Thiokol's Jack Buchanan recalled the ambience: "It was
a fairly lax time when everybody sort of pushed back and got real
comfortable because I think everybody knew that ten minutes was
not reasonable. We were figuring on 20 minutes or 30 minutes, which
it turned out to be."[127] McDonald, Mulloy, and Reinartz continued to
review the data, discussing the inconsistencies and disputing the
kinds of conclusions that could be drawn. McDonald and Reinartz
discussed the qualification of the motor between 40°F and 90°F.
McDonald agreed that these qualification temperatures were incompatible with Thiokol's recommendation. He noted that those tests
were done before he was assigned to the project, but that he had seen
documents that the motor was qualified from 40°F to 90°F. He said,
however, that he could not recall seeing any static test results showing what condition the O-rings were in under those temperatures.[128]
    Everyone assumed Thiokol was reworking their technical argu-
ment, doing further calculations, and would return with a stronger
engineering analysis that supported their stance. Mulloy reflected, "I
thought they were going to come back with a temperature floor that
took into account all the data."[129] Al McDonald recalled:

> I personally thought that the reason it was taking so long was that
> our engineers made up that presentation only with the data that
> showed why we could not launch [at less than 53°F] and not any
> data that may have supported that it could be launched at lower
> temperature, and they were looking for that data, or that they were
> indeed trying to make a new calculation based on what they did
> know and decide where they could come off at 53 or some other
> number, and that is why it was taking so long. I didn't find out
> until about two days later that that was all the data that there was,
> and there wasn't anymore, and they didn't have any more, and they
> did not run any calculations to try to stand by that data. I didn't
> know that at all. I was very surprised.[130]

At Marshall. People also talked informally about the data while waiting for Thiokol to come back on line.[131] Neither Powers nor Coates,
who both supported Thiokol's position, expressed their views.
Instead, the conversation was filled with affirmations of redundancy,
and agreement that the shuttle could be launched at much lower
temperatures than 53°F.[132] There was some conversation about the
program and that the program could not afford to limit flights in the

winter months to avoid cold weather. If that turned out to be true, they speculated that heaters and heating blankets to warm the joints on the launch pad would be necessary.[133] However, at Marshall, as at Kennedy, there also seemed to be consensus that Thiokol would stick to their no-launch position.[134] So far, they had not wavered. Marshall's Larry Wear was speculating to Thiokol's Boyd Brinton that Thiokol might opt for a maneuvering strategy that would create a no-launch rationale everyone could support: a recommendation not to fire below the qualification temperature of 40°F. This strategy, although it did not take into account the 30°F data, took away many of the contradictions in their analysis and conformed to a rule everyone acknowledged. It would amount to delaying the launch on a technicality to give themselves more time to analyze the effects of cold temperatures. Brinton agreed that the motor would perform adequately at that temperature and that Thiokol might opt for that strategy.[135] Thiokol's Kyle Speas, who witnessed that conversation, said: "I don't think that he [Wear] felt that 40 degrees was the lowest temperature that could be launched at. He was looking at it more from a maneuvering position–type standpoint. Thiokol had reservations, you know, instead of having them conform to the whole thing, [they could] put the monkey back into somebody else's pocket, saying that we haven't qualified for that temperature, giving Thiokol more time to analyze the situation."[136]

George Hardy discussed the data with some of the people near him, hearing "no disagreement or differences in our interpretation of what we believed the data was telling us."[137] However, he and Lovingood were acting on the assumption that Thiokol would stick to the recommendation. Ben Powers, who was sitting across from Hardy and next to Lovingood, heard the conversation. Powers stated: "They were already talking about who they were going to call on the phone to shut it down. They were already lining up the phone calls in their mind. 'These are the guys we have to notify because we're not going to go.' You know, that's what they were doing. As soon as we get off of this, we've got to notify this guy and this guy and this guy that we can't launch tomorrow. They were already headed down that road. It was no-launch at that time."[138]

The off-line caucus over, the second teleconference resumed. The people at Kennedy and Marshall came back on line. Thiokol's Joe Kilminster announced that Thiokol had reconsidered. They reversed their first position, recommending launch. Kilminster read the revised engineering analysis supporting the recommendation. Mulloy requested that Kilminster fax a copy of Thiokol's flight-readiness

rationale and launch recommendation to both Kennedy and Marshall. The teleconference ended about 11:15 P.M. EST. At Kennedy, McDonald argued on behalf of the Thiokol engineering position. Subsequently, Mulloy and Reinartz telephoned Aldrich, discussing the ice that had formed on the launch pad and the status of the recovery ships. They did not inform Aldrich about the teleconference and Thiokol engineers' concerns about the effects of cold temperature on the O-rings. The Kennedy Space Center meeting broke up about midnight.

Kilminster stated Thiokol's position, reading from the notes he had roughed out: Thiokol acknowledged that the O-rings on SRM-25 (STS 51-L) would be about 20° colder than SRM-15 (STS 51-C). Although temperature effects were admittedly a concern, the data predicting blow-by were inconclusive. Thiokol acknowledged that the cold would increase the O-ring's hardness, taking it longer to seal the gap, and thus more hot ignition gases may pass the primary seal. But erosion tests indicated that the primary O-ring could tolerate three times more erosion than experienced on SRM-15, the previous worst case. And further, even if the primary O-ring did not seal, the secondary would be in place and would fulfill its function as a backup, sealing the joint. When Kilminster finished, Hardy, looking around the table at Marshall, asked, "Has anybody got anything to add to this?" People were either silent or said they had nothing to add. Reinartz, as Shuttle Projects Manager, asked all parties to the teleconference whether there were any disagreements or any other comments concerning the Thiokol recommendation. No one at any of the three locations said anything.

As a consequence, people at Marshall and Kennedy, precluded from hearing what happened during the caucus at Thiokol by the audio mute, were not aware that anyone at Thiokol still objected to launch.[139] Marshall's John Q. Miller remembered: "I could detect no dissension. Kilminster was speaking for Thiokol. He speaks for the corporation in a case like this, and I heard nothing and got no indication that people out there were not happy other than the concerns they had expressed during the first part [of the teleconference]."[140]

In keeping with NASA procedures, Mulloy then asked Joe Kilminster to send a copy of his flight-readiness rationale and recommendation to Marshall and Kennedy.[141] McDonald said that he would not sign it; Kilminster did. No discussion occurred because, as Thiokol's Jack Buchanan explained, getting Kilminster's signature was the norm: "I represent Thiokol at the Kennedy Flight Readiness Reviews, and every contractor including me has a sheet that they have to sign saying we're ready. Every contractor has a signature page that they present to the KSC. That would be Martin Company, USBI, Thiokol, Lockheed and any other contractor and the payload community, even other NASA organizations. They all have signed sheets. The launch pad's ready. The ships are ready. There's many of these signed sheets. The Air Force has a signed sheet. The range is ready. Safety has a signed sheet. Public Affairs even has one. That is a common thing. It is normal for Kilminster to sign as Program Director, and I think everybody expected him to sign."[142]

The teleconference ended about 11:15 P.M. EST, but conversation about it continued.

At Marshall. People did not linger, but dispersed quickly because it had been a long day and many of them had to be back very early to support the launch.[143] Thiokol's Brinton and Speas stayed behind to wait for the fax from Thiokol. While waiting, Brinton called the Utah home of his close friend Arnie Thompson in order to speak to Thompson's wife, who was Brinton's secretary. Meanwhile, Thompson returned home, and Brinton asked him how he felt about the launch. Thompson said that they did not have any data at this temperature, so he felt that it was certainly more conservative to wait to fly until the afternoon, when the temperature was predicted to be higher. He observed that Marshall was usually extremely conservative, and he wondered why they were not in this situation, too, but that he had no data that said that they were going to have any trouble.[144]

At Thiokol. People also scattered quickly. The working engineers described their feelings after the teleconference. Boisjoly recalled, "I left the room feeling badly defeated, but I felt I really did all I could to stop the launch."[145]

Larry Sayer described their conversation as they left the teleconference room: "I told him, I said, Roger, you can't feel bad about this because we did as engineering what we were supposed to do, and this is if there's a concern we bring that concern to management. I told him I felt we didn't have a real strong position. We had a lot of, you know, we had a lot of feelings that we were concerned about those temperatures, but we didn't have a solid position that we could quantify. And he was very somber."[146]

O-Ring Task Force Chair Don Ketner recalled: "I hadn't changed my opinion, but I still had the feeling which I had had as soon as the first questions came up, which was that we just didn't have enough conclusive data to convince anyone. And I felt that we had presented as much information as we had. I knew you couldn't prove from that data, because of the inconclusiveness of it: that is, you could argue that you shouldn't launch at 75°."[147]

Jerry Burn recalled:

Well, we weren't working on the assumption that it would fail. We were working under the idea that we would have problems, and how extensive those problems would be were unknown. . . . I do know that when people left the room in the telecon, I felt there would be [O-ring] damage. I was hoping there wouldn't be, but I felt there would be. Roger Boisjoly was 100 percent convinced there would be damage. He went by my desk as we were leaving and he mentioned to make sure that I document the damage well because we'd have a new data point. You know, we'd gotten to the point where we knew that the decision had been reversed, but as far as detrimental damage, I don't believe that anybody there felt—I don't know if it came across that way, but we were just afraid of going outside the data base. I was afraid of it, I think everybody was still afraid of it. It's not like—I mean, we had high concern, but it's not like we knew that explosion would occur or that damage of a complete joint would occur.[148]

Thiokol's Larry Sayer: "I guess you have to look at it very hard after what happened, but at that time with the circumstances we had, I would have probably made the same decision that Mr. Lund made. It wasn't something that really bothered me afterwards or caused me any discomfort relative to emotional stability relative to that subject. In fact, the next day I was not watching the launch. I was in my office working, and someone came in after the launch and poked their head in and said that it blew up, and I said, 'You're kidding.'"[149]

Brian Russell said: "We were not asked as the engineering people. It was a management decision at the vice presidents' level, and they had heard all that they could hear, and I felt there was nothing more to say, and also felt in my mind that I didn't see a dangerous concern. I knew we were entering into increased risk, and I didn't feel comfortable doing that."[150]

Russell and Boisjoly also had a conversation. Boisjoly said he felt the decision had been jammed down his throat. Russell observed that the situation that night was different from usual: instead of proving that it was safe to fly, they had to prove it was unsafe.[151] Boisjoly agreed.[152] Boisjoly, Maw, and Ebeling rode home together. Maw recalled that both Boisjoly and Ebeling made the statement that they did not get to vote. According to Maw, Boisjoly stated that "he had done everything he could and he felt like he had a clear conscience. He—I don't think he expected something to go wrong, but he felt like he had done what he could. If something did go wrong, he felt like he had a clear conscience, like he had done everything he could."[153]

When Ebeling arrived home, he told his wife that he thought a grave mistake was made. He was angry and did not sleep well.[154] On the other hand, Thiokol's Brinton stated, "Let me tell you that I went home and slept well that night. I was not concerned."[155]

After the others left, Russell remained in the teleconference room to help Kilminster and Lund with the fax because he was the only person present who knew how to operate it. He helped with the wording of the final recommendation chart containing the flight-readiness rationale (see fig. 12). Consistent with the flight rationale Kilminster gave during the teleconference, it acknowledged remaining concerns, admitted more blow-by was likely, and asserted redundancy. Russell, commenting on the process, said that Lund and

FINAL THIOKOL LAUNCH RECOMMENDATION

MTI Assessment of Temperature Concern on SRM-25 (51L) Launch

o Calculations show that SRM-25 o-rings will be 20° colder than SRM-15 o-rings

o Temperature data not conclusive on predicting primary o-ring blow-by

o Engineering assessment is that:

   o Colder o-rings will have increased effective durometer ("harder")

   o "Harder" o-rings will take longer to "seat"

      o More gas may pass primary o-ring before the primary seal seats (relative to SRM-15)

         o Demonstrated sealing threshold is 3 times greater than 0.038" erosion experienced on SRM-15

   o If the primary seal does not seat, the secondary seal will seat

      o Pressure will get to secondary seal before the metal parts rotate

         o O-ring pressure leak check places secondary seal in outboard position which minimizes sealing time

o MTI recommends STS-51L launch proceed on 28 January 1986
   o SRM-25 will not be significantly different from SRM-15

Joe C. Kilminster, Vice President
Space Booster Programs

Morton Thiokol Inc.
Wasatch Division

Information on this page was prepared to support an oral presentation and cannot be considered complete without the oral discussion

Fig. 12

Kilminster seemed convinced that they were not going significantly outside of the experience range and their decision would not result in any catastrophe.[156] About 9:45 P.M. MST, Russell typed the flight rationale and faxed it to Kennedy and Marshall.

At Kennedy. When the teleconference was over, Thiokol's Al McDonald took a strong stand against launch. McDonald was surprised not only at the Thiokol reversal, but that the Thiokol recommendation contained no temperature limit. He told Mulloy and Reinartz that he was amazed that NASA accepted the recommendation, based on the earlier conversation that the motor was qualified from 40°F to 90°F. Recalled McDonald: "That is when I said directly to them, I don't understand, because I certainly wouldn't want to be the one that has to get up in front of a board of inquiry if anything happens to this thing and explain why I launched the thing outside of what I thought it was qualified for."[157] McDonald then said that if engineering concerns for the effect of cold on the O-rings were not sufficient to recommend not launching, there were two other concerns. First, the launch pad was being covered with ice because of a combination of strong winds, the dropping temperature, and a makeshift freeze protection plan to keep pipes from freezing by allowing water to flow slowly out of the water systems on the pad. Second, high seas had forced back the ships that were sent out to recover the SRBs when they separated from the Orbiter and fell in the sea. Thus, the ships would not be in position at launch time, risking loss of the costly SRBs. Mulloy responded that NASA and the appropriate contractors were aware of both situations. The Mission Management Team was aware of and monitoring both launch pad ice and recovery area weather. Mulloy told McDonald he planned a call to Arnold Aldrich, Level II Program Manager and head of Mission Management, to advise him that recovery area weather did not now meet Launch Commit Criteria.[158]

McDonald left to wait for Thiokol's flight rationale in the fax room. Cecil Houston called Acting Launch Director Gene Sestile to report the outcome of the Thiokol teleconference, as promised, and to get an update on the recovery ships. Sestile then set up a teleconference between himself, Houston, Reinartz, Mulloy, and Aldrich, who was at his motel. Mulloy reported that the weather had forced the recovery ships toward shore. Aldrich agreed that action was appropriate, asking how the launch would be affected by the ships not being able to recover the SRBs immediately. Mulloy stated that they would lose the parachutes, and they could afford that cost, but they would not lose the SRBs because they would float and could be recovered later. The Thiokol teleconference was not mentioned.[159] Mulloy testified: "I did not discuss with Mr. Aldrich the conversations that we had just competed with Morton Thiokol. At that time, and I still consider today, that was a Level III issue, Level III being an SRB element or an external tank element or Space Shuttle Main Engine element or an Orbiter. There was no violation of Launch Commit Criteria. There was no waiver required in my judgment at that time and still today. And we work many problems at the Orbiter and the SRB and the External Tank level that never get communicated to Mr. Aldrich or Mr. Moore. It was clearly a Level III issue that had been resolved."[160]

McDonald distributed copies of the Thiokol flight rationale. Around midnight, as the people at Kennedy were leaving, McDonald overheard Reinartz and Mulloy discussing whether they ought to tell Bill Lucas about the teleconference outcome as soon as they got back to the hotel. The comment he overheard was: "I don't think I want to wake the old man and tell him this right now because we did not change the decision to proceed with the launch. If we would have decided not to, then yes, we would have to do that."[161]

At 1:30 A.M. on January 28, the Ice/Frost Inspection Team assessed the ice on the launch pad. NASA alerted Rockwell International, prime contractor for the Orbiter in Downey, California, where specialists began investigating possible effects of ice on the Orbiter. Reporting for launch at 5:00 A.M., Mulloy and Reinartz told Marshall Director Bill Lucas and James Kingsbury, Director of Marshall's Science and Engineering Directorate, of Thiokol's concerns about temperature and the teleconference resolution. At approximately 7:00 A.M., the ice team made its second launch pad inspection. On the basis of their report, the launch time was slipped to permit a third ice inspection. At 8:30, the *Challenger* crew were strapped into their seats. At 9:00, the Mission Management Team met with all contractor and NASA representatives to assess launch readiness. They discussed the ice situation. Rockwell representatives expressed concern that the acoustics at ignition would create ice debris that would ricochet, possibly hitting the Orbiter or being aspirated into the SRBs. Rockwell's position was that they had not launched in conditions of that nature, they were dealing with the unknown, and they could not assure that it was safe to fly. The Mission Management Team discussed the situation and polled those present, who voted to proceed with the launch.

At 1:30 A.M., the ambient temperature had dropped to 29°F. Two teams were dispatched to the launch pad. One repaired a hardware problem on a ground fire detector, interrupting the tanking operation and slipping launch time from 9:38 to 10:38 A.M. The other, the Ice/Frost Inspection Team, assessed launch pad ice.[162] The team reported that, despite antifreeze, ice covered the shuttle's huge fixed service tower. The strong wind and cold transformed the water trickling through its drains into 1/8" to 3" ice on the floors and icicles, some as long as 18", which hung from the multilevel, 235' structure. The walkway used in case of emergency flight crew evacuation from the Orbiter was coated with ice. The ice team began ice removal. Hearing the ice team report at 3:00 A.M., Charles Stevenson, Leader of the Ice/Frost Inspection Team, became concerned that at liftoff, ice debris from the fixed service tower would strike the Orbiter, damaging its heat tiles. The cold weather had produced another unprecedented situation. No one knew what the effect of acoustics at liftoff would be on the ice on the service tower, which was distant from the launch vehicle. He reported to the Launch Director, recommending they not launch. The Launch Director asked the Orbiter contractor, Rockwell International, to research the threat of ice to the Orbiter. Shortly after midnight Pacific Standard Time at its Downey facility, a Rockwell aerodynamicist and debris people began an analysis for the Mission Management Team meeting scheduled for 9:00 A.M. EST at Kennedy.

At Kennedy and Marshall, people arrived early to support the launch. The teleconference was not a focus of conversation or a source of lingering controversy at either place. An exception occurred at Marshall, where Thiokol's Boyd Brinton bumped into Hardy, who asked for a copy of Thiokol's final chart. They discussed the teleconference, how the O-rings worked, and what they had been talking about. Standing in a hallway beside a blackboard, they "drew some pressure time traces and discussed O-ring sealing in general."[163] Brinton told Hardy about his phone conversation with Arnie Thompson after the teleconference. Brinton concluded that Hardy had no second thoughts or lingering concerns and appeared very comfortable with the decision.[164]

At Kennedy, as Lucas and Kingsbury arrived at the Launch Control Center at 5:00 A.M., Mulloy and Reinartz reported the outcome of the Thiokol teleconference. According to Lucas, he was told that Thiokol had a concern about the weather, and that it was discussed and "concluded agreeably that there was no problem."[165] The crew was awake by 6:00 A.M. After a leisurely breakfast, they received a weather briefing.[166] They were informed of the temperature, now at 24°F, and of the ice on the pad, but not of the teleconference. They were told to expect some hold time on the pad due to the ice problem and the Launch Commit Criterion that the ambient temperature be 31°F for the shuttle to be launched. The seven crew members got into their flight gear, boarded *Challenger*, and were strapped into their seats at 8:36 A.M. At 7:30 A.M., Casablanca, the alternate TAL site, had been declared NO-GO because of precipitation and low ceilings. This decision had no effect on the countdown, however, because the primary TAL site, Dakar, had acceptable weather.[167] Moreover, Dakar had lighting facilities for a night landing, meaning they could launch later that day if the ice situation deferred the launch to the secondary launch window, which opened at 3:48 P.M. and closed at 6:48 P.M.

From 7:00 to 8:45 A.M., the ice team inspected the launch pad a second time.[168] The ambient temperature rose from 26°F to 30°F during their inspection. The water system had continued to trickle, the winds had continued to blow, and some of the stairwells in the fixed service tower looked like "something out of Dr. Zhivago."[169] The ice team snapped off the icicles; 95 percent of the ice in the water troughs was removed.[170] However, sheet ice still firmly adhered to the service tower. One ice team member measured temperatures on the SRBs, External Tank, Orbiter, and launch pad with an infrared pyrometer. The right SRB, sitting in shadow, measured about 8°F near the aft region, while the left SRB, in the direct sun, was about 25°F. The ice team, which did not routinely report surface temperatures, did not include this finding in its report because there were no Launch Commit Criteria on surface temperatures.[171]

Meanwhile, Rocco Petrone, President of Rockwell's STS Division, arrived at the Downey facility at 4:30 A.M. PST and spoke with the Rockwell personnel who had been working on the ice problem. Petrone arranged a teleconference between the Downey people and two Vice Presidents/Program Managers and others at the Cape to clarify Rockwell's position on launch. The Downey staff concluded that ice on the service tower would travel only about halfway to the vehicle, so it was not a threat. They had seen an aspiration effect on previous launches, by which things were pulled into the SRB after ignition, but never from as far away as the fixed service tower. Some ice might land on the Mobile Launch Platform, however, and they did not know what effect that would have. This was, as they said, "a concern." The Chief Engineer summarized what they knew, concluding: "It's still a bit of Russian roulette; you'll probably make it. Five out of six times you do playing Russian roulette. But there's a lot of debris. They could hit [the Orbiter] direct, they could be kicked up later by the SRBs, and we just don't know how to clear that."[172] Rockwell's Director of Technical Integration in Downey conveyed the Downey position to Rockwell personnel at Kennedy: "Our data base does not allow us to scientifically tell you what's going to happen to the ice. Therefore, we feel we're in a no-go situation right now."[173]

Ice/Frost Team Leader Stevenson reported to Arnold Aldrich just before the Mission Management Team meeting, apparently not informing Aldrich of his earlier recommendation to delay launch because of the ice.[174] After hearing Stevenson's report, however, Aldrich himself decided to delay launch until 11:38 A.M. to allow the sun to melt more ice on the pad and to allow the ice team to perform a third ice assessment at launch minus 20 minutes and report back.[175]

Aldrich convened the Mission Management Team meeting at 9:00 A.M. He asked representatives from the Ice/Frost Inspection Team, Rockwell, Kennedy Director of Engineering Horace Lamberth, and Johnson's Director of the Orbiter Project Richard Colonna for their positions on the threat of ice to the Orbiter. Kennedy Engineering had quantitative data. They had calculated ice trajectories, concluding that even if the wind increased, the fixed service tower was so distant that ice would not come in contact with the Orbiter. If ice debris were aspirated into the SRBs, the pieces would be the size of ice cubes or smaller and would not cause problems. Colonna reported similar conclusions from Houston by the Mission Evaluation Team. He also noted that even if the calculations were in error, the ice would be lightweight and hit at an angle, having a low probability of causing damage.

Next, Aldrich telephoned Tom Moser, a member of the Mission Evaluation Team in Houston, whose calculations confirmed what had already been said. Aldrich then asked Rockwell Vice President Bob Glaysher for the contractor analysis and position on launch. Glaysher discussed aspiration effects, the possible ricochet of ice from the tower, and what ice resting on the Mobile Launch Platform would do at ignition. He stated that the ice was an unknown condition and they were unable to predict where it would go or what damage it could do to the Orbiter. When asked by Aldrich for his view, Martin Cioffoletti, Rockwell's other Vice President at Kennedy, said, "We felt that we did not have a sufficient data base to absolutely assure that nothing would strike the vehicle, and so we could not lend our 100 percent credence, if you will, to the fact that it was safe to fly."[176]

Following this discussion, Aldrich asked all present for their position regarding launch. Colonna, Lamberth, and Moser all recommended proceeding. When asked, Glaysher said, "Rockwell cannot assure that it is safe to fly."[177]

From Glaysher's statement, Aldrich concluded: "He did not disagree with the analysis the JSC [Johnson Space Center] and KSC had reported, that they would not give an unqualified go for launch as ice on the launch complex was a condition which had not previously been experienced, and thus this posed a small additional, but unquantifiable, risk. [They] did not ask or insist that we not launch, however. At the conclusion of the above review, I felt reasonably confident that the launch should proceed."[178] Lamberth testified that he also did not interpret Rockwell's position as a no-launch recommendation: "It just didn't come across as the normal Rockwell no-go safety of flight issues come across."[179]

Aldrich then heard reports from all other contractors and the ice team. Ice on the shuttle itself was not a concern. However, the north and west sides of the fixed service tower were still encrusted with ice and icicles. The ice team's assessment was that ice would not strike the Orbiter during liftoff or ascent because of the horizontal distance between the service tower and the Orbiter. The decision was made to launch pending a final ice team inspection to again clear the area and assess any changes. Subsequently, all parties were asked to notify Aldrich of launch objections due to weather. Rockwell made no phone calls or further objections to Aldrich or other NASA officials after the 9:00 Mission Management Team meeting or after the resumption of countdown.[180]

While the Mission Management Team meeting was in progress, Thiokol people were arriving at work in Utah. Although not all the feelings and actions that morning of the Thiokol engineers who participated in the teleconference are known, the information available shows the same diversity of views as they indicated the night before. Bob Ebeling told the people in his car pool about the teleconference, saying that "according to our engineering people, the 51-L was going to blow up on the pad."[181] Boisjoly reported going straight to his office, having no conversations with anyone. Russell recalled discussing the teleconference and the fact that Thiokol had been concerned with some people he bumped into.[182] Bill Macbeth answered a phone call from Thiokol's Boyd Brinton, who called from the Cape about another matter. Macbeth told Brinton he was "pretty ashamed and embarrassed at our performance last night. I feel bad that Mulloy and Hardy had to point out to us we had done a poor job." Brinton agreed.[183]

At the Cape, the final ice team inspection was made between

10:30 and 10:55 A.M. EST. The deck ice had started to melt in areas getting direct sun. Icicles were falling from the fixed service tower. The team swept up the debris and again cleared the ice from the water troughs. Ambient temperature was 36°F when the final ice inspection was complete. At the final Mission Management Team meeting, the team reported no significant change. All members of the Mission Management Team gave their "go" for launch.

A little after 11:25 A.M., the terminal countdown began. STS 51-L was launched at 11:38 A.M. EST. The ambient temperature at the launch pad was 36°F. The mission ended 73 seconds later as a fireball erupted and the *Challenger* disappeared in a huge cloud of smoke. Fragments dropped toward the Atlantic, nine miles below. The two SRBs careened wildly out of the fireball and were destroyed by an Air Force range safety officer 110 seconds after launch. All seven crew members perished.

Chapter Nine

CONFORMITY AND TRAGEDY

Juxtaposing a stereotyped version of the eve of the launch against a thick description of events that night illuminates a more complex version of reality. The resulting imagery does not so readily identify schedule-oriented, amorally calculating managers who violated safety rules in response to production pressure as an explanation for the launch decision. Some examples:

• The availability of a secondary launch window in the afternoon of January 28, when the temperature was going to be in the 40s and 50s, neutralizes arguments that a failure to launch that morning would have serious political and scheduling repercussions.

• The posttragedy imagery of solidary opposition of managers versus engineers is not borne out: most participants were both managers *and* engineers, but more interesting is the little-known fact that the working engineers did not all share the same view. When Thiokol engineers were chosen to testify before the Presidential Commission and the House Committee, those who held views that differed from Boisjoly and Thompson were not called. Thus one point of view was presented, and an image of engineering consensus was created.

• With the exception of Thiokol's Arnie Thompson, all teleconference participants acknowledged that Mulloy's criticisms of the Thiokol engineering argument were accurate and well taken, agreeing that Thiokol's data analysis was filled with inconsistencies and inadequately supported their no-launch recommendation.

• Contradicting impressions that Marshall was determined to launch no matter what, George Hardy, who stated he was "appalled," also said in the next breath that he "would not recommend launch against the contractor's objection," and later, "For God's sake, don't let me make a dumb mistake." While Thiokol caucused, Marshall's Lovingood and Hardy discussed who needed to be called in order to cancel the launch. Everyone at Marshall and Kennedy expected Thiokol to hold to their original position, yet the conversation was unemotional and banal in those sites: no strategizing about how to derail a no-launch recommendation, no secret conversations, no phone calls to or from the outside.

• Contradicting speculation after the accident that managers opposed Thiokol's launch recommendation because a 53°F low-temperature limit would have a dramatic effect on the program by restricting flights at Kennedy and limiting future use of Vandenberg Air Force Base for the shuttle, Marshall personnel apparently believed the resolution was simple and not disruptive to program goals. During the caucus, when the conversation at Marshall turned to the consequences of a possible 53°F Launch Commit Criterion for the SRBs, necessary changes they discussed were heat sensors, heaters, and heating blankets to wrap the joints at the launch site to protect the O-rings from cold.

These and other details of the eve of the launch never became accessible to the public. Still, the thick description in the last chapter does not answer all questions about the effects of production pressure on the launch decision. Many were addressed in chapter 2 (and I will not repeat those still-relevant points now), but what was missing from that discussion was microanalysis looking for possible connections between the economic strain the agency was experiencing and the actions of NASA managers that night. Drawing on the revisionist history presented in the last chapter, this chapter too challenges conventional understandings. We begin by reconsidering the historically accepted explanation that linked production pressure to rule violations and managerial wrongdoing, examining the views of teleconference participants and management actions. Instead of rule violations, we find that managers' decision making was rule-based: managers abided by all internal NASA rules and norms for flight-readiness assessments.[1] Then, analyzing the teleconference, we find that the launch decision resulted not from managerial wrongdoing, but from structural factors that impinged on the decision making, resulting in a tragic mistake.

Consistent with the conventional explanation, NASA's environment of competition and scarcity and the production pressure that resulted had a powerful impact on the teleconference proceedings. However, revising history, it did not operate as traditionally assumed. First, production pressure was institutionalized, affecting all participants' decision making prerationally. Second, production pressure alone does not explain what happened. All the factors that

contributed to the normalization of deviance in the history of decision making were reproduced in the teleconference exchange: the
work group's production of culture, the culture of production, and
structural secrecy. This event becomes a microcosm enabling us to
see how these same factors combined in a single disastrous decision.

THE AMORAL CALCULATOR
HYPOTHESIS RECONSIDERED

Official investigators asked all participants whether any "special"
pressures from outside existed or were exerted on NASA to get this
particular launch up on schedule. All denied it. Further, although all
acknowledged NASA's ambitious launch schedule, all who were
asked by official investigators (not all were asked) denied that NASA
managers had put schedule before safety. Most compelling on this
issue are the unsolicited statements of engineers who opposed the
launch. For example, Marshall's Leon Ray said: "That didn't happen
to us, all that smoke filled rooms stuff. Cost and schedule didn't
drive the *Challenger*."[2] Thiokol's Jack Kapp stated: "[Schedule] was
not a major factor, although I must in all truthfulness say everybody
is aware of the fact we had some ambitious schedules to meet. But
given a situation that was recognized at the time as being dangerous—schedule wouldn't have driven that."[3]

Thiokol's Brian Russell felt their recommendation for delay would
have minor consequences for the schedule: "I felt that if we could get
up to the temperatures within our experience base, then the same
analyses that had justified the previous flights could continue to justify any flights in the future. And so I didn't see it as a program stop-
per by delaying that launch in my own mind."[4] Marshall's Keith
Coates, who also opposed the launch, said: "I am concerned with the
way the investigation is being perceived by people. As engineers, we
recognize that we have risks, and that when we have a failure, that
we do our best to resolve that failure and get on. The investigation is
being perceived as a fault-finding, and that bothers me because, as I
said, while I disagreed technically with the decision to fly, I have a lot
of respect for the people who made the decision to fly, and if I had
been in their shoes, I'm not sure what I would have done; nor do I
think anyone else would know what they would have done unless
they had been in their shoes."[5]

Instead, participants asserted that the *Challenger* launch decision
was the outcome of a "strictly technical" discussion.[6] All partici-

pants stated that Marshall managers proceeded with the launch because the data led them to that conclusion.

Official investigators also asked all participants whether knowledge of NASA's ambitious launch schedule was behind Thiokol management actions. No one testified that NASA's schedule concerns drove Thiokol senior administrators, not even the engineers most upset by the reversal. If Thiokol managers were amorally calculating costs and benefits that night, significant costs existed for Thiokol in case of failure, which Marshall's Larry Mulloy outlined: "I cannot understand why they would be motivated to take a risk that they thought was a safety of flight risk, because they have absolutely no incentive to do that. Our contract incentive is set up in the opposite direction. Thiokol has a production incentive contract. They don't have an award fee that they get depending upon how I think they respond to my wishes. They have a $10 million penalty if an SRM causes a Crit. 1 failure. That $10 million can escalate into much more money than that, given the loss of the mission success fee that is on a block of vehicles, which is another 12.5 percent over the fee otherwise earned. They have negative incentives to take a risk."[7] Marshall's Larry Wear concurred: "That [penalty clause] was not an easy negotiation. They didn't just roll over and play dead and volunteer to have a $10 million penalty clause. That was a tough negotiation when that was established, three or four years ago. . . . But, whatever consideration that I can imagine that might have passed through their brains that night, that had to be much smaller than the issue at stake. Probably an even more important issue was if you do something dumb here, you've just made your condition worse."[8]

The postaccident revelation that NASA had been looking for a second contractor to supply SRMs was apparently behind the Presidential Commission's conclusion that Thiokol executives reversed the original recommendation "in order to accommodate a major customer."[9] However, Thiokol was to remain the sole source supplier for two more NASA hardware purchases: one was for 60 flights, the other for 90 flights, so the threat was not imminent.[10] Also, teleconference participants denied that second-source bidding had shaped actions by Thiokol officials. Thiokol's Boyd Brinton said it was "absolutely not even a consideration" and that he had "never seen a consideration like that in any of the technical decisions."[11] Marshall's Ben Powers said: "I knew there was talk of a second source, but I wouldn't have thought it would have made any difference, because they were our only contractor and we need shuttle motors, so, so what? We've got to get them from somewhere. And to get an alternate source you'd have to have somebody that has demonstrated and qualified. There is no one else that can produce that motor at this time. I don't think that had anything to do with this at all. It was strictly dealing with the technical issues."[12] Thiokol's Bob Lund pointed out that "as far as second source on this flight decision, gadzooks, you know, the last thing you would want to do is have a failure."[13]

Still, all participants acknowledged that there was pressure that night. Opinion about the nature of that pressure was divided. Many managers and engineers from both Thiokol and Marshall said that it was no different from the pressure surrounding every important decision.[14] However, many of the engineers in Utah who opposed the launch disagreed.[15] They said they *experienced* more pressure than usual. Consistent with their insistence that the discussion was a "strictly technical discussion," however, they linked the pressure to the technical discussion itself: specifically, Marshall managers' criticisms of Thiokol's data analysis and conclusions that "shifted the burden of proof."[16] To Thiokol's Brian Russell, "the statements that seemed to me to put pressure or at least to make me feel the pressure were the responses of being appalled at our recommendations, the responses of when we should launch, how long do we have to wait, April? And also there was argument about what sort of rationale we could have for 53°, and what temperature could we make a strong statement on. And the data couldn't tell us which temperature other than 53°. . . . I had the feeling that we were in the position of having to prove that it was unsafe instead of the other way around, which was a totally new experience."[17]

Boisjoly also identified the technical discussion as the source of the pressure:

> I felt personally that [Thiokol] management was under a lot of pressure to launch and that they made a very tough decision, but I didn't agree with it. One of my colleagues [Russell] that was in the meeting summed it up best. This was a meeting where the determination was to launch, and it was up to us to prove beyond a shadow of a doubt that it was not safe to do so. This is in total reverse to what the position usually is in a preflight conversation or a Flight Readiness Review. It is usually exactly opposite that.
> DR. WALKER: Do you know the source of the pressure on management that you alluded to?
> ROGER BOISJOLY: Well, the comments made over the [teleconference] net is what I felt. I can't speak for them, but I felt it—I felt the tone of the meeting exactly as I summed up, that we were being put in a position to prove that we should not launch rather than being

put in the position and prove that we had enough data to launch. And I felt that very real.[18]

"Shifting the burden of proof" was perceived by the engineers as a deviation from normal practice. However, neither NASA norms nor rules were violated on the eve of the launch. Although Marshall personnel did assert their adherence to rules and procedures throughout, I am not basing my conclusion solely on the statements of those who would have an obvious interest in presenting themselves as rule-abiding. My review of rules, procedures, and norms used in FRR decision making for the SRB group in previous launches and by other work groups responsible for other parts of the shuttle allowed me to identify the same norms, rules, and procedures in operation on the eve of the launch. The finding that Marshall managers conformed to prescribed NASA practices is affirmed by the House Committee.[19] Although volume 1 of the Presidential Commission's report describes procedural irregularities by Marshall managers as a contributing cause of the disaster,[20] the technical appendices in volume 2 contain the report of the Commission's "Pre-Launch Activities" investigative team, which states that "the launch decision followed established management directives" and that "the management tools described above did not adequately protect against the STS 51-L mishap."[21]

A RULE-BASED DECISION

That night, teleconference participants faced several unprecedented situations. First was the record cold temperature. Although armed with engineering principles that functioned as universalistic rules and with their own particularistic, experience-based rules developed as they learned about joint behavior, they had no rule about how the O-rings would function at the predicted temperatures. Thiokol engineers thought the cold would reduce resiliency, which in turn would affect redundancy. They had data on redundancy; they had data on resiliency. But they had no data on how resiliency affected redundancy.[22] The result was no temperature Launch Commit Criterion for the SRB joints. Second, although teleconferences occurred "before launch every hour of the day and night,"[23] the SRB work group had never had a technical concern that required a teleconference discussion between the two engineering communities after the L-1 FRR.[24] Finally, in all the years of association between Thiokol and Marshall, this was the first time the contractor ever had come forward with a no-launch recommendation.[25] It is doubtful that any work group had previously had this experience: recall that most launches are stopped, not by an FRR-type recommendation from the contractor, but at the Cape because of some "hardware glitch" at the launch site.

Mileti notes that "in the face of uncertainty, people's preferences take over."[26] At NASA, however, the intrusion of preference into technical decision making was constrained by numerous rules. The 8:15 teleconference was not part of FRR, which had occurred two weeks earlier. It was an ad hoc meeting. But in a situation of uncertainty about both the technical issue and the procedures, Marshall managers enacted the procedures appropriate to a Level III FRR. The format for the evening was established during the preliminary afternoon teleconference and Marshall managers followed it throughout the evening teleconference. The unprecedented nature of the teleconference gave no basis for comparison, which led many participants who opposed the launch to view "shifting the burden of proof" and other managerial actions as irregular. It was not only that no one had been in an eve-of-launch, "no-fly" teleconference before, but some participants, including some of the working engineers, had never been in a Level III FRR before. The perception of some managerial actions as irregular was also fueled, in some cases, by the positions people held in the organization, which as we know limits access to information, ability to understand it, and meaning ascribed to it. Even participants who attended FRRs frequently were not familiar with all the procedures and rules that were customarily used in NASA Level III managers' decision making. Disinterested about decision-making rules prior to the tragedy, many questioned them afterward.

Two weeks earlier in the 51-L FRR, Thiokol had certified in writing that the SRBs were flight-ready. That determination had been presented and certified at every level of FRR. Any observation or change arising between FRR and launch called for a risk assessment. Thiokol was now reversing it original certification, presenting an engineering analysis that altered their previous recommendation to Level III. The teleconference proceeded according to the protocol for a formal Level III FRR. Thiokol's Brian Russell said: "In my own mind, it was very much like a Flight Readiness Review. In fact, that's what we were doing, was discussing the readiness of that vehicle to fly under the conditions that we anticipated."[27]

Briefly summarized, the procedures (described in detail in chaps. 3, 4, and 5) were: (1) For any new observation or change, a risk assessment was made, then weighed against the existing technical rationale for establishing flight safety in order to determine risk acceptability. (2) The engineering analyses, conclusions, and flight-readiness assessment were presented in formal charts, accompanied by oral explanation. (3) The presentation was subjected to adversarial "probes" and "challenges" to test the technological rigor of the hazard assessment. (4) According to the Delta review reporting concept, if the issue was resolved at Level III with no change in the existing technical rationale for flight readiness, the technical issue was not relayed up the hierarchy to Levels II and I for further consideration. On the other hand, if the change or the new evidence altered the existing rationale, that change was conveyed up the hierarchy to other levels, where it would be presented by the Level III Project Manager and scrutinized by Level II and I personnel.

The predicted cold temperature was a signal of potential danger that set in motion the Acceptable Risk Process. Thiokol engineers reviewed the evidence, prepared their data analysis, conclusions, and recommendation, then made the formal presentation. Marshall's Larry Mulloy took charge, as he normally did for an SRB Project Level III review. In taking an adversarial stance toward Thiokol's data analysis and conclusions, Mulloy and Hardy were enacting their usual roles in Level III FRR as Project Manager and Deputy Director of the S&E Directorate, respectively. Marshall's Larry Wear testified, describing the culture of FRR:

> I have seen a lot that has been written and a lot that the Commission has asked or has implied in their response which, to me, infers a lack of understanding of the climate that we live and work in. It is absolutely normal for an engineer to present data, to present what he concludes from the data, and to have other engineers disagree with that, or not understand that, or need to understand various facets of it because they don't draw the same conclusion. That is normal. It is normal for a contractor engineer or one of the NASA engineers to present his data and his conclusions and have various people either disagree with what he is saying or test him, to see if he really has good reason for believing in what he says, or whether he just has a very shallow conclusion that he has drawn and reached. That goes on all day long. That is the world we live in.
>
> . . . So, with the participants that were in that meeting that night, the level of questioning that was raised, and whatever challenges may have been thrown out or inquiry may have been raised, there was nothing in that meeting that was abnormal from a hundred other meetings that those participants had been involved in on previous occasions. In our system, you are free to say whatever you wish, to recommend whatever you wish. But you've got to be able to stand the heat, so to speak, based on what you have said.[28]

Boisjoly confirmed that harsh criticism by Level III management was the norm, stating that he had been in FRRs at which "we took a lot of flak, a major amount of flak. . . . We are always challenged and asked to present the data base, present the proof, present the information that supports the statement that you are making at the time. . . . We were always put in a position, and quite frankly in many respects nitpicked, to prove that every little thing that we had was in proper order and had the proper engineering rationale and data to back it up in order to fly. And in this instance we were being challenged in the opposite direction."[29]

Both Mulloy and Hardy testified that they challenged the rigor of every engineering launch position. However, they denied Boisjoly's assertion that they were "shifting the burden of proof." Mulloy stated:

I found their conclusion without basis and I challenged its logic. Now, that has been interpreted by some people as applying pressure. I certainly don't consider it to be applying pressure. Any time that one of my contractors or, for that matter, some of Mr. Hardy's [S&E] people who come to me with a recommendation and a conclusion that is based on engineering data, I probe the basis for their conclusion to assure that it is sound and that it is logical. And I assure you, sir, that there was no reversal of the tradition of NASA which says prove to me why you can't fly versus prove to me why you can. To me, it doesn't make any difference. If somebody is giving me a recommendation and a conclusion that is based upon engineering data, I am going to understand the basis for that recommendation and conclusion to assure that it is logical. I think that has been interpreted, when one challenges someone who says, "I don't have anything too quantitative, but I'm worried," that that is pressure, and I don't see it that way.[30]

Hardy stated:

I am likely to probe and sometimes even challenge either a pro position or a con position, or sometimes even both . . . and the objective of this is just simply to test the data, test the degree of understanding of the individual that is presenting the data, test not only his engineering knowledge, but his engineering assessment of that data. So that is characteristic of the way that I do, and I think anybody that knows me would realize that that is not interpreted as coming on strong or applying pressure. Thiokol will also remember that on more than one occasion that I have rejected their proposal to fly. . . .

I categorically reject any suggestion that the process was prove to me it isn't ready to fly as opposed to the traditional approach of prove that this craft is ready to fly. I have no responsibility for schedules, I have no responsibility for manifest. I couldn't tell you with any degree of certainty what the next launch cargo was on the next flight. I couldn't tell you with any high degree of certainty even what date it was scheduled to fly. As I said, I have no responsibility for that, and I have occasion to know that at times, but it is not my primary job. I would hope that simple logic would suggest that no one in their right mind would knowingly accept increased flight risk for a few hours of schedule.[31]

What was the source of these diametrically opposed perceptions about "shifting the burden of proof"? In every FRR the contractor engineers for all shuttle projects had to prove that *their data supported their engineering conclusions and the launch recommendation*. The Thiokol engineers believed that the burden of proof was different than in the past because *they had never come forward with a no-launch recommendation in an FRR before.* Marshall managers and Thiokol proponents of the no-launch recommendation were both partially correct: managers did not behave differently, but the burden of proof did shift. It shifted because the contractor position shifted. There was not one set of practices for determinations of "go" and another for determinations of "no-go": the same adversarial strategies and tests of scientific validity of engineering arguments were called for in both situations. Thiokol was challenged in the "opposite direction" because their position was in the opposite direction. Since their position deviated from the norm, the burden of proof deviated from the norm.

In the following, Marshall's Ben Powers, who opposed the launch, corrects the investigator's question, carefully reframing it to reflect the culturally accepted FRR practice of testing to assure the adequacy of technical rationale behind all readiness assessments.

> MR. MOLESWORTH: That was, was it not, an unusual position for a contractor to be in—to prove that his system could not function?
> BEN POWERS: I don't think you can use those words, "to prove that it would fail." I think he was put in a position to where he had to explain—and I'm not defending Hardy nor myself nor anybody else, I'm trying to be as accurate as I can—I think that your words should be chosen to say Thiokol was being put into a position *to explain why their rationale was "No fly."*[32] (Emphasis added)

Other controversial Marshall managerial actions also conformed to the Level lll guidelines usually employed. Official investigators questioned why Mulloy and Reinartz did not report the teleconference to Arnold Aldrich, Level II in the FRR reporting chain, either before the teleconference or when they phoned him afterward to discuss the weather in the recovery ship area.[33] When questionable organizational actions become public, middle-level managers often state that top administrators were not informed, shielding them from blame. Official investigators pushed hard on the issue of "who knew what when," for in light of the tragic outcome it appeared that if Levels II and I had been informed of the teleconference they would have intervened and the astronauts would not have died. The Presidential Commission stated: "The decision to launch the *Challenger* was flawed. . . . If the [Level II and I] decisionmakers had known all the facts, it is highly unlikely that they would have decided to launch 51-L on January 28, 1986."[34] Normally, however, Level II was brought in only after the issue was addressed at Level III, and then only if there had been a change in the previous engineering rationale for flight readiness. The teleconference produced no change in the three-factor technical rationale behind the 51-L FRR recommendation to accept risk. Therefore, not to report conformed to the Delta review reporting protocol. When Mulloy testified about that postteleconference call to Aldrich, he was not asked for the rules, but the rules were his answer:

> There was in the Launch Criteria a statement relative to recovery area, and that was that if there is a possibility that the boosters cannot be recovered, that is an "Advisory Call." It does not say you cannot launch under those circumstances. I took that to be my responsibility to advise Mr. Aldrich that we possibly would not be in a position to recover the Solid Rocket Boosters because they would be some 40 miles from the impact area. . . . The O-ring and other special elements of a Level III were considered in the management system to be a delegation to the Level III Project Manager to make disposition on any problems that arose on those. Our judgment was that there wasn't any data that was presented that would change the rationale that had been previously established for flying with the evidence of blow-by . . . [therefore] there would not be any requirement to have that rationale approved by Level II or Level I. There was no violation of Launch Commit Criteria. There was no waiver required in my judgment at that time and still today [either would have required him to report to Level II]. . . . It was clearly a Level III issue that had been resolved.[35]

Similarly, Reinartz "did not perceive any clear requirement for interaction with Level II, as the concern was worked and dispositioned with full agreement among all responsible parties [Levels III and IV] as to that agreement."[36]

Their testimony that they did not report the teleconference because it was a Level III matter is consistent with how they reported technical issues from Level III in all previous FRRs. Recall Mulloy's description of Marshall culture in the years preceding the tragedy: "We were absolutely relentless and Machiavellian about following through on all the required procedures at Level III."[37] Acting Launch Director Gene Sestile and Cecil Houston, who were on the postteleconference phone hookup between Mulloy, Reinartz, and Aldrich, affirmed that Aldrich was not told. We can also see the Delta review reporting concept operating in their postteleconference discussion about telling Lucas. Recall that as they left Kennedy that night, McDonald overheard Mulloy tell Reinartz: "I don't think I want to wake the old man and tell him this right now because we did not change the decision to proceed with the launch. If we would have decided not to [launch], then yes, we would have to do that."[38] Mulloy and Reinartz testified they did tell Lucas in the morning, an action conforming to the Marshall FRR requirement that everything be reported to the Center Director. The question with Lucas, for them, was not "if" but "when."

But "what" also matters to us, for we want to know, if people were told, what they were told. Lucas, who testified he was informed the next morning of Thiokol's initial below-53°F no-launch recommendation and shown the final recommendation, said they told him "that an issue had been resolved, that there were some people at Thiokol who had a concern about the weather, that that had been discussed very thoroughly by the Thiokol people and by the Marshall Space Flight Center people, and it had been concluded agreeably that there was no problem, that he had a recommendation by Thiokol to launch and our most knowledgeable people and engineering talent agreed with that. So from my perspective, I didn't have—I didn't see that as an issue."[39] Apparently, Mulloy and Reinartz, who at the end of the teleconference did not know Thiokol engineers still objected, accurately transmitted its conclusion to Lucas, omitting the details of the engineering analysis: they "statused" it, which was the normative treatment for "resolved problems."

The House Committee stated no reporting rules were violated that night. They pointed out that "Marshall management used poor judgment in not informing the (Level II) NSTS [National Space Transportation System] Program Manager" but concluded that there was "no evidence to support a suggestion that the outcome would have been any different had they been told."[40] In the absence of witnesses to these oral exchanges, we still must wonder who knew what. We have seen in the past how structural secrecy, not individual secrecy, operated as the social construction of risk developed in the SRB work group was passed up the hierarchy, becoming the col-

lective construction of risk. Official organizational practices, including problem resolution and reporting rules that reduced the amount
of information going to the top, operated to systematically censor
what was relayed upward on the eve of the launch. The teleconference was not a secret meeting: in addition to the participants, many
others at Kennedy knew about it, providing a control on anything
that Marshall managers might have said or not said. If Aldrich had
been notified, he probably would have been given the same information as Lucas, drawing the identical conclusion. What we know
about structural secrecy—hierarchical reporting rules, specialization, information dependencies, the influence of Project Managers,
and uncertainty absorption as information is passed upward—
affirms the House Committee's conclusion: if told, Level II would
not have intervened.

Other controversies about Level III personnel and rule violations
existed:

• During the teleconference, Marshall participants challenged
Thiokol's position, referring to a NASA document authorizing
launch for the entire shuttle system at ambient temperatures of 31°F
to 99°F degrees. Recall that after the teleconference, McDonald challenged Mulloy and Reinartz's decision to launch outside the SRM
qualification temperature range of 40°F to 90°F. Here again, we see
the effects of structural secrecy. Marshall managers were going by
the book, but McDonald's position denied him familiarity with the
relationship of the first rule to second. After the tragedy, McDonald
testified that the 31°F lower limit was a "higher-level" temperature
specification from Johnson Space Center. The higher-level specification was incorporated in the "lower-level spec," Thiokol's 40°F–90°F
specification document. McDonald commented: "The way the system is supposed to work is, you're supposed to comply with your
own specification plus any higher level specifications that may
involve the entire shuttle system. And I was unaware of that, frankly,
that that criteria was in there."[41]

• Mulloy's willingness to accept a final launch recommendation
from Thiokol that contained no specified temperature limit, which
McDonald questioned, also was rule-based. The rule was that the
Launch Director made temperature decisions about the shuttle system, using the existing Launch Commit Criteria establishing a 31°F
lower limit, unless a Launch Commit Criterion for a particular component contradicted it.

• Mulloy's request for a copy of the final launch recommendation
signed by Thiokol Vice President Joe Kilminster also was rule-based.

In every Level III FRR, Mulloy received a signed launch recommendation from the contractor, and although this was an ad hoc meeting, he followed the FRR protocol.[42] In hindsight, with the tragic outcome known and no experience in a no-launch FRR available for comparison, the request for a signature appeared ominous to many of the participants, and hence to the official investigators. This perception was reinforced by McDonald's statement on the teleconference that he would not sign. Later, McDonald said that he was not sure what the "normal" procedure was: Kilminster usually signed in a formal FRR, but if anything came up afterward, it usually was handled orally. But the public view that he was overridden by his supervisor was in error. "I think that [statement] has been misinterpreted, at least by the press. They said that I was overruled by my supervisor. That is not true at all. I chose not to sign that. He didn't overrule me. I felt that that decision was an engineering decision made by the people that understood the problem the best, that had all of the data and facts, and they are the ones who should recommend it. And that is why I made that. It wasn't that I was overruled."[43]

Among posttragedy accounts, the rule-based nature of the teleconference discussion was noted only by the House Committee. Mulloy reflected, "With all procedural systems in place, we had a failure."[44] If schedule-oriented, amorally calculating middle managers violating rules in pursuit of organization goals were not responsible for the launch decision, what happened?

THE PRODUCTION OF CULTURE, THE CULTURE OF PRODUCTION, STRUCTURAL SECRECY, AND DISASTER

Production pressure had a powerful impact on the teleconference discussion. But to attribute the controversial actions of Marshall managers on the eve of the launch to production pressure alone is to assume, wrongly, that the launch decision was the outcome of that single cultural imperative. The Marshall culture was, in fact, dominated by pressure to achieve three cultural imperatives: political accountability, bureaucratic accountability, and the standards of the original technical culture. And we find all three expressed in Mulloy's explanation of his "April" comment, which affirms his preoccupation with Marshall standards for "scientific" engineering analyses (original technical culture), procedural rules for risk assessment (bureaucratic accountability), and production goals (political accountability):

There are currently no Launch Commit Criteria for joint temperature [bureaucratic accountability]. What you are proposing to do is to generate a new Launch Commit Criteria on the eve of launch, after we have successfully flown with the existing Launch Commit Criteria 24 previous times [bureaucratic accountability; political accountability]. With this LCC, do not launch with a temperature less than 53°F, we may not be able to launch until next April [political accountability]. It is all in the context, again, with challenging your interpretation of the data, what does it mean and is it logical, is it truly logical [original technical culture] that we really have a system that has to be 53 degrees to fly [political accountability]?[45]

But these concerns were not the exclusive domain of Marshall managers. The cultural imperatives of bureaucratic accountability, political accountability, and the original technical culture were part of the frame of reference all participants brought to the teleconference. Their interaction that night was scripted in advance by the same cultural understandings that shaped their choices in the years preceding January 27, 1986. In combination, the imperatives of the culture of production produced a mistake of tragic proportions. McCurdy and Romzek and Dubnick stressed that the infusion of bureaucratic and political accountability into the NASA culture did not eliminate the space agency's original technical culture, but these new pressures made its tenets more difficult to carry out.[46] On the eve of the launch, we see how true this was.

That night, a core component of worldview that many at Marshall, Kennedy, and Thiokol, the key managers among them, brought to the teleconference was the belief in redundancy. The belief in redundancy was the product of the work group culture and the incremental accretion of history, ideas, and routines about the booster joints that began in 1977. It was based on a scientific paradigm in the Kuhnian sense: agreed-upon procedures for inquiry, categories into which observations were fitted, and a technology including beliefs about cause-effect relations and standards of practice in relation to it. These traits, reinforced by the cultural meaning systems that contributed to its institutionalization, gave the belief in redundancy the sort of obduracy Kuhn remarked upon. That night, the SRB work group culture fractured: the Thiokol engineers tried to overturn the belief in redundancy that they had created and sustained in the preceding years.[47] A scientific paradigm is resistant to change. For those who adhere to its tenets, alteration requires a direct confrontation with information that contradicts it: a signal that is too clear to misperceive, too powerful to ignore.

After the *Challenger* tragedy, it appeared that Thiokol engineers had given exactly that sort of signal. Describing the engineering opposition to the launch, Marshall's Leon Ray said, "We lay down in the bucket."[48] The "bucket" he referred to is a trench under the launch pad that diverts the rocket flame and exhaust flow away from the vehicle. It forces the flame to make a 90° turn into a water trench, which keeps it from rebounding from the concrete and immolating the vehicle. The imagery of engineers "lying down in the bucket" suggests a strong signal: an all-out engineering effort to stop launch by throwing their bodies under the rocket. For Thiokol engineers, this effort took two forms: the formal presentation of charts and verbal opposition. The initial Thiokol engineering recommendation, do not launch unless O-ring temperature is 53°F or greater, was an unprecedented Thiokol stance—itself a deviant event that should have constituted a signal of potential danger to management. Further, Thompson and Boisjoly and other engineers who opposed the launch believed they had done everything they could to argue vigorously against the launch.

But the information exchange that night was undermined by structural secrecy. As teleconference participants groped to find the appropriate rule about resiliency, redundancy, and temperature in an unprecedented situation, all enacted the "rules of appropriateness" drawn from the culture of production.[49] The imperatives of the original technical culture, bureaucratic accountability, and political accountability *interfered with the production and exchange of information so essential to hazard assessment.* The charts the engineers prepared contained patterns of information that included mixed, weak, and routine signals; thus the engineers failed to make their case because their engineering argument *deviated from the rigorous scientific standards of the original technical culture for FRR risk assessments.* On the other hand, their verbal opposition was impaired by bureaucratic and political accountability. Participants *conformed to cultural understandings about hierarchical relations and functional roles* that, in combination with *production pressure,* affected both the production and exchange of information. The result was missing signals. Altogether, it was the wrong stuff to overturn an entrenched scientific paradigm.

CHARTS AND THE ORIGINAL TECHNICAL CULTURE

In the NASA culture, the Acceptable Risk Process required that every flight-readiness risk assessment weigh new conditions, tests, or engineering calculations against the documented technical rationale of record that supported a component's status as an acceptable risk. On the eve of the launch, the Thiokol engineers were arguing that the weather would affect O-ring resiliency, which in turn would affect redundancy. Thus, they were challenging the three-factor technical rationale supporting the booster joint's redundancy.

The Thiokol engineers constructed their charts from the information at hand. In an unprecedented situation typical of unruly technology, they examined the available data on temperature and created a practical rule: do not launch unless O-ring temperature is 53°F or greater. However, the charts contained routine, mixed, and weak signals, undermining the scientific credibility of their position. As a consequence of the latter two, the launch recommendation and analysis violated norms of the original technical culture about what constituted a technical argument acceptable in FRR. While to outsiders after the tragedy, Marshall managers appeared to be engaging in deviance and misconduct, ironically it was the Thiokol engineering analysis that was deviant because it did not conform to established norms for "engineering-supported" technical positions at NASA. Thus, it was inadequate as a challenge to the documented risk rationale supporting redundancy.[50]

Routine signals. Remember that during the teleconference the engineering charts arrived at Kennedy and Marshall one at a time and out of sequence. The Thiokol conclusions and recommendations charts were the last to arrive at Marshall and Kennedy, so the data analysis charts that came before them were read without knowledge of the specifics of Thiokol's final position. The data analysis charts presented no new data; the information in them was known, was accepted, and had been repeatedly presented in other situations *when Thiokol recommended proceeding with launch on the basis of redundancy.* Several of the charts themselves were old. They were pulled from previous presentations and so had been seen before. The resiliency data had been presented at a special Marshall briefing by Arnie Thompson during the previous summer and also had been included in the August 19 NASA Headquarters briefing in Washington. The House Committee found "no evidence that new data were presented during the January 27th teleconference that were not available to Thiokol at the time of the [STS 51-L] Flight Readiness Review. Moreover, the information presented was substantially the same as that presented at the August 19th briefing at which time they had recommended that it was safe to fly as long as the joints were leak checked to 200 psi, were free from contamination in the seal area and met O-ring squeeze requirements. No mention was made of a temperature constraint at that time or anytime between then and the January 27th teleconference."[51]

Consequently, as the charts arrived, they were viewed as routine signals, with a taken-for-granted meaning: they affirmed the cultural belief in redundancy for many. Mulloy stated, "Everything they had said up to that time [the arrival of the conclusions and recommendations] was simply a repeat of what we already knew."[52] What participants "already knew" was that those data had been taken into consideration on every previous launch decision. The outcome of those considerations was the development of the three-factor technical rationale affirming SRB joint redundancy and repeated recommendations by Thiokol to accept risk and proceed with launch, even after the January 1985 cold-weather launch. Said Marshall's Larry Wear: "In my hearing their data and their recommendation, I had been very conscious of the fact that the secondary seal was there, was in place. If you go back through our FRRs, where we have discussed this particular subject on numerous occasions, if you go back and read the CIL through the eyes of an engineer, it is very evident in that one. So, it is in our litany, all the way through. It's in our thought process all the way through."[53]

Mixed signals. The Thiokol presentation gave mixed signals as well. First, the 53°F launch recommendation contradicted launch recommendations in the past. At the next FRR after the January 1985 launch, Thiokol came forward with a recommendation for acceptable risk based on their calculations that the joint would seal with three times the damage that had occurred at an O-ring temperature of 53°F. Moreover, several times in the past the ambient temperature had dropped below 53°F for a scheduled launch and Thiokol had not raised any concerns. The temperature that very morning had been in the 40s. As retired Marshall chemist John Schell said, "It kind of surprised me that one day they were going to launch at 40 degrees and the next they were saying [it had to be] 53°."[54]

At no time previously had Thiokol tried to initiate a temperature Launch Commit Criterion. The House Committee found that "the efforts of Thiokol engineers to postpone the launch commendable; however, Thiokol had numerous opportunities throughout the normal flight readiness review process following flight 51-C in January 1985 to have the new minimum temperature criteria established."[55] Ironically, the reason they raised temperature as an issue on the eve of the launch was in response to an inquiry by Marshall's Larry Wear.

The House Committee noted that "the concerns and recommendations of the Thiokol engineers [on the eve of the launch] were solicited by NASA, and in as much as they had not come forth with the recommendation for a higher minimum temperature criterion on earlier occasions when it was planned to launch at temperatures below 53 degrees, it is unlikely that this recommendation would have been made on this occasion without the specific inquiry by NASA."[56]

Second, the charts themselves contained mixed signals—or, as Marshall engineer Ben Powers put it, they were "talking out of both sides of their mouth."[57] Thiokol argued for a correlation between cold and O-ring blow-by. But, in fact, the existence of such a correlation was contradicted by other data in their charts.[58]

1. The first statement among the conclusions (fig. 11.12) acknowledged that other factors had contributed to blow-by in the past. Blow-by had occurred on other missions, independent of temperature.
2. The summary of blow-by tests (fig. 11.9) indicated no leakage [blow-by] in tests at both 75°F and 30°F.
3. The history of O-ring temperatures (fig. 11.11) showed that two developmental motors and two qualifying motors successfully fired at temperatures below the 53°F limit Thiokol recommended.
4. Thiokol's argument for correlation between cold temperature and blow-by was based on the comparison of the only two launches with field joint blow-by, shown in figure 11.10. Although SRM-15 (STS 51-C), launched with an O-ring temperature of 53°F, had the most blow-by, SRM-22 (STS 61-A), launched with an O-ring temperature of 75°F, had experienced "the next worst blow-by."[59] The data on SRM-22 Thiokol cited to prove their point actually contradicted it.
5. The hardness/temperature chart (fig. 11.7) showed O-ring hardness increased to 90 at 30°, but the blow-by tests (fig. 11.9) showed no leakage at that temperature.
6. Thiokol's proposed 53°F limit contradicted other temperature guidelines that teleconference participants knew: the manufacturer's data showing Viton resiliency at 25°F, the SRM qualification firings at propellant mean bulk temperatures of 40°F–90°F with no failures, and the Launch Commit Criterion specifying an ambient temperature of 31°F–99°F for the entire Space Shuttle.

Noting the mixed signals, the House Committee observed that "Thiokol's advice and recommendations to NASA were inconsistent, and therefore, the arguments presented during the January 27th teleconference might not have been as persuasive at the time as they now appear to be in hindsight."[60]

Weak signals. In NASA's original technical culture, engineering analysis attained credence because risk assessments followed scientific, rule-bound engineering standards. Further, all new tests, flight experience, and calculations were weighed against the preexisting engineering rationale for flight when determining risk acceptability. To be integrated into or to alter the developing knowledge base for a component, an engineering analysis had to meet certain criteria. Three aspects of the Thiokol presentation violated normative expectations about acceptable technical arguments, thus making the no-launch recommendation a weak signal.

First, although practical experience is the most important data an engineer has, two astronaut Presidential Commissioners noted that Thiokol's use of the *experience base* deviated from traditional practice. Thiokol posed a launch decision floor of 53°F, using the flight experience base for cold temperature. This position made good sense to many after the tragedy. As Gouran et al. put it, "Why the absence of a systematic correlation within the available data base would justify taking a risk outside the data base, however, is not intuitively obvious."[61] However, according to the two astronauts, sticking with the *flight* experience base was unusually restrictive. Astronaut Neil Armstrong, Vice Chair of the Presidential Commission, said to Thiokol's Lund: "The recommendation was to stay within your experience base, but I find that to be a peculiar recommendation in the operation of any kind of system because normally you say, 'From our experience base, our data points, and our analysis, and our extrapolations, we would be willing at any time to go beyond our experience point out this far' as a next step. And the only reason you would say 'I would stay within my experience base' is that you had a problem at that point that said you dare not go any farther."[62]

However, the latter was not the case because after the January 1985 launch, Thiokol engineers told NASA that they *could* go farther: the safety margin indicated the rings would tolerate three times the erosion possible under the worst-case conditions. Wondering why they had not used the 40°F qualification temperature as the cut-off, Presidential Commission member Astronaut Sally Ride pointed out that it was accepted practice to go beyond the flight experience base, using data points from qualification testing to establish a guideline: "Normally, when you are trying to extrapolate beyond your flight experience, you rely on your qualification testing program, and a system or a subsystem is qualified to fly within a certain regime or a certain envelope. That would include environmental effects like temperature."[63]

The violations of two other criteria of the original technical culture were considered perhaps more significant lapses at NASA, however. A necessary element of an acceptable engineering argument was known in NASA-speak as the standard of *logical consistency:* an engineering recommendation about flight readiness had to take into account all available data, which had to be consistent with the recommendation. This standard is what Mulloy referred to when he testified, "It is all in the context, again, with challenging your interpretation of the data, what does it mean and is it logical, is it truly logical, that we really have a system that has to be 53 degrees to fly?"[64] Thiokol's 53°F limit did not take into account all the data Thiokol presented: test data showing the joint worked at lower temperature and data showing problems at higher temperatures. Thus, the final recommendation violated the standard of logical consistency that was a tenet of NASA's original technical culture.

Clarke writes about the operation of a "disqualification heuristic" in decision making about risk.[65] It is a mind-set that disqualifies disconfirming information and critical data and viewpoints, while highlighting confirming information. It protects experts from seriously considering the likelihood of catastrophe. Here we find an interesting reversal. Thiokol engineers, convinced of the threat, selectively drew from all available data that which supported their point that it was unsafe to fly. Said Russell: "[Mulloy's] argument was rational and logical in my mind. I didn't agree with it necessarily . . . because I felt that the data of the soot blow-by which was extensive on 51-C and the resiliency tests were telling us something. Namely, that the colder the temperature, the higher the risk. And the other data, though they didn't support that, I tended not to give them as much value in my own mind."[66]

Third, acceptable technical arguments in FRR had to meet the quantitative standards of scientific positivism.[67] A strong norm supporting quantification, an aspect of the original technical culture, operated at both NASA and Thiokol. Endorsement of it was found consistently across interview transcripts. Observational data, backed by an intuitive argument, were behind all engineering analysis. But subjective, intuitive arguments required lab work and tests before they were considered admissible evidence in FRR. Otherwise, they remained untested hypotheses that were unacceptable in NASA's science-based, rule-bound prelaunch FRR. At Marshall in particular, where the Von Braun technical culture originated, "solid technical arguments" were required and people were grilled extensively in FRR to uncover weak analyses that might obscure a threat to flight safety. Thiokol was trying to establish a correlation based on observed blow-by on two missions. The basis of Boisjoly's argument was that the *quality* of the damage was worse on the January 1985 launch because the *putty looked different* than in other instances of blow-by. Not only was this an intuitive argument according to NASA standards (because it was based on observation), but Thiokol's comparison of SRM-15 and SRM-22—the linchpin of their position—did not support a correlation: blow-by occurred on the one motor at 53°F and on the other at 75°F.

A weak signal is one conveyed by information that is informal and/or ambiguous, so that its significance—and therefore the threat to flight safety—is not clear. Thiokol engineers themselves acknowledged that their position was not adequately supported by their data, saying their argument was "subjective," "inconclusive," "qualitative," "intuitively correct," and based on "engineering 'feel.'" Thus, when Marshall challenged Thiokol to "prove it" by "quantifying their concerns," they were asking Thiokol to conform to the standards of the original technical culture by supplying quantitative data. When Boisjoly replied, "I couldn't quantify it. I had no data to quantify it, but I did say I knew it was away from goodness in the current data base," his reference to "away from goodness in the current data base" was known in NASA culture as "an emotional argument."[68] "Real" technology conformed to the norms of quantitative, scientific positivism. Any FRR engineering analysis that did not meet these standards was, in effect, a weak signal in the NASA system.

On the morning of the *Challenger* launch, Aldrich also invoked the same standards for technical arguments when he called a Mission Management Team meeting to assess ice. Of the three reports given to NASA, two included quantitative data about the distance the ice would fall and rebound and the probability of resultant damage to the vehicle. These statements were accompanied by recommendations to accept risk and proceed with the launch. From the Rockwell people, he received no presentation of quantitative data. They discussed the possibilities, stating that the evidence was "inconclusive," that the condition was an unknown, and that they were unable to predict and therefore unable to assure that it was safe to fly. Their position, when weighed against the standards of NASA's original technical culture, was a subjective, intuitive argument (i.e., a weak signal). Absent consensus, Aldrich made a management risk decision, proceeding with launch based on the quantitative data that had been presented.

These parallels with the eve-of-launch teleconference notwithstanding, both official investigations concluded that on the morning of the launch, Aldrich did the correct thing.[69] Based on the language Rockwell representatives used to express their position (e.g., "there's a probability in a sense that it was probably an unlikely event, but I could not prove that it wouldn't happen"; "we can't give you 100 percent assurance"; and "I could not predict the trajectory that the ice on the mobile launch platform would take at SRB ignition") both the Commission and House Committee observed that Rockwell's launch recommendation was ambiguous—in effect, a weak signal.[70] The Commission found "it difficult, as did Mr. Aldrich, to conclude that there was a no-launch recommendation."[71] Noting that Rockwell "expressed concern" without actually saying, "stop the flight, it is unsafe", the House Committee observed the dubious benefits that befall contractors when giving weak signals: "If the odds favor a successful flight they do not have to be responsible for canceling, yet if the mission fails they are on record as having warned about potential dangers."[72]

TALK AND BUREAUCRATIC ACCOUNTABILITY

Bureaucratic accountability affected the teleconference outcome by creating missing signals. When Leon Ray said, "We lay down in the bucket," he referred not just to the charts that made up the formal presentation but to what he and, later, the public, perceived as a signal sufficiently strong to overturn management's definition of the situation: vigorous verbal opposition to the launch by the working engineers in the SRB work group, led by Boisjoly and Thompson. But, like the charts, vocal opposition by engineers was not as strong at the time as it seemed in retrospect. First, chapter 8 shows clearly that consensus was absent: not all engineers shared the view represented by Boisjoly and Thompson. Second, and more important, many of those who *were* opposed to launching, who had doubts, or who even had information that might have supported the Thiokol position either did not speak at all or conveyed their views in a manner that did not make their position clear to others. These missing signals distorted conclusions drawn by people on both sides of the argument, greatly affecting the Thiokol engineers, who were unaware of the extent of support for their position, and the participants at Marshall and Kennedy, who after the caucus were unaware that opposition to the launch remained.

The ability to give a strong signal through verbal opposition was undermined because *all* participants conformed to cultural understandings about talk and bureaucratic accountability in FRR that silenced people. All were disadvantaged at the outset because physical separation was a source of structural secrecy: missing signals were an inevitable aspect of a three-location teleconference. With no video transmission, communication between settings depended entirely on words, inflection, and silences, for many visual cues that normally aid interpretation—such as gestures, facial expressions, body posture, activity—were unavailable. Communication depended on individual willingness to speak to an unseen audience. But missing signals were also a consequence of recreating the practices of a formal Level III FRR in an ad hoc meeting. Unlike other engineering discussions, FRR was manifestly hierarchical, guided by many rules and procedures about the form, content, and structure of information exchange. Rather than the original technical culture's professional accountability that gave deference to those at the bottom of the organization with the greatest technical expertise, the bureaucratic accountability that developed in the 1970s prevailed. In his study of the Tenerife air disaster, Weick observed that under conditions of stress, communication—even among equals—tends to turn hierarchical.[73] Giddens and others have noted how the rules people use as resources for interaction are both the medium for interaction and its outcome.[74] Probably for both these reasons, the participants reproduced the social inequalities of the engineering profession and the organization. The result was missing signals: information that was not communicated to all participants of the teleconference.

Participants were both empowered and disempowered to speak by their position in the hierarchy. The people who dominated the teleconference discussion were those in the SRB work group whose position exposed them to the most—and the most recent—information. Information empowered people, making them central to the conversation; lack of it marginalized them. One of the consequences of imposing the rigors of FRR on this ad hoc meeting was that participants abided by the stringent FRR standards about what was admissible as evidence. In order to speak, participants felt they needed "solid technical data." Lacking it, people abided by functional roles and hierarchical relations at the cost of more missing signals. Participants repeatedly affirmed that their ability to speak was limited because they either (1) were unable to interpret the information they had, (2) were able to interpret it but were limited in their ability to understand its relevance to the whole by their specialization, or (3) did not define themselves as having the authority to speak. For any or all of these reasons, they deferred to hierarchy, depending on others for information and its interpretation.[75] To show how bureaucratic accountability affected teleconference talk, we divide the 8:15 EST teleconference into three periods: precaucus, caucus, and postcaucus.

Precaucus. Several specialists not in the SRB work group but asked to participate under the matrix system did not speak over the teleconference network because they did not feel that their position gave them sufficient expertise on the SRB joints.[76] Some, who had been in the work group but transferred to a new position, did not speak because they did not feel their knowledge of the technical issue was current enough. For example, Marshall's John Q. Miller, transferred to a new program several months earlier, said he was "no longer able to maintain total technical knowledge."[77] He stated that he did not understand the discussion about the resiliency data because those data were gathered after his transfer. He assessed the Thiokol argument on the basis of its engineering logic, concluding that the Thiokol data did not support a temperature correlation. He explains his reluctance to talk and his dependence on others for interpretation of the technical issues: "If everybody else had agreed to launch, I would not have stood up and said, 'No, it's outside the experience base, we can't launch.' However, if Thiokol had recommended, which they did, not to launch, I would not have objected to that recommendation. . . . I did not have a firm position because I was not as current on all of the test data that had been generated in the last 6 to 12 months."[78]

Throughout the evening, those who did have something to say conformed to cultural understandings about hierarchical reporting relationships, functional roles, and the kind of data admissible as evidence in FRRs. In several instances, people relayed information to their supervisors, then said nothing more. The supervisor did not relay the information on the network, so it did not become part of the discussion. This pattern was a source of many missing signals that night. In FRR, the managers become the "voice."[79] Once engineers present their supervisor with the engineering assessment and recommendation, the engineering responsibility is complete. The supervisor decides what to do with that information. Powers stated: "I'm in a supporting role to Mr. McCarty and when I tell him something he can take it and throw it in the trash can if he wants to. That's his call, not mine."[80]

Missing signals occurred because, while the superior depended on engineers for data and its interpretation, engineers depended on their superior to speak on their behalf in FRR. By virtue of position, supervisors had access to information that crossed disciplinary boundaries and so allegedly knew "more," but without hands-on knowledge of lab work, they also knew "less." Unless they received a thorough briefing or made a concerted effort to explore the issue on their own, they were not empowered to speak. In briefing their supervisors, engineers were limited because they had no more information than Thiokol had. So the managers they supported either received abbreviated statements that fell into the category of "engineering feel," or bits and pieces of information without an interpretation of their relationship to the whole. Under cultural assumptions about what was acceptable data, supervisors did not enter an FRR conversation with intuitive or poorly understood information. So on the eve of the launch, they were not empowered to speak by the information they had; they also deferred to hierarchy. They were silent, relying on Project Managers who did have first-hand knowledge.

Project Managers had both the authority vested in the Project Manager position and the influence in technical decision making accorded expertise that crosscut NASA's disciplinary divisions. They received information from all specializations related to their project, and they interpreted it for those both above and below them in the hierarchy, as Mulloy indicated when he referred to his position as "the neck of the hour glass." The formal authority of their position was reinforced by expert knowledge. The Level III and IV Project Managers in the SRB work group were "hands-on" managers. They were the main repository for the specialized knowledge of the various labs that worked for them, *and* the position required that they integrate and master this knowledge in order to make recommendations about risk acceptability to their superiors in adversarial FRR challenges. Their expertise and authority created *patterned informational dependencies:* managers above and below them depended on them for the last word in technical analyses.

Both sides of the argument that night were affected by missing signals resulting from cultural assumptions about hierarchical reporting relationships, functional roles, and admissible evidence. A scenario from Marshall shows how data supporting launch was excluded from the conversation. Marshall's John Schell, the chemist who came out of retirement for the night because only he could locate information about Viton and cold temperature, had data that contradicted the Thiokol position. He said nothing on the network because he was an "outsider" who had spent the last five years away from the job and because he was asked to "get data" not "give a recommendation." Schell passed the information to Riehl. As a representative of the Materials and Properties Lab, Riehl understood well the effect of temperature on the rings, but he did not understand joint dynamics, the specialty of the other main lab, Structures and Propulsion.[81] Here was the matrix system's weakness: information from both labs was necessary to understand how resiliency affected redundancy. Schell and Riehl gave their data on cold temperature and Viton to their superior Schwinghamer. The three were silent during the teleconference because Schwinghamer had, in turn, relayed what they knew to his superior, Wayne Littles. Schwinghamer explained his silence:

> If it's colder, the O-ring will be slower to respond from the pure resiliency aspect, but I just don't know what that means in terms of pressure sealing, you know. . . . I had a little reservation about it because I knew it was going to be slower, stiffer, but you know, that pressure sealing of the gas effect was supposed to take care of all that. And I thought it did, frankly. At that point, I would have to *defer to some structures mechanics types because once you get out of the material and its properties, then if I venture an opinion, it is only speculation because I am really not that kind of an expert.* We had some data on Viton, and we knew that Viton, like plain rubber, as it gets colder, it gets stiffer. *That's just common sense. But that ain't very quantitative, frankly.* I didn't take a position because there's no way on a basis of just simple resilience—as complicated as that joint is, you have to understand more than that to appreciate how it works, and I really didn't take a position. . . .[82]
>
> Then I showed it [the data] to Wayne, and I felt that, well, *he has the information, and if he thinks it is important enough, I figured he knew more about the operation of that joint than I did. . . . I thought he would certainly bring it up if he felt uncomfortable about it. I just knew the materials end.* We are highly disciplinarily oriented here, and we usually don't get much out of our field, and I just couldn't take that data about the hardness of the O-ring and do anything with it in terms of how does the joint function.[83] (Emphasis added)

Littles did not report their data on the network. When asked why he did not bring it up himself, Riehl cited the tenet of the original technical culture: engineering statements based on "solid technical data" were necessary to circumvent hierarchy in FRR. Marshall's Bill Riehl: "I don't normally like to break the chain of command. However, it depends on how strongly I feel technically about something. First I'm going to have to have technical data to support something when I go. That's the way I am, solid technical data. . . . I've got to be able to substantiate it with data, because you don't go up [the hierarchy in FRR] on subjective feelings, because if you don't have data, if

I feel like I don't have data to back me up, the boss' opinion is better than mine."[84]

A second scenario, at Thiokol, shows how data supporting Thiokol's "no-fly" recommendation was kept from the conversation by these same factors. Marshall's Cecil Houston and Thiokol's Jack Buchanan, both Resident Managers at Kennedy, participated at the Cape in order to take care of any teleconference-related administrative tasks. Neither commented on the network because they felt they did not have good comprehension of the technical issues.[85] Yet Houston and Buchanan's positions at Kennedy gave both important information that no one else had. Both knew that rainwater had been found in the joint when STS-9 was removed from the launch pad and destacked in 1983. Since cold turns water to ice and ice might alter the movement of the rings, this information would have been important to the Utah engineers. But neither Houston nor Buchanan understood how the joint worked, and so neither could interpret the significance of what they knew. Houston said nothing; Buchanan whispered the information to Al McDonald, whom he saw as the Thiokol representative at Kennedy with the appropriate expertise to evaluate this information for input into the teleconference. McDonald did not mention this over the network, assuming that Thiokol engineers knew. They did not.

Effect on Thiokol participants. Of the people at Marshall and Kennedy silenced by these cultural understandings, three supported Thiokol's position: Marshall's Coates and Powers and Thiokol's McDonald. None of them expressed support for Thiokol's position over the network. Keith Coates, a former S&E engineer in the SRB work group, told George Hardy prior to the teleconference that he was concerned about the cold because he recalled the blow-by and erosion on STS 51-C.[86] But he said nothing during the teleconference because he did not understand all the data being presented, not having worked on the SRMs for eight or nine months,[87] and he had only observational data, which in the NASA culture is not the stuff of "real" technology. Asked why Hardy did not report his concerns during the teleconference, Coates said: "I think—I have got to be fair here—I put it in the context of I can't quantify that concern. So that, in the atmosphere of an engineering decision, may not be worth a whole lot. [It's] not scientific, something achieved through scientific process."[88]

Powers said nothing on the network. However, near the end of the precaucus part of the teleconference, we again see the reproduction of hierarchy. As Hardy began his infamous "appalled" comment, SRB Chief Engineer Jim Smith got up from his seat to consult with John McCarty, Deputy Director of the Structures and Propulsion Lab, in order to get that laboratory's position on Thiokol's recommendation. Smith, whispering so as not to interfere with Hardy's statement, asked McCarty for the lab position. McCarty then asked Powers, his support person, who whispered that he thought it was too cold and that he supported the contractor's position 100 percent. Then the Thiokol caucus request broke up the meeting, and their conversation.[89] Powers said nothing more because he had told his boss his view, and because "we [at Marshall] expected Thiokol would come back with 'Hey, we're sticking to our guns. It's no-launch, folks.'" Also, Thiokol engineers were making the argument he would have made, and he had no more data.[90] When he told Smith and McCarty that he supported the Thiokol position because it was too cold, that too was a "subjective argument" in FRR culture.

As a result of these missing signals, participants in Utah heard no support for their position prior to the caucus. Thiokol's Brian Russell recalled: "I didn't know who was down at Marshall Space Flight Center, and I thought to myself where are the Leon Rays and the Jerry Peoples and the Ben Powers and those people? Are they there? When we have technical interchanges on the telephone or personal visits, those are the people that we sit across the table with and argue every detail, or at least present our data, talk about it, sometimes argue it, in very great detail. . . . So I wondered what was on their minds and I wondered if they were hearing what we were trying to say or at least what we felt."[91]

What Thiokol participants *did* hear were statements in favor of launch from three Project Managers—Hardy, Mulloy, and McDonald—the three participants with the greatest influence in technical decision making. Mulloy and Hardy had both aggressively challenged the Thiokol engineering conclusions. Although Hardy had moved from the SRB Project to S&E, and thus no longer held the Project Manager position, he maintained his influence through his past reputation as Project Manager and through a continuing "hands-on" relationship with the SRB Project. As Thiokol's Jack Kapp said of Mulloy and Hardy, "These are knowledgeable, brilliant men."[92] Thiokol engineers heard respected managers in positions of power tear apart their data analysis before their own bosses and others on the network, then respond in language and tone that many thought harsh; hierarchy, functional roles, and information simultaneously empowered the talk of some while silencing others. The verbal opposition from Project Managers must have made it difficult for some engineers to express their views, in particular the working engineers who did not regularly interact with Mulloy and Hardy in FRRs. Thiokol's Jerry Mason acknowledged the impact of their remarks on those at the bottom of the FRR hierarchy: "And the comments that they made, in my view, probably got more reaction from the engineer at the lower level than they would from the manager, because we deal with [those] people, and managers, all the time. Those guys are higher level managers to some of the engineers than they are to us, who we're dealing with on a peer basis."[93]

Of perhaps even greater significance to Thiokol's definition of the situation (although less publicized) was Al McDonald's sole comment on the teleconference net. Made just before the caucus, McDonald was heard by all to make a prolaunch statement: Thiokol's own Project Manager apparently agreed with Mulloy and Hardy that the joint would be redundant. McDonald's statement dumbfounded the Thiokol engineers. They thought that he, of all their upper management, would be most likely to understand. He worked closely with the working engineers and was always their "voice" when the engineers went to Huntsville for FRRs.[94] As Mulloy's counterpart, McDonald's views carried great weight with the Thiokol Vice Presidents as well. Circumstances perhaps made Mason and Wiggins more susceptible than usual to the views of all three Project Managers. Recall that Mason and Wiggins arrived late at the preteleconference meeting, witnessing only the preparation of the conclusions and recommendations charts. Once the teleconference began, they heard Hardy and Mulloy point out inconsistencies in the Thiokol data analysis. Then they heard all three Project Managers take the position that despite the cold, the joints would be redundant. For Mason and Wiggins, whose positions did not put them in "the seat of understanding" about the engineering analysis, these were strong signals indeed. Small wonder, then, that Thiokol senior administrators called for a caucus.

Caucus. The caucus demonstrates even more powerfully how bureaucratic accountability undermined the professional accountability of the original technical culture, creating missing signals. Mason, the top official present, took charge. As Senior Vice President of Thiokol's Wasatch Operations, Mason did not normally attend FRRs. He testified that he would "normally get a copy of the handout and leaf through it from the standpoint of maintaining just a general awareness of what the issues were."[95] Not only was he poorly informed about the technical issue, but he was the administrator present who was the least practiced in running an engineering discussion. He was also a virtual stranger to the working engineers. Mason's first actions reproduced the hierarchical, adversarial strategies of FRR that had guided the teleconference proceedings so far. By taking charge, by suggesting they go back over the engineering argument and consider Mulloy's critique of it, and by reiterating Mulloy's points (which he saw as "new data"), Mason conformed to the FRR script for managers, which was to probe and challenge engineering arguments in order to eliminate the flaws and make them tighter. Although we do not want to lose sight of the fact that in the eyes of everyone present he also aligned himself with the three Project Managers' positions on risk, we raise a different point.

When Mason followed his cultural script, taking an adversarial stand in relation to his own engineers, he elicited their traditional cultural script for FRR: they defended their position, responding to Mason instead of interacting with each other to find new data or to rework the data they had. Thompson and Boisjoly rehashed the Thiokol charts and the arguments already presented. With no opportunity for the engineers to consult among themselves or do new calculations, Thompson and Boisjoly still could not quantify the correlation that, on the basis of observational data, the engineers believed existed. So the caucus produced no new insights that converted Thiokol's "intuitive argument" into a "solid technical position" that could delay the launch.

With the exception of Boisjoly and Thompson, none of the engineers stated his position in the caucus. Again, the authority vested in the speakers by virtue of position influenced who spoke and who did not. Mason, Boisjoly, and Thompson dominated the conversation. Mason was empowered to speak solely by his authority as senior official present. In contrast, Boisjoly and Thompson's authority came from hands-on experience. Still, we must wonder what empowered their continued protest in the face of management opposition. In a powerful demonstration of how position affected the engineering argument by giving access to information and ability to interpret it, Boisjoly and Thompson had direct access to information that other teleconference participants had not seen first-hand. Both had been at the Cape for the disassembly of STS 51-C, the January 1985 cold-temperature launch. They had seen the blow-by.

NASA standards about "solid technical arguments" notwithstanding, the qualitative difference they saw in the blow-by on that launch had alarmed them. Their direct observation was behind their willingness to deviate from accepted standards by making an "intuitive" argument. It was behind the formulation of the Thiokol formal presentation and, now, empowered them to continue arguing for delay, even after opposition by three Project Managers and now Mason. Said Boisjoly: "And I discussed the SRM-15 [STS 51-C] observations, namely, the 15-A [left motor] had 80 degrees arc black grease between the O-rings, and make no mistake about it when I say black, I mean black like coal. It was jet black. And 15-B [right motor] had a 110 degree arc of black grease between the O-rings. . . . I tried one more time with the photos. I grabbed the photos, and I went up and discussed the photos once again and tried to make the point that it was my opinion from actual observations that temperature was indeed a discriminator and we should not ignore the physical evidence that we had observed."[96]

Boisjoly noted the difference in the power of information to persuade when an observation is made "first-hand" rather than in photographs: "What I saw at the Cape made an impression. I don't think the photos made the same impression because the area was small and flat."[97]

Although a few of the others at Thiokol made comments of a factual nature, most were silent. In another display of bureaucratic accountability and structural secrecy, they deferred to others who had more authority to speak than they. Some said they were silent because their views were already being stated, either by Mason or by Boisjoly and Thompson, and they had no new information or insights to add to what was being said. Engineer Joel Maw, the thermal expert, indicated that specialization limited his understanding and caused him to defer to others: "I agreed that we didn't have a correlation between cold weather and blow-by, but I didn't feel like I was in a position to make that judgment. Mr. Boisjoly and Mr. Thompson were the experts as far as the structure of the O-ring."[98]

There was yet another critical moment in the caucus when norms about hierarchy and talk created missing signals. The contractor was required to give a company position in FRR. When a single engineering opinion was required and consensus was absent, a "management risk decision" was the norm, so Mason's decision to exclude engineers from the decision making was consistent with precedent. However, when Mason began polling the managers, no one in the room really knew how all the engineers stood. The silence was interpreted by people on both sides of the issue as support for their own stance. Jack Kapp, who opposed the launch, said: "I guess I did not know how the other principals there stood, but I did not hear anybody in the engineering complex, the design engineering complex, those ten that weren't in the upper management position, I did not hear any of them capitulate in any way concerning their recommendation not to fire."[99] While Bill Macbeth, who favored launch, recalled that "it became obvious that we were not going to get a 100 percent unanimous technical expression. We still had two people that were not comfortable with launching, and other than that I don't think there was any strong dissension against launch in that room. . . . I heard nothing in the remarks from anybody other than Boisjoly and Thompson after the original Mulloy-Hardy challenge to indicate to me that anybody had any real strong feelings that a launch was an absolute mistake."[100]

Mason began the managers' discussion, again repeating the data supporting redundancy and taking a position in favor of launch. When it was Lund's turn to vote, the prolaunch votes of Mason, Wiggins, and Kilminster proceeded his. Lund hesitated, then Mason made his "hat" comment. Russell indicated that, rather than being deviant, Mason's comment conformed to expectations about management risk decision making and Lund's functional role. Affirming legitimacy of bureaucratic accountability in technical decisions, Russell said: "He [Lund] made some statements of unsurety, and it was a couple minutes into that period that Mr. Mason said, 'Take off your engineering hat and put on your management hat.' And there were some people that were upset at that comment, I guess. You know, that's a comment that tends to come up: Let's make a decision here. And all right, it would be somewhat of a pressure comment, but in my own mind he [Lund] is the Engineering Manager. He's the number one manager over Engineering, so I kind of interpreted that as essentially Bob's decision: the managers make decisions, and the Engineering Manager makes the engineering decisions."[101]

Lund did not voice support for the original engineering position that he helped to construct. He was disempowered, first, by his position in the voting order and in the hierarchy, facing three prolaunch votes, two by people who were his superiors. Second, he had no new information or insights that would counter the arguments raised by Marshall, now reiterated by Mason, that would convert the Thiokol analysis into a solid, technical argument that conformed to normative standards for engineering arguments at NASA. Third, he received no further social support from his engineers. Larry Sayer observed: "He [Lund] turned around and looked at the people that support him that put the engineering data together, and no one said anything. I think he was really looking for help—is there anything else we can do from an engineering point of view that would make our original recommendation strong enough that we could stand behind it?"[102]

But Lund got nothing but silence from the people who were closest to the SRB joint problem. The engineers were disempowered from speaking for the same reasons as Lund: senior personnel had taken prolaunch positions, and they were lowest in the pecking order; they had no new information or insights, necessary to contest management risk decisions; and, separated from their Project Manager and Powers and Coates at Marshall by the mute button and from each other by Mason's hierarchical, adversarial strategy, they also lacked support. Brian Russell, who opposed the launch, reflected on the impact of these missing signals: " . . . times in the past when I've been overruled of my initial position, I have changed my position in my own mind. I see, yes, okay, they're right. I'll go along. And so I wasn't sure how much of that had happened in everybody else's minds. . . . The decision was to be made, and a poll was to be taken. And I remember distinctly at the time wondering whether I would have the courage, if asked, what I would do and whether I would be alone. I didn't think I would be alone, but I was wondering if I would be, and if I would have the courage, I remember that distinctly, to stand up and say no."[103]

But he never found out how his colleagues felt because there were more missing signals: the engineers were not asked. They sat quietly and listened to what the Vice Presidents were doing, themselves reproducing hierarchical relations by conforming to the norms of management risk decisions: engineers give technical input to managers; managers make the decisions. Larry Sayer noted: "We're normally making engineering positions to management, and we try to do the best job we can to make our recommendations, and then management, using that data plus whatever knowledge they then have as far as schedule, as far as risk and other things that make a management risk decision. So our position always is if it's something that's going to really vitally affect the corporation, then we take it up through our management chain both within the engineering and through the program management side to make sure that they're aware of what our concerns are, what our recommendation is, and then after a decision is made then we normally support that decision and go work towards that."[104]

They felt empowered to continue to challenge management only when they had the data to back up their position. As Riehl said, "If I feel like I don't have data to back me up, the boss' opinion is better than mine."[105] They had no more data.

Postcaucus. After the caucus, agreed-upon assumptions about bureaucratic accountability continued to affect talk by creating missing signals. Thiokol's Kilminster, Vice President of the Booster Program, was responsible for final certification of flight readiness of the SRMs in all FRRs. When the teleconference resumed, he relayed only the revised conclusions and launch recommendation. Again conforming to norms about hierarchical relations and letting their boss be the "voice," those in Utah who still opposed launch said nothing. Typically, NASA did not ask for a vote count when a contractor's senior official presented the company position in FRR.[106] Again, Marshall management conformed, with enormous consequences. Separated by the physical locations of the teleconference sites and "on mute" during the Thiokol caucus, no one at Marshall and Kennedy knew how the reversal came about, nor did they know that many engineers still opposed the launch. These missing signals generated others: any voices that might have been raised in support of the Thiokol engineers were silenced. Thiokol's McDonald, at Kennedy, recalled: "I had no way of understanding what caused that change, whether they found some data that was indeed at lower temperatures that made that all inconclusive, or they just drew it from the same data we discussed that didn't appear conclusive. . . . I certainly didn't want to make it [the engineering decision] because I didn't have all the data, but from what data I had seen I didn't feel I could make that recommendation, and they clearly had more information than I did."[107]

Those opposed to the launch had a final opportunity to speak on the teleconference. After Kilminster read the recommendation, Reinartz first asked Marshall's George Hardy for his view. In doing so, Reinartz was following the chain of command. Normally, as Shuttle Projects Office Manager, Reinartz required a recommendation from the Project Manager (Mulloy) and the lead person from S&E (Hardy).[108] Hardy asked the Marshall group, "Has anyone got anything to add to this?" Although he addressed the group and he looked around the entire table, he said the names of only two or three of the managers, but not the others, among whom were Coates and Powers.[109] No one commented or expressed dissent. Then Reinartz asked all parties on the network whether there were any disagreements or any other comments concerning the Thiokol recommendation. No one voiced disagreement or tried to keep the discussion going.

With no visual cues and no verbal dissent and uninformed that engineers at Thiokol still objected, Marshall managers assumed unanimity. Mulloy said: "Nobody at KSC made any comment at that point. There were no other comments made from Marshall, where there were a large group of engineers assembled, and there were no further comments from Utah, so my assumption was that there was no dissent to that decision."[110]

Why did no one opposed to the launch speak out when Reinartz and Hardy asked for comments at the end of the teleconference? To the penetrating questions raised by official investigators about this silence, all participants responded that everyone had a chance to speak that night. The freedom to speak your piece was a belief asserted by both managers and engineers in other contexts and described in previous chapters as a taken-for-granted part of NASA culture. Both Marshall and Thiokol managers affirmed that the conduct of the teleconference was true to the "full disclosure" principle, saying it was an "open meeting," the engineers had "freedom of speech," there were "no restrictions," "no requirements to stick to chain of command," and it was a "very, very open discussion."[111] More significantly, the working engineers in the SRB group who opposed launch also asserted they had freedom of speech and that there were no restrictions. However, even as they gave verbal allegiance to a culture of full disclosure, the SRB work group engineers articulated the FRR norms about hierarchy, functional roles, and admissible data that had permeated the teleconference interaction to this point. Unanimously, they affirmed that position in the organization determined who would speak, by virtue of both authority and information. They affirmed the legitimacy of management risk decisions and the superior ability of managers to decide. Absent new data or insights, they could say no more. They knew their place.

Thiokol's Roger Boisjoly:

I did not agree with some of the statements that were being made to support the decision. I was never asked nor polled, and it was clearly a management decision from that point. I must emphasize, I had my say, and I never [would] take [away] any management right to take the input of an engineer and than make a decision based upon that input, and I truly believe that. I have worked at a lot of companies, and that has been done from time to time, and I truly believe that, and so there was no point in me doing anything any further than I had already attempted to do.[112]

Thiokol's Brian Russell:

> There was no negative response at that point. There was no one who said, I feel that we're doing the wrong thing. I didn't say it, Roger didn't say it, and any of the others who might have thought it didn't say it. And my reason for that is, here again we had tried everything we could do, at least I thought, to come up with a "no" recommendation. We were unsuccessful. And once the decision had been made, with all of the arguments presented on the table, I didn't feel it was even appropriate that I talk with Joe [Kilminster] a little bit after it to say, hey, Joe, I think you made the wrong decision. And so I didn't say that, even though I was still not in agreement with the decision. But they were capable, I felt, of understanding what we had to say and of making the decision accordingly.[113]

Thiokol's Jack Kapp:

> I was disappointed in the decision. I never did change my mind. At the same time, I clearly understood that [reversal]. I wished we could have quantified our assessments. I wish we could have been a lot more specific. . . . I remember feeling in my own mind that, gosh, I wish it hadn't gone that way because I don't feel good about it, but a lot of smart men have heard the argument, men that have directed the program for years and years. I've been right before; I've been wrong before. I felt like management has made a decision now. I'm not going to holler about it or anything else. We'll just go on with it.[114]

Marshall's Ben Powers:

> I can say anything I want to, but it was just useless. Once the contractor has said, "Hey, I'm reversing my position, you know, I took my data and threw it in the trash can and I'm ready to go, I'm ready to fly," you're pretty well committed. I did not know that Thiokol was going to reverse themselves. If I knew they were going to do that, then I would have been really trying to twist some arms. But I thought, hey, this is their show. They're going to get the job done. They are going to get it stopped, and I'm for it. And that's what it looked like. And Hardy and Lovingood were already talking about how they were going to shut it down. And when they [Thiokol] came back with the reversal, I just couldn't believe it. And if I look back on it now, what I should have done is I should have done everything within my power to get it stopped. I should have taken over the meeting and all that. But, you know, really I'm not of that grade structure or anything.[115]

Marshall's Keith Coates:

I personally opposed Thiokol's reversal, but I did not adamantly state that. I did not stand up and lay down on the tracks. . . . I have felt that I could speak out whenever I wished to speak out. I don't know whether others feel different. But I feel like I can tell my boss. I feel like I could tell George Hardy, Larry Mulloy, but as I said, it was their call. They were in the position to make the decision, and I had nothing further to add to the data that they had been presented.[116]

And at Kennedy, Project Manager Al McDonald, who himself sometimes made management risk decisions, gave two reasons for not speaking, both demonstrating taken-for-granted assumptions about hierarchy, information, and talk:

I certainly didn't want to make it [the engineering decision] because I didn't have all the data, but from what data I had seen I didn't feel I could make that recommendation. . . . I knew at that point in time that if I was going to have any impact on this thing, I was going to have to deal with Mr. Mulloy and Mr. Reinartz, because I felt that my management, if I had talked to them, would say, Well, have you talked to Mr. Mulloy and Mr. Reinartz about that?[117]

As a final missing signal, after the teleconference and unheard by his Utah colleagues, McDonald spoke out against launch for the first time, protesting to Mulloy and Reinartz at the Cape. Not only was McDonald's protest a missing signal as far as Thiokol participants were concerned, it was a weak signal to those at the Cape because he added no information to the redundancy/resiliency issue. Clearly unhappy about the outcome, he expressed his inability to understand how they could accept a recommendation outside the 40°F qualification temperature; he questioned launching with ice on the launch pad and rough weather in the recovery area. The House Committee noted that his remarks lacked "clear, unambiguous language" because he did not say that the launch should not be undertaken below 53°F.[118]

Responding to McDonald, Mulloy asserted rules about functional roles and bureaucratic accountability: these issues were in neither McDonald's nor Mulloy's domain and would be handled by others. Since Thiokol data did not support a new Launch Commit Criterion for the SRBs, the decision about temperature was no longer a Level III decision. It would be made by the Launch Director at the Cape, who would make a decision based on the 31°F Launch Commit Criterion for the shuttle system. Mulloy told McDonald that his latter two concerns were being handled by the appropriate people. McDonald said nothing more.

TALK AND POLITICAL ACCOUNTABILITY

NASA's environment of competition and scarcity made political accountability a part of NASA culture in the 1970s. It resulted in production pressure that became institutionalized, having a subtle, insidious effect on the outcome of the teleconference. Contrary to conventional interpretations of the eve of the launch, production pressure influenced the actions of all participants. We have already seen some of its effects in the preceding section: as Weick noted, hierarchy becomes more salient for people in stressful situations.[119] Now we see additional effects on both the production of information in the preteleconference preparations and exchange of information during the teleconference, adding to structural secrecy and missing signals.

Teleconference preparation. Thiokol engineers chose to schedule the second teleconference for 8:15 P.M. EST. Rather than creating their presentation and notifying NASA when they were ready, they created a deadline for themselves. The engineers were used to deadlines and working under time pressure, and this deadline demonstrates the internalized taken-for-granted nature of production concerns in the work place. Setting the deadline demonstrates their understanding of the need to coordinate a meeting between people located in three places and three time zones, some of whom had already gone home for the day. It also reflects a concern with cost/efficiency. Everyone knew that the ground crew would begin filling the External Tank with hydrogen and oxygen sometime after midnight, and while all insisted (and launch history verifies) that the tanking operation never deterred a full discussion of a technical issue and never prevented a launch delay in the past, everyone knew that a no-launch recommendation reached before tanking began would save money and ground-crew time and energy.

So Thiokol engineers set the deadline, then hurried to meet it, never requesting more time (though they did take some more, while Marshall and Kennedy waited on line for the teleconference to begin). When they set the deadline, the Thiokol engineers believed that an 8:15 start gave them adequate time—even a half hour more than they first suggested—to prepare their formal presentation. They were confident about what they were going to present, since the data analysis had been underway since 12:30 MST, and that they had enough time to do the necessary calculations and to make up the charts. And setting the start of the teleconference at 8:15 does not appear to have limited the time spent in discussion. No one reported feeling hurried; everyone reported that they said everything they wanted to say. In fact, the discussion could have continued for two more hours before the crew began filling the External Tank.

In hindsight, however, we see—and they realized—that the Thiokol presentation contained inconsistencies that weakened their argument. The deadline clearly affected Thiokol's preparation time, which may have affected the data analysis, thereby having a possible detrimental effect on both the charts and the talk—for it was "solid technical arguments" that gave engineers their voice at NASA. As it turned out, the preparations took longer than the Thiokol engineers thought. Facing the self-imposed deadline and acting on cultural assumptions about hierarchy and specialization as well, they began by dividing the work assignments along functional roles. The engineers separated to prepare individual charts; Lund and those remaining created the conclusions and recommendations charts without benefit of the data; observations were made in small conversation groups that were not made to the group as a whole; no opportunity existed for the full group collectively to review all charts before the presentation began. We do not know whether, given more preparation time, they would have thought of assessing all the charts collectively. If they had, we do not know whether they would have created a different kind of presentation that converted the mixed, weak, and routine signals into an engineering analysis that met the standards of the original technical culture. But the deadline clearly blocked the opportunity.

Precaucus. All participants acknowledged pressure that night. Several Thiokol engineers stated that the pressure was unusual. Boisjoly and Russell identified its source as the "tone" of the meeting, referring to the attacks on their engineering position and the "appalled" and "April" comments of Hardy and Mulloy. And although many engineers reported that the "tone" was normal, others felt that Marshall's criticisms of the Thiokol argument were unduly strong; Maw, who had not had much contact with Hardy and Mulloy, felt "intimidated." For some of the engineers present, the tone of the meeting may have affected their willingness to speak, creating more missing signals.

Given the unprecedented circumstances of the teleconference, we can assume that pressure was considerable. Although Mulloy and

Hardy were powerful in relation to the other participants, they were middle-level managers in the NASA system, accountable first to Marshall Center Director Bill Lucas, then to Levels II and I. As middle-level managers, Hardy and Mulloy were subject to performance pressure, of which production pressure was but one component. They were also responsible for ensuring that the work group met the standards of the original technical culture and bureaucratic accountability. The situation was unprecedented for Mulloy and Hardy in three ways that aligned with all three cultural imperatives in their domain: (1) A safety-of-flight issue materialized in a no-launch recommendation after the L-1 FRR. (2) Thiokol's recommendation was not supported by a "solid engineering analysis." (3) The 53°F criterion for flight had long-run implications for schedule. Mulloy, in particular, as Project Manager, was personally responsible for the outcome of this teleconference, as he was every other FRR. He would be the person to defend the technical rationale and no-launch recommendation to Level II. Hardy, too, would be accountable, as the senior Marshall official present, a former Project Manager, and Deputy Director of S&E (top official among the "cops" that monitored the contractors' performance).

Yet all managers testified that pressure was normal—evidence of the extent to which production pressure had become institutionalized. That it was taken-for-granted did not mean managers did not experience the pressure, however. Quite the contrary, as Marshall's Larry Wear told me:

> Pressure exists in everything we do. So to deny that there was pressure in that room would be false. There is always pressure in the room when you are discussing any of this sort of data, whether you are going to run a test tomorrow, whether you're trying to decide whether it's all right to load propellant in a motor that's got a deviation in insulation that you're going to vary. Before a launch, it is a form of pressure because you've got a time constraint and you've got to consider something here and you've got to say yes or no in X timeframe or you're going to delay something. Also, there's a pressure you put on yourself because you've got to make a decision and, you know, you're involved in it. If, for no other reason, your own pride is involved, and if I've got to make a decision, I need to make a good decision, I need to make a right decision. In this particular case, it's a launch decision, it's a very, very serious one, because it is a safety issue they've raised and it's very serious. So you've got to make the right decision. There is always an element of pressure when one has to make a decision, especially if that decision is a very serious decision.

I hate to equate it to normal, everyday activities. But it is very difficult to have a session with an object on the table that has big importance, whether it is dollars, or a schedule, or success of a test, or whatever, without there being some manner of pressure. So there was no atmosphere of pressure, if you want to call it that, that all of us weren't conditioned with twenty years ago. It's with you. It's a way of life. To call off a launch, there would be a certain degree of trauma involved in that. But it wouldn't be the end of the world, or the end of your job, or the end of the agency. It's just a thing you do when you have to do it. So I think it's difficult for people outside the January 27 meeting to understand that whatever sense of pressure was present there really isn't any different from 24 other meetings, even though this was a very serious topic.[120]

If pressure was normal, as managers said, what accounts for many engineers' experience of it as unusual?

1. *Marshall managers' harsh criticism of the Thiokol position.* Thiokol engineers experienced the Marshall criticism as unusual pressure because the format of an FRR had been mixed in with an ad hoc engineering discussion. Adversarialism and harsh criticism were the norm in FRRs, as everyone attested. However, normally in FRRs, Thiokol's Project Manager Al McDonald presented and discussed all the charts, including the recommendations and conclusions. McDonald was the "voice," buffering interaction between Marshall and his own engineers. When particular technical issues came up, the engineers McDonald had assembled as his "support staff" responded with what they knew from tests, calculations, and post-flight analysis to support certain points. But the Project Manager was the one who interacted most with Marshall representatives. He was the spokesperson and thus responsible for responding to all the "probes" and "challenges." Normally, he, not the engineers, absorbed most of the criticism.

On the eve of the launch, McDonald was at the Cape. The engineers, in Utah, presented all the charts (with the exception of the conclusions and recommendation, which Lund gave) and responded to all questions themselves, as was customary in any ad hoc engineering discussion. Although the Thiokol engineers who participated in the teleconference had witnessed the adversarialism in FRR (though not all had, because not all had been called previously to support FRRs), never before had they been responsible for the full presentation and thus the direct recipients of all Marshall's adversarialism. Add to this the fact that Marshall identified flaws in the engineering analysis that the engineers themselves recognized and for which they had no more supporting data, and their experience of unusual pressure is understandable.

2. *The "appalled" and "April" comments.* NASA's institutional history of competition, scarcity, and elite political bargains led to cultural change that then "trickled down" to affect talk in the teleconference, creating missing signals. When we analyze Mulloy's answer to a question about his strong responses during the teleconference, his words reveal that he experienced performance pressure from all three cultural imperatives: production concerns that flowed from the agency's political accountability, concerns about the rigor of the technical argument and consequent safety implications indicative of the original technical culture, and concerns about reporting up the hierarchy in a system strong on bureaucratic accountability. Mulloy explained:

> I was frustrated. It [Thiokol's analysis] was dumb. The data said one thing, the recommendation another. If there was a temperature problem, a 53° ambient limit did not solve the technical issue, which was what temperature did the joint need to be. At what temperature was resiliency affected? If we set the LCC at 53°, the joint temperature could be much higher or much lower. 53° told us nothing about risk [original technical culture]. If they had come back with no-launch, we would've called Arnie [Aldrich] to cancel the launch. Telling him that MTI had a concern about the cold temperature and the joints would've been enough to stop it.
>
> Then I would have taken the no-launch recommendation forward to a full-fledged Level II the next morning. I would've felt naked. I couldn't have defended it [original technical culture; bureaucratic accountability]. I probably would've said that MTI was no-go because of concerns about the effect of cold temperature on the O-rings, and then had McDonald presenting, since he was at the Cape. Arnie would've challenged MTI, the same as I did. Arnie's sharp. He'd seen the data, had heard it many times. McDonald would have been in a real pickle. It was a bad story, a bad rationale, not well thought out [original technical culture]. And it was a real box, because we would shut down the program based on the temperature criteria that they had [political accountability].[121]

NASA managers did certainly want to launch. All of them—managers and engineers alike—always did, because launching shuttles was their job. But in addition, Mulloy was anticipating the next stage in the launch decision chain, in which political accountability, bureaucratic accountability, and the original technical culture came together for him personally. As the "neck of the hour glass," he would, like Thiokol engineers that night, have to stand before higher authority and be accountable for the SRB work group's choice. And, like Thiokol engineers, he would have to face adversarial challenges, publicly assuming responsibility for launch delay and an engineering analysis that was indefensible ("I would've felt naked") according to normative FRR standards.

We know that the teleconference was unprecedented in another way that, combined with this performance pressure, forms a plausible explanation of the managerial comments that set the meeting's tone. Normally, Level III FRRs were face to face, not by teleconference. The lack of face-to-face contact may have affected Mulloy and Hardy's talk, allowing them to vent their frustrations in uncustomary ways. In face-to-face encounters, we tend to respond in ways that allow the other person to save face.[122] Although Level III FRR was normally an aggressive, adversarial, no-holds-barred exchange between participants ranked differently in the hierarchy (managers vs. engineers, as well as Level III vs. Level IV), the teleconference made people in other locations invisible. The exchange was more impersonal. Hardy and Mulloy attacked the engineering position ("My God, *Thiokol,* when do you want me to launch, next April?"), not individual engineers. The invisibility of Thiokol's Utah participants allowed the performance pressure on Marshall managers to be expressed in abnormal ways.

Again we see the relationship between position, information, and structural secrecy. People in other settings could not see Hardy and Mulloy. With no visual cues to help them interpret those comments, the words and inflection constituted the entire meaning. In fascinating testament to the power of visual information, neither Thiokol's Brinton or Speas nor Marshall's Powers or Coates (both opposed to launch), who were with Hardy at Marshall, perceived his comments as pressure. They could see him (Powers noted that he "had his sleeves rolled up" and was "working hard to understand") and heard statements that did not carry over the network to Utah and the Cape ("For God's sake, don't let me make a dumb mistake" and Hardy's conversation with Lovingood about phone calls to cancel the launch). Those opposed to the launch who were in the same setting as Hardy had more information that altered interpretation. For many of those opposed to the launch in Utah, however, Hardy's "appalled" comment had a different effect, apparently obliterating the significance of his subsequent "but I will not agree to launch against the contractor's recommendation," which those opposed at Marshall interpreted as supportive of the Thiokol position.

In sum, the tone of the meeting was set outside the teleconference by political elites who took actions that added political accountability and bureaucratic accountability to NASA's original technical culture. These imperatives trickled down to NASA managers who, experiencing performance pressure, openly expressed their frustrations about upholding these standards. The separate locations and the absence of video transmission added an impersonality to the exchange that allowed managers to vent their feelings. Further, people in other locations did not have access to other visual and verbal information that would have provided interpretive context to the words they heard. Coming from "respected, brilliant men" above them in the hierarchy, these comments greatly affected listeners at the bottom who were defending a position publicly exposed as having analytic inadequacies.

Caucus. In the hurried return of Mason and Wiggins to Thiokol, we see in bold relief the subtlety with which political accountability was carried into the setting, affecting talk. They rushed back from Brigham City to participate in an unprecedented delay recommendation, an act that emphasized production pressure and the importance of the Thiokol stance for contractor-government relations. No engineers suggested Mason used an intimidating tone during the caucus, and Wiggins apparently did not enter into the discussion. However, Mason did not normally participate in technical discussions with working engineers. The presence of the Senior Vice President (apart from anything he said) was surely a reminder of production concerns that must have added to the difficulties of arguing a position already limited by "subjective" observational data.

Brian Russell was the only person who, in interview transcripts, articulated the effect of political accountability on talk. First asserting the silencing effects of having no data to add to the discussion and hearing no support from engineers in other places, he said:

> And the pressure I felt was not pressure from our management or even from the NASA people directly to make me be a little bit afraid to get up and say no, we shouldn't launch, if I were asked. It's just that the pressure of stopping the National Space Program or delaying it, which you have a big ball rolling down the hill and you're going to stand up and say stop, and that's no small thing.
>
> I did feel that kind of pressure, although I tended to try to focus on the 51-L issue because also in my mind was the safety of that flight first, and I was willing to slow down, if need be, to help be more sure of safety. I didn't see in my mind that, oh, my, gosh, we're going to for sure have a catastrophe like we had. I did not see that. But the risk I thought was increased toward that, and that's what worried me.[123]

On January 27, 1986, the five-step decision sequence was enacted once again. The predicted cold weather was a signal of potential danger, creating uncertainty about the relationship between O-ring resiliency and redundancy. Arranging the teleconference was an official act acknowledging escalated risk. There followed a review of the evidence, culminating in an official act indicating the normalization of deviance: accept risk. The decision was followed by the destruction of STS 51-L, a signal that did what the Thiokol engineers could not do: it overturned the scientific paradigm on which the launch decision was based.

The structural secrecy that prevailed that night was destroyed with the *Challenger*. Teleconference participants grappled with the reality of what they had done, said, and left unsaid. As the unexpressed opposition to the launch was made public, other teleconference participants expressed surprise. Marshall's John Miller compared the eve of the launch with other launch delay decisions:

> There was nothing that strongly expressed during the course of the telecon. There was no lay down in the bucket type approach, I don't think. I didn't interpret it as being an absolute no. I detected a concern and I'm sure everyone has concern for every launch, and this was no exception. It was just an added concern because of the low temperature, but the conclusion and recommendation charts, you know, nowhere on the charts did they say we recommend that you do not fly, period. That is the way that I interpreted them, that they had a strong concern, strong enough to put it on a chart, but no diehards. . . . I was shocked when I heard two or three days later that there were so many Thiokol engineering people who opposed the launch. I never got any indication.[124]

Marshall's Wayne Littles recalled the final minutes of the teleconference when Hardy and Reinartz asked for additional comments:

> . . . the kind of meetings we have, you know, there is no restriction on people voicing opinions, and particularly the group of people that were there. For somebody like Ben Powers, for instance, not to raise an objection is out of character with Ben Powers. I would have thought that if Ben Powers had a serious concern that he would have raised an objection or told somebody. He normally does that. That's the mode he operates in. I have no problem with that. But no, there was no one there who voiced a contrary opinion. There was no one on the telecon at Thiokol who voiced a contrary opinion. There was nobody at the Cape who voiced a contrary opinion. I understand since that McDonald made some comments to Larry Mulloy at the Cape, subsequent to the telecon, after we went off line. He didn't say that on the loop, whatever he said down there. I've also been told by Joe Kilminster that he talked to Al McDon-

ald twice within the hour following the telecon, and that Al didn't voice any objection to him either time and raised no concern. I understand where we are now. The event is tragic beyond belief. But it does seem to me that there are some people who are reacting at this point beyond what they did that night. There was just no indication, from my standpoint.[125]

On the eve of the launch, Boisjoly felt he had "done everything he could." So did the others who opposed the launch. It is true that within the nested cultural context of the teleconference, they did literally everything "in their power." And so did managers. The engineers were willing to maintain the silence imposed by bureaucratic and political accountability because, despite their opposition, apparently *they did not believe the SRB joints would fail.* They believed that they were entering into an area of increased risk, that more damage to the O-rings might occur, but with the exception of Bob Ebeling (who did not state a position during the teleconference), no one thought that a complete ring burn-through was possible. As Thiokol's Jerry Burn said: "We weren't working on the assumption that it would fail. We were working under the idea that we would have problems, and how extensive those problems would be was unknown."[126] And after the teleconference, even Boisjoly acted as if he expected the mission to return. He stopped by Burn's office, telling him to be sure to document the damage well in the postflight analysis of STS 51-L because they would have a new data point enabling them to gain more insight about cold-temperature effects.[127]

The work group followed all the cultural imperatives, accepting the possibility of increased damage once again. Conforming to rules diminishes uncertainty by permitting a sense of control over uncertain situations. The "can do" attitude had played an important role in the SRB work group's normalization of deviance in the past. In contrast to the devil-may-care, overly confident attitude of NASA senior administrators alluded to in the press after the tragedy, the "can do" attitude in the SRB work group was grounded in conformity: rigorous adherence to the engineering methods, routines, and lessons of Marshall's original technical culture under Von Braun and to the Lucas-created strictness in conforming to organizational norms, rules, and bureaucratic procedures governing decisions about flight readiness. Work group members asserted that in the past their belief in "people and procedures," "the adversarial process," "first-hand engineering knowledge," and "going by the book" assured them of the correctness of their technical analyses.[128] On the eve of the launch, the very fact of their conformity to engineering routines and methods, and to the organizational rules, norms, and procedures for risk assessment, may have contributed to their silence at the end of the teleconference by reinforcing their belief that the joints would not fail.

Afterward, with the clarity of vision that hindsight usually brings, the engineers wished they had done more. Boisjoly tortured himself with thoughts that he might have been able to stop the launch by calling the newspapers. Russell said, "If I had foreseen the catastrophe, I think I'd have made some other calls and even circumvented the system."[129] And Bob Ebeling said: "When Larry Mulloy mentioned to Thiokol, 'When do you want me to launch, next April?' to this day I'm sorry I didn't stand up. Larry knows me and he knows my voice, and I'm sorry to this day I didn't stand up and say, well, if that's when it warms up, Larry, that's when it will have to be launched. I'm really sorry I didn't do that."[130]

But we must wonder. Ebeling, the one person who said he believed at the time a catastrophe was imminent, said nothing, and as Powers said: "If I look back on it now, what I should have done is I should have done everything within my power to get it stopped. I should have taken over the meeting and all that. But, you know, really I'm not of that grade structure or anything."[131]

Within the culture, the resource that circumvented the structural barriers to talk was information, and no one had any more. Many of the Thiokol engineers, Larry Sayer in particular, said afterward that if they had had more data, the outcome would have been different:[132]

> We didn't have a lot of the data that we have now. We didn't have a lot of the quantitative things that we can use to say we have to have these things happen, and the temperatures have to be here, and so forth. . . . My feeling is if you have a solid engineering position, you're backed up by analysis. You're backed up by test data and everything else. And if we had had that kind of presentation and that kind of data where we had said, you know, 50 degrees, below 50 degrees we have a high probability at this point of it not working, then I don't have any question in my mind that our management, Marshall people, and everyone would have said, "No, we'd better not launch." But we were in a position of not having a real strong engineering position, then it came down to a matter of judgment.[133]

But with hindsight, we too have clear vision. They could have delayed the launch with the information they had.

1. Thiokol could have opted for the maneuvering strategy Marshall's Wear discussed with Brinton during the caucus: recommend delay until the temperature was 40°F or greater. This recommendation would have met the criteria for an "engineering-supported" state-

ment; it eliminated the unusually restrictive reliance on the experience base that Commissioners Ride and Armstrong noted; and it satisfied the technical standard of logical consistency.[134] It would have held up under the scrutiny of Marshall managers, who could have defended the recommendation to their superiors. The strategy would have achieved delay by using a technicality, rather than data supporting a relationship between cold temperature and O-ring damage. The working engineers would have gained time to do more research.

2. More stunning is the observation that they *did* have the pertinent data. There were charts they did not imagine and did not construct that, if created, would have provided the quantitative correlational data required to sustain their position. After the disaster, two members of the Presidential Commission's investigative staff, Alton G. Keel, Jr., its Executive Director, and Randy R. Kehrli, a Department of Justice attorney—two nonengineers doing some Monday morning quarterbacking—put together two charts for the Commission report that demonstrate how Thiokol might have arrayed the data to create a strong signal. The tragedy itself made these charts possible because it made temperature a salient variable. It was fairly easy, knowing the technical cause, to go back and reconstruct the data in ways that made sense of the outcome.

Figure 13.1 simulates some of the thinking that night. It shows that when only flights with field joint anomalies are examined in relation to temperature, no correlation is evident. Thiokol engineers made no systematic presentation of temperature data for all launches. Furthermore, no participant at any of the three locations asked for an examination of all data points. Instead, as Thiokol engineers made their presentation, people at both Marshall and Thiokol recalled other missions with field joint anomalies at temperatures warmer than that for STS 51-C. They were, in effect, taking a mental sample, making a mental chart like figure 13.1. They were talking from loosely remembered data of all flights with field joint anomalies, which led them to the same conclusion as this chart: no temperature correlation. Figure 13.2 is a trend analysis. It shows that *when all missions are taken into account, a correlation between O-ring anomalies and temperature appears. Of the flights launched above 65°F, three out of seventeen, or 17.6 percent had anomalies. Of the flights launched below 65°F, 100 percent had anomalies.* Since all test evidence only approximates flight experience, flight experience itself is considered superior evidence in engineering risk assessment. Thiokol was posing a correlation between temperature and O-ring damage, but in fact the data on the charts were inadequate for assessing that important question: no Thiokol chart plotted temperature against O-ring damage for each launch, showing what happened when temperature varied. Such a chart would have been a strong signal indeed.[135]

Why did Thiokol engineers not have this idea?[136] Again, we must

POSTACCIDENT TEMPERATURE ANALYSIS

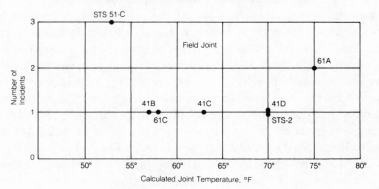

Plot of flights with incidents of O-ring thermal
distress as function of temperature

Fig. 13.1

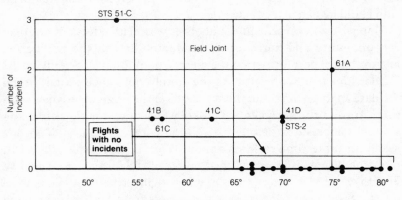

Plot of flights with and without incidents of O-ring
thermal distress

***

NOTE: Thermal distress defined as O-ring erosion, blow-by,
or excessive heating

Fig. 13.2

consider the structural and cultural forces, the complexity of the task environment, and the propensity of the past to shape outcomes in the present. In the years before the *Challenger* launch, cold temperature never became a major concern because it required three consecutive days of record cold to lower the O-ring temperature on STS 51-C. The engineers believe that a recurrence was an unlikely event, and in fact, cold temperature was never central to the work of the O-Ring Seal Task Force. At the time, other problems were more important than a run of cold temperatures that was unlikely to materialize. So, for the same reason that Thiokol engineers did not scramble to run resiliency tests for a complete range of temperatures, they did not systematically calculate O-ring temperatures for each launch. On the eve of the launch, in what ultimately may have been the most consequential effect of position and structural secrecy, Thiokol's Utah location left contractor engineers desensitized to the possibility of trend data concerning temperature and O-ring performance. According to Roger Boisjoly, the division of labor at Thiokol prevented any of the working engineers from developing an awareness that there was a performance trend in which temperature might be important.[137] For each launch disassembly, different Thiokol engineers went to the Cape. No Utah engineer was there for all postflight analyses, blocking the opportunity for intuiting a trend based on observational data across all launches in the setting where temperature could be personally experienced.

The past was present in yet another way. The effects of political accountability and bureaucratic accountability on the production and exchange of information are clear. Could it have been different? If a deadline had not been set in the preteleconference period, would the data analysis have met the standards of the original technical culture? Apparently all people at Thiokol were content with their analysis and conclusions before the teleconference began, and no one asked for more time for preparation. What if, instead of a Level III protocol, Marshall managers had followed the format of a Level IV FRR in which the assumptions guiding engineering discussions were different: managers and working engineers from both Marshall and Thiokol engaged in what was described as a "sleeves rolled up" informal exchange of information that was typically cooperative rather than divisive; engineers talked to engineers; adversarial tensions, bureaucratic rigor, and managerial performance pressures of the Level III FRR were reduced.

Would substituting a more collaborative exchange for the adversarial formalism of a Level III FRR have changed the outcome? If managers and engineers across all three locations had put their heads together to see what they could make of the data, would they have come up with the trend analysis in figure 13.2? Would Cecil Houston or Jack Buchanan have mentioned the rainwater in the joint of STS-9, and would it have become a factor converting the no-launch recommendation to an engineering-supported statement? Would those engineers who held back because they had "no real data" have expressed their opposition? If Mason had begun the caucus by giving the engineers time to talk among themselves and rethink their analysis and recommendation, rather than beginning in an adversarial manner that caused them to defend what they had already stated, would they have put together all data points on O-ring damage and temperature? If the caucus had not been on mute, or if all three locations had been polled, would the outcome have been different? We do not know.

What we do know is that on the eve of the launch, facing unprecedented conditions, neither managers nor engineers suggested that they proceed a different way. The professional accountability of the original technical culture that gave deference to technical experts at the bottom of the hierarchy was alive, but not well. We know this because the rules that forced the Thiokol engineers' concerns to be considered worked: the teleconference happened. But in the interaction leading up to the teleconference and the teleconference exchange itself, professional accountability took a back seat. The opportunity to put together an engineering analysis in keeping with the standards of the original technical culture was blocked by shared understandings about bureaucratic accountability and political accountability. Cultural explanations often leave us adrift when it comes to individual choice. They can suggest the menu of choices available to those in a particular socio-politico-historical niche;[138] but when it comes to explaining a particular choice, we must look at what people say and do to find out what institutional forces are expressed and taken into account.[139] In the teleconference discussion, we can see the macro-micro connection: the triumvirate of cultural imperatives affected talk, and therefore the launch decision.

On the surface, the teleconference appeared to be an informal discussion that was disorganized—we might even conclude after reading chapter 8 that it was poorly organized. Participants said that they normally found teleconferences to be disorderly affairs, and this one was no exception. But underlying the experience of disorder on the eve of the launch was an underlying order. Underneath the adversarial and divisive interaction, the launch decision was a collaborative effort: all participants enacted their cultural scripts, reproducing and affirming the nested cultural meaning systems in which the interaction occurred.[140] Howard Becker writes that culture is people acting in concert to the same set of rules.[141] Culture predates their activities, providing common understandings that provide rules for appropriate action. Writing about the *Challenger* disaster, Starbuck and Milliken noted that "people acting on the basis of habits and obedience are not reflecting on the assumptions underlying their actions."[142] However, facing new conditions, people do improvise, remaking and changing cultural understandings. But that did not happen here. It can truly be said that the *Challenger* launch decision was a rule-based decision. But the cultural understandings, rules, procedures, and norms that always had worked in the past did not work this time. It was not amorally calculating managers violating rules that were responsible for the tragedy. It was conformity.

Chapter Ten

LESSONS LEARNED

> The sorrow of mistakes is sometimes very diffuse and some-
> times very pointed. It is sometimes the sorrow of failed
> action and sometimes the sorrow of failed conduct. The sor-
> row of mistakes has been expressed as *the too-lateness of
> human understanding* as it lies along the continuum of
> time, and as a wish that it might have been different both
> then and now.
> Marianne A. Paget, *The Unity of Mistakes*

In *American Technological Sublime,* David Nye argues that the American reverence for technology is such that we have invested technological masterworks with transcendent, near-religious significance.[1] Almost at its inception, NASA's space program became a cultural icon. For many citizens, it represented and glorified American enterprise, cutting-edge technology and science, pioneering adventurism, and national and international power. Each awesome space achievement was a celebration of technological advance that reaffirmed the spiritual and moral values that supposedly underlay it. Keeping our eyes on the skies got us through hard times. The achievements of the space program offered hope during the tumultuous, cruel 1960s and 1970s, as the country was torn by racial conflict, the assassinations of John Kennedy, Martin Luther King, Jr., and Robert Kennedy, protests against the Vietnam war and for the rights of women and minorities, Watergate. In an era of despair for heroes lost, silver-suited astronauts were endowed with heroic stature.

During the 1980s, the symbolic meaning of NASA and its programs sustained the myth of American superiority through times of urban unrest, homelessness, violent crime, drug wars, and the decline of the cities, education, and international power. But skepticism and opposition grew. Spending money for a space station to explore other planets when the needs of this one were great seemed ludicrous. Even for NASA supporters, some of the excitement waned. Apparently, the public bought the definition of routine space flight that was conveyed when the Space Shuttle was declared operational by top NASA administrators and the White House in 1982. Daily work routines were no longer disrupted for space missions as they had been earlier in NASA history, when the first astronauts orbited the earth and set foot on the moon. This change did not substantially alter the symbolic meaning of the space agency and its programs: every vehicle launched still carried the extra weight of the American Dream going awry. But for many citizens, shuttle launches had become routine and taken-for-granted.

Perhaps more than any mission for years, the *Challenger* mission captured the attention of the public. Diverse in race, sex, ethnicity, and religion, the *Challenger* crew was a symbolic reminder that the all-too-elusive possibilities of equal opportunity and unity from diversity could still be attained. Christa McAuliffe was included to give this mission symbolic meaning of another sort. As a woman and professional teacher, she was chosen not because children learned more about space from a teacher orbiting the earth, but because her presence reinforced NASA's message that Space Shuttle flight was routine and the President Reagan image as proeducation at a time when education funds were being slashed. Symbolism notwithstanding, NASA believers, skeptics, and opponents alike developed a personal interest in this mission. Because Christa McAuliffe was among the crew, the media informed us not only of their names, but about their families and their histories. We witnessed their training. When the disaster occurred, we knew these astronauts. Americans had to come to grips with both personal and symbolic loss: the deaths of the astronauts and the anguish of their families as well as the destruction of all the space program represented.

We needed an explanation. The immediate appointment of a Presidential Commission, headed by a politician, to investigate the tragedy underscored the very real political and economic consequences of this event. This action, which usurped and upstaged the oversight responsibility of the House Committee on Science and Technology, stands in stark contrast to reaction to the deaths of three astronauts in the 1967 Apollo launch pad fire, after which NASA conducted its own investigation. In 1967, NASA was a powerful agency with a powerful leader able to stave off an outside inquiry. In 1986, a weakened NASA, essentially leaderless, could not. A Presidential Commission was appointed because of the greater visibility and public attention this event attracted. Christa McAuliffe made this mission special. An explanation had to be found for the deaths of the *Challenger* crew and corrective actions implemented in order to restore the legitimacy of NASA and guarantee the survival of the space program.

In moving language, the reports of both the Presidential Commission and the House Committee emphasized the need to get the program going again. The Commission wrote: "Each member of the Commission shared the pain and anguish the nation felt at the loss of seven brave Americans in the *Challenger* accident on January 28, 1986. The nation's task now is to move ahead to return to safe space flight and to its recognized position of leadership in space. There could be no more fitting tribute to the *Challenger* crew than to do so."[2] The House Committee wrote: "We are at a watershed in NASA's history and the Nation's space program. NASA's 28-year existence represents its infancy. We must use the knowledge and experience from this time to insure a strong future for NASA and the US space program throughout the 21st century. . . . Though we grieve at the loss of the *Challenger* crew, we do not believe that their sacrifice was in vain. They would not want us to stop reaching into the unknown. Instead, they would want us to learn from our mistakes, correct any problems that have been identified, and then once again reach out to expand the boundaries of our experience in living and working in outer space."[3]

To their great credit, both investigations made it clear that the disaster was not merely a technical failure; the NASA organization was implicated. Moreover, both were critical of the economic and political environment that were the structural origins of disaster. But the distillation of the complex and far-reaching findings of these two official investigations in more popular accounts focused public attention—and blame—on the middle managers who made the decisions and allegedly concealed technical problems from top administrators. This focus created the impression that the outcome could have been different had middle managers behaved differently. This historically accepted explanation not only masked the complex structural causes of the disaster, it obscured the fact that individual responsibility spanned hierarchical levels. First, it deflected attention from the powerful elites, far removed from the hands-on risk assessment process, who made decisions and took actions that compromised both the shuttle design and the environment of technical decision making for work groups throughout the NASA-contractor system. Congress and the White House established goals and made resource decisions that transformed the R&D space agency into a quasi-competitive business operation, complete with repeating production cycles, deadlines, and cost and efficiency goals. While both the Presidential Commission and the House Committee emphasized pressure as a contributing cause of the tragedy, only the House Committee located political elites outside the space agency as the source: "The Committee, the Congress, and the Administration have played a contributing role in creating this pressure. Congressional and Administration policy and posture indicated that a reliable flight schedule with internationally competitive flight costs was a near-term objective."[4]

Second, the conventional explanation disguised NASA's own top administrators' contribution to the tragedy. In order to maintain their own power and that of their threatened institution, they made bargains that altered the organization's goals, structure, and culture. These changes had enormous repercussions. They altered the consciousness and actions of technical decision makers, ultimately affecting the *Challenger* launch deliberations. Also, NASA top administrators responded to an environment of scarcity by promulgating the myth of routine, operational space flight. Perpetuating that myth, they established a policy that allowed nonastronauts on shuttle missions. Neither official investigation pursued the question of who actually decided that a teacher should fly on the *Challenger* mission. By limiting their agenda to the technical failure and decision making about the technology, the investigations effectively prevented the NASA–White House negotiations that culminated in the Teacher in Space mission from becoming a matter of public concern.

Finally, the focus after the disaster on managerial wrongdoing also removed from the public spotlight the difficulty of making engineering decisions about the shuttle technology. Although both official investigations repeatedly and publicly emphasized that the technology was developmental, not operational, the attention to managers marginalized the role of skilled working people, daily engineering routine, and the technology. Thus, the public remained largely ignorant of the difficulty of decision making about a joint that was invisible once the segments were stacked, the complexity of the sealing dynamics, and the inability of known tests to replicate the environmental forces on the booster joints during a mission. Little publicity was given the conclusion shared by the Presidential Commission and the House Committee on Science and Technology that no one at NASA nor Thiokol fully understood the operation of the joint prior to the accident.[5] The public did not learn that qualified technical experts doing the risk assessments, following rules and using all the usual precautions, made a mistake.

It was, as Presidential Commission member Astronaut Neil Armstrong remarked, "a tender design."[6] After the tragedy, Solid Rocket

Booster work group members continued to learn about the technology. Previously convinced that they had understood the joint, now they were forced to agree that they had not. It was not just temperature effects that they did not understand. The failure analysis showed that other factors, previously not taken into account, had contributed to the technical failure: the potential for ice in the joint, putty behavior, and effects of violent wind shear.[7] As the postdisaster analysis taught them more and more, explanations about the technical cause changed. They learned that the O-rings had been badly burned, but the charred material had sealed the joint, avoiding a launch pad debacle. The seal might have held had not unpredicted, unprecedented wind shear buffeted the vehicle, dislodging the material and allowing the hot gases to penetrate the joint. More testing was done. They discovered that overcompression of the O-rings was a contributing cause.[8] Prior to STS 51-L, they had worried about undercompression, shimming the joints to increase it. The outside referee tests, authorized in 1985 to settle the seven-year dispute between Thiokol and Marshall engineers about the size of the gap during joint rotation, were completed. The referee testing found a gap size even smaller than Thiokol's measurement, affirming Thiokol's assertion that the joint was fully redundant throughout the ignition transient, and therefore the design ought to have remained in the C 1R category.[9]

Most serious was the discovery that misunderstanding about the most critical technical rule used had contributed to a flawed decision. The temperature specifications had been misread, misinterpreted, and misused by both Thiokol and Marshall for years. The 31°F–99°F ambient temperature requirement for launching the entire shuttle system was created early in the program by other engineers. In a palpable demonstration of structural secrecy, many teleconference participants had jobs that did not require them to know Cape launch requirements and so were unaware that this temperature specification existed.[10] More profoundly disturbing, however, was that for Mulloy, Hardy, and many others who used it regularly in launch decision making at the Cape, this launch decision rule had become dissociated from its creators and the engineering process behind its creation. They had followed it repeatedly, taking for granted the interpretative work that other engineers had done.[11] They did not know that Thiokol had not tested the boosters to the 31°F lower limit.[12] Instead, Thiokol (with NASA engineering oversight) had established the limit using the military temperature specification for Viton O-rings.[13] Pondering the 31°F lower limit, which he and others believed was a legitimate decision rule on the eve of the launch, Mulloy later testified, "I was referencing a non-existent data base."[14] Worse, a close reading of that complex original temperature document showed that launching based on ambient temperature alone was wrong: the temperature of the boosters had to be calculated each time, taking into account heat from the sun and proximity to the External Tank.[15] None of the teleconference participants understood that prior to the tragedy.

When asked why all the factors influencing joint behavior were not known prior to STS 51-L, Mulloy responded: "I guess it is difficult for me to answer this way, but I was not smart enough to know it before hand. The people, Morton Thiokol and the [S&E] engineers that we have, had been looking at this problem over the last 7 to 8 years. They were looking at the observations and making judgments and making recommendations to continue flying, based on those data, and they were not smart enough to recommend the additional testing, and the people who reviewed that at levels above me were not smart enough to say that we need to do more than what we are doing. It is tough for me to say that, but I don't know any other way in hindsight."[16]

All causal explanations have implications for control. The benefit of explanations that locate the immediate cause of organizational failure in individual decision makers is that quick remedies are possible. Responsible individuals can be fired, transferred, or retired. New rules that regulate decision making can be instituted. Having made these changes, the slate is clean. Organizations can go on. In *The Limits of Safety*, Scott Sagan alerts us to the politics of blame.[17] Both for easy public digestion and for NASA's survival, the myth of production-oriented, success-blinded middle managers was the best of all possible worlds. It removed from public scrutiny the contributions to the disaster made by top NASA officials, Congress, and the White House; it minimized awareness of the difficulty of diagnosing the risky technology. Locating blame in the actions of powerful elites was not in NASA's interest. And focusing attention on the fact that, after all this time, the technology still could defy understanding would destroy the NASA-cultivated image of routine, economical spaceflight and with it the Space Shuttle Program. The myth of managerial wrongdoing made the strategy for control straightforward: fix the technology and change the managerial cast of characters, implement decision controls, and proceed with shuttle launches.

Invariably, the politics of blame directs our attention to certain individuals and not others when organizations have failures. Invariably, the accepted explanation is some form of "operator error," isolating in the media spotlight someone responsible for the hands-on work: the captain of the ship, a political functionary, a technician, or middle-level managers.[18] To a great extent, we are unwitting participants because without extraordinary expenditure of time and energy we cannot get beyond appearances. But we are also complicitous, for we bring to our interpretation of public failures a wish to blame, a penchant for psychological explanations, an inability to identify the structural and cultural causes, and a need for a straightforward, simple answer that can be quickly grasped. But the answer is seldom simple. Even when our hindsight is clear and we acknowledge the players omitted from the media spotlight, as long as we see organizational failures as the result of individual actions our strategies for control will be ineffective, and dangerously so. Consequently, it becomes important to examine the lessons we can draw from the *Challenger* tragedy.

Perhaps the most obvious lesson is about the manufacture of news and the social construction of history in an age when most people are distanced from events and depend on published accounts for information. Even when incredible resources are brought to bear on understanding a public failure involving an organization, the explanation is likely to be more tangled and complex than it appears. Both the consumers and producers of information about public events should beware of the retrospective fallacy.[19] Retrospection corrects history, altering the past to make it consistent with the present, implying that errors should have been anticipated. Understanding organizational failure depends on systematic research that avoids the retrospective fallacy by going beyond secondary sources and summaries, relying instead on personal expertise based on original sources that reveal all the complexity, the culture of the task environment, and the meanings of actions to insiders at the time.

Beyond this general caveat are many lessons from the case itself. The point of this book is to provide a revisionist history of this event and, at the same time, to explain it. That done, we now consider what else it teaches us. Anthropologist Clifford Geertz reminds us that having gained access to an unfamiliar universe, put ourselves in touch with the lives of strangers, and considered what the knowledge thus attained tells us about that society, cultural analysts must extrapolate the significance for social life as such, drawing "large conclusions from small, but very densely textured facts."[20] Embedded in the *Challenger* case are general lessons, both theoretical and practical. In order better to explore them, I first will bring into sharp focus the causes of the tragedy. I then consider what the case teach-

394   CHAPTER TEN

es us about science, technology, and risk; decision making in organizations; and organizational deviance and misconduct. Finally, I demonstrate the significance of the new territory it introduces: the normalization of deviance and its implications for mistake, mishap, and disaster.

THE SOCIAL ORGANIZATION OF A MISTAKE

The *Challenger* disaster was an accident, the result of a mistake. What is important to remember from this case is not that individuals in organizations make mistakes, but that mistakes themselves are socially organized and systematically produced. Contradicting the rational choice theory behind the hypothesis of managers as amoral calculators, the tragedy had systemic origins that transcended individuals, organization, time, and geography. Its sources were neither extraordinary nor necessarily peculiar to NASA, as the amoral calculator hypothesis would lead us to believe. Instead, its origins were in routine and taken-for-granted aspects of organizational life that created a way of seeing that was simultaneously a way of not seeing. The normalization of deviant joint performance is the answer to both questions raised at the beginning of this book: Why did NASA continue to launch shuttles prior to 1986 with a design that was not performing as predicted? Why was the *Challenger* launched over the objections of engineers? The salient environmental condition was NASA's institutional history of competition and scarcity. We want to know how it affected decision making. Official launch decisions accepting more and more risk were products of the production of culture in the SRB work group, the culture of production, and structural secrecy. What is important about these three elements is that each, taken alone, is insufficient as an explanation. Combined, they constitute a nascent theory of the normalization of deviance in organizations.

Production of culture. In the history of decision making, we saw a decision sequence whose repetition indicated the development of a work group culture. The decision-making process itself was a key factor in the normalization of technical deviation. As the Level III and IV Marshall and Thiokol managers and engineers assigned to work on the SRB joints interacted about booster joint performance, they developed norms, values, and procedures that constituted a scientific paradigm. That paradigm supported a belief that was central to their worldview: the belief in redundancy. It developed incrementally, the product of learning by doing. It was based on operating standards consisting of numerous ad hoc judgments and assumptions that they developed in daily engineering practice. Each time a signal of potential danger occurred, it challenged the scientific paradigm and the cultural construction of risk that was its product. Each time, the work group analyzed the new information, weighing it against their definition of the situation and the scientific paradigm that underlay it. Each time they elaborated the original paradigm, based on tests and mathematical models that advanced their understanding, coming forward with engineering determinations of acceptable risk and official recommendations to fly.

The technology mattered. An unprecedented joint design and the uncertainty that accompanied it were fundamental to the production of the work group culture. More accurately, it was the repeated cycles of uncertainty/certainty that were fundamental. Signals of potential danger sent the working engineers scrambling to unravel the evidence and compare tests and flight experience. Prior to each launch, uncertainty had to be converted to certainty—and it was. But it was always temporary, for indubitably new information surfaced from tests or flight experience that was again a signal of potential danger, challenging the prevailing construction of risk and calling for renegotiation. The five-step decision sequence would again follow. With no formal rules to guide them initially about how to respond to the deviant performance of the SRB joints, the SRB work group evolved a set of solutions to a problem they faced in common: norms, beliefs, and procedures relevant to their specific task. Initially without guidelines, the work group evolved an unobtrusive normative structure that was reinforced over time and did, in fact, guide. A fundamental sociological notion is that choice creates structure, which in turn feeds back, influencing choice. In the history of decision making about the SRBs, we see this principle at work, as work group participants created an official cultural construction of risk that, once created, influenced subsequent choices.

The formal aspects of their decision making stabilized their definition of the situation. Of course, using engineering principles and practice to convert uncertainty to certainty was their job. But in addition, each time a launch recommendation was required, diverse opinions had to be pulled together into a collective public position. Creating documents that solidified the group stance committed group members to that stance. Their technical analysis and corrective actions were then subjected to the often painfully adversarial scrutiny of FRR. The challenges raised in FRR drove them back to the labs for further work. Engineering positions that withstood the challenges, as the risk assessments of the SRB joints did, gained in stability primarily because of the engineering analysis behind them, but also because of social affirmation: attacked and defended in a public forum, they received official certification. The capstone, of course, was the feedback from the postflight engineering analysis that showed that their prelaunch corrective actions and performance predictions in response to the most recent anomaly had been correct. Alone, however, the social affirmation and commitment generated as work group decisions were processed are insufficient to explain the normalization of deviance. The culture of production and structural secrecy were environmental and organizational contingencies that caused the work group culture to persist.

Culture of production. Cultural beliefs in the environment affirmed the work group's definition of the situation, informing their sensemaking in common directions. These cultural scripts were part of the worldview that work group participants brought to decision making, providing taken-for-granted sets of invisible rules about how to act in the situations that they faced. Their actions conformed to the culture of production in which they worked; thus, they were acceptable and not deviant in that context. As trained engineers, the situations they faced were consistent with what they understood to be normal technology in large-scale technical systems. Interpretive flexibility, disagreement, crafted decision rules, recurring anomalies, the maxim "Change is bad," and the discipline and methods of scientific positivism all were aspects of the engineering worldview that endowed their decision making with legitimacy. Similarly, their actions conformed to the culture of a profession customarily located in bureaucratic technical production systems. Throughout the history of decision making, the work group's behavior demonstrated a belief in the legitimacy of bureaucratic authority relations and conformity to rules as well as the taken-for-granted nature of cost, schedule, and safety satisficing.

What made this professional culture so powerful as a legitimating force was its nested quality: it was reproduced in the NASA organization and Marshall Space Flight Center, reinforcing the normalcy of their decisions. The workplace was dominated by three cultural imperatives: the original technical culture, bureaucratic accountability, and political accountability. We saw how NASA's environment changed, so that the original technical culture struggled to survive amid institutionalized production concerns and bureaupathology. We saw how the performance pressure at Marshall mandated conformity to all three cultural imperatives, and that the work group did, in fact, conform to all three. Not only did the correspondence of their actions with these cultural scripts normalize their actions, in their view, but their *awareness* of their conformity had a separate effect. The fact that they did everything they were supposed to do reinforced the technical choices they made. For these technical experts at the bottom of the launch decision chain, conformity was the source of the NASA "can do" attitude.

Structural secrecy. To understand why the SRB work group's construction of risk persisted, a final factor is necessary. The culture of production existed for all work groups in the NASA-contractor system. Yet many times work groups did not normalize signals of potential danger: the outcome was a decision of unacceptable risk. Why, in this case, did a work group routinely normalize the technical deviation? Structural secrecy not only prevented a reversal of the scientific paradigm, it perpetuated their view that the O-rings were an acceptable risk: patterns of information, organizational structure, processes, and transactions, and the structure of regulatory relations contributed to a collective organizational construction of risk.

Signals of potential danger were embedded in patterns of information that affected the work group's definition of the situation. Signals lost their salience as a result of the risk assessment process. Signals were mixed: information indicating trouble was interspersed with and/or followed by information signaling that all was well. Signals were weak: the initial threat posed by an anomaly was neutralized as its consequences for performance were measured and understood. Signals were routine: recurring anomalies that were within predictions assured work group participants that they understood the joint and that it was safe to fly. As the work group conveyed information up the hierarchy, organizational structure, processes, and transactions perpetuated the construction of risk created in the work group. Many mechanisms were built into the process so that people outside the work group could (and did) challenge the group's risk assessment. Nonetheless, limiting informational dependencies remained that restricted challenges. The ability of others to intervene, altering the work group's paradigmatic worldview, was inhibited by systematic censorship of information: its patterned reduction due to official organizational practices, specialization, and the tendency of top decision makers to rely on signals when unable to discriminate in decision-making situations.

Finally, structural secrecy originating in the structure of regulatory relationships interfered with safety regulators' capability to contradict the work group's cultural construction of risk. Autonomy and interdependence were its sources. Even Safety, Reliability, & Quality Assurance, the one regulatory unit with personnel who worked closely with the NASA and Thiokol engineers on a daily basis, did not challenge the work group's definition of the situation. Their job was to review, not to produce, data and conduct tests as an independent check. Dependent on the work group for information and its interpretation, they became enculturated. They reviewed the engineering analysis and agreed. Similarly situated in the culture of production and exposed to the engineering analysis, incremental decision making, patterns of information, decision streams, commitment process, and paradigm development, they also normalized joint performance that deviated from expectations.

On the eve of the *Challenger* launch, the production of culture, the culture of production, and structural secrecy combined in epitome of the history of decision making. The teleconference was a microcosm through which we watched these patterns of the past reproduced in a single, dynamic exchange. In a situation of perhaps unparalleled uncertainty for those assembled, all participants' behavior was scripted in advance by the triumvirate of cultural imperatives that shaped their previous choices. The preexisting scientific paradigm and its dominant ideology, the belief in redundancy, formed the all-important frame of reference for the information that was presented. This decision frame was neither latent nor arbitrary, but manifest and in keeping with the tenets of the original technical culture: the Acceptable Risk Process required that all new information be weighed against the preexisting technical rationale for flying with a component. The original technical culture also was manifest in the selection of people from various specializations to attend and in the stringent FRR standards for adversarial challenges and solid, science-based, engineering analyses.

The engineering discussion was undermined by structural secrecy at the outset because participants were in three locations. As the teleconference got underway, the original technical culture, bureaucratic accountability, and political accountability exacerbated this structural secrecy by affecting both the production and exchange of information. Instead of an irrefutable signal that matched the engineers' sense of "lying down in the bucket," the result was weak and missing signals incapable of altering the scientific paradigm supporting the belief in acceptable risk. The charts the Thiokol engineers prepared contained patterns of information that consisted of mixed, weak, and routine signals; thus the engineering argument failed because it did not meet the standards of scientific excellence prescribed by the original technical culture. In cultural terms, the engineering rationale for delay was a weak signal. Bureaucratic and political accountability were responsible for missing signals. Verbal opposition was undermined because all participants conformed to common understandings about talk and bureaucratic accountability in FRR that silenced people. Political accountability, and the production pressure that was its result, had its impact on both the production of information and its exchange. Again guided by unspoken shared understandings, the Thiokol engineers set a time limit for their preparations, possibly sabotaging their own opportunity to put together an engineering presentation capable of reversing the scientific paradigm. And managers, themselves responsible for bureaucratic, political, and technical accountability, experienced performance pressure in all three areas that resulted in actions that silenced people. The scientific paradigm prevailed.

The historically accepted explanation of the launch that has production goals dominating managerial actions, resulting in rule violations and individual wrongdoing, gives way to a complex imagery of bounded rationality exacerbated by additional environmental and organizational contingencies. What is compelling is how structures of power, history, processes, and layered cultures that affected all participants' behavior at a subtle, prerational level combined to produce the outcome. Although the discussion itself was adversarial, in keeping with the tradition of FRR, the outcome was a cooperative endeavor: all participants followed agreed-upon, taken-for-granted understandings that dictated how they were to proceed. Like the history of decision making, the eve of the launch was characterized by conformity, not deviance. Socially organized and history-dependent, it is unlikely that the decision they reached could have been otherwise, given the multilayered cultures to which they all belonged.

Having made explicit the social organization of mistake that resulted in the *Challenger* launch decision and the theory of the normalization of deviance that it generates, we turn to the lessons the case holds about science, technology, and risk; decision making in organizations; and organizational deviance and misconduct.

SCIENCE, TECHNOLOGY, AND RISK

Although much research has examined how the riskiness of technological products is constructed for and by the public, much less attention has been placed on the construction of risk in the workplace by technical experts with official responsibility to do so. We have examined scientific practice, how workers assign meanings to and make decisions about technology, and how individuals and organizations, as a consequence, construct risk. Keep in mind the ways this case differs from other studies of risk assessment. First, in contrast to the well-known laboratory studies of individual perceptions of risk by psychologists Tversky, Kahneman, and their colleagues,[21] the case takes us out of the laboratory into a real-world example of risk assessment in an organization. Second, the risky decisions we examined were made by workers engaged in predictive work prior to the operation of the boosters. They were not monitoring and manipulating an operating technology to prevent failures, like astronauts and NASA personnel in Mission Control during a spaceflight, the crew on an aircraft carrier flight deck, or workers in a nuclear power plant, who are confronted with baffling signals from technology in action. Third, risk assessment was (1) a manifest goal of the organization, (2) continuous, (3) systematic, formalized, and regulated, and (4) openly negotiated between individuals and groups in order to produce an official collective organizational construction of risk.

In keeping with the urgings of Clarke, Douglas, Freudenberg, Heimer, Short, and others who have argued for what Short identified as the "social transformation" of risk analysis,[22] this case shifts attention from the psychological to the social, organizational, and institutional influences on risk assessment. It demonstrates the inseparability of social and technical factors in understanding the scientific practice of engineers and the determinations of risk that result.[23] It illuminates culture as supremely important in shaping risk assessments in the workplace.[24] Moreover, the case sheds some light on how scientific paradigms develop, operate, and persist. Although the work group's scientific paradigm is hardly of the scope of the scientific communities and dominant paradigms described in Kuhn's path-breaking *The Structure of Scientific Revolutions*, it is Kuhn in microcosm. The stability of scientific paradigms and their resistance to change is Kuhn's theme. It is also ours.

To understand how workers constructed risk, we opened up the "black box" of scientific practice, showing the ambiguity, the conflicting standards, the negotiations, and the disagreements about the nature of evidence and its meaning.[25] Interpretive flexibility existed and knowledge was socially constructed, but it was bounded by the real, constraining material elements of the technology in question. Each time a launch decision had to be made, the technical experts snatched certainty from the jaws of uncertainty, pulling together a coherent technical analysis. Scholars who study scientific practice in other laboratory settings find that transforming disorder into order is basic to scientific work, which Leigh Star describes as "the representation of chaos in an orderly fashion."[26] We watched as, over time, the work group developed a scientific paradigm that incrementally expanded to include recurring anomalies, becoming more stable in the process. Resolution of anomalies is what Kuhn identifies as the business of "normal science." In a paragraph that reads like a description of work group practice, he writes: "Discovery commences with the awareness of anomaly, i.e., with the recognition that nature has somehow violated the paradigm-induced expectations that govern normal science. It then continues with a more or less extended exploration of the area of anomaly. And it closes only when the paradigm theory has been adjusted so that the anomalous has become the expected."[27]

Kuhn explains the stubborn persistence of paradigms by identifying several factors that apply to this case. First, a paradigm endures because it is more successful than its competitors in solving a few problems that a group of practitioners identifies as acute. It never solves all the problems, thus generating disagreement, but the absence of standard interpretations does not prevent a paradigm from guiding research. Second, normal science *is* normative. Normal science is "puzzle solving" *within* the paradigm: the paradigm determines what is a problem and what rules and methods will be used to examine it. Practitioners whose research is based on a shared paradigm are committed to the same rules and standards for scientific practice; these rules limit both the nature of acceptable solutions and the steps by which they are obtained.[28] A new theory meets resistance because it implies a change in the rules that governed puzzle solving in the past, as well as the dismantling of the scientific work successfully completed.[29] Third, the first received paradigm usually accounts for most of the observations and experiments readily available to practitioners. However, further development entails specialization: elaborate equipment, esoteric vocabulary and skills, and refinement of concepts. Specialization narrows worldview, creating rigidity and resistance to paradigm change.

In language that staggers in its seeming prescience about events at NASA, Kuhn describes normal science as "an attempt to force nature into the preformed and relatively inflexible box that the paradigm supplies. No part of the aim of normal science is to call forth new sorts of phenomena; indeed those that will not fit the box are often not seen at all."[30] According to Kuhn, the recalcitrance of a scientific paradigm is in the worldview it engenders; it is overturned when a crisis arises that precipitates a transformation of that worldview. The result is a scientific revolution that changes the paradigm and the rules of normal science. Paradoxically, paradigms themselves are indispensable to scientific revolutions because anomalies are only apparent against the background provided by the paradigm.[31] However, the *Challenger* case proves that a move from anomaly to "scientific revolution" must circumvent still other constraints. It reminds us that paradigm obduracy is embedded in institutional forces in environment and organization that go beyond scientific communities and the received wisdom of scientific practice. Most science and technology is done in socially organized settings that can hardly be described as neutral. The professions and organizations that generate scientific and technical work struggle to survive in an often turbulent environment, experiencing competition and scarcity, and thus decisions are governed by other cultural imperatives that coexist uncomfortably with those of scientific practice. This case shows how these imperative also become part of practitioners' worldview, providing additional constraints, blocking opportunities for innovation.

Further, this case shows how resource commitments, bureaucratic procedures, and routines in organizations restrict search, resulting in bounded rationality that focuses attention in some directions at the expense of others. Finally, we saw how the very process of producing scientific and technical knowledge can contribute to paradigm persistence. Producing analysis in documentary form, following the rules, making results public, and receiving affirmation committed the work group to lines of action that were difficult to reverse without evidence sufficiently strong that it overturned the scientific paradigm. But the case also reminds us that the success of a paradigm revolution can depend not only on converting tacit knowledge that is subjective and intuitive into a form that is acceptable, under the prevailing rules of the game, but also—like most successful revolutions—on power. At NASA, the crisis that precipitated a transformation of worldview and resulted in a paradigm shift was not the teleconference, but the *Challenger* disaster.

DECISION MAKING IN ORGANIZATIONS

Study of the *Challenger* launch decision stands with research on other famous decisions with historic consequences, such as Graham Allison's *Essence of Decision*, which explains the 1962 Cuban missile crisis, and Janis's *Groupthink*, which explores U.S. policy decisions that dealt with the Bay of Pigs, North Korea, Pearl Harbor, and Vietnam.[32] However, what we learn about decision making in this case must be tempered by awareness that it differs in important ways from both of the above. Both Janis and Allison look at policy decisions by elite policymakers, working in secrecy. The *Challenger* case shows bottom-up decision making that was conveyed to people throughout the hierarchy. In addition, the decision making here consisted of a repeating choice about the same technical object in a highly regulated, open decision process. While elite policymakers played a vital role, they were absent from the technical decision making. The focus necessarily is on technical experts doing engineering work. And though the outcome of their decision making can be called policy, the participants called it risk assessment, flight readiness, or simply engineering. Beneath its unique historic significance and the risky technology that was NASA's business, the *Challenger* launch decision is a story about routine decisions in the workplace. Thus, the case also contains lessons about decision making in organizations, regardless of their enterprise.

The conventional explanation of the launch decision rested on a rational choice theory of decision making: amorally calculating managers, experiencing production pressures, weighing costs and benefits, violating safety rules, then launching. Rational choice theory assumes that decisions are made by informed, calculating individuals working toward clear sets of goals. Decision making, in this view, is an intentional and outcome-oriented activity, so research typically focuses on individual decision makers and outcomes. In contrast, naturalistic studies focus on the decision process.[33] Sensemaking, in this view, is about contextual rationality, so the task is to expose the constraints, both hidden and explicit, both informal and formal, that act on decision makers.[34] This case takes a naturalistic approach, relying on historical ethnography and thick description of decision processes to understand what happened at NASA. It reminds us that actions are not necessarily the outcome of intent, conscious choice, or planning, even though the outcomes may, to some extent, be predictable.[35] It underscores the importance of exploring both decision context and the interpretive work of the people making choices. Moreover, it suggests the importance of history as a context for sensemaking. Mayer Zald observed that organizations exist *in* history, deeply embedded in institutional environments; simultaneously, they exist *as* history, products of accumulated experience over time.[36] The *Challenger* incident demonstrates a decision context in which both converged, affecting the interpretation of information.

The case is, to borrow from Douglas, an exemplar of *How Institutions Think*.[37] It falls squarely into that branch of organization theo-

ry known as the "new institutional school of organizational analysis."[38] The new institutionalism emphasizes the way in which non-
local environments—industries, professions, and the like—penetrate
organizations, creating a frame of reference, or worldview, that individuals bring to decision making and action.[39] The distinctiveness of
the institutional approach is its message about what happens at the
local level. Yet the connection between the environment and individual action has seldom been explored in the context of real-world
decision making. Organizations are a mediating link between environment and individual action. What sets the *Challenger* case apart
is that it makes explicit the linkages between environment, organization, and individual action.

In the history of SRB decision making, we saw how cultural
beliefs originating in the environment—the engineering profession,
the aerospace industry, and the NASA organization—affected worldview and thus actions and outcomes at the local level. On the eve of
the launch, these same nested cultures shaped group dynamics.
Many posttragedy accounts concluded that Janis's theory of groupthink—perhaps the leading theory of group dynamics and decision
making—was responsible for the launch decision.[40] But many of the
elements of groupthink were missing, and those that were present
have explanations that go beyond the assembled group to cultural
and structural sources.[41] To a great extent, group dynamics during
the teleconference—and the outcome—reflected key assumptions in
the environment and the workplace that played themselves out in
interaction. Perhaps most alarming is the way that production pressures and bureaupathology became institutionalized and taken-for-
granted, having a profound impact on the proceedings.

But individuals were not passive recipients of culture. The case
shows how people at all hierarchical levels actively participated in its
production and reproduction. The institutional approach acknowledges the role of powerful elites who take advantage of the environment to further their own interests as well as those of the organization.[42] Brint and Karabel point out that organizational interests do
not simply "reflect in a mirrorlike fashion the distribution of power
in the larger society," but are given a distinctive stamp by people with
the power to do so.[43] We saw that historic actions by elites gave a particular character to NASA culture. But the work group's own historic
actions brought a culture into existence. The discovery of this work
group culture expands our notions to include the idea of a dynamic
organizational culture that can transcend organizational structure,
developing as people interact about tasks, disappearing when the task

ends, perhaps reappearing in other forms as the group members become part of other groups. This example can help us understand new forms of organization in which people gather and disband around changing tasks.[44]

By linking environment, organization, and decision making, the case alerts us to the subtlety of definitional processes, affirming the power of historic actions in organizations that become solidified into norms, standard operating procedures, and a shared worldview that shapes future choices. It displays the forces in culture and social structure that simultaneously set limits to and present possibilities for rationality in human affairs. It extends what we know about bounded rationality, showing how taken-for-granted assumptions, predispositions, scripts, conventions, and classification schemes figure into goal-oriented behavior in a prerational, preconscious manner that precedes and prefigures strategic choice. It conveys a stunning message about the influence of these preconscious schema on the production, exchange, and interpretation of information in organizations. It shows that organization and interorganization structure and patterns of information can play a decisive role in the persistence of worldview. It identifies the influence of cultural understandings, developed in the workplace as employees interact about a common task, on choice.

Consequently, the case supplies the agency missing from the new institutionalism. It affirms a theory of practical action that links institutional forces, social location, and individual thought and action.[45] For at its essence, the case is a picture of individual rationality irretrievably intertwined with position in a structure.[46] Position in the engineering profession, the aerospace industry, and the various organizations made up the labyrinthine NASA-contractor network was a key determinant of individual and collective determinations of risk. Position determined social mission. Position determined access to information. Position determined responsibility for acting on information and the actions that legitimately could be taken. Position contributed to ability to interpret information and to the worldview brought to that interpretation. Perhaps most important, position determined power to shape opinions and outcomes in one's own and other organizations.

ORGANIZATIONAL DEVIANCE AND MISCONDUCT

By concluding that the decision to launch the *Challenger* was a mistake, I am not dismissing the relevance of this case for our understanding of organizational misconduct. Far from it. The factors commonly associated with organizational misconduct, described in chapter 2, were alive and well at NASA: the competitive environment, organization characteristics—structure, processes, and transactions—and the regulatory environment.[47] But I also found other factors, previously unacknowledged, that played an important role. What is new is how they combined to normalize technical deviation in joint performance. Although allegations of managerial rule violations were proved false, it is possible that the processes of deviance normalization revealed in this case may play a role in facilitating rule violations and misconduct when they do occur in other organizations. Consequently, what happened at NASA has important implications for the unresolved question of "good" people and "dirty" work raised earlier: How is it that people who often appear to be pillars of the community (educated, employed, scout leaders, or churchgoers) violate rules, regulations, and laws on behalf of their organizations that result in harmful outcomes?

Perhaps the most dramatic insights are about the link between culture and individual choice.[48] Consistently, scholars have assigned great importance to culture as a cause of organizational deviance and misconduct.[49] Two observations are generally held to be true. First, in industries and organizations where misconduct occurs, normative environments exist that conflict with those of the outside world: what society defines as illegal, deviant, or unethical comes to be defined in the industry and organization as normative.[50] Second, these normative environments have a unidimensional character. Reflecting competitive pressures from the environment, organizational culture emphasizes production goals, which are presumed to be the driving force behind calculated decision making that culminates in violative behavior.[51] This case takes us in important new directions on both these issues.

On the first, the case is a vivid example of these conflicting normative standards: after the tragedy, outsiders perceived that continuing to fly under the circumstances that existed at NASA was deviant; yet as the problem unfolded, insiders saw their behavior as acceptable and nondeviant. However, it develops our insights further by showing the manifold aspects of culture and how it shaped the meaning of actions to insiders. In the history of decision making, we saw the development of work group norms encouraging behavior that came to be considered acceptable practice and nondeviant to many people within the organization: launching when equipment was not operating as expected, with peculiar variations in O-ring performance from launch to launch. We also saw the process by which the boundaries defining acceptable behavior incrementally widened, incorporating incident after aberrant incident.

On the second issue, the case affirms the important effects of production pressure on decision making but shows these effects to be much more subtle than previously surmised. Moreover, rather than a unidimensional culture, the case shows that a multilayered, multidimensional, and dynamic culture affected technical decision making at NASA.[52] We saw decision making dominated by conformity not to a single cultural imperative about production goals, but to three imperatives. Accepting the deviant performance of the SRB joint was legitimate because the work group conformed to the norms and rules of the many cultures to which they belonged. The case suggests the role of conformity in choice, affirming the insight that for participants in the culture in question, it is not deviance that is learned but *conformity* to group rules and norms.[53] We see evidence of the role of conformity in the deviant outcomes described in such works as Hannah Arendt's *Eichmann in Jerusalem* and Herbert Kelman and Lee Hamilton's *Crimes of Obedience,* although neither work is helpful in illuminating the longitudinal microlevel processes and dynamic interactions in which participants created a definition of the situation.[54]

The case also introduces risk as a concept useful for understanding the deviant and unlawful acts of organization members. Research has been preoccupied with the deterrent effect of punishment and how people's calculations of the costs and benefits of their actions figure into decisions to violate laws, rules, and regulations.[55] Incorporated into this rational choice model is the probability of getting caught and being punished. Consequently, the social construction of risk— the risk of incurring costs versus rewards—looms as a crucial but often unarticulated factor in decisions to violate. But this particular piece of the deterrence puzzle—individual *perceptions* of rewards and punishments and the probability that they will be incurred—has seldom been unraveled in a real-life workplace situation.[56] This case allows us a first glimpse of internal organizational dynamics that affect the individual perceptions of risk that are the linchpin of deterrence theory. At NASA, the construction of SRB joint risk prevented the work group from fully acknowledging the harm (or costs) to others, to their organizations, and to themselves of their decisions concerning the O-ring problem. To the contrary, the definition of the situation that prevailed led them to believe that launches would be successful, resulting in rewards.

Although we cannot conclude that what happened at NASA applies to other incidents of alleged organizational misconduct, neither can we conclude that amoral calculation is always at the bottom of them. We need research that examines decision making within both organizational and environmental contexts, and when that is impossible, we must still wonder how the subtle interaction of competitive environment, organizational characteristics, and regulatory systems operate to influence the actions of people in organizations, shaping their worldview and their interpretation of information. Even when the case for amoral calculation appears convincing (such as the decision to continue manufacturing the flawed Ford Pinto, with documents that verify administrators estimated the cost of repairing the flaw against the probable estimated costs of loss of life when they decided not to alter the design),[57] we must be curious about the dynamics of decision making. We must wonder how the organizational context shaped the construction of risk held by key decision makers: the risk of injury or death to Pinto passengers, the risk that costs would befall the corporation, and the risk that those who decided to continue manufacturing the unchanged Pinto would be held personally responsible for any harmful outcomes. We must wonder about processes that normalize deviance, possibly allowing organization members honestly to view their actions as normal, rather than deviant.

The answer to the question of "good" people and "dirty" work suggested by this research is that culture, structure, and other organizational factors, in combination, may create a worldview that constrains people from acknowledging their work as "dirty." Thus, rather than contemplating or devising a "deviant" strategy for achieving the organization's goals and then invoking techniques of neutralization in order to proceed with it or rationalize it afterward,[58] they may never see it as deviant in the first place. How influential can the deterrent effects of punishment and costs be when environmental contingencies, cultural beliefs, and organizational structures and processes shape understandings so that actors do not view their behavior as unethical, deviant, or having a harmful outcome?

Finally, the case also demonstrates why it is so difficult for legal and normative systems to assign responsibility properly when organizations have harmful outcomes.[59] It is well known that division of labor in organizations obfuscates responsibility for organizational acts.[60] Specialization means that many people throughout the hierarchy are likely to participate in making a decision or product. Some may unknowingly participate in deviant, illegal, or unethical acts because their position leaves them unaware of how their action connects to the actions of others, what the outcome will be, or that legal or normative structures are in jeopardy. It is also well known that top administrators play a powerful but invisible role by determining goals, setting deadlines, determining sanctions, and otherwise influencing the environment of decision.[61] But when social control agents attempt to identify responsible individuals, middle managers are most likely to be held accountable because they made the decisions—or failed to make the decisions—that seemed temporally connected to the harmful actions.[62] Charges of intent or negligence find a ready resting place on middle managers because of their position and role in the organization. Although no legal system was involved, the NASA case follows the classic pattern for organizational misconduct: middle managers were assigned normative responsibility and left "twisting in the wind," while more powerful administrators—some outside the NASA organization, who had acted years earlier in ways that influenced the outcome—were not. This case sets forth many factors that help explain why people at both the top and the bottom of the hierarchy escape the attention of legal and normative systems when organizations have harmful outcomes. Organizational complexity provided the usual obstacles to attributing responsibility. But the analysis shows that the issue of accountability was disguised by history, geography, and culture as well.

ON MISTAKE, MISHAP, AND DISASTER: THE NORMALIZATION OF DEVIANCE IN ORGANIZATIONS

Embedded in the diversity of the three foregoing topics is a common theme: the formation of worldview and how it affects the interpretation of information in organizations. Possibly the most significant lesson from the *Challenger* case is how environmental and organizational contingencies create prerational forces that shape worldview, normalizing signals of potential danger, resulting in mistakes with harmful human consequences. The explanation of the *Challenger* launch is a story of how people who worked together developed patterns that blinded them to the consequences of their actions. It is not only about the development of norms but about the incremental expansion of normative boundaries: how small changes—new behaviors that were slight deviations from the normal course of events—gradually became the norm, providing a basis for accepting additional deviance. No rules were violated; there was no intent to do harm.

Yet harm was done. Astronauts died. Although the explanation of the *Challenger* launch is not a story of harm done through amoral calculation, we can find no relief in the tale that has been told in these pages. It is a story that illustrates how disastrous consequences can emerge from the banality of organizational life. It is a story of rather ordinary influences on decision making that operate inconspicuously but with grave effect. No fundamental decision was made at NASA to do evil; rather, a series of seemingly harmless decisions were made that incrementally moved the space agency toward a catastrophic outcome.

The forces of formal organization that generate routinization and collective action are legendary, acknowledged in classic writings on bureaucracy by Max Weber, Frederick Taylor, Chester Barnard, Michel Crozier, Peter Blau, and Robert Merton, and visually represented in such films as Chaplin's *Modern Times*. Less acknowledged are the nested cultural influences that, although invisible, are equally influential in molding common understandings in organizations. Although the case concerns a technical artifact and statistical deviation, it suggests how conformity to rules and norms, incrementalism, precedent, patterns of information, organizational structure, and environmental conditions congeal in a process that can create a change-resistant worldview that neutralizes deviant events, making them acceptable and nondeviant. While the normalization of deviance can be functional for an organization, reducing uncertainty, allowing coherence, and creating continuity between past, present, and future, it also can lead to mistake resulting in mishap and, as in the *Challenger* incident, disaster. We must wonder to what extent the same general factors normalizing technical deviation at NASA contributed to other well-known incidents, now publicly acknowledged as mistakes: for example, the Hubble telescope, the downfall of IBM, and U.S. involvement in the Vietnam War.[63] These, too, had long incubation periods punctuated by signals of potential danger; the responsible organizations proceeded as if nothing was wrong in the face of evidence that something was wrong.

It would be a serious mistake, on our part, to simply chalk the *Challenger* tragedy up as the inevitable outcome of the size, complexity, and political conflicts of a huge, technocratic bureaucracy. Research suggests that deviance is normalized in other organizations—smaller, less complex, less formalized, nontechnocratic—contributing to mistake with portentous human consequences. Barry M. Turner, in *Man-Made Disasters*, investigated accidents and social disasters, seeking any systematic patterns that might have preceded these events.[64] Examining formal and complex organizations (none of which matched the size, complexity, and formalization of NASA), Turner found that the responsible organizations had "failures of foresight" in common.[65] The disasters had long incubation periods characterized by a number of discrepant events signaling potential danger. Turner notes that typically these events were overlooked or misinterpreted, accumulating unnoticed. Among the contributing factors he identified were norms and culturally accepted beliefs about hazards. Although Turner did not have access to data that allowed him to examine decision making within its social context, his finding suggests the existence of culturally based worldviews that affected the interpretation of information.

We also have some research evidence on members of professional associations. These professionals were not regular members of a workforce, but made decisions under more autonomous circumstances, less governed by bureaucratic formalisms, in which mistake also had awesome human consequences. How professional worldview affects scientists' interpretive work is a theme in several of the books of biologist and paleontologist Stephen Jay Gould.[66] In his first—and classic—major work that explicitly claims this agenda, *The Mismeasure of Man*, Gould examined the great IQ debate on racial differences in mental worth.[67] He reanalyzed and replicated the eighteenth- and nineteenth-century classical data sets in craniometry and intelligence testing that established the mental inferiority of oppressed and disadvantaged groups—races, classes, or sexes—thus proving they deserved their deprived status. However, in his reanalyses, Gould continually found "a priori prejudice, leading scientists to invalid conclusions from adequate data, or distorting the gathering of data itself."[68]

Gould attributed this finding to a societal culture affirming the correctness of racial ranking—whites at the top, blacks at the bottom, and other races distributed in between. This societal belief, he asserts, affected the worldview of professional scientists, distorting research on heredity, intelligence, and race. He notes that some distortions were outright cases of fraud, but most were mistakes that reflected "unconscious biases that record subtle and inescapable constraints of culture."[69] In many cases, he found scientists, working honestly, had "averted anomalies," disregarding or misinterpreting obvious and clear discrepant information that contradicted their worldview.[70] Their profound belief in the correctness of racial ranking probably precluded these scientists from seeing any possible harmful human consequence of their work, so they probably did not view their decisions as risky decisions (unless we count their own career risk). But the human consequences were far-ranging. Gould observes, "Few tragedies can be more extensive than the stunting of life, few injustices deeper than the denial of an opportunity to strive or even to hope, by a limit imposed from without, but falsely identified as lying within."[71]

Whereas Gould relied on historical data, sociologist Stephen J. Pfohl watched professionals making decisions. In *Predicting Dangerousness*, Pfohl recounts how he and his research associates observed teams of mental health professionals assessing the potential dangerousness of inmates being reviewed for possible release from the Lima State Hospital for the Criminally Insane, as mandated by Ohio law.[72] Human behavior being what it is and with no fail-safe method for evaluating the potential for harm of a given individual (hospitalized or not), predicting the dangerousness of these inmates certainly qualifies as risk determination under conditions of uncertainty. Each team was composed of a social worker, a psychiatrist, and a psychologist. Pfohl and his colleagues found that each of these three professional types had distinctive worldviews—theoretical dispositions developed from training and experience—that channeled themselves into specific diagnostic interactions: selective structures for question asking and answer hearing.

Each professional brought to the interviews previous patient experience and a stock of clinical interpretations typical of his or her profession. This led to "preinterview theorizing" that selectively focused team members' expectations about the person they were going to interview.[73] In the hearings, team members asked patients "theory-pushing" questions geared toward confirming an impression they had of patients on the basis of patient records.[74] They selectively heard and misheard responses, fitting patient statements and reactions into professional theoretical categories prescribing symptoms of mental illness.[75] Often information that Pfohl and his associates defined as normal and a possible signal of patient mental health (an aggressive response, inarticulateness, or withdrawal in a public examination with enormous personal implications) was interpreted by the hearing officials as a signal of potential danger because it fit preexisting professional diagnostic schema. In this hospital for the criminally insane, evidence of sanity was deviant, discrepant information. It contradicted the professionals' worldview and was interpreted as a signal of potential danger requiring continued hospitalization.

Pfohl concludes that these practices convey an image of psychiatric decision making as "an inherently contextbound process, rooted more in the sense-making practices of clinicians than in the discernible behavior of patients."[76] Each professional was predicting the dangerousness of the patients by assessing information in relation to his or her own professional frame of reference. Because a collective decision was necessary, these three definitions had to be negotiated. The final construction of risk upon which the decision to release or not to release was based was a product of social interaction and reflected—not surprisingly—the differential distribution of power between the risk assessors. Typically, the definition given by the psychiatrist on the team won.

Even the smallest organizations we create—intimate relationships—are vulnerable. When we marry or live with another person, we develop a division of labor, share some goals, compete for scarce resources, and socialize new members. We come and go regularly, making decisions daily, creating precedents, decision streams, culture, and history. Like their larger and more formal counterparts, these small organizations are also subject to failures with human consequences. My research, *Uncoupling: Turning Points in Intimate Relationships*, shows that when intimate relationships end, the normalization of signals of potential danger has contributed to the outcome.[77] Despite the enormous disparity in size, complexity, and function from NASA, *Uncoupling* is in some ways a better demonstration of the general process than the foregoing examples because, like the NASA case, it allows us to watch repeating decisions about the same object over time and gives comparable data about the effect of patterns of information on the definition of a situation. To exploit this opportunity, I will focus on structural secrecy here. Bear in mind, however, that cultural beliefs in the environment and the production of culture also play a role in the normalization of signals of potential danger in intimate relationships.

When relationships end, the typical pattern is that one person, whom I call the initiator, begins leaving the relationship socially and psychologically before the other. By the time the still-committed partner realizes that the relationship is in serious jeopardy, the initiator has already left in a number of ways, so that efforts to save it are likely to fail. What we want to know is, How it is possible, when two people live together, for one to slip so far away without the other's noticing and acting to avert what is so often experienced as a personal disaster? The answer is in signals of potential danger, patterns of information, and the worldview that the partner brings to the interpretation of information. Initiators display their discontent with weak signals, mixed signals, and signals that become routine.

For many reasons, initiators do not give irrefutable signals that would immediately get their partners' attention until late in the uncoupling process. In the beginning, of course, signals are usually weak, but even as discontent grows, initiators may say, "I'm unhappy in this relationship," but they do not say, "I'm unhappy—and I'm seeing someone else." They also send mixed signals: they may complain bitterly, then withdraw into silence; they may threaten to leave, then repent with a declaration of love. In both cases, the partner takes the second signal as affirmation that the upset was temporary and now all is well. Repeated signals—threats to leave or constant complaints—which were at first an attention-getting deviation from the routine, upon repetition become routine.

As was true for the SRB work group, context, gradualism, and incrementalism also matter. Signals of potential danger tend to lose their salience because they are interjected into a daily routine full of established signals with taken-for-granted meanings that represent the well-being of the relationship. A negative signal can sometimes become simply a deviant event that momentarily mars the smoothness of the ongoing routine. As the initiator's unhappiness grows, the number, frequency, and seriousness of signals of potential danger increase. While they would surely catch the partner's attention if they all came at once, this is not the case when new signals are introduced slowly amid others that indicate stability. A series of discrepant signals can accumulate so slowly that they become incorporated into the routine; what began as a break in the pattern becomes the pattern. Small changes—new behaviors that were slight deviations from the normal course of events—gradually become the norm, providing a basis for accepting additional deviance.

To the interpretation of this information, the partner brings a worldview, or frame of reference, that includes integrated sets of assumptions, expectations, and experiences about the relationship and his or her position in it. As in the NASA case, these beliefs are cultural, originating in the environment and the organization, and they include expectations about what is normal in relationships in general and in this one in particular. When new signals do not fit, the partner sifts through the mixed array, selectively incorporating signals consistent with the ongoing frame of reference and normalizing those that challenge it. Averting disaster becomes an infinitely receding possibility for the partner. The normalization of deviance is cemented by the partner's commitment to a line of action that is difficult to reverse—the decision to couple, to stay in the relationship—for to reverse it not only contradicts cultural expectations about relationships but calls into question past decisions. Only when finally confronted with a signal so direct, painful, and undeniable that it cannot be normalized—for example, the initiator leaves or asks the partner to leave; the divorce is final—is the partner in the position of the teleconference participants after the *Challenger* tragedy. Only then does the partner clearly see the signals of potential danger that existed all along.

THE INEVITABILITY OF MISTAKE

The foregoing examples affirm that mistakes are systematic and socially organized, built into the nature of professions, organizations, cultures, and structures. Collectively, they are chilling in their suggestion that the normalization of deviance creates the potential for mistake in organizations large and small. We are left with a disturbing question: If the normalization of deviance neutralizes signals of potential danger in intimate relationships—two decision makers, unencumbered by complex hierarchy, technology, and "blizzards of paperwork"—how can we expect to control it in larger organizations that deal in risky technology? In *Normal Accidents*, Perrow concludes that accidents are normal, or inevitable, in certain technological systems.[78] He identifies the source of dangerous accidents as the system, not its component parts. When a technical system has many parts that interact and also are tightly coupled, it is capable of generating unfamiliar, unexpected sequences that are not visible or not immediately comprehensible. Because tightly coupled technical systems have little slack, or "give," they offer few opportunities to recover when something begins to go wrong. The *Challenger* disaster can justifiably be classed as a normal accident: an organizational-technical system failure that was the inevitable product of the two complex systems. But this case extends Perrow's notion of system to include aspects of both environment and organization that affect the risk assessment process and decision making. Increasing the basic pessimism of the original model of normal accidents, we learn that even when technical experts have time to notice and discuss signals of potential danger in a well-attended meeting prior to putting the technology into action, their interpretation of the signals is subject to errors shaped by a still-wider system that includes history, competition, scarcity, bureaucratic procedures, power, rules and norms, hierarchy, culture, and patterns of information.

An obvious advantage of the workplace over the settings of the other examples is that formal organizations can create rules, structures, and processes to regulate risky decision making. High-reliability theorists take the position that accidents can be prevented through good organizational design and management.[79] They examine organizations with good safety records in order to find out what conditions promote safety. Among their recommendations for achieving high reliability are: safety as the priority objective of organizational elites; effective trial-and-error learning from accidents; redundancy built into both organization and technology to enhance safety; decentralized decision making that enables quick, flexible responses to field-level surprises; and a "culture of reliability," built on a military model of intense discipline and training, that maximizes uniform and appropriate responses by those closest to the risky technology.

It goes without saying that managing organizations better can reduce the possibility of mistake leading to mishap and disaster, and that implementing these suggestions can be beneficial. Every possible avenue should be pursued. However, we should be extremely sensitive to the limitations of known remedies. While good management and organizational design may reduce accidents in certain systems, they can never prevent them. Their potential may vary, depending on the characteristics of the technology, the organization, and the task. We might surmise that the greater the complexity of technology, organization, and task, the greater the probability of an accident. The causal mechanisms in this case (although certainly distinctive on all three scores) suggest that technical system failures may be more difficult to avoid than even the most pessimistic among us would have believed. The effect of unacknowledged and invisible social forces on information, interpretation, knowledge, and—ultimately—action are very difficult to identify and to control.

From our luxurious retrospective position, NASA appeared to have used its penchant for rule making to create a decision-making structure absolutely suited to preventing the normalization of deviance. Those who designed the FRR process wanted to assure information exchange. They designed it to circumvent the problems of hierarchy, specialization, and information dependence associated with communication in large organizations. Yet those same factors were reproduced in FRR, mirroring the structural secrecy found in the parent organization. NASA even had the foresight to incorporate rules and procedures designed to produce signals that would challenge developing scientific paradigms and entrenched worldviews: open meetings, a "fishbowl" environment that was adversarial both in theory and in practice, and a matrix system of mandated participation by people whose specializations and dissociation from work groups maximized the diverse perspectives brought to bear on risk assessments.[80] NASA had a well-conceived system for regulating decisions about risk that worked in every case but one. Moreover, the eve-of-launch teleconference was exactly the sort of decentralized decision process in response to crisis that high-reliability theorists call for. We saw how unseen history, political bargains, culture, and structure undermined decision making that night.

Reconsider, for a moment, the significance of Richard Cook, the Resource Analyst at NASA Headquarters who wrote a memo to his boss in July 1985 warning that the joint problem was compromising flight safety. Cook was an outsider—not a trained engineer, not located at Marshall, not a member of the work group. Prior to writing his memo, he had worked at NASA Headquarters only a short time. He did not talk to engineers in the SRB work group.[81] He saw the trend of accepting more and more technical deviation and, not being privy to engineering analysis and the technical rationale, saw the trend itself as a signal of potential danger. Not ensnared by the multilayered cultures of decision making, he immediately defined as deviant and risky what people in the launch decision chain defined as normal and acceptable.

From Cook's insight, we might conclude that one solution to the normalization of deviance in organizations is to rely more on outsiders who, uncoopted by membership in the system, can introduce contradictory signals that challenge entrenched worldviews: in NASA's case, supplement the various safety regulatory structures that require engineering expertise with people who are outside the launch decision process and/or nonengineers. While this strategy is worthwhile and may reduce accidents, there are good reasons to think its potential would be limited. As a nonengineer, Cook lacked the expertise and authority to persuade. But even when outsiders were given authority, as they were in NASA's Space Shuttle Crew Safety Panel and the Aerospace Safety Advisory Panel, autonomy and interdependence affected their ability to spot developing problems and to effect change. Alternatively, we might conclude that making rules to increase the clarity of insider signals would be effective: for example, contractors must say "go" or "no-go" (not "we cannot give 100 percent assurance that it is safe to launch"); launch decisions made after the formal FRR must be face to face, not by teleconference, so participants get all visual and verbal signals. Improving clarity of signals is useful and must be done. Still, we must not be deceived. Strategies for control must target the cause. But when we focus on the ability of individuals to interject clear, alternative signals that challenge a paradigmatic worldview, we target only one aspect of an intricate, massive causal system that spanned environment, organization, and individual choice.

The scope and interconnectedness of this causal system make mistakes inevitable. Identifying the systemic causes of a mistake is difficult because, in the wake of a tragedy, organizational learning is, at best, partial. Not all contributing causes will surface. Like the Presidential Commission, both internal and external accident investigations will be affected by structural secrecy resulting from the autonomy and interdependence that systematically undermine regulatory efforts. Most likely to remain undetected are the imperceptible interwoven cultures and structures that affect decision making. So veiled are their effects that even decision makers are unaware of their operation. The traditional response to organizational failure is to first identify and then change personnel in key decision-making positions. Admittedly, changing the cast of characters is much easier than an all-encompassing organizational analysis. But environment, organization, and individual choice are inseparable. We must be aware that personnel replacements will be subject to the effects of the same cultures and structures. In fact, any remedy that targets only individuals misses the structural origins of the problem.[82] Furthermore, we must be concerned about the ironies of social control: attempts to change an organizational system tend to complicate it, adding to the very problems they are meant to prevent.

Take culture, for example. Even if we can sort out multidimensional and dynamic organizational cultures so that we can see the role they played in a failure, tinkering with culture can have unintended system consequences that are hard to predict. An aspect of culture may simultaneously enhance safety in some situations and undermine it in others. If we try to change it in response to an accident, are we undermining some part of the system that formerly was functioning in an acceptable manner? For example, Marshall's Lucas-inspired culture sounds very like the "culture of reliability" that high-reliability theorists call for to assure uniform responses and predictability in time of crisis.[83] Military discipline, clear authority structure, formalization, and rigor were its hallmarks. We might even conclude that prior to the tragedy, this culture was working well: problems were discovered, launches were delayed while fixes were implemented, and disasters were averted. However, the *Challenger* teleconference is stunning testimony to what Sagan calls "the dark side of discipline."[84] The previously shared values about rule following, authority relations, and technical rigor that participants automatically invoked on the eve of the launch did not work in the best interest of safety.

Consider the uneasy tension between bureaucratic accountability, political accountability, and the original technical culture at NASA. An obvious change dictated by the *Challenger* launch decision is to use rules to create a more democratic culture: empower people closest to the technology to speak, to collectively consult, to cast the deciding votes. The importance of democratic processes to technical decision making notwithstanding, we are left to wonder about the possible undetectable safety ramifications of change. Normal-accident theorists point out the inevitable strain between centralization and decentralization in organizations that produce and use hazardous technologies. To what extent can democracy thrive in one part of NASA's sprawling bureaucracy, which requires centralization over all to construct and launch the shuttle? While new rules might regulate effectively to the advantage of lower participants in formal launch decision making, would conflict between democratic and hierarchical authority relations erupt in a condition of unstructured crisis, creating ambiguity and inertia? Recall, also, that standards for solid technical arguments undermined the engineering argument on the eve of the launch. The tacit knowledge that was the essence of engineering as a craft—and thus of risk assessments—was incompatible with NASA demands for explicit knowledge that could be readily conveyed to others. What happens in flight-readiness decisions if the intuitive and subjective are acceptable in risk assessments? What happens if bureaucratic mandates are less rigorous? How can we truly explore the consequences of cultural tinkering in advance of such changes in order to minimize the ironies of social control?

Structural changes in a system can also backfire. Corrective actions that make the system more complex—as many do—also create new ways it can fail.[85] Two steps taken to improve on the sending and receiving of signals after the *Challenger* tragedy come to mind. Both official investigations questioned the meaningfulness of a C 1 designation at NASA when 748 shuttle items bore that designation. So NASA reviewed the status of all C 1 and C 1R items. After reevaluation, the number of C 1 items nearly doubled. The good news was that more components received the extra FRR attention accorded a C 1 item; the bad was that the designation itself became an even more routine signal. In a second corrective, NASA created more redundancy in reporting practices to increase the flow of information up the hierarchy. Sagan's concerns about the relationship between increased redundancy, system complexity, and unintended negative safety consequences are affirmed.[86] Already overburdened with paperwork, engineers now spend more time away from the technology in order to do desk work. And while those at the top may get more information, the increased information is likely to increase structural secrecy rather than knowledge and sensitivity about individual problems. Oral communication in data dumps, meetings, teleconferences, and FRR remain the most important means of exchanging technical information at NASA.

Perhaps the most troubling irony of social control demonstrated by this case is that the rules themselves can have unintended effects on system complexity and, thus, mistake. The number of guidelines—and conformity to them—may increase risk by giving a false sense of security.[87] But in addition, a proliferation of rules regulating an industry, a task, or information exchange may create confusion, defy mastery, or result in some regulations being selectively ignored. Also, large numbers of rules and procedures create monitoring difficulties for safety regulators, increasing the workload and reducing the possibility of detecting problems. Even the characteristics of the rules themselves can undermine their ability to regulate, as we saw both in the history of decision making and on the eve of the launch:

1. Recency: The date of origin may affect the legitimacy of a rule or procedure or knowledge of its existence, undermining its effectiveness.
2. Relevance: The degree to which a rule or procedure is defined as relevant to a particular task or to organizational goals may influence willingness to abide by it.
3. Complexity: A rule or procedure having many interrelated parts may be difficult to interpret, generating mistake out of misunderstanding.
4. Vagueness: When a rule is stated in general or indefinite terms or unclearly expressed, mistake can result.

We also must take seriously the difficulty of learning from small samples.[88] When failure is rare, like the shuttle disaster, we probe to find the cause, to prevent a recurrence. But disasters are produced by complicated combinations of factors that may not congeal in exactly the same way again. Complex organizational systems and complex technical systems are susceptible to many kinds of failures. Attempts to understand, predict, and control notwithstanding, the bottom line is this: organizations can never create rules to cover all organizational and technical conditions. Even rule-burdened NASA had no rules for the unprecedented situation on the eve of the launch. When teleconference participants automatically invoked Level III

FRR procedures, this self-imposed structure still left room for the unplanned, the capricious, the unsystematic. Marshall's best-informed joint expert, Leon Ray, was wrongly believed to be out of town and so was not invited; instead of starting the risk assessment at Thiokol when first informed of the cold weather, Ebeling waited until after everyone had lunch, shortening the preparation time; Brian Russell, one of Thiokol's leading joint experts, was frequently out of the teleconference room faxing the charts because the support staff normally responsible had gone for the day.

Finally, there is the unruly nature of the technology itself. Sociologist Everett C. Hughes, in "Mistakes at Work," tells us that mistakes are indigenous, systematic, normal by-products of the work process.[89] Every occupation has its mistake calculus—a calculus of the probability of making mistakes, which depends on skills, frequency of performance, and the nature of the task itself. Medical work, for example, is very like engineering in large-scale technical systems. It requires risky decisions about a complex system for which failure has human consequences. Marianne Paget, in her book *The Unity of Mistakes,* calls medical work "error-ridden activity" because mistakes are indigenous to the work process. She writes: "My characterization undermines the semantic sense of mistakes as uncommon, aberrant, or culpable acts. In saying this, I do not wish to imply that medical mistakes are never aberrant, culpable, or uncommon. Rather, it is the whole activity that is exceptional, uncommon, and strange because it is error-ridden, inexact, and uncertain and because it is practiced on the human body."[90]

Risk assessments at NASA went on under circumstances that made risk fundamentally incalculable: large, complex organizations regularly and systematically fail because their parts interact in unpredictable ways that defy planning; large-scale technical systems cannot be tested to environmental conditions, because all environmental conditions cannot be foreseen. Prior to the tragedy, NASA's SRB work group members were well aware that their work was error-ridden activity. When corrections to the SRB joints had been completed after the disaster, Marshall's Leon Ray, who believed it was a job well done, told me, "We'll blow another one, but it won't be the Solid Rocket Booster that does it." And as I left an interview with Marshall's SRB Chief Engineer Jim Smith, his closing remark was, "Keep your fingers crossed for us."

The managers and engineers in the SRB work group knew that every flight was risky. They did not put Christa McAuliffe on the shuttle. They worked in a system created by political leaders in Congress and the administration who cut resources and NASA leaders who, in response, accelerated the flight rate, altered the culture, and exploited the symbolic and political potential of the Space Shuttle by taking citizens for rides. High-reliability theorists point out, and correctly, that organizational elites need to make safety *the* priority. However, this presumes that organizations can resist environmental forces that affect culture. This case requires us to remain cynical about this possibility.[91] After the *Challenger* disaster, both official investigations decried the competitive pressures and economic scarcity that had politicized the space agency, asserting that goals and resources must be brought into alignment. Steps were taken to assure that this happened. But at this writing, that supportive political environment has changed. NASA is again experiencing the economic strain that prevailed at the time of the disaster. Few of the people in top NASA administrative positions exposed to the lessons of the *Challenger* tragedy are still there. The new leaders stress safety, but they are fighting for dollars and making budget cuts.[92] History repeats, as economy and production are again priorities. The lingering uncertainty in the social control of risky technology is how to control the institutional forces that generate competition and scarcity and the powerful leaders who, in response, establish goals and allocate resources, using and abusing high-risk technical systems.

Appendix A

Cost/Safety Trade-offs?
Scrapping the Escape Rockets
and the SRB Contract Award
Decision

Three types of past NASA decisions were alleged to be cost/safety trade-offs after the *Challenger* tragedy: (1) budget cuts affecting the operation of the safety program (e.g., reduced budget allocations to officially designated safety regulatory organizations), (2) cost cutting by reduced hardware testing, and (3) cost cutting in technical design decisions. Tracing reductions in the safety program and hardware testing was difficult for me because the implications for mission safety were not clear. Technical design decisions, however, were a different matter: they had extensive paper trails that described the implications for mission safety. I selected the decision to scrap escape rockets and the award of the SRB contract to Morton Thiokol. The decision to eliminate the escape rockets had generated little public controversy at the time it was made, but became controversial after the *Challenger* tragedy. The decision to award the SRB contract to Morton Thiokol, Inc., was controversial at the time it was made and was revived after the tragedy. When the seven astronauts lost their lives, these design decisions appeared in hindsight to be very, very wrong. After examining the factors that went into them at the time these decisions were made, however, I concluded that these were not the cost/safety trade-offs they appeared to be after the tragedy.

Scrapping the escape rockets. NASA decided on solid fuel rocket boosters for the shuttle. Solid fuel is costly, but the rockets that use it have fewer moving parts than those using liquid fuel and so are less expensive to operate. A major safety concern, however, arises from the inability to shut down a solid fuel rocket once it has been ignited.[1] NASA Administrator James C. Fletcher (head of the agency in the 1970s and again after the *Challenger* tragedy) announced that this danger would be compensated for by a system of escape rockets.

These escape rockets would make possible crew escape in emergency situations during the dangerous "first stage of ascent"—the first two minutes of flight during which the SRBs burned to thrust the shuttle into orbit. Escape rockets had been used on all previous NASA spacecraft. But the shuttle's Orbiter was enormous—the size of a DC-9—and the escape rockets required to get the vehicle away from the flaming SRBs would, by necessity, weigh more than the shuttle's payload.[2] Weight was a concern. NASA had already made several design decisions to reduce the weight of the Orbiter in order to increase payload capacity. After Fletcher's announcement, a decision was made to scrap the two escape rockets on the Orbiter in order to reduce weight.

After the disaster, analysts cited this as an example of the powerful priority of commercial success at NASA. Although those who made this observation presented no discussion of safety implications, the nature of the decision—jettisoning the escape rockets to reduce weight to increase payload capacity—allowed readers to infer that economic interests were the priority and safety was being traded away.[3] But a close look at the factors that influenced this decision indicates that such an interpretation is incorrect.[4] NASA's elimination of the escape rockets to save the crew was accompanied by a full exploration of all possible first-stage crew escape systems. Nothing could save the crew if a failure occurred while the boosters were thrusting; any method of ejecting the crew from the Orbiter would fail because the astronauts would be pulled into the flame behind the boosters. And if the vehicle blew up on the ground because of SRB joint failure, it would occur at the instant of ignition. There would be no warning. All methods considered were eliminated because of one or more of the following: "limited utility, technical complexity, lack of reaction time and appropriate cues, and/or performance and mission objectives impact."[5] Consequently, NASA's first-stage design rationale was as follows:

> The best approach to maximizing the recovery of the Orbiter and minimizing the risk to the crew was to ensure first stage ascent through the process of conservative design, analysis, and certification. The residual, time critical, identified failures for first stage ascent were of a nature such that there were no known measurement parameters that would forecast, with sufficient action time, a pending catastrophic event requiring abort. Usually, the event itself was the only indication of the failure, such as attach structure failure or SRB burnthrough. In addition, *there were no practical means available that would give complete coverage during first stage ascent with any significant increase to crew survivabil-*

ity. Therefore, the program adopted the position that first stage ascent must be assured.[6] (Emphasis added)

Economic interests certainly were a consideration in the decision to jettison the escape rockets. Budgetary constraints and the inverse relationship between Orbiter weight and payload capacity played a role in this particular design decision, but safety was not the trade. Since the escape rockets could not save the crew, their exclusion from the design lightened Orbiter weight with no safety trade-off. When the *Challenger* tragedy occurred, NASA's first-stage design rationale, stated above, had not changed: the goal was to assure reliable first-stage ascent. The factors influencing that rationale remained the same. "All of the possible first stage crew escape enhancements would require a method to detect the impending failure in time to take successful action, a requirement that currently has no practical solutions for all the possible scenarios."[7] The House Committee on Science and Technology, investigating the provision for crew safety in case of in-flight emergencies, concluded: "Crew escape options were considered when the Shuttle was originally designed and the basic situation has not changed. Many initially attractive options do not significantly reduce risk to the crew either because they may not reduce exposure to the principal hazards or because they add risks of their own. . . . Crew escape during the ascent phase appears infeasible. . . . Launch abort during SRB burn appears impossible but it may be possible to decrease risk to the crew after SRB separation, primarily through mission design."[8]

The SRB contract: segmented versus monolithic boosters. In contrast to the previous example, NASA's awarding of the SRB contract to Morton Thiokol was extremely controversial, not only after the *Challenger* disaster but at the time it was made. Among the proposals that NASA reviewed when deciding on the SRB design was Aerojet's proposal for a seamless monolithic structure. NASA chose Thiokol's segmented booster design. After the *Challenger* disaster, the *Los Angeles Times* stated that NASA's choice of the segmented design was a cost/safety trade-off: "The National Aeronautics and Space Administration decided years ago against buying seamless Solid Rocket Boosters for the Shuttle that would have precluded potential failure associated with joints and seals because the segmented rockets offered by Morton Thiokol, Inc. were cheaper, according to a 1973 NASA document."[9]

The "1973 NASA document" to which the *Los Angeles Times* referred was the report to NASA administrators from the Source

Evaluation Board (SEB), which assessed the four proposals submitted for the SRB contract. The *Times* reported correctly. The SEB report acknowledged that "the monolithic case design proposed by Aerojet precluded the potential failure modes associated with joints and seals and, thus, contributed to a highly reliable design."[10]

Soon after the *Times* article appeared, Richard F. Cottrell, retired president of Aerojet, contested NASA's 1973 choice of a segmented SRB in a memorandum to the Presidential Commission.[11] Cottrell charged that NASA's SRB contract decision sacrificed safety for economy: "The Fletcher-NASA management organization had its mind made up to use segmented rockets. Their determination was based on the spurious conclusion that the complicated segmented devices were a vehicle for competition (if not for Space), which automatically would result in lower cost."[12]

The 1973 Aerojet proposal had warned NASA of the hazards of a segmented design, stating that the rubber seals necessary between the joints of a segmented design and the frequent handling necessary to assemble the segments jeopardized safety.[13] In a detailed comparison of monolithic versus segmented SRM designs, Aerojet's proposal concluded, "The relative simplicity of a conventional monolithic SRM results in significantly fewer failure modes as compared with a segmented SRM."[14] Arguing for the superiority of Aerojet's proposed monolithic structure, Cottrell's 1986 memorandum stated that NASA had knowingly traded cost savings for safety.

Both the *Times* article and Cottrell's memorandum argued that a seamless rocket booster—one having no joints—would have precluded potential failure associated with the joints of segmented rockets. Who could find fault with this statement? If *Challenger* had no joints, the joints would not have failed. But the truism becomes "more true" in retrospect; because the flawed joint was the technical cause of the accident, the information that a monolithic structure was rejected because a segmented rocket was cheaper looked like "amoral calculation" writ large. I found the following, however.

Four firms bid for the contract: Aerojet Solid Propulsion Co., Lockheed Propulsion Company, Morton Thiokol, and United Technologies Center (UTC). Three of the four firms submitted designs for segmented rocket motors. As customary, NASA set up a SEB to examine the proposals and report its findings to NASA's then-Administrator, James C. Fletcher. The SEB evaluated all four proposals on the basis of "mission suitability," which included (1) design, development, and verification, (2) manufacturing, refurbishment, and product support, and (3) management.[15]

Out of a possible 1,000 points, the SEB scored the four firms bid-ding as shown in table A1.[16] Note that according to column (1), Lock-heed's segmented motor design was ranked best of the four bidders; Thiokol's design was ranked lowest. However, the text of the SEB report stated that the technical problems with Thiokol's design were minor and readily correctable. Also, Thiokol's cost proposal was about $122 million lower than Lockheed's cost, which was evaluat-ed second lowest.[17] The SEB report concluded, "The Lockheed Mis-sion Suitability advantage over Thiokol and UTC was minimal; the Thiokol cost advantage was relatively significant in each area of con-sideration; the Aerojet Mission Suitability score was significantly below the scores of the three higher ranked firms."[18]

Fletcher, who as Source Selection Officer was responsible for the decision, chose Thiokol. His rationale, stated publicly, emphasized cost: "We reviewed the Mission Suitability factors in the light of our judgment that cost favored Thiokol. We concluded that the main criticisms of the Thiokol proposal in the Mission Suitability evalua-tion were technical in nature, were readily correctable, and the cost to correct did not negate the sizeable Thiokol cost advantage. Accordingly, we selected Thiokol for final negotiations."[19]

What neither the *Times* article nor Cottrell's memorandum to the Presidential Commission mentioned was that the SEB's report stated that Aerojet's monolithic design was ranked low because it was inad-equate in several ways: "The strength of the case was found inade-quate for the prelaunch bending moment loads and was not designed with an adequate safety factor for water impact loads. An inert weight adjustment would be required to correct for the load inadequacy.

TABLE A1. CONTRACTOR PROPOSALS FOR SOLID ROCKET MOTOR: SOURCE EVALUATION BOARD'S SUMMARY OF MISSION SUITABILITY

| Manufacturer | Score | Overall Adjective Rating | Mission Suitability (rank) | | |
			Design, Development, and Verification (1)	Manufacturing, Refurbishment, and Product Support (2)	Management (3)
Lockheed	714	Very good	1	1	4
Thiokol	710	Very good	4	2	1
UTC	710	Very good	2	3	2
Aerojet	655	Good	3	4	3

Aerojet proposed ablative materials in the nozzle which were not developed or characterized. The use of these materials offered a potential saving in program cost, but with attendant technical and program development risk. The proposed thickness of the ablative materials was insufficient to meet required safety factors. The propellant formulation proposed by Aerojet was selected on the basis of data from 30-gallon mixes, which was a high technical risk."[20]

This paragraph is a good example of the sort of technical material that is part of this decision history. The precise meaning may escape the reader, but the point is clear: after the disaster, a monolithic design may have been superior in principle, but in practice, the one Aerojet submitted was found wanting. Challenging this SEB assessment in his memorandum to the Commission, former Aerojet President Cottrell stated, "It degrades the Aerojet 'Mission Suitability' on the basis of alleged design deficiencies totally unsupported by the facts in the proposal."[21]

I am not able to judge these conflicting technical arguments. However, I found some support for the SEB's evaluation of Thiokol's design from three sources: investigations conducted by the Government Accounting Office (GAO), Marshall engineer W. Leon Ray, member of the SEB, and Thiokol engineer Roger M. Boisjoly. In 1973, when NASA announced the contract award, Lockheed filed an official protest of NASA's decision with the GAO. Lockheed maintained that the SEB report on which the NASA decision was based had miscalculated costs, biasing the decision against Lockheed and in favor of Thiokol.[22]

The GAO investigated the bid protest. The 98-page investigation report included a reassessment of *both* the technical and cost aspects of the Lockheed and Thiokol designs. The GAO report (1) confirmed that the choice of segmented rocket boosters would not sacrifice performance quality and (2) agreed with Fletcher's statement that shortcomings in Thiokol's design were not major and were readily correctable.[23] The GAO review noted that NASA's reliance on cost did not represent "an unreasonable exercise of discretion."[24] However, the report stated that a "more reasonable evaluation of certain fuel costs would have reduced the difference between Lockheed's and Thiokol's cost proposals by about $68 million, or from approximately $122 million to about $56 million."[25]

With reevaluation, Thiokol's cost proposal was still about $56 million less than Lockheed's. As a result of its investigation, the GAO recommended that Fletcher, as Source Selection Officer, determine whether the selection decision should be reconsidered, in light of the finding that the difference between Lockheed's and Thiokol's cost proposal was significantly less than what had been reported by the SEB.[26] Note that the GAO recommended reconsideration because costs had been miscalculated, *not* because mission safety was jeopardized by an inferior Thiokol design. On June 24, 1974, two days after the GAO's announcement of the bid protest decision, NASA's Fletcher announced a decision to proceed with Thiokol, based on the conclusion that the initial rationale for the selection of Thiokol remained valid.[27] My review of the history of the contract award decision diminishes the power of Aerojet's argument that NASA's choice of a segmented SRB was a cost/safety trade-off. But in the process of tracking it down, I uncovered other information that indicated possible misconduct of another sort.

After the *Challenger* tragedy, several sources argued that Thiokol was chosen over top-ranked Lockheed because Fletcher had long-standing connections to Utah, where Morton Thiokol's Wasatch operations were located.[28] Moreover, Bell and Esch reported that two former Thiokol employees held positions with the SEB.[29] Because of these allegations, Senator Albert Gore, Jr. (D-Tenn.), requested that the GAO review again NASA's 1973 decision to grant Thiokol the contract for the SRBs. This time, the focus of investigation was (1) did NASA's response to GAO's bid protest decision comply with procurement regulations (i.e., did NASA reevaluate costs as requested), and 2) did NASA staff involved in the contracting decision violate Executive Order 11222, which provides standards for federal employees to follow in avoiding conflicts of interest.[30]

On the first question, the GAO found that NASA properly responded to the bid protest decision that it reconsider the Thiokol selection.[31] On the second, the GAO found no evidence to indicate violations of conflict-of-interest statutes or regulations by NASA Administrator Fletcher that would bear upon the Thiokol decision.[32] Paradoxically, the GAO report revealed that Fletcher had a vested interest in an Aerojet pension fund at the time of his participation in the SRM contractor selection process.[33] Also, the GAO verified that two former Thiokol employees served on the SEB, as Bell and Esch reported, but found that one had no financial ties with Thiokol and the other had a Thiokol pension plan that was independent of Thiokol's earnings. Thus, neither were disqualified from participating in the selection process.

Conflict-of-interest investigations, like the GAO's, try to find identifiable formal financial ties. Informal connections, like friendships, political loyalties, and personal allegiances are harder to ferret out and harder to prove if charges are brought. A March 1986 article in the *Akron Beacon Journal* quoted a source who stated, "There's no question that one of the main reasons Thiokol got the award was because Senator Moss (D-Utah) was chairman of the Aeronautical and Space Sciences Committee and Jim Fletcher was the administrator of NASA."[34] Fletcher denied that politics played any role in the selection.[35] Perhaps the contract decision involved misconduct of another sort: not a cost/safety trade-off based on assessing technology and the cost to produce it, but a political alliance with its own costs and benefits to be calculated. The political considerations of the SRB contract decision remain elusive. Nonetheless, I learned that in NASA's choice of Thiokol (1) cost played a major role in the decision and (2) both the first GAO report and the SEB report stated that segmented SRBs would not sacrifice performance quality, and that the shortcomings in Thiokol's design were readily correctable.[36] Even if political considerations did affect the decision to give the contract to Thiokol, I had not found any evidence that, when the decision was made, it was a cost/safety trade-off.

This conclusion was verified by both Marshall's Leon Ray and Thiokol's Roger Boisjoly. The top NASA administrators were not the ones making the engineering evaluation of the contractor designs. Engineering opinion about the relative merits of the designs of Lockheed, Aerojet, and Thiokol should carry the day in resolving this issue. About Aerojet's monolithic design, Ray, a member of the SEB, said that when the contract decisions were made, "everyone" was using segmented boosters—an important aspect of the social context of decision. Boisjoly agreed, noting that a monolithic structure like Aerojet's design might have reduced risk of a disaster due to SRB joint failure, but a monolithic design had other risks. Both types of designs had to be preloaded with solid fuel propellant and shipped to the Cape. Boisjoly pointed out that shipment was accompanied by "great risk of explosion": the larger the segments, the more propellant, the greater the risk. He said: "A monolithic structure contains 1.1 million pounds of propellant. It's bad enough shipping it in segments, which contain less."[37]

As for the choice between the segmented designs of Lockheed and Thiokol, Marshall's Ray said: "We (SEB members) all went into it with the very best knowledge that we could gather, and we at that time thought it (Thiokol's) was an acceptable design. We thought from a technical standpoint that probably Lockheed had a little better design, but that's yet to be proven. It's strictly speculation."[38]

Since the technical cause of the *Challenger* accident was a flawed joint in the segmented booster design, Aerojet's arguments about the dangers of a segmented SRB took on a relevance after the tragedy that they did not have at the time this decision was made. The social context in which Aerojet's original proposal was judged was one where segmented SRBs routinely were used for other purposes. In fact, Thiokol engineers had based their design for the shuttle boosters on the Air Force's Titan III rocket, a segmented booster from United Technologies viewed by the industry as one of the most reliable ever produced.[39] The social context of Cottrell's 1986 memorandum was one punctuated by an entirely different experience with segmented SRBs: the destruction of the *Challenger.*

Investigating the SRB contract award decision and the scrapping of the escape rockets throws doubt on speculation about a history of cost/safety trade-offs at NASA. Both examples illustrate the seductiveness of the retrospective fallacy. Most certainly, I did not have access to all the information, and certainly both decisions could have been probed more deeply than I did. But what I found did not affirm either decision as an example of organizational misconduct and amoral calculation on the part of NASA senior administrators.

Appendix B

Supporting Charts and Documents

FINAL TELECONFERENCE PARTICIPANTS

NASA Marshall Space Flight Center	Morton Thiokol Wasatch Division
1. George B. Hardy, Deputy Director, Science and Engineering, MSFC	1. Jerald Mason, Senior Vice President, Wasatch Operations, MTI
2. Judson A. Lovingood, Deputy Manager, Shuttle Projects Office, MSFC	2. Calvin Wiggins, Vice President and General Manager, Space Division, MTI
3. Leslie F. Adams, Deputy Manager, SRB Project, MSFC	3. Joe C. Kilminster, Vice President, Space Booster Programs, MTI
4. Lawrence O. Wear, Manager, SRM Project Office, MSFC	4. Robert K. Lund, Vice President, Engineering, MTI
5. John Q. Miller, Technical Assistant, SRM Project, MSFC	5. Larry H. Sayer, Director, Engineering and Design, MTI
6. J. Wayne Littles, Associate Director for Engineering, MSFC	6. William Macbeth, Manager, Case Projects, Space Booster Project Engineering, Wasatch Division, MTI
7. Robert J. Schwinghamer, Director, Material and Processes Laboratory, MSFC	7. Donald M. Ketner, Supervisor, Gas Dynamics Section and Head Seal Task Force, MTI
8. Wilbur A. Riehl, Chief, Nonmetallic Materials Division, MSFC	8. Roger Boisjoly, Member, Seal Task Force, MTI
9. John P. McCarty, Deputy Director, Structures and Propulsion Laboratory, MSFC	9. Arnold R. Thompson, Supervisor, Rocket Motor Cases, MTI
10. Ben Powers, Engineering Structures and Propulsion Laboratory, MSFC	10. Jack R. Kapp, Manager, Applied Mechanics Department, MTI
11. James Smith, Chief Engineer, SRB Program, MSFC	11. Jerry Burn, Associate Engineer, Applied Mechanics, MTI
12. Keith E. Coates, Chief Engineer, Special Projects Office, MSFC	12. Joel Maw, Associate Scientist, Heat Transfer Section, MTI
13. John Schell, Retired Engineer, Materials Laboratory, MSFC	13. Brian Russell, Manager, Special Projects, SRM Project, MTI
	14. Robert Ebeling, Manager, Ignition System and Final Assembly, SRB Project, MTI

Present at KSC

14. Cecil Houston, MSFC Resident Manager, at KSC
15. Stanley R. Reinartz, Manager, Shuttle Projects Office, MSFC
16. Lawrence B. Mulloy, Manager, SRB Project, MSFC

Present at MSFC

15. Boyd C. Brinton, Manager, Space Booster Project, MTI
16. Kyle Speas, Ballistics Engineer, MTI

Present at KSC

17. Allan J. McDonald, Director, SRM Project, MTI
18. Jack Buchanan, Manager, KSC Operations, MTI

Fig. B1

RELEVANT ORGANIZATION CHARTS: NASA AND MORTON THIOKOL

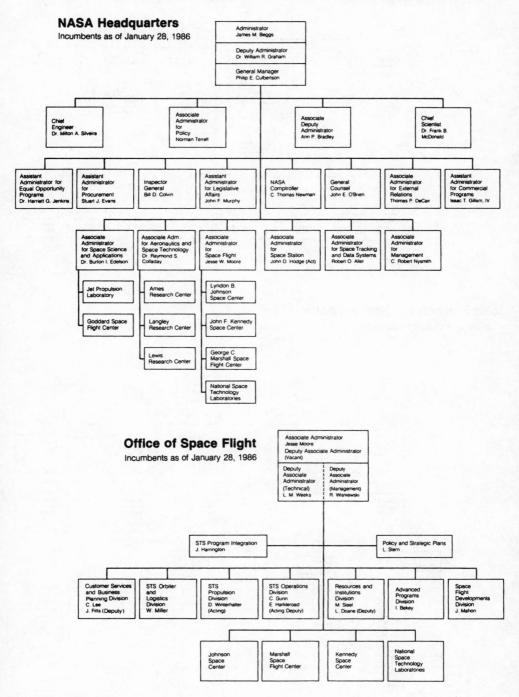

Fig. B2.1

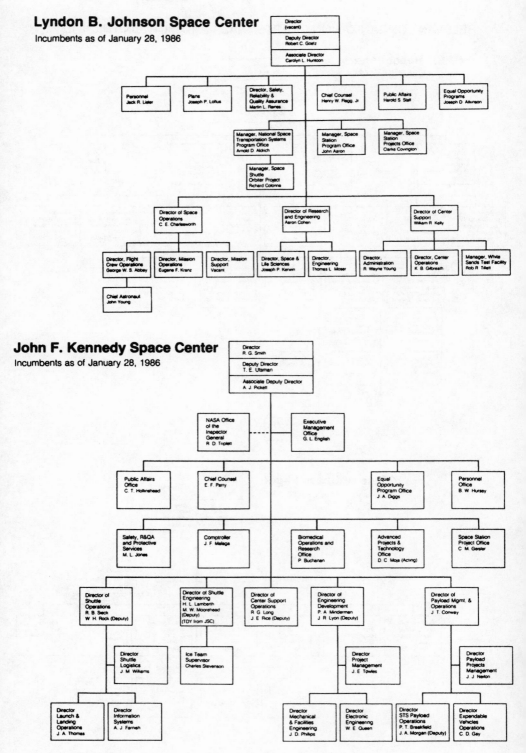

Lyndon B. Johnson Space Center

Incumbents as of January 28, 1986

John F. Kennedy Space Center

Incumbents as of January 28, 1986

Fig. B2.2

George C. Marshall Space Flight Center Organization Charts

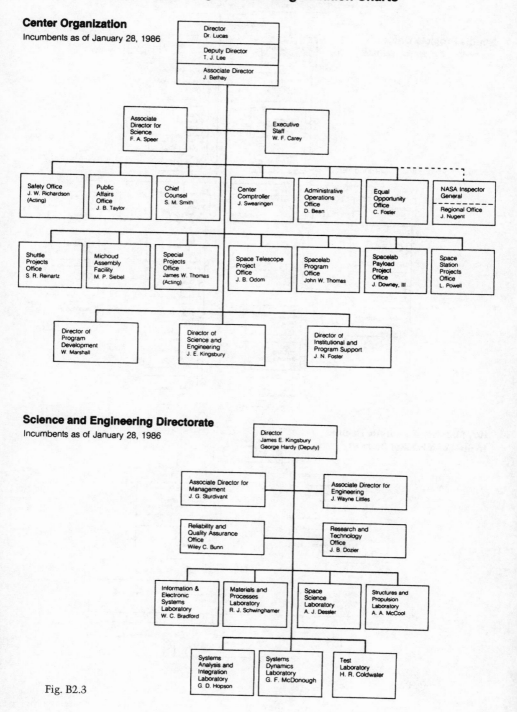

Fig. B2.3

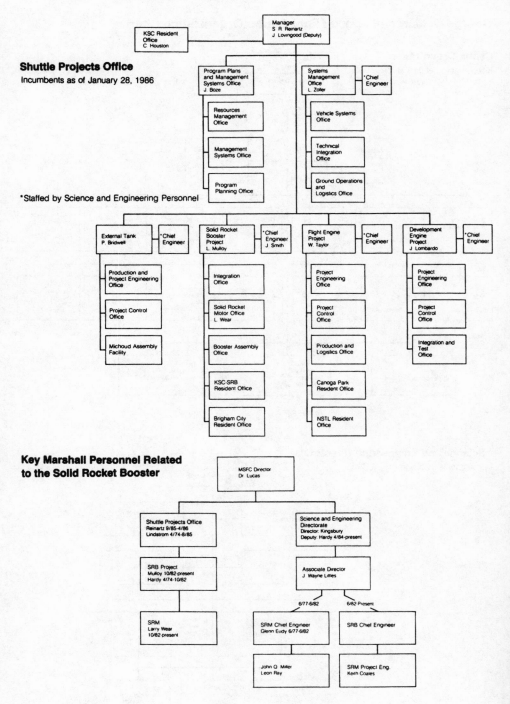

Shuttle Projects Office

Incumbents as of January 28, 1986

*Staffed by Science and Engineering Personnel

**Key Marshall Personnel Related
to the Solid Rocket Booster**

Fig. B2.4

**Morton Thiokol 27 Jan 1986
Meeting Participants**

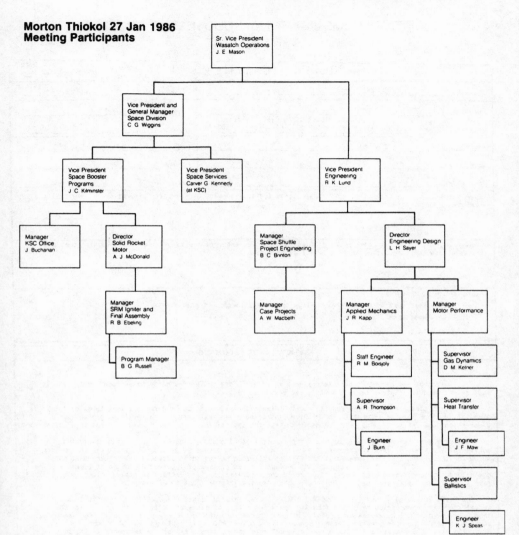

Fig. B2.5

CRITICAL ITEMS LIST ENTRY, SOLID ROCKET BOOSTER, CRITICALITY 1R

127-

SRB CRITICAL ITEMS LIST

of SE 019-127-2H   Sheet 1 of 2

Subsystem: SOLID ROCKET MOTOR	Criticality Category 1R Reaction Time: Immediate to Sec.
Item Code. 10-01-01	Page: A-35
Case, P/N 1U50129, 1U50131, 1U50130, 1U50169, 1U50185,	
Item Name: 1U51473 (Joint Assys, Factory P/N 1U51768, Field: 1U50747)	Revision:
No. Required: 1 (11 segments)	Date: November 24, 1980
FMEA Page No.: A-4 of MSFC-RPT-724	Analyst: Hoskins
Critical Phases: Boost	Approved:

Failure Mode & Causes: Leakage at case assembly joints due to redundant O-ring seal failures or primary seal and leak check port O-ring failure.

Failure Effect Summary: Actual Loss - Loss of mission, vehicle, and crew due to metal erosion, burnthrough, and probable case burst resulting in fire and deflagration.

Redundancy Screens & Measurements

1) Fail - Leak test does not verify integrity of leak check port seal.

2) Fail - Not tested.

3) Pass - No known credible causes.

RATIONALE FOR RETENTION

A. DESIGN

- Each O-ring of the redundant pair is designed to effect a seal. The design is based upon similar single seal joints used in previous large diameter, segmented motor cases.

A small MS port leading to the annular cavity between the redundant seals permits a leak check of the seals immediately after joining segments. The MS plug, installed after leak test, has a retaining groove and compression face for its O-ring seal. A means to test the seal of the installed MS plug has not been established.

The surface finish requirement for the O-ring grooves is 63 and the finish of the O-ring contacting portion of the tang, which slides across the O-ring during joint assembly, is 32. The joint design provides an OD for the O-ring installation, which facilitates retention during joint assembly. The entry portion of the tang provides 0.125-inch standoff from the O-rings contact portion of the tang during joint assembly. The design drawing specifies O-ring lubricant prior to the installation. The factory assembled joints (dwg. 1U51768) have an additional seal provided by the subsequently applied internal case insulation.

The field assembled joints (Dwg. 1U50747) and factory assembled joints (Dwg. 1U51768) benefit from the increased O-ring compression resulting from the centering effect of shims of .032-.036 inches between the tang O.D. and clevis I.D. of the case joint. However, redundancy of the secondary field joint seal cannot be verified after motor case pressure reaches approximately 40% of MEOP. It is known that joint rotation occurring at this pressure level with a resulting enlarged extrusion gap causes the secondary O-ring to lose compression as a seal. It is not known if the secondary O-ring would successfully re-seal if the primary O-ring should fail after motor case pressure reaches or exceeds 40% MEOP.

The O-ring for the case joint and test port are mold formed of high temperature, compression set resistant, fluorocarbon elastomer. The design permits five scarf joints for the case joint seal. The O-ring joint strength must equal or exceed 40% of the parent material strength.

B. TESTING

A full scale clevis joint test verified the structural strength of the case and pins (TWR-10547). A hydroburst life cycle test (TWR-11664) demonstrated the primary seal's ability to withstand four times the flight requirement of one pressurization cycle and the secondary seal's ability to continue to seal under repeated cycling (54 cycles) with the primary seal failed. The joint seals withstood ultimate pressure of 1483 psi during the burst tests, yielding a safety factor of 1.58. The Structural Test Article (STA-1) verified the seals capability under flight loads and further verified the redundancy of the secondary seal.

The joint seals have performed successfully in four developmental and three qualification motor static firings.

Fig. B3.1

CRITICAL ITEMS LIST ENTRY, SOLID ROCKET BOOSTER,
CRITICALITY 1R *(continued)*

SRB CRITICAL ITEMS LIST

Sheet 2 of 2

Subsystem: SOLID ROCKET MOTOR	Criticality Category 1R Reaction Time Immediate to Sec.
Item Code: 10-01-01	Page: A-36
Case, P/N 1U50129, 1U50131, 1U50130, 1U50169, 1U50185, Item Name: 1U51473 (Joint Assys, Factory, P/N 1U51768, Field: 1U50747)	Revision:

RATIONALE FOR RETENTION (CONT'D)

A lightweight case joint verification test (TWR-12690) has demonstrated the secondary seal performance with a purposely pre-failed primary O-ring and demonstrated three pressure cycles on the primary seal with one cycle to 1.40 times maximum expected operating pressure.

C. INSPECTION

The tang -A- dia. and clevis -C- dia. are measured and recorded. These diameters control the radial spacing between tang and clevis. The depth, width and surface finish of the O-ring grooves are verified. The segment finish of the tang is also verified. The O-ring seal mating surfaces of the forward and aft segments are verified for flatness and surface finish. The following characteristics are inspected on each O-ring to assure conformance to the standards.

- o Surface voids and inclusions
- o Mold flashing
- o Scarf joint mismatch or separation
- o Cross section
- o Circumference

Each assembled joint seal is tested per STW7-2747 via pressurizing the annular cavity between seals to 50 ± 5 psi and monitoring for 10 minutes. A seal seating pressure of 220 psi, with return to 0 psig, may be used prior to the test. A pressure decay of 1 psig or greater is not acceptable. Following seal verification by QC, the leak test port plug is installed with QC verifying installation and torquing.

D. FAILURE HISTORY

No known record of failure due to case joint seal leakage on segmented 156" or Titan IIIC motors.

No failures in the four development and three qualification SRM motor test firings.

Fig. B3.2

CRITICAL ITEMS LIST ENTRY, SOLID ROCKET
BOOSTER, CRITICALITY 1

SRB CRITICAL ITEMS LIST

Sheet: 1 of 2

Subsystem SOLID ROCKET BOOSTER

Criticality Category 1   Reaction Time Immediate to Sec.

Item Code 10-01-01

Page A-6A

Item Name *Case, P/N (See Retention Rationale)
(Joint Assys, Factory P/N 1U50147 Field: 1U50747

Revision:

No. Required 1 (11 segments, 3 field joints, 7 plant joints)

Date: December 17, 1982

FMEA Page No. A-4 of MSFC-RPT-724

Analyst: Garber

Critical Phases: Boost

Approved: [signature]

Failure Mode & Causes: Leakage at case assembly joints due to redundant O-ring seal failures or primary seal and leak check port O-ring failure.

NOTE: Leakage of the primary O-ring seal is classified as a single failure point due to possibility of loss of sealing at the secondary O-ring because of joint rotation after motor pressurization.

Failure Effect Summary: Actual Loss - Loss of mission, vehicle, and crew due to metal erosion, burnthrough, and probable case burst resulting in fire and deflagration.

RATIONALE FOR RETENTION

Case, P/N 1U50129, 1U50131, 1U50130, 1U50185, 1U50147, 1U50715, 1U50716, 1U50717
1U51473

A. DESIGN

The SRM case joint design is common in the lightweight and regular weight cases having identical dimensions. The SRM joint uses centering clips which are installed in the gap between the tang O.D. and the outside clevis leg to compensate for the loss of concentricity due to gathering and to reduce the total clevis gap which has been provided for ease of assembly. On the shuttle SRM, the secondary O-ring was designed to provide redundancy and to permit a leak check, ensuring proper installation of the O-rings. Full redundancy exists at the moment of initial pressurization. However, test data shows that a phenomenon called joint rotation occurs as the pressure rises, opening up the O-ring extrusion gap and permitting the energized O-ring to protrude into the gap. This condition has been shown by test to be well within that required for safe primary O-ring sealing. This gap may, however, in some cases, increase sufficiently to cause the un-energized secondary O-ring seal to lose compression, raising question as to its ability to energize and seal if called upon to do so by primary seal failure. Since, under this latter condition only the single O-ring is sealing, a rationale for retention is provided for the simplex mode where only one O-ring is acting.

The surface finish requirement for the O-ring grooves is 63 and the finish of the O-ring contacting portion of the tang, which slides across the O-ring during joint assembly, is 32. The joint design provides an OD for the O-ring installation, which facilitates retention during joint assembly. The tang has a large shallow angle chamfer on the tip to prevent the cutting of the O-ring at assembly. The design drawing specifies application of O-ring lubricant prior to the installation. The factory assembled joints have NBR rubber material vulcanized across the internal joint faying surfaces as a part of the case internal insulation subsystem.

A small MS port leading to the annular cavity between the redundant seals permits a leak check of the seals immediately after joining segments. The MS plus, installed after leak test, has a retaining groove and compression face for its O-ring seal. A means to test the seal of the installed MS plug has not been established.

The O-rings for the case joints are mold formed and ground to close tolerance and the O-rings for the test port are mold formed to net dimensions. Both O-rings are made of high temperature, low compression set fluorocarbon elastomer. The design permits five scarf joints for the case joint seal rings. The O-ring joint strength must equal or exceed 40% of the parent material strength.

B. TESTING

To date, eight static firings and five flights have resulted in 180 (54 field and 126 factory) joints tested with no evidence of leakage. The Titan III program using a similar joint concept has tested a total of 1076 joints successfully.

Fig. B4.1

CRITICAL ITEMS LIST ENTRY, SOLID ROCKET
BOOSTER, CRITICALITY 1 *(Continued)*

SRB CRITICAL ITEMS LIST		Sheet 2 of 2

		Immediate
Subsystem: SOLID ROCKET BOOSTER	Criticality Category __1__	Reaction Time to Sec.
Item Code: 10-01-01	Page:	A-68
Item Name: *Case, P/N (See Retention Rationale) (Joint Assys, Factory P/N 1U50747 Field: 1U50747	Revision:	

RATIONALE FOR RETENTION (CONT'D)

A laboratory test program demonstrated the ability of the O-ring to operate successfully when extruded into gaps well over those encountered in this O-ring application. Uniform gaps of 1/8-inch and over (TWR-13486) successfully withstood pressures of 1600 psi. The Hydroburst Program (TWR-11664) and the Structural Test Program (STA-1) for the standard weight case (TWR-12051) and the Lightweight Case Joint Certification Test (TWR-12829) all have shown that the O-ring can withstand a minimum of four pressurizations before damage to the ring can permit any leakage.

Further demonstration of the capability of joint sealing is found in the hydro-proof testing of new and refurbished case segments. Over 540 joints have been exposed to liquid pressurizations at levels exceeding motor MEOP with no leakage experienced past the primary O-ring. The only occasions where leakage was experienced was during refurbishment of STS-1 where two stiffener segments were severely damaged during cavity collapse at water impact.

A more detailed description of SRM joint testing history is contained in TWR-13520, Revision A.

C. INSPECTION

The tang -A- diameter and clevis -C- diameter are measured and recorded. The depth, width and surface finish of the O-rings grooves are verified. The surface finish of the tang is also verified. Characteristics are inspected on each O-ring to assure conformance to the standards to include:

   o Surface conditions
   o Mold flashing
   o Scarf joint mismatch or separation
   o Cross section
   o Circumference
   o Durometer

Each assembled joint seal is tested per STW7-2747 via pressurizing the annular cavity between seals to 50 ± 5 psi and monitoring for 10 minutes. A pressure decay of 1 psig or greater is not acceptable. Following seal verification by QC, the leak test port plug is installed with QC verifying installation and torquing.

D. FAILURE HISTORY

No failures have been experienced in the static firing of three qualification motors, five development motors and ten flight motors.

Fig. B4.2

POSTACCIDENT TREND ANALYSIS
Anomalies, Leak Check Pressure, and Joint Temperature

O-Ring Anomalies Compared with Joint Temperature and Leak Check Pressure

Flight or Motor	Date	(Solid Rocket Booster)	Joint/ O-Ring	Pressure (in psi)		Erosion	Blow-by	Joint Temp °F
				Field	Nozzle			
DM-1	07/18/77	–	–	NA	NA	–	–	84
DM-2	01/18/78	–	–	NA	NA	–	–	49
DM-3	10/19/78	–	–	NA	NA	–	–	61
DM-4	02/17/79	–	–	NA	NA	–	–	40
QM-1	07/13/79	–	–	NA	NA	–	–	83
QM-2	09/27/79	–	–	NA	NA	–	–	67
QM-3	02/13/80	–	–	NA	NA	–	–	45
STS-1	04/12/81	–	–	50	50	–	–	66
STS-2	11/12/81	(Right)	Aft Field/Primary	50	50	X	–	70
STS-3	03/22/82	–	–	50	50	–	–	69
STS-4	06/27/82	unknown: hardware lost at sea		50	50	NA	NA	80
DM-5	10/21/82	–	–	NA	NA	–	–	58
STS-5	11/11/82	–	–	50	50	–	–	68
QM-4	03/21/83	–	Nozzle/ Primary	NA	NA	X	–	60
STS-6	04/04/83	(Right)	Nozzle/ Primary	50	50	(¹)	–	67
		(Left)	Nozzle/ Primary	50	50	(¹)	–	67
STS-7	06/18/83	–	–	50	50	–	–	72
STS-8	08/30/83	–	–	100	50	–	–	73
STS-9	11/28/83	–	–	100²	100	–	–	70
STS 41-B	02/03/84	(Right)	Nozzle/ Primary	200	100	X	–	57
		(Left)	Forward Field/ Primary	200	100	X	–	57
STS 41-C	04/06/84	(Right)	Nozzle/ Primary	200	100	X	–	63

Dash (–) denotes no anomaly.
NA denotes not applicable.
NOTE: A list of the sequence of launches (1-25), identified by STS mission designation, is provided on pages 4 thru 6.

¹ On STS-6, both nozzles had a hot gas path detected in the putty with an indication of heat on the primary O-ring.

² On STS-9, one of the right Solid Rocket Booster field joints was pressurized at 200 psi after a destack.

Fig. B5.1

POSTACCIDENT TREND ANALYSIS
Anomalies, Leak Check Pressure, and Joint Temperature
(Continued)

Flight or Motor	Date	(Solid Rocket Booster)	Joint/ O-Ring	Pressure (in psi) Field	Nozzle	Erosion	Blow-by	Joint Temp °F
STS 41-C (cont'd)		(Left)	Aft Field/ Primary	200	100	(3)	—	63
		(Right)	Igniter/ Primary	NA	NA	—	X	63
STS 41-D	08/30/84	(Right)	Forward Field/Primary	200	100	X	—	70
		(Left)	Nozzle/ Primary	200	100	X	X	70
		(Right)	Igniter/ Primary	NA	NA	—	X	70
STS 41-G	10/05/84	—	—	200	100	—	—	78
DM-6	10/25/84	—	Inner Gasket/ Primary	NA	NA	X	X	52
STS 51-A	11/08/84	—	—	200	100	—	—	67
STS 51-C	01/24/85	(Right)	Center Field/Primary	200	100	X	X	53
		(Right)	Center Field/ Secondary	200	100	(4)	—	53
		(Right)	Nozzle/ Primary	200	100	—	X	53
		(Left)	Forward Field/Primary	200	100	X	X	53
		(Left)	Nozzle/ Primary	200	100	—	X	53
STS 51-D	04/12/85	(Right)	Nozzle/ Primary	200	200	X	—	67
		(Right)	Igniter/ Primary	NA	NA	—	X	67
		(Left)	Nozzle/ Primary	200	200	X	—	67
		(Left)	Igniter/ Primary	NA	NA	—	X	67
STS 51-B	04/29/85	(Right)	Nozzle/ Primary	200	100	X	—	75
		(Left)	Nozzle/ Primary	200	100	X	X	75

Dash (−) denotes no anomaly.
NA denotes not applicable.
NOTE: A list of the sequence of launches (1-25), identified by STS mission designation, is provided on pages 4 thru 6.

3 On STS 41-C, left aft field had a hot gas path detected in the putty with an indication of heat on the primary O-ring.

4 On a center field joint of STS 51-C, soot was blown by the primary and there was a heat effect on the secondary.

Fig. B5.2

POSTACCIDENT TREND ANALYSIS
Anomalies, Leak Check Pressure, and Joint Temperature
(Continued)

Flight or Motor	Date	(Solid Rocket Booster)	Joint/ O-Ring	Pressure (in psi) Field	Pressure (in psi) Nozzle	Erosion	Blow-by	Joint Temp °F
STS 51-B (cont'd)		(Left)	Nozzle/ Secondary	200	100	X	-	75
DM-7	05/09/85		Nozzle/ Primary	NA	NA	X	-	61
STS 51-G	06/17/85	(Right)	Nozzle/ Primary	200	200	X[5]	X	70
		(Left)	Nozzle/ Primary	200	200	X	X	70
		(Left)	Igniter/ Primary	NA	NA	-	X	70
STS 51-F	07/29/85	(Right)	Nozzle/ Primary	200	200	([6])	-	81
STS 51-I	08/27/85	(Left)	Nozzle/ Primary	200	200	X[7]	-	76
STS 51-J	10/03/85		-	200	200	-	-	79
STS 61-A	10/30/85	(Right)	Nozzle/ Primary	200	200	X	-	75
		(Left)	Aft Field/Primary	200	200	-	X	75
		(Left)	Center Field/ Primary	200	200	-	X	75
STS 61-B	11/26/85	(Right)	Nozzle/ Primary	200	200	X	-	76
		(Left)	Nozzle/ Primary	200	200	X	X	76
STS 61-C	01/12/86	(Right)	Nozzle/ Primary	200	200	X	-	58
		(Left)	Aft Field/Primary	200	200	X	-	58
		(Left)	Nozzle/ Primary	200	200	-	X	58
STS 51-L	01/28/86			200	200			31

Dash (−) denotes no anomaly.
NA denotes not applicable.
NOTE: A list of the sequence of launches (1-25), identified by STS mission designation, is provided on pages 4 thru 6.

[5] On STS 51-G, right nozzle had erosion in two places on the primary O-ring.

[6] On STS 51-F, right nozzle had hot gas path detected in putty with an indication of heat on the primary O-ring.

[7] On STS 51-I, left nozzle had erosion in two places on the primary O-ring.

Fig. B5.3

ENGINEERING MEMORANDA WRITTEN AFTER
O-RING CONCERN ESCALATES

NASA

National Aeronautics and
Space Administration

Washington, D.C.
20546

JUL 17 1985

MPS.

TO: M/Associate Administrator for Space Flight

FROM: MPS/Irv Davids

SUBJECT: Case to Case and Nozzle to Case "O" Ring Seal Erosion
           Problems

As a result of the problems being incurred during flight on both
case to case and nozzle to case "O" ring erosion, Mr. Hamby and I
visited MSFC on July 11, 1985, to discuss this issue with both
project and S&E personnel. Following are some important factors
concerning these problems:

A.   **Nozzle to Case "O" ring erosion**

There have been twelve (12) instances during flight where there
have been some primary "O" ring erosion. In one specific case
there was also erosion of the secondary "O" ring seal. There
were two (2) primary "O" ring seals that were heat affected (no
erosion) and two (2) cases in which soot blew by the primary
seals.

The prime suspect as the cause for the erosion on the primary "O"
ring seals is the type of putty used. It is Thiokol's position
that during assembly, leak check, or ignition, a hole can be
formed through the putty which initiates "O" ring erosion due to
a jetting effect. It is important to note that after STS-10, the
manufacturer of the putty went out of business and a new putty
manufacturer was contracted. The new putty is believed to be
more susceptible to environmental effects such as moisture which
makes the putty more tacky.

There are various options being considered such as removal of
putty, varying the putty configuration to prevent the jetting
effect, use of a putty made by a Canadian Manufacturer which
includes asbestos, and various combination of putty and grease.
Thermal analysis and/or tests are underway to assess these
options.

Thiokol is seriously considering the deletion of putty on the QM-
5 nozzle/case joint since they believe the putty is the prime
cause of the erosion. A decision on this change is planned to be
made this week. I have reservations about doing it, considering
the significance of the QM-5 firing in qualifying the FWC for
flight.

Fig. B6.1

ENGINEERING MEMORANDA WRITTEN AFTER
O-RING CONCERN ESCALATES *(Continued)*

It is important to note that the cause and effect of the putty varies. There are some MSFC personnel who are not convinced that the holes in the putty are the source of the problem but feel that it may be a reverse effect in that the hot gases may be leaking through the seal and causing the hole track in the putty.

Considering the fact that there doesn't appear to be a validated resolution as to the effect of putty, I would certainly question the wisdom in removing it on QM-5.

B.   Case to Case "O" Ring Erosion

There have been five (5) occurrences during flight where there was primary field joint "O" ring erosion. There was one case where the secondary "O" ring was heat affected with no erosion. The erosion with the field joint primary "O" rings is considered by some to be more critical than the nozzle joint due to the fact that during the pressure build up on the primary "O" ring the unpressurized field joint secondary seal unseats due to joint rotation.

The problem with the unseating of the secondary "O" ring during joint rotation has been known for quite some time. In order to eliminate this problem on the FWC field joints a capture feature was designed which prevents the secondary seal from lifting off. During our discussions on this issue with MSFC, an action was assigned for them to identify the timing associated with the unseating of the secondary "O" ring and the seating of the primary "O" ring during rotation. How long it takes the secondary "O" ring to lift off during rotation and when in the pressure cycle it lifts are key factors in the determination of its criticality.

The present consensus is that if the primary "O" ring seats during ignition, and subsequently fails, the unseated secondary "O" ring will not serve its intended purpose as a redundant seal. However, redundancy does exist during the ignition cycle, which is the most critical time.

It is recommended that we arrange for MSFC to provide an overall briefing to you on the SRM "O" rings, including failure history, current status, and options for correcting the problems.

Irving Davids

cc:
M/Mr. Weeks
M/Mr. Hamby
ML/Mr. Harrington
MP/Mr. Winterhalter

Fig. B6.2

ENGINEERING MEMORANDA WRITTEN AFTER
O-RING CONCERN ESCALATES *(Continued)*

MORTON THIOKOL. INC.   COMPANY PRIVATE
Wasatch Division

Interoffice Memo

31 July 1985
2870:FY86:073

TO: R. K. Lund
           Vice President, Engineering

CC: B. C. Brinton, A. J. McDonald, L. H. Sayer, J. R. Kapp

FROM: R. M. Boisjoly
           Applied Mechanics – Ext. 3525

SUBJECT: SRM O-Ring Erosion/Potential Failure Criticality

This letter is written to insure that management is fully aware of the seriousness of the current O-Ring erosion problem in the SRM joints from an engineering standpoint.

The mistakenly accepted position on the joint problem was to fly without fear of failure and to run a series of design evaluations which would ultimately lead to a solution or at least a significant reduction of the erosion problem. This position is now drastically changed as a result of the SRM 16A nozzle joint erosion which eroded a secondary O-Ring with the primary O-Ring never sealing.

If the same scenario should occur in a field joint (and it could), then it is a jump ball as to the success or failure of the joint because the secondary O-Ring cannot respond to the clevis opening rate and may not be capable of pressurization. The result would be a catastrophe of the highest order – loss of human life.

An unofficial team (a memo defining the team and its purpose was never published) with leader was formed on 19 July 1985 and was tasked with solving the problem for both the short and long term. This unofficial team is essentially nonexistent at this time. In my opinion, the team must be officially given the responsibility and the authority to execute the work that needs to be done on a non-interference basis (full time assignment until completed).

Fig. B7.1

ENGINEERING MEMORANDA WRITTEN AFTER
O-RING CONCERN ESCALATES *(Continued)*

R. K. Lund   31 July 1985

It is my honest and very real fear that if we do not take immediate action to
dedicate a team to solve the problem with the field joint having the number
one priority, then we stand in jeopardy of losing a flight along with all the
launch pad facilities.

R. M. Boisjoly

R. M. Boisjoly

Concurred by:

J. R. Kapp

J. R. Kapp, Manger
Applied Mechanics

                         COMPANY PRIVATE

Fig. B7.2

ENGINEERING MEMORANDA WRITTEN AFTER
O-RING CONCERN ESCALATES *(Continued)*

MORTON THIOKOL. INC.
Wasatch Division

Interoffice Memo

2871:FY86:141
22 August 1985

TO: S.R. Stein,
        Project Engineer

CC: J.R. Kapp, K.M. Sperry, B.G. Russell, R.V. Ebeling, H.H. McIntosh,
        R.M. Boisjoly, M. Salita D.M. Ketner

FROM: A.R. Thompson, Supervisor
        Structures Design

SUBJECT: SRM Flight Seal Recommendation

The O-ring seal problem has lately become acute . Solutions, both long and
short term are being sought, in the mean time flights are continuing. It is
my recommendation that a near term solution be incorporated for flights
following STS-27 which is currently scheduled for 24 August 1985. The near
term solution uses the maximum possible shim thickness and a .292 +.005/-.003
inch dia O-ring. The results of these two changes are shown in Table 1. A
great deal of effort will be required to incorporate these changes. However,
as shown in the Table the O-ring squeeze is nearly doubled for the example
(STS-27A). A best effort should be made to include a max shim kit and the
.292 O-ring as soon as is practical. Much of the initial blow-by during
O-ring sealing is controlled by O-ring squeeze. Also more sacrificial O-ring
material is available to protect the sealed portion of the O-ring. The added
cross-sectional area of the .292 dia O-ring will help the resilience response
by added pressure from the groove side wall.

Several long term solutions look good; but, several years are required to
incorporate some of them. The simple short term measures should be taken to
reduce flight risks.

A.R. Thompson

ART/jh

TC 2018 (REV 7-84)

Fig. B8

ENGINEERING MEMORANDA WRITTEN AFTER
O-RING CONCERN ESCALATES *(Continued)*

MORTON THIOKOL INC.
Wasatch Division

Interoffice Memo

1 October 1985
E150/RVE-86-47

TO: A. J. McDonald, Director
            Solid Rocket Motor Project

FROM: Manager, SRM Ignition System, Final Assembly, Special
            Projects and Ground Test

CC: B. McDougall, B. Russell, J. McCluskey, D. Cooper,
            J. Kilminster, B. Brinton, T. O'Grady, B. MacBeth,
            J. Sutton, J. Elwell, I. Adams, F. Call, J. Lamere,
            P. Ross, D. Fullmer, E. Bailey, D. Smith, L. Bailey,
            B. Kuchek, Q. Eskelsen, P. Petty, J. McCall

SUBJECT: Weekly Activity Report
            1 October 1985

EXECUTIVE SUMMARY

HELP! The seal task force is constantly being delayed by every possible
means. People are quoting policy and systems without work-around. MSFC
is correct in stating that we do not know how to run a development
program.

GROUND TEST

1. The two (2) GTM center segments were received at T-24 last week.
Optical measurements are being taken. Significant work has to be done
to clean up the joints. It should be noted that when necessary SICBM
takes priority.

2. The DM-6 test report less composite section was released last week.

ELECTRICAL

As a result of the latest engineering analysis of the V-1 case it
appears that high stress risers to the case are created by the phenolic
DFI housings and fairings. As it presently stands, these will probably
have to be modified or removed and if removed will have to be replaced.
This could have an impact on the launch schedule.

Fig. B9.1

ENGINEERING MEMORANDA WRITTEN AFTER
O-RING CONCERN ESCALATES *(Continued)*

A. J. McDonald, Director
1 October 1985
E150/RVE-86-47
Page 2

FINAL ASSEMBLY

One SRM 25 and two SRM 26 segments along with two SRM 24 exit cones were
completed during this period. Only three segments are presently in
work. Availability of igniter components, nozzles and systems tunnel
tooling are the present constraining factors in the final assembly area.

IGNITION SYSTEM

1. Engineering is currently rewriting igniter gask-o-seal coating
requirements to allow minor flaws and scratches. Bare metal areas will
be coated with a thin film of HD-2 grease. Approval is expected within
the week.

2. Safe and Arm Device component deliveries is beginning to cause
concern. There are five S&A's at KSC on the shelf. Procurement,
Program Office representatives visited Consolidated Controls to discuss
accelerating scheduled deliveries. CCC has promised 10 A&M's and 30
B-B's no later than 31 October 1985.

O-RINGS AND PUTTY

1. The short stack finally went together after repeated attempts, but
one of the o-rings was cut. Efforts to separate the joint were stopped
because some do not think they will work. Engineering is designing
tools to separate the pieces. The prints should be released tomorrow.

2. The inert segments are at T-24 and are undergoing inspection.

3. The hot flow test rig is in design, which is proving to be
difficult. Engineering is planning release of these prints Wednesday or
Thursday.

4. Various potential filler materials are on order such as carbon,
graphite, quartz, and silica fiber braids; and different putties. They
will all be tried in hot flow tests and full scale assembly tests.

5. The allegiance to the o-ring investigation task force is very
limited to a group of engineers numbering 8-10. Our assigned people in
manufacturing and quality have the desire, but are encumbered with other
significant work. Others in manufacturing, quality, procurement who are
not involved directly, but whose help we need, are generating plenty of
resistance. We are creating more instructional paper than engineering
data. We wish we could get action by verbal request but such is not the
case. This is a red flag.

R. V. Ebeling

Fig. B9.2

ENGINEERING MEMORANDA WRITTEN AFTER
O-RING CONCERN ESCALATES *(Continued)*

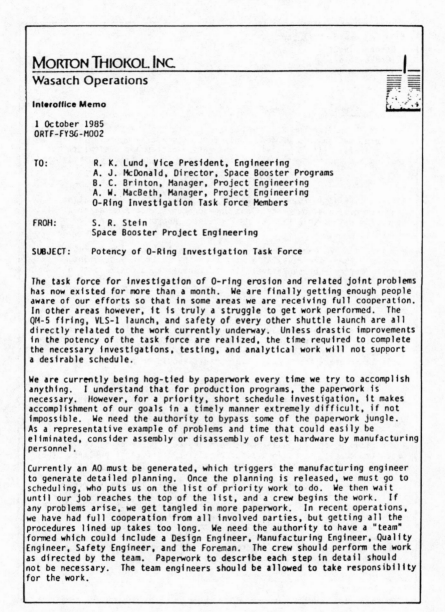

MORTON THIOKOL. INC.

Wasatch Operations

Interoffice Memo

1 October 1985
ORTF-FY86-M002

TO: R. K. Lund, Vice President, Engineering
          A. J. McDonald, Director, Space Booster Programs
          B. C. Brinton, Manager, Project Engineering
          A. W. MacBeth, Manager, Project Engineering
          O-Ring Investigation Task Force Members

FROM: S. R. Stein
          Space Booster Project Engineering

SUBJECT: Potency of O-Ring Investigation Task Force

The task force for investigation of O-ring erosion and related joint problems
has now existed for more than a month. We are finally getting enough people
aware of our efforts so that in some areas we are receiving full cooperation.
In other areas however, it is truly a struggle to get work performed. The
QM-5 firing, VLS-1 launch, and safety of every other shuttle launch are all
directly related to the work currently underway. Unless drastic improvements
in the potency of the task force are realized, the time required to complete
the necessary investigations, testing, and analytical work will not support
a desirable schedule.

We are currently being hog-tied by paperwork every time we try to accomplish
anything. I understand that for production programs, the paperwork is
necessary. However, for a priority, short schedule investigation, it makes
accomplishment of our goals in a timely manner extremely difficult, if not
impossible. We need the authority to bypass some of the paperwork jungle.
As a representative example of problems and time that could easily be
eliminated, consider assembly or disassembly of test hardware by manufacturing
personnel.

Currently an AO must be generated, which triggers the manufacturing engineer
to generate detailed planning. Once the planning is released, we must go to
scheduling, who puts us on the list of priority work to do. We then wait
until our job reaches the top of the list, and a crew begins the work. If
any problems arise, we get tangled in more paperwork. In recent operations,
we have had full cooperation from all involved parties, but getting all the
procedures lined up takes too long. We need the authority to have a "team"
formed which could include a Design Engineer, Manufacturing Engineer, Quality
Engineer, Safety Engineer, and the Foreman. The crew should perform the work
as directed by the team. Paperwork to describe each step in detail should
not be necessary. The team engineers should be allowed to take responsibility
for the work.

Fig. B10.1

ENGINEERING MEMORANDA WRITTEN AFTER O-RING CONCERN ESCALATES *(Continued)*

Distribution
1 October 1985
Page two

I know the established paperwork procedures can be violated if someone with enough authority dictates it. We did that with the DR system when the FWC hardware "Tiger Team" was established. If changes are not made to allow us to accomplish work in a reasonable amount of time, then the O-ring investigation task force will never have the potency necessary to resolve the problems in a timely manner.

Scott R Stein
S. R. Stein

Fig. B10.2

ENGINEERING MEMORANDA WRITTEN AFTER
O-RING CONCERN ESCALATES *(Continued)*

ACTIVITY REPORT

The team generally has been experiencing trouble from the business as usual attitude from supporting organizations. Part of this is due to lack of understanding of how important this task team activity is and the rest is due to pure operating procedure inertia which prevents timely results to a specific request.

The team met with Joe Kilminster on 10/3/85 to discuss this problem. He wanted specific examples which he was given and he simply concluded that it was every team members responsibility to flag problems that occurred to organizational supervision and work to remove the road block by getting the required support to solve the problem. The problem was further explained to require almost full time nursing of each task to insure it is taken to completion by a support group. Joe simply agreed and said we should then nurse every task we have.

He plain doesn't understand that there are not enough people to do that kind of nursing of each task, but he doesn't seem to mind directing that the task never-the-less gets done. For example, the team just found out that when we submit a request to purchase an item, that it goes through approximately 6 to 8 people before a purchase order is written and the item actually ordered.

The vendors we are working with on seals and spacer rings have responded to our requests in a timely manner yet we (MTI) cannot get a purchase order to them in a timely manner. Our lab has been waiting for a function generator since 9-25-85. The paperwork authorizing the purchase was finished by engineering on 9-24-85 and placed into the system. We have yet to receive the requested item. This type of

Fig. B11.1

ENGINEERING MEMORANDA WRITTEN AFTER
O-RING CONCERN ESCALATES *(Continued)*

example is typical and results in lost resources that had been planned
to do test work for us in a timely manner.

I for one resent working at full capacity all week long and then
being required to support activity on the weekend that could have been
accomplished during the week. I might add that even NASA perceives that
the team is being blocked in its engineering efforts to accomplish its
tasks. NASA is sending an engineering representative to stay with us
starting Oct 14th. We feel that this is the direct result of their
feeling that we (MTI) are not responding quickly enough on the seal
problem.

I should add that several of the team members requested that we be
given a specific manufacturing engineer, quality engineer, safety
engineer and 4 to 6 technicians to allow us to do our tests on a
non-interference basis with the rest of the system. This request was
deemed not necessary when Joe decided that the nursing of the task
approach was directed.

Finally, the basic problem boils down to the fact that ALL MTI
problems have #1 priority and that upper management apparently feels
that the SRM program is ours for sure and the customer be damned.

Roger Boisjoly 10/4/85

Fig. B11.2

Appendix C

On Theory Elaboration, Organizations, and Historical Ethnography

The *Challenger Launch Decision* is the second in a series of three books that develop and refine a method for theorizing about organizations while simultaneously using this method to solve substantive problems. In this appendix, I briefly describe the method, its use in the *Challenger* case, and the Mertonian framework that underpins this analysis. I then make some observations about the process of historical ethnography for organizational analysis. My goal here is simply to make readers sensitive to essential theoretical and methodological issues not visible in the text. For those wishing greater detail, I have written about theory elaboration elsewhere, and the third book is in progress.[1]

Theory elaboration: an overview. The roots of this method are in Simmel's formal sociology.[2] He wrote that certain features of concrete phenomena can be extracted from reality, allowing us to compare activities and events that appear radically different in content, yet have common structural arrangements. Thus the sociologist could identify marital conflict and martial conflict as essentially similar interactive forms. Building on Simmel's ideas, theory elaboration develops general theory by qualitative case comparison of organizational forms that vary in size, complexity, and function. By theory, I mean theoretical tools in general (theory, models, and/or concepts). By elaboration, I mean the process of refining these theoretical tools in order to specify more carefully the circumstances in which they do or do not apply.

We choose a case because we have good theoretical and empirical reasons to think it might be an example of *x*. The analysis can be developed from ethnography, interviews, original documents, or secondary analysis. Some theoretical logic is always part of our selection, either implicitly or explicitly, so first we want to make it explicit, then we use it as an heuristic device to guide the analysis and organize insights. Perhaps the guiding theory is better thought of as an analytic framework that opens possibilities at the same time that it focuses the research. The opening of possibilities is guaranteed, first, by the use of Lindesmith's method of analytic induction,[3] which forces us to reconsider and adjust theory every time we encounter a piece of disconfirming evidence, and second, by the requirement to fully explain the case, which forces us to take into account and integrate all data that bear on the incident or activity in question. The data can contradict the theory, reveal previously unseen inadequacies, confirm parts of it but not all, and/or indicate new hypotheses. Also, if we stick to the twin requirements of analytic induction and full case analysis, we (1) protect against force fitting the data to the theory and (2) generate new theories we would not have predicted in the beginning.

The possibility of discovery is enhanced by another aspect of this method. Many people tend to specialize in research, restricting their inquiries to one organizational form: the world system, business organizations, educational institutions, prisons. Varying organizational form (groups, simple formal organizations, complex organizations, subunits, networks) to examine a particular phenomenon can give us qualitatively different information and, thus, new insights. So, for example, we might want to examine tight and loose coupling, concepts regularly used to describe relations in complex organizations, to find their relevance in families or friendships. Sometimes shifting the unit of analysis also shifts the level of analysis. Because of the different sorts of data available from macro- and microlevel analysis, comparing organizational forms that vary in size, complexity, and function can help us explore the relationship between environment, organization, and individual behavior.

The Challenger case: Mertonian underpinnings. The NASA/*Challenger* case began as part of a project using this method to elaborate a theory of organizational misconduct. When the disaster occurred, I was already working on two cases: police misconduct and family violence. I was looking for a third case involving a large complex organization that was not a corporation, in order to avoid the business firm bias of most previous theorizing and develop a general theory that applied to a variety of organizations. The *Challenger* incident seemed the ideal complement to the other two. The major elements of the theory of organizational misconduct on which the three-case comparison is based are:

1. The competitive environment (competition, scarce resources, and norms), which generates pressure on organizations to violate laws and rules in order to attain goals;[4]
2. Organizational characteristics (structure, processes, and transactions), which provide opportunities to violate;[5]
3. The regulatory environment (autonomy and interdependence), which is affected by the relationship between regulators and the organizations they regulate, frequently minimizing the capacity to control and deter violations, consequently contributing to their occurrence.[6]

Each is related to misconduct, but what is interesting is how they are interrelated, so that misconduct results from the three in combination. This explanatory scheme was, itself, a product of theory elaboration. In my initial use of this method, I applied Merton's social structure and anomie theory (SSAT) to organizations as the units of analysis, which resulted in a reconceptualization of SSAT that was the basis for the first environmental element above.[7] Because Cloward and Ohlin's work is so often used in conjunction with SSAT, this reconceptualization quite naturally led me to analyze the applicability of their ideas about opportunity structures to legitimate organizations.[8] Thus, I arrived at the second element. With these two reconceptualized theories joined, the role of the regulatory environment in shaping patterns of individual choice became clear, completing the explanatory scheme. Of course, many other theories, models, and concepts developed by scholars over the years are included and essential. This method differs from, and hence adds to, the extant literature by (1) emphasizing these three elements as the major theoretical building blocks and (2) posing a linkage between them.

In the *Challenger* case, I used the three building blocks and their subconcepts to organize the data and conceptualize. This case was attractive because the data made it possible to explore the macro-micro connection, as I mentioned in chapter 2. As it turned out, the Mertonian-based structural theory (as elaborated by this case) and the microlevel processes that reflect and reinforce it are the explanation of this case: the theory of the normalization of deviance is a grounded theory showing the nexus of the three structural conditions. However, the data and the need to explain the case drove how the story was told. Consequently, the three elements and the linkage between them may not be apparent to specialist readers. (I contributed further to this invisibility by omitting a discussion of the framework per se in the text in order to tell the *Challenger* story lucidly and minimize the kind of detail that would be mind-numbing to nonspecialists.) The analysis of the competitive environment as structural impetus (competition, scarce resources, and norms) appears in chapters 1 and 6. The analysis of organization characteristics that provide opportunities (structure, processes, and transactions) appears in chapters 3 through 7. The regulatory environment (autonomy and interdependence) and its connection to decision making appears near the end of chapter 7. Then, following the description of the eve of the launch in chapter 8, these same factors combine to form the analysis in chapter 9.

Much remains to be said about the use of theory elaboration in this analysis; for now, I emphasize two points. First, the case demonstrates that this method fulfills its promise to protect against unconscious bias that might lead a researcher to find exactly what he or she had hypothesized. Analytic induction and the requirement to fully explain the event or activity in question contradicted my beginning typological assumption that the *Challenger* launch decision was an example of organizational misconduct, as traditionally defined, causing me to start over. Further, the many contradictions I found (and experienced as "mistakes" throughout the project) forced me to shift theoretical gears again and again, incrementally pushing me toward the final, inductively developed explanation of the case. Second, this method also fulfills its promise to reject and refine aspects of the guiding theory, add to others that were not the central focus at the outset, and initiate new theory. In the final chapter, I described how the case elaborated theories about science, technology, and risk, decision making in organizations, and deviance and misconduct. I then discussed the main innovation: a nascent conceptual scheme about signals of potential danger and the normalization of deviance. Implications for the Mertonian-based theory of organizational misconduct will be addressed in future writing.[9]

Historical ethnography and organizational analysis: the research process. Among my primary data sources were over 122,000 pages of documents gathered by the Presidential Commission that are catalogued and available at the National Archives, Washington, D.C. I have relied on volumes 1, 2, 4, and 5 of the 1986 *Report of the Presidential Commission on the Space Shuttle Challenger Accident*, volumes 4 and 5 of which contain 2,500 pages of hearing transcripts from both closed and open Commission sessions; the 1986 report of the Committee on Science and Technology, U.S. House of Representatives, which includes two volumes of hearing transcripts; and the published accounts of journalists, historians, scientists, and others. Using the Freedom of Information Act, I secured copies of additional documents from the warehouse at Johnson Space Center, where originals of all NASA documents relevant to the *Challenger* incident are stored. I drew on materials available at the National Archives Motion Picture and Video Library, NASA History Office, and NASA Media Library, Washington, D.C., as well as at the Marshall Space Flight Center History Library, Huntsville, Alabama. While in Huntsville, I took the official bus tour of Marshall and a personal tour that included the room where all teleconferences are held, a demonstration of the teleconference communication system, and the Huntsville Operations Support Center, where all countdowns, launches, orbiting flights, and landings are monitored.

I conducted interviews, in person and by telephone. I also used transcripts from 160 interviews conducted by 15 experienced government investigators who supported Commission activities, totaling approximately 9,000 pages stored at the National Archives. Nearly 60 percent of those interviewed by government investigators never testified before the Presidential Commission or the House Committee on Science and Technology. The investigators took depositions, using interview guides. They questioned interviewees on two topics: the history of SRB decision making and events on the eve of the launch. If people participated in both, they were interviewed twice, resulting in two transcripts, one for each topic. However, those who testified extensively before the Commission about the eve of the launch were only questioned by these investigators about the history of decision making.

The research process is more difficult to discuss briefly but is important to record in light of recent forceful addresses given by two leading organization theorists to specialists in organization studies: Mayer Zald, who argued for the systematic development of an anthropology of organizations, and Charles Perrow, who urged that more research attention be directed toward the role of organizations in social problems.[10] As a historical ethnography of a modern complex organization responsible for a failed technology that caused irreversible damage—lives lost, a social fabric weakened by anguish and cynicism, millions of dollars in costs to taxpayers—this case fits both those categories. In contrast to organization studies, other areas in sociology have a tradition of ethnographic organizational analysis that targets social problems, extending from classics such as Whyte's *Street Corner Society* and Burawoy's *Manufacturing Consent* to recent compelling work that includes Diamond's *Turning Gray Gold*, Ostrander's *Money for Change*, Duneier's *Slim's Table*, and Chambliss's *Beyond Caring*.[11] What distinguishes the NASA/*Challenger* case from these others is the conjunction of history, ethnography, and complex organization: I was not a participant observer inside the organization examining structure and process, but an outsider retrospectively reconstructing both, relying heavily on documents. What was gained and lost by this situation is hard to weigh, but positioning myself at NASA in 1977 to watch the process of deviance normalization that began then and culminated in the 1986 tragedy was, of course, impossible.

Previously, I have described how the very characteristics of organizations that we study—structure, specialization, culture, language, interorganizational relations—are obstacles to understanding in field work.[12] These obstacles were more difficult to surmount than in my previous experience because of NASA's size, complexity, and core technology. Moreover, experienced in ethnography and interviewing, I had no experience doing history. The research focus was the intersection of structure with definitional processes that spanned 1977–86, and the body of available archival data was enormous. Thus, many of my normal research strategies did not apply. For example, three weeks of photocopying only the interview transcripts at the National Archives filled a four-drawer file. I began coding them, but soon realized not only that their coding could take a year or more, but also that this strategy was flawed: aggregating statements from all interviews by topic (a practice I had often used) would extract parts of each interview from its whole. Each person was giving a chronological account of (1) the history of decision making and (2) the eve of the launch; hence, the whole stream of actions was absolutely essential context for understanding insider definitions of the situation. Also, within each chronology were many incidents, decisions, and events about which I needed other participants' views. So the method I evolved was to proceed chronologically, examining turning points decision by decision, event by event, first examining documents created at the time, identifying the people who participated in a decision or event and others who knew about it, then reading the whole interview transcripts and hearings testimony of those people, which I could compare with the documentary record. Because I wanted to see how interpretations varied depending on a person's position, this strategy was complicated by the number of NASA and Thiokol employees across the hierarchy who were pulled together in these decisions by NASA's matrix system and FRR.

My interviewing was similarly driven by the historical chronology, and so ebbed and flowed throughout the project. The interviewees, subject matter, and timing were dictated by the gradually unfolding picture and the questions that arose. Some examples: in 1988, I interviewed 18 people connected to safety regulation at NASA Headquarters and throughout the organization because regulation was what I was investigating at the time; when I had questions about the Presidential Commission's investigation, I contacted a Presidential Commission member (anonymous), Emily Trapnell, Randy Kehrli, and Jack Macidull, who were members of the Commission's investigative staff, and Chairman William Rogers; when I was reconstructing decisions that required evaluating testimony about cold temperature at the Cape (STS 51-B, January 1985, and the eve of the launch), I contacted the National Climatic Data Center to secure temperature records for the Cape Canaveral launch site. Thiokol's Roger Boisjoly was an important source of information in the early years, and we maintained a telephone interviewing relationship through most of the project, as I did with others. However, I deferred interviews with Marshall Space Flight Center employees until June 1992, when I finished reconstructing the history of decision making for 1977–85. By then, I felt that I could evaluate the validity of what they told me because I had come to an understanding of events based on the archival records and had mastered NASA procedures, structure, language, and the technology. I did open-ended interviews designed to test my understanding of these aspects of culture, the correctness of my analysis, and to lead in new directions. I maintained telephone interviewing relationships with them for the duration of the project.

Analytic induction drove the explanation in unexpected directions. The analysis outgrew my first idea for a chapter in a book of three case comparisons, outgrew my second idea for a slender volume. I can only conclude that a full case analysis that makes macromicro connections by incorporating both organization *in* history and organization *as* history inevitably leads to a big book.[13] However, the salient overarching message about the research process of historical ethnography for organizational analysis may best be conveyed by this Indian story related by anthropologist Clifford Geertz: "An Englishman who, having been told that the world rested on a platform which rested on the back of an elephant which rested in turn on the back of a turtle, asked (perhaps he was an ethnographer; it is the way they behave), what did the turtle rest on? Another turtle. And that turtle? "Ah, Sahib, after that it is turtles all the way down."[14] Geertz tells the story to point out that cultural analysis is necessarily incomplete, and the more deeply it goes, the less complete it is. History, too, is an unfolding process, difficult to capture. When history, ethnography, and complex organizations combine, it is guaranteed to be, as the Indian said, "turtles all the way down." That was the essence of my research experience, even at the end. But we do not need to get through all the turtles. At some point, we can predict the next turtle and we know that more turtles will not change the essence of what we have learned by looking under the platform, under the elephant, and under a number of turtles. What matters in developing an anthropology of organizations is that we go beyond the obvious and grapple with the complexity, for explanation lies in the details.

Acknowledgments

Acknowledging the sources of support for my research again invokes the themes of this book, for it requires the reconstruction of a nine-year history revealing the incrementalism behind the interpretative work and decision making that resulted in this manuscript. My initial debt is to Sim Sitkin, who in December 1985, asked me to participate in a panel the following August on secrecy in organizations. This invitation, coming a few weeks prior to the *Challenger* tragedy, hastened my first conceptualization and writing (Sim and his colleague Jack Brittain then creatively used the data from that paper as the basis for their well-known "Carter Racing Case," now used in many business schools). By that August meeting, my initial foray into the case had already raised more questions than answers. Still, a book might never have materialized had I not been the fortunate recipient of two fellowships (Centre for Socio-Legal Studies, Wolfson College, Oxford, 1986–87; American Bar Foundation, Chicago, 1988–89) that gave me the resources necessary to delve deeply into archival materials as well as lively intellectual environments in which to air my work-in-slow-progress. In Chicago, I also benefited from continuing conversations with colleagues in the Department of Sociology, Northwestern University, and my year-long participation in John Padgett's Workshop on Organizations, University of Chicago.

I am indebted to historians and archivists who helped me find my way through the enormous documentary record of this incident: in particular, David Paynter, National Archives, Washington, D.C.; the staff at the Motion Picture and Video Library, National Archives; Michael Wright, Marshall Space Flight Center History Library, Huntsville, Alabama; Constance Moore, NASA Still Photo Library, Washington, D.C.; Nena Jones, NASA Still Photo Library, Johnson

Space Center, Houston, Texas; and Lee Saegesser, NASA History Office, Washington, D.C. I owe much to the many people who generously gave hours of their time for interviews throughout this project. I am especially grateful to the few to whom I turned repeatedly: Thiokol's Roger Boisjoly and Marshall's Larry Mulloy, Leon Ray, Jim Smith, and Larry Wear. I want to pay special tribute to Leon Ray, the engineer and "organization gadfly" who was "way, way down" in the Marshall hierarchy. Ray died suddenly in 1994. Although he was the person at NASA who knew the most about the Solid Rocket Booster joints, he was virtually ignored by postaccident interviewers. For his understanding and love of technology, for all he taught me, I thank him.

My major intellectual debts are to Robert K. Merton, whose conceptualization of social structure, cultural structure, and deviance provided the theoretical foundation for my analysis; to Peter Berger and Thomas Luckmann for their compelling insights on the linked processes and social organization that underpin the social construction of reality; and to Charles Perrow and James F. Short, whose innovative research and theorizing about organizations, complex technical systems, and risk opened a new area for sociologists. Over nine years, I accumulated many other intellectual debts. The hundreds of hours studying engineering reports, drawings, and Flight Readiness Review presentations together with the continual discoveries requiring both major and minor refinements in the analysis often left me anxious and ill at ease with a project that refused to hold still so I could understand it. Experiencing the loneliness of the long distance writer, I worked and reworked my ideas in continuing conversations with valued colleagues and friends Howard Becker, David Karp, Susan Ostrander, and Albert J. Reiss, Jr.. The development of my analysis also benefited from the thoughtful responses of colleagues to early papers and seminar presentations: John Braithwaite, Albert K. Cohen, Paul J. DiMaggio, Robert Emerson, Gilbert Geis, Keith Hawkins, Eiko Ikegami, John Logsdon, Robert F. Meier, Robert K. Merton, Charles Perrow, Shula Reinarz, Susan Shapiro, Karl Weick, and Brian Wynne.

I did not circulate book chapters for review, as many do, because each chapter is linked to the next and cannot be understood without the whole. To those who read the completed manuscript, I am deeply grateful. I asked a variety of specialists to criticize it because the book deals with a well-known public event and a complex technology and organization, and also falls within several major areas in sociology. This manuscript was reviewed by people at NASA and Morton Thiokol, and by Jack Macidull (FAA investigator, engineer, high-performance-aircraft pilot, and member of the Presidential Commission's investigative team). In addition, I received scrupulously thorough reviews from expert sociologists in the key theoretical areas the manuscript addresses: Steven Brint, Daniel F. Chambliss, Lee Clarke, Lewis Coser, Albert J. Reiss, Jr., and Robert N. Stern. Busy with their own writing, they made time for mine. Not only did their comments help me clarify my arguments, but the rich conversations that resulted were the most rewarding aspect of the research.

I thank my research assistant, Daniel F. Egan, who intelligently and carefully created a newspaper and magazine article archive, carried out extensive library research, and did bibliographic editing. I thank my computer guru, Joe Scavuzzo, whose wizardry speeded the book's completion, and Stephen Vedder, Mary Binell, and Dawn Vaughan at the Boston College Audio-Visual Department, who digitalized photographs and documents for reproduction. To John Tryneski, my editor at the University of Chicago Press, and the entire staff associated with the book's design and production, I convey my appreciation for the quality and speed of the group effort that successfully met an impossible deadline. I thank my friends, all with expertise in areas different from mine, who passed the extreme test of friendship by listening, with interest, to details about this project that only an author could love: Deborah Brown, Betty Goldstein, Sandra Gradman, Jill Kneerim, Lorna McKenzie-Pollock, Michele Salzman, and Annebelle Scavuzzo. Finally, I thank my support team that has endured through all the books, through all the events, over the years: my dearest loves, my very best work, my children.

Notes

PREFACE

1. Bruno Latour and Steve Woolgar, *Laboratory Life: The Construction of Scientific Facts* (Beverly Hills, Calif.: Sage, 1979); Bruno Latour, *Science in Action* (Cambridge: Harvard University Press, 1987).

2. Paul J. DiMaggio and Walter W. Powell, "Introduction," in *The New Institutionalism in Organizational Analysis*, ed. Walter W. Powell and Paul J. DiMaggio (Chicago: University of Chicago Press, 1991), 15–27.

3. See appendix C. See also Diane Vaughan, "Theory Elaboration: The Heuristics of Case Analysis," in *What Is a Case? Exploring the Foundations of Social Inquiry*, ed. Charles C. Ragin and Howard S. Becker (Cambridge: Cambridge University Press, 1992); Diane Vaughan, "The Macro-Micro Connection in White-Collar Crime Theory," in *White-Collar Crime Reconsidered*, ed. Kip Schlegel and David Weisburd (Boston: Northeastern University Press, 1992).

CHAPTER 1

1. Gene Gurney and Jeff Forte, *Space Shuttle Log: The First Twenty-Five Flights* (Blue Ridge Summit, Pa: AERO, 1988), 260.

2. Presidential Commission on the Space Shuttle *Challenger* Accident, *Report to the President by the Presidential Commission on the Space Shuttle Challenger Accident* (Washington, D.C.: Government Printing Office, 1986), 1:1.

3. Ibid. 4:1–178.

4. Ibid., 94.

5. Ibid., 95.

6. Ibid., 97. Both Moore and Arnold Aldrich, Manager, National Space Transportation Systems Program, Johnson Space Center, testified that temperature had been considered and was not viewed as a problem (ibid., 20, 57).

7. Although the testimony concerning security at Kennedy and Vandenberg was excised from the verbatim transcription of this closed session, the remainder of the testimony was included in the Commission's official report (ibid., 214–37).

8. Philip M. Boffey, "NASA Had Warning of a Disaster Risk Posed by Booster," *New York Times*, 9 February 1986, sec. A, 1.

9. Ibid., 32.

10. Ibid.; Richard C. Cook to Michael Mann, "Problem with SRB Seals," memorandum, 23 July 1985, reproduced in Presidential Commission, *Report* 4:391.

11. Boffey, "NASA Had Warning," 32.

12. Transcripts of both closed and open Commission hearings are in volumes 4 and 5 of the Commission's report. The transcripts of this closed Commission hearing are found in Presidential Commission, *Report* 4:243–326.

13. Ibid. 1:206.

14. Presidential Commission on the Space Shuttle *Challenger* Accident, STS 51-L Data and Design Analysis Task Force, "Views of *Challenger* Space Shuttle: Explosion, Tank Failure, etc.," 13 June 1986, Motion Picture and Video Library, National Archives, Washington, D.C.

15. Presidential Commission, *Report* 1:72.

16. Ibid., 122.

17. Ibid., 104.

18. U.S. Congress, House, Committee on Science and Technology, *Investigation of the Challenger Accident: Report* (Washington, D.C.: Government Printing Office, 1986); U.S. Congress, House, Committee on Science and Technology, *Investigation of the Challenger Accident: Hearings*, 2 vols. (Washington, D.C.: Government Printing Office, 1986). The House Committee on Science and Technology has full oversight responsibilities for NASA. The Committee annually authorizes NASA funds, endorsing the programs and activities of NASA. As a part of its oversight role, the Committee reviewed the work of the Presidential Commission and of NASA in investigating the causes of the accident and similarly reviewed the recommendations to resume safe flight. The Committee's investigation encompassed a review of the Presidential Commission report, its appendices, and the investigation team reports. In addition, the Committee held 10 formal hearings involving 60 witnesses, as well as interviewing many outside the formal hearing setting (U.S. Congress, House, *Investigation: Report*, 1–3).

19. U.S. Congress, House, *Investigation: Report*, 4–5. The Presidential Commission acknowledged (but did not assert as the fundamental problem) the fact that neither NASA nor Thiokol "fully understood the mechanism by which the joint sealing action took place" and their failure to fix the problem (Presidential Commission, *Report* 1:148).

20. Presidential Commission, *Report* 1:3; Malcolm McConnell, *Challenger: A Major Malfunction* (Garden City, N.Y.: Doubleday, 1987); Richard C. Cook, "The Rogers Commission Failed: Questions It Never Asked, Answers It Didn't Listen To," *Washington Monthly*, November 1986, 13–21; Richard S. Lewis, *Challenger: The Final Voyage* (New York: Columbia University Press, 1988).

21. David Ignatius, "Did the Media Goad NASA into the Challenger Disaster?" *Washington Post*, 30 March 1986, sec. D, 1.

22. Gurney and Forte, 259.

23. Ignatius.

24. Ibid., 5.

25. Cook, "Rogers Commission"; McConnell.

26. Cook, "Rogers Commission"; McConnell, 168–71; Walter V. Robinson, "NASA Blamed for Shuttle Disaster," *Boston Globe*, 10 June 1986, sec. A, 1; Ed Magnuson, "Fixing NASA," *Time*, 9 June 1986, 14–25.

27. Presidential Commission, *Report* 1:176.

28. Cook, "Rogers Commission."

29. Presidential Commission, *Report* 1:176–77.

30. Ibid., 176.

31. Ibid.

32. Trudy E. Bell and Karl Esch, "The Fatal Flaw in Flight 51-L," *IEEE Spectrum*, February 1987, 48.

33. Richard P. Feynman, *What Do You Care What Other People Think? Further Adventures of a Curious Character* (New York: Norton, 1988), 194.

34. Transcript, Federal News Service, Washington, D.C.

35. Presidential Commission, *Report* 1:176.

36. McConnell, 131.

37. See, e.g., ibid., 22–23. Shirley Green, interview transcript, 2 April 1986, NASA Headquarters files, National Archives, Washington, D.C.

38. McConnell, 168.

39. Green, interview transcript, 2 April 1986, 6.

40. Presidential Commission, *Report* 2:J10.

41. Ibid., J20.

42. R. Jeffrey Smith, "Experts Ponder Effect of Pressures on Shuttle Blowup," *Science* 231 (28 March 1986): 1495.

43. See, e.g., Presidential Commission, *Report* 1; U.S. Congress, House, *Investigation: Report*; McConnell; Joseph J. Trento, *Prescription for Disaster: From the Glory of Apollo to the Betrayal of the Shuttle* (New York: Crown, 1987).

44. McConnell, 23.

45. Feynman, *What Do You Care What Other People Think?* 216–17.

46. Brian Woods, "A Sociotechnical History of the United States Space Shuttle" (masters thesis, University of Edinburgh, 1995).

47. Barbara S. Romzek and Melvin J. Dubnick, "Accountability in the Public Sector: Lessons from the Challenger Tragedy," *Public Administration Review* 47 (1987): 227–38.

48. Walter A. McDougall, *And the Heavens and the Earth: A Political History of the Space Age* (New York: Basic Books, 1985), 8.

49. Ibid., 420–35.

50. Romzek and Dubnick, 231.

51. Lewis.

52. Alex Roland, "The Shuttle: Triumph or Turkey?" *Discover*, November 1985, 29.

53. McConnell, 32.

54. National Aeronautics and Space Administration, *NASA Pocket Statistics* (Washington, D.C.: Office of Headquarters Operations, Code DA, January 1991), C-26–C-27.

55. Dennis Overbye, "Success amid the Snafus," *Discover*, November 1985, 54; James A. Van Allen, "Myths and Realities of Space Flight," *Science* 232 (30 May 1986): 1075–76.

56. James A. Van Allen, "Space Science, Space Technology, and the Space Station," *Scientific American* 254 (January 1986): 35.

57. Romzek and Dubnick.

58. John M. Logsdon, "The Space Shuttle Program: A Policy Failure?" *Science* 232 (30 May 1986): 1102.

59. Ibid., 1100.

60. McDougall, 108–11, 118–21.

61. Roland, 40–41; Trento.

62. Logsdon, 1101.

63. Roland, 35–38; McDougall, 383–86.

64. Roland, 40.

65. Logsdon, 1101; Roland, 35.

66. Roland, 35.

67. NASA received $3.2 billion for fiscal year 1972, restricting shuttle development allocations to $5–$6 billion, with a peak annual budget of $1 billion (Logsdon, 1101).

68. Donald MacKenzie and Judy Wajcman, eds., *The Social Shaping of Technology: How the Refrigerator Got Its Hum* (Milton Keynes: Open University Press, 1985).

69. Brian Woods, "A Sociotechnical History of the United States Space Shuttle" (Ph.D proposal paper, Department of Sociology, University of Edinburgh, March 1995), 2.

70. Roland, 38.

71. Logsdon, 1101–2; Lewis, 54. For a detailed account of how political, economic, and social factors influenced how shuttle technologies were chosen and others dismissed, see Woods, "Sociotechical History" (masters thesis).

72. Logsdon, 1101. See also Roland, 38.

73. Logsdon, 1102.

74. James Fletcher to Casper Weinberger, letter, 6 March 1972, quoted ibid., 1101–2.

75. McConnell, 40.

76. Logsdon, 1101.

77. Ibid.

78. Roland, 38; Gregg Easterbrook, "Big Dumb Rockets," *Newsweek*, 17 August 1987, 53; McConnell, 39; Trento, 107.

79. McConnell, 39.

80. Ibid., 35.

81. Van Allen, "Space Science," 36.

82. Roland, 48.

83. For a thorough assessment of how shuttle technology affected organization structure, see Woods, "Sociotechnical History" (masters thesis).

84. Presidential Commission, *Report* 1:165.

85. Logsdon, 1105.

86. Presidential Commission, *Report* 1:4.

87. Stuart Diamond, "NASA Cut or Delayed Safety Spending," *New York Times*, 24 April 1986, sec. A, 1.

88. Eliot Marshall, "The Shuttle Record: Risks, Achievements," *Science*, 231 (14 February 1986): 664–66.

89. Presidential Commission, *Report* 1:164; Roland, 35–38.

90. Presidential Commission, *Report* 1:164–65.

91. Ibid. 2:J31.

92. Johnson Space Center, JSC-07700, cited in Presidential Commission, *Report* 2:J-31; Norman Schulze, interview transcript, 15 April 1986, NASA Headquarters files, National Archives, Washington, D.C., p. 15.

93. Presidential Commission, *Report* 1:5.

94. Gurney and Forte, 38–48; Presidential Commission, *Report* 1:164.

95. Presidential Commission, *Report* 1:164.

96. Roland, 38.

97. Van Allen, "Space Science," 36.

98. Logsdon, 1100.

99. McConnell, 41.

100. U.S. Congress, House, *Investigation: Report*, 120.

101. Van Allen, "Space Science," 36. As early as February 1982 (five months before the shuttle was declared operational), a General Accounting Office (GAO) report recommended that NASA reconsider its payload pricing policy because of Space Transportation System cost increases. Between the time NASA established payload prices and the GAO examination, projected average costs to fly a shuttle mission increased 73 percent. GAO advised that unless prices were adjusted, the NASA budget was going to bear a major portion of the increased costs (General Accounting Office, Mission Analysis and Systems Acquisition Division, Report no. MASAD-82-15:B-202664, 23 February 1982, National Archives, Washington, D.C.).

102. Roland, 48.

103. Ibid.

104. Presidential Commission, *Report* 1:164.

105. McConnell, 62–64.

106. Overbye, 71–72.

107. McConnell, 64.

108. Ibid.

109. Roland, 40–41.

110. Eliot Marshall, "DOE's Way-Out Reactors," *Science* 231 (21 March 1986): 1357–59.

111. Roland, 45.

112. Presidential Commission, *Report* 1:171–72.

113. Gurney and Forte, 116–27, 152–59.

114. Presidential Commission, *Report* 1:171–72.

115. McConnell, 98.

116. Ibid., 14.

117. James Fletcher to T. F. Rogers, memorandum, 16 November 1976.

118. Study Team to James Fletcher, "Unique Personality for Space Shuttle Flights: Preliminary Report," memorandum, 8 June 1976, NASA Headquarters, Washington, D.C., p. 1.

119. Ibid.

120. Alan Ladwig, "The Space Flight Participant Program: Taking the Teacher and Classroom into Space" (paper IAF-85-438 presented at 36th Congress of International Astronautical Federation; Stockholm: Pergamon, 1985), 2.

121. McConnell, 100–102.

122. Ladwig, 3.

123. National Aeronautics and Space Administration, News and Information Branch, "Background Information on Teacher in Space Project" (Washington, D.C., 1 December 1985), 1.

124. Presidential Commission, *Report* 1:168.

125. Ibid.

126. Ibid., 164.

127. Ibid., 170; 2:J31.

128. Ibid. 1:165.

129. Ibid., 173–74.

130. Ibid., 166–70.

131. Ibid., 168–69.

132. U.S. Congress, House, *Investigation: Report*, 126.

133. Presidential Commission, *Report* 1:170.

134. McConnell, 30.

135. Van Allen, "Space Science," 35.

136. NASA, *Pocket Statistics*, C-26–C-27.

137. Presidential Commission, *Report* 1:173.

138. The 24 flights planned for 1987 would have been impossible because inventory was insufficient (U.S. Congress, House, *Investigation: Report*, 116–17).

139. NASA, News and Information Branch, "Teacher in Space Project," 2.

140. Presidential Commission, *Report* 1:176; 2:J30.

141. Ibid. 1:171.

142. Ibid., 164.

143. McConnell, 67.

144. Presidential Commission, *Report* 1:170–71; U.S. Congress, House, *Investigation: Report*, 127.

145. McConnell, 65.

146. Ibid.

147. Van Allen, "Space Science."

148. Ibid., 165.

149. U.S. Congress, House, *Investigation: Report*, 3, 123.

CHAPTER 2

1. Presidential Commission, *Report* 1:103, 123, 139, 159; U.S. Congress, House, *Investigation: Report*, 144, 146, 245.

2. See appendix C. These three factors are combined in a theory of organizational misconduct that I first developed in Diane Vaughan, *Controlling Unlawful Organizational Behavior: Social Structure and Corporate Misconduct* (Chicago: University of Chicago Press, 1983), chaps. 4–6.

3. James G. March and Zur Shapira, "Managerial Perspectives on Risk and Risk Taking," *Management Science* 33 (1987): 1404–18.

4. Charles Perrow, *Normal Accidents: Living with High-Risk Technologies* (New York: Basic Books, 1984).

5. Dorothy Nelkin and Michael Brown, *Workers at Risk: Voices from the Workplace* (Chicago: University of Chicago Press, 1984).

6. Francis T. Cullen, William J. Maakestad, and Gray Cavender, *Corporate Crime under Attack: The Ford Pinto Case and Beyond* (Cincinnati: Anderson, 1987).

7. Robert A. Kagan and John T. Scholz, "The 'Criminology of the Corporation' and Regulatory Enforcement Strategies," in *Enforcing Regulation*, ed. Keith Hawkins and John M. Thomas (Boston: Kluwer-Nijhoff, 1984), 69–72.

8. Ibid., 70.

9. Keith Hawkins, "The Uses of Legal Discretion: Perspectives from Law and Social Science," in *The Uses of Discretion*, ed. Keith Hawkins (Oxford: Clarendon, 1992), 24.

10. Jeffrey Pfeffer and Gerald R. Salancik, *The External Control of Organizations: A Resource Dependence Perspective* (New York: Harper and Row, 1978), 40–54.

11. Vaughan, *Controlling Unlawful Organizational Behavior*, 59.

12. Everett C. Hughes, definitive contributor to the Chicago school of sociology, first raised this question in his exploration of the lack of response of German citizens to the Nazi regime's extermination of the Jews. See Hughes, "Good People and Dirty Work," *Social Problems* 10 (1962): 3–11.

13. See, e.g., Marshall B. Clinard and Peter C. Yeager, *Corporate Crime* (New York: Free Press, 1980); Edwin H. Sutherland, *White-Collar Crime* (New York: Dryden, 1949); Barry M. Staw and Eugene Swajkowski, "The Scarcity Munificence Component of Organizational Environments and the Commission of Illegal Acts," *Administrative Science Quarterly* 20 (1975): 345–54; Sally S. Simpson, "The Decomposition of Antitrust: Testing a Multi-Level Longitudinal Model of Profit-Squeeze," *American Sociological Review* 51 (1986): 859–75; Jacob Perez, "Corpo-

rate Criminality: A Study of the One Thousand Largest Industrial Corporations in the USA" (Ph.D. diss., University of Pennsylvania, 1978).

14. See, e.g., Gilbert Geis, "The Heavy Electrical Equipment Antitrust Cases of 1961," in *Criminal Behavior Systems*, ed. Marshall B. Clinard and Richard Quinney (New York: Holt, Rinehart and Winston, 1967), 131–51; Cullen, Maakestad, and Cavender.

15. Robert Jackall, *Moral Mazes: The World of Corporate Managers* (New York: Oxford University Press, 1988).

16. Peter N. Grabosky, *Wayward Governance: Illegality and Its Control in the Public Sector* (Woden: Australian Institute of Criminology, 1989).

17. See, e.g., Staw and Swajkowski; Kagan and Scholtz, 71–72; Clinard and Yeager; Vaughan, *Controlling Unlawful Organizational Behavior.*

18. Kagan and Scholtz; Vaughan, *Controlling Unlawful Organizational Behavior.*

19. March and Shapira, 1407–8.

20. For elaborations and variations of this theme, see Peter K. Manning, "'Big Bang' Decisions: Notes on a Naturalistic Approach," in *The Uses of Discretion*, ed. Keith Hawkins (Oxford: Clarendon, 1992); Michael Hechter, ed., *The Microfoundations of Macrosociology* (Philadelphia: Temple University Press, 1983); Constance Perin, "How Organizational, Technical, and Cultural Processes Work Together for Safety" (paper presented at Safety Culture in Nuclear Installations meeting of American Nuclear Society, Vienna, 24–28 April 1995).

21. Aaron Wildavsky, "Choosing Preferences by Constructing Institutions: A Cultural Theory of Preference Formation," *American Political Science Review* 81 (March 1987): 3–21; DiMaggio and Powell, "Introduction," 9–10.

22. DiMaggio and Powell, "Introduction," 10–11.

23. This brief summary refers to the Carnegie school of organization theory. Among the foundational works are Herbert A. Simon, *Administrative Behavior: A Study of Decision-Making Processes in Administrative Organizations*, 3d ed. (New York: Free Press, 1976); James G. March and Herbert A. Simon, *Organizations* (New York: Wiley, 1958); Richard M. Cyert and James G. March, *A Behavioral Theory of the Firm* (Englewood Cliffs, N.J.: Prentice-Hall, 1963).

24. March and Simon.

25. Simon, *Administrative Behavior.*

26. Karl E. Weick, *The Social Psychology of Organizing* (Reading, Mass.: Addison-Wesley, 1979).

27. James G. March and Johan P. Olsen, *Ambiguity and Choice* (Bergen: Universitetsforlaget, 1979); Charles E. Lindblom, "The Science of Muddling Through," *Public Administration Review* 19 (1959): 79–88.

28. For a wonderful example of how the media created history from Feynman's ice water demonstration, see Thomas F. Gieryn and Anne E. Figert, "Ingredients for a Theory of Science in Society: O-Rings, Ice Water, C-Clamps, Richard Feynman, and the *New York Times*," in *The-*

ories of Science in Society, ed. Susan Cozzens and Thomas F. Gieryn (Bloomington: Indiana University Press, 1990).

29. W. Leon Ray, telephone interview by author, 13 October 1992.

30. Presidential Commission, *Report* 4:811.

31. Ibid., 793.

32. Ibid. 2:H68–H95.

33. McConnell, 52–53.

34. Bell and Esch, 39; National Aeronautics and Space Administration, Source Evaluation Board, "Selection of Contractor for Space Shuttle Program Solid Rocket Motors," 19 November 1973, National Archives, Washington, D.C., pp. 20–21.

35. William J. Broad, "NASA Chief Might Not Take Part in Decisions on Booster Contracts; Accusations of Bias in '73 Award to Thiokol Prompt His Statement," *New York Times,* 7 December 1986, sec. A, 1.

36. "The Shuttle Inquiry: Business Aspects. Shuttle's Builders Have Varying Stakes in Program's Future," *New York Times,* 30 January 1986, sec. A, 5; Steven Greenhouse, "Thiokol: Contractor at Biggest Risk," *New York Times,* 3 February 1986, sec. A, 11.

37. Presidential Commission, *Report* 1:140.

38. Ibid., 1.

39. Cook, "Rogers Commission," 13–21. See also Richard C. Cook, "The Challenger Report: A Critical Analysis of the Report to the President of the Presidential Commission on the Space Shuttle *Challenger* Accident" (1986, mimeograph; for copies, contact Richard C. Cook, Rt. 1, Box 1182, King George, VA 22485).

40. Phillip M. Boffey, "Panel Asks NASA for All Records on Booster Risks: Space Agency to Comply," *New York Times,* 10 February 1986, sec. B, 10.

41. Presidential Commission, *Report* 1:1.

42. Cook, *"Challenger* Report"; Philip Shenon, "NASA Accused of Cover-up in Shuttle Deaths," *New York Times,* 14 November 1988, sec. B, 6; Linda Greenhouse, "NASA Told to Release *Challenger* Disaster Tape," *New York Times,* 30 July 1988, sec. A, 31.

43. William E. Schmidt, "Reporters Use New Technology to Thwart NASA's Secrecy," *New York Times,* 20 March 1986, sec. A, 24; Evan Thomas, "Painful Legacies of a Lost Mission," *Time,* 24 March 1986, 28–29.

44. Michael Tackett, *"Challenger* Cabin, Crew Found: Shuttle's Gear May Yield Clues," *Chicago Tribune,* 10 March 1986, sec. A, 1.

45. Phillip M. Boffey, *"Challenger* Crew Knew of Problem, Data Now Suggest; NASA Reverses Its Stand," *New York Times,* 29 July 1986, sec. A, 1.

46. Excerpts from transcription of flight-deck tape recording appears in Boffey, *"Challenger* Crew Knew of Problem." Complete transcription appears in Lewis, 15–16.

47. Kerwin's report concluded that the astronauts perished either because of loss of air due to cabin decompression or upon impact with

478 NOTES TO PAGES 44-49

the water. The cause of death could not be accurately determined (Boffey, "*Challenger* Crew Knew of Problem").

48. Presidential Commission, "Views of *Challenger* Space Shuttle."

49. Cook, "*Challenger* Report"; "NASA Cover-up of *Challenger* Data Alleged," *Boston Globe,* 13 November 1988, sec. A, 1. According to Commission investigator Jack Macidull, whether death was painful or instantaneous was a factor relevant in determining the amount of insurance claims against NASA (Jack Macidull, telephone interview by author, 5 April 1995).

50. Presidential Commission, *Report* 4:378.

51. Cook to Mann, "Problem With SRB Seals."

52. Storer Rowley and Michael Tackett, "Internal Memo Charges NASA Compromised Safety," *Chicago Tribune,* 9 March 1986, sec. A, 8; Robert Reinhold, "Astronauts' Chief Says NASA Risked Life for Schedule: 'Awesome' List of Flaws," *New York Times,* 9 March 1986, sec. A, 1.

53. The list was prepared by Stephen G. Bales, Chief of the Johnson Space Center's System Division, in response to a request from the Johnson National Space Transportation System Program Office for a review of all flight safety critical items in the Shuttle Program. For the list, see "Flight Safety Critical Items Analyzed in Wake of Accident," *Aviation Week and Space Technology,* 17 March 1986, 87–95.

54. Excerpt from memo text, *New York Times,* 9 March 1986, sec. A, 36.

55. Phillip M. Boffey, "Engineer Who Opposed Launching *Challenger* Sues Thiokol for $1 Billion," *New York Times,* 28 January 1987, sec. A, 1. See also *Boisjoly v. Morton Thiokol, Inc.,* 706 F.Supp. 795 (D.Utah 1988).

56. See *United States ex rel. Boisjoly v. Morton Thiokol, Inc.,* Civil Action no. 87-209 (D.D.C.), 3.

57. Ibid., 8.

58. Presidential Commission, *Report* 1:148; 5:1446. See also Richard P. Feynman, "Personal Observations on Reliability of Shuttle," ibid. 2: appendix F.

59. In no uncertain terms, however, the Commission concluded that commercial interests were directly linked to Thiokol management's decision making the night before the launch. Thiokol was in the midst of negotiation over a renewal of its multibillion-dollar NASA contract to continue producing the SRBs (ibid. 1:139; McDougall, 438–40). In the official report, the Commission stated that Thiokol management reversed its first position and recommended launch at the urging of Marshall managers and contrary to the views of its own engineers *in order to* "accommodate a major customer" (Presidential Commission, *Report* 1:104; 2:F4).

60. R. L. Banks and S. C. Wheelwright, "Operations vs. Strategy: Trading Tomorrow for Today," *Harvard Business Review,* May/June 1979, 112–20; Cyert and March.

61. "Shuttle Disaster Puts Spotlight on Safety: Trade-off for Performance Is a Critical Question," *Wall Street Journal,* 4 February 1986, 26.

62. Presidential Commission, *Report* 5:1531.

63. Carol A. Heimer, "Social Structure, Psychology, and the Estimation of Risk," *Annual Review of Sociology* 15 (1988): 503.

64. Presidential Commission, *Report* 5:867.

65. See, e.g., Stuart Diamond, "NASA Wasted Billions, Federal Audits Disclose," *New York Times*, 23 April 1986, sec. A, 1; Diamond, "NASA Cut or Delayed Safety Spending." NASA rebuttal to these articles can be found in David W. Garrett, Press Release no. 86–52, in *NASA News* (Washington, D.C.: NASA Headquarters, 25 April 1986). See also Trento; McConnell; Easterbrook, 46–60.

66. For a cogent discussion of the very limited abort possibilities even after SRB burn, see Karl Esch, "How NASA Prepared to Cope with Disaster," *IEEE Spectrum*, March 1986, 32–36. Significantly, the Esch article was based on an interview assessing astronaut survival that was conducted before the accident, it is thus untinged by retrospection. But recommendations for escape mechanisms *after* the accident were of equally limited potential. See "Johnson Team Urges Parachutes, Rockets for Shuttle Crew Escape," *Aviation Week and Space Technology*, 14 July 1986, 141. See also Presidential Commission, *Report* 2:J24.

67. Marshall engineer W. Leon Ray was a member of the Source Evaluation Board that assessed contractor designs; see W. Leon Ray, interview transcript, 25 March 1986, Marshall Space Flight Center files, National Archives, Washington, D.C., p. 5. See full analysis in appendix A of this volume.

68. Erving Goffman, *The Presentation of Self in Everyday Life* (Garden City, N.Y.: Anchor, 1959); Jack Katz, "Cover-up and Collective Integrity: On the Natural Antagonisms to Authority Internal and External to Organizations," *Social Problems* 25 (1979): 3–17.

69. Data from STS 51-L were secured in accordance with the Contingency Operations Plan that required all data tape originals to be impounded and copied. All flight controller logs and hardcopies also were impounded. Copies were made, inventories were created, and access was controlled (Presidential Commission, *Report* 2:J23).

70. Judson A. Lovingood, interview transcript, 23 April 1986, Marshall Space Flight Center files, National Archives, Washington, D.C.; Presidential Commission, "Views of *Challenger* Space Shuttle."

71. Jack Macidull, telephone interview by author, 28 July 1994.

72. Perrow, *Normal Accidents.*

73. Macidull, telephone interview, 28 July 1994. John F. Clark, who served on the failure review board investigating Apollo 13 in 1970 and at the time of this writing was vice president of public policy at the American Institute of Aeronautics and Astronautics, verified the constraint that fault tree analysis puts on the release of information. He wrote in a letter to the *New York Times* that much of what was construed as cover-up after the accident was NASA's unwillingness to release information until they had completed an analysis of the data, which requires three to six months ("NASA Has Been Telling the Truth All Along," letter, *New York Times*, 18 August 1986, sec. A, 16).

74. "NASA Cover-up of *Challenger* Data Alleged," 18.

75. William R. Freudenberg, "Nothing Recedes Like Success? Risk Analysis and the Organizational Amplification of Risks," *Risk* 3 (1992): 1–35. For extensive confirmation of this assertion, see publications list of the Disaster Research Center, University of Delaware, which specializes in the organizational response to large-scale community crises.

76. Bruce Murray, *Journey into Space: The First Thirty Years of Space Exploration* (New York: Norton, 1989), 225.

77. Bunda L. Dean, telephone interview by author, 20 June 1992.

78. Alex S. Jones, "Withheld Shuttle Data: A Debate over Privacy," *New York Times*, 27 January 1987, sec. C, 1.

79. Ibid. NASA based its refusal to release on Exemption 6 of the FOIA (passed by Congress in 1977), which allows exemption from mandatory disclosure of personnel and similar files if disclosure would constitute a "clearly unwarranted invasion of privacy."

80. When journalists sued, the space agency's right to withhold the information was affirmed in the courts ("NASA Cover-up of *Challenger* Data Alleged").

81. Roger Boisjoly, personal interview by author, 15 February 1988.

82. John Walton, "Making the Theoretical Case," in *What Is a Case? Exploring the Foundations of Social Inquiry*, ed. Charles C. Ragin and Howard S. Becker (New York: Cambridge University Press, 1992), 125.

83. Presidential Commission, *Report* 1:125–27.

84. Ibid., 159.

85. See photocopy of documents submitted and approved for the waiver of the requirements for the SRM seal (ibid. 1:239–44). Waivers will be discussed in detail in chap. 4.

86. Mulloy's lifting of the Launch Constraint will be discussed in detail in chap. 5.

87. Most often, the term is used to indicate a negative evaluation, but being a genius or being a serial killer will each violate the norms of many groups.

88. Out of the country for the academic year 1986–87, I had written to teleconference participants about my research. In August 1986, I mailed copies of a preliminary paper I had written to 38 NASA and Thiokol participants in order to get feedback and pave the way for future interviews. During the fall, I received 10 responses at the Centre for Socio-Legal Studies, Wolfson College, Oxford, where I was working, when a letter arrived stating that the writer would like to help me, but since astronauts' families recently had sued, he did not feel free to give out any information. I received no more replies.

89. For details, see appendix C.

90. Robert Darnton, *The Great Cat Massacre and Other Episodes in French Cultural History* (New York: Basic Books, 1984), 6.

91. AFter the disaster, much attention was given to rule violations throughout the Space Shuttle Program. In a two-article series in the *New York Times*, Stuart Diamond reported many rule violations at NASA

("NASA Wasted Billions"; "NASA Cut or Delayed Safety Spending"). My inquiry only focused on rules and procedures allegedly violated by Level III and IV managers and engineers in the SRB Project work group.

92. For a review of research substantiating my position, see Heimer, "Social Structure," 500–501.

93. Peter L. Berger and Thomas Luckmann, *The Social Construction of Reality* (New York: Doubleday, 1966); Peter Marris, *Loss and Change* (London: Routledge and Kegan Paul, 1974), 5–22; Perrow, *Normal Accidents,* 321–24; Diane Vaughan, *Uncoupling: Turning Points in Intimate Relations* (New York: Oxford University Press, 1986).

94. Marris, 9–11; Gerald Zaltman, "Knowledge Disavowal" (paper presented at conference on Producing Useful Knowledge for Organizations, Graduate School of Business, University of Pittsburgh, October 1982).

95. Zaltman, 3–5.

96. Vaughan, *Uncoupling,* 62–102.

97. William R. Freudenberg, "Perceived Risk, Real Risk: Social Science and the Art of Probabilistic Risk Assessment," *Science* 242 (October 1988): 44–49.

98. Ibid.

99. John Van Maanen and Steve Barley, "Cultural Organization: Fragments of a Theory," in *Organizational Culture,* ed. Peter J. Frost, Larry F. Moore, Meryl Ries Louis, Craig C. Lundberg, and Joanne Martin (Beverly Hills, Calif.: Sage, 1985).

100. Kathleen L. Gregory, "Native-View Paradigms: Multiple Cultures and Culture Conflicts in Organizations," *Administrative Science Quarterly* 28 (1983): 359–76; Meryl Reis Louis, "An Investigator's Guide to Workplace Culture," in *Organizational Culture,* ed. Peter J. Frost, Larry F. Moore, Meryl Ries Louis, Craig C. Lundberg, and Joanne Martin (Beverly Hills, Calif.: Sage, 1985); Harrison M. Trice and Janice M. Beyer, *The Cultures of Work Organizations* (Englewood Cliffs, N.J.: Prentice-Hall, 1993); Van Maanen and Barley.

101. The many facets of organizational culture are examined in Joanne Martin, *Cultures in Organizations: Three Perspectives* (New York: Oxford University Press, 1992). This logic is also demonstrated empirically in research on structure and process. See, e.g., Karl E. Weick, "Educational Organizations as Loosely Coupled Systems," *Administrative Science Quarterly* 21 (1976): 1–19; John Van Maanen and Edgar H. Schein, "Toward a Theory of Organizational Socialization," in *Research in Organizational Behavior,* ed. Barry Staw and L. L. Cummings (Greenwich, Conn.: JAI, 1979).

102. The three-factor technical rationale confirming SRB joint redundancy will be explained in the chronological history of decision making in chaps. 3–5. This rationale for accepting risk consisted of the experience base, safety margin, and self-limiting aspects of joint sealing.

103. Maury Silver and Daniel Geller, "On the Irrelevance of Evil: The Organization and Individual Action," *Journal of Social Issues* 34 (1978): 125–35.

104. Dorothy E. Smith, "The Social Construction of Documentary Reality," *Sociological Inquiry* 44 (1974): 257–67.

105. William H. Starbuck and Frances J. Milliken, "Executives' Perceptual Filters: What They Notice and How They Make Sense," in *The Executive Effect*, ed. Donald Hambrick (Greenwich, Conn.: JAI, 1988).

106. Ibid., 38.

107. Barry Turner, *Man-Made Disasters* (London: Wykeham, 1978); Barry Turner, "The Organizational and Interorganizational Development of Disasters," *Administrative Science Quarterly* 21 (1976): 383, 392–93. Perrow found that in the accident at the nuclear power plant at Three Mile Island, warning signals before the accident were seen as "background noise"; only in retrospect did they become signals of danger to insiders. Charles Perrow, "The President's Commission and the Normal Accident," in *The Accident at Three Mile Island: The Human Dimensions*, ed. David Sills, Charles Wolf, and Vivian Shelanski (Boulder, Colo.: Westview, 1981).

108. Lawrence B. Mulloy, telephone interview by author, 19 March 1993.

109. Lawrence B. Mulloy, "Level I Flight Readiness Review, 51-E," Reel .098, 21 February 1985, Motion Picture and Video Library, National Archives, Washington, D.C.

110. For explanation, see U.S. Congress, House, *Investigation: Hearings* 1:410–12.

111. For acknowledgment of the videotape and a transcript, see Presidential Commission, *Report* 2:H2, H42.

112. For other examples and precedent for ethnographic studies of context, meaning, and its effects on decisions and action in organizations, see Peter K. Manning, *The Narc's Game* (Cambridge: MIT Press, 1980); Peter K. Manning, *Symbolic Communication: Signifying Calls and the Police Response* (Cambridge: MIT Press, 1988); Peter K. Manning, *Organizational Communication* (New York: Aldine de Gruyter, 1992). For a similar strategy examining risk assessments, see Constance Perin, "Organizations as Contexts: Implications for Safety Science and Practice," *Industrial and Environmental Crisis Quarterly* 8 (in press); Jean-Michel Saussois and Herve Laroche, "The Politics of Labeling Organizational Problems: An Analysis of the *Challenger* Case," *Knowledge and Policy: The International Journal of Knowledge Transfer* 4 (1991): 89–106.

113. Barbara W. Tuchman, *Practicing History: Selected Essays* (New York: Ballantine, 1981), 75.

114. See, e.g., Stephen Jay Gould, *The Mismeasure of Man* (New York: Norton, 1981); Thomas S. Kuhn, *The Structure of Scientific Revolutions* (Chicago: University of Chicago Press, 1962).

CHAPTER 3

1. Presidential Commission, *Report* 1:120.

2. Gary Alan Fine, *With the Boys* (Chicago: University of Chicago

Press, 1987); Clifford Geertz, *The Interpretation of Cultures* (New York: Basic Books, 1973), 17.

3. Louis, 91.

4. Clifford Geertz, "Thick Description: Toward an Interpretive Theory of Culture," in *The Interpretation of Cultures* (New York: Basic Books, 1973), 14–17.

5. Ibid., 14.

6. Thomas P. Hughes, "The Evolution of Large Scale Technical Systems," in *The Social Construction of Technological Systems: New Directions in the Sociology and History of Technology,* ed. Wiebe Bijker, Thomas P. Hughes, and Trevor J. Pinch (Cambridge: MIT Press, 1987).

7. J. B. Hammack and M. L. Raines, *Space Shuttle Safety Assessment Report,* Johnson Space Center, Safety Division, 5 March 1981, National Archives, Washington, D.C., p. 10.

8. Lawrence O. Wear, interview transcript, 12 March 1986, Marshall Space Flight Center files, National Archives, Washington, D.C., p. 27.

9. See also Peter Whalley, *The Social Production of Technical Work* (Albany: State University of New York Press, 1986).

10. Hammack and Raines, 11–12.

11. Feynman, "Personal Observations on Reliability of Shuttle," F1–F5. See also how these phrases were singled out for attention in "Flight Readiness Review Treatment of the O-Ring Problems," Presidential Commission, *Report* 2:H1–H3.

12. I began uncovering the five-step decision sequence in 1987–88 and did not find the document describing the Acceptable Risk Process until working on the project at the National Archives in 1989. It was not until 1991 that I realized that in addition to revealing the norms, beliefs, and procedures of the work group, my discovery of this sequence was independent proof that work group decision making conformed to NASA rules for hazard assessment, as described in the *Shuttle Safety Assessment Report.*

13. I analyzed FRR documents for all levels of FRR for four launches: STS 51-I (August 27, 1985), 51-J (October 3, 1985), 61-A (October 30, 1985), and 61-B (November 26, 1985).

14. Johnson Space Center, JSC-14814, vols. 1-4, cited in Hammack and Raines.

15. U.S. Congress, House, *Investigation: Report,* 123–26; Presidential Commission, *Report* 4:178–83.

16. For each mission, a preliminary launch date is established when payloads are assigned. Optimally, the launch would occur at the end of the 15 months, but changes in the flight manifest and technical issues needing additional time to resolve typically extended the 15 months to more, calling for change in the scheduled launch date. Thus, each launch has two dates, the preliminary one and a final one established after mission preparation is well underway. For example, STS 51-L, the *Challenger* mission, was first scheduled for July 2, 1985, then rescheduled for January 23, 1986. See Presidential Commission, *Report* 2:J7.

17. "Space Shuttle Flight Readiness Reviews," NASA Program Directive, Space Shuttle Operations, Integration Division, SPO-PD 710.5A, 26 September 1983, cited in Presidential Commission, *Report* 1:82, reproduced in U.S. Congress, House, *Investigation: Report*, 380–83.

18. Pfeffer and Salancik, 40–54.

19. James D. Smith, personal interview by author, 8 June 1992.

20. Ibid.

21. Lawrence O. Wear, personal interview by author, 2 June 1992.

22. W. Leon Ray, telephone interview by author, 27 January 1993.

23. These changes were identified in FRR documents as "Class 1 changes." Many Class 1 changes were reported in FRR for each flight.

24. Lawrence B. Mulloy, telephone interview by author, 27 January 1993.

25. Keith E. Coates, interview transcript, 25 March 1986, Marshall Space Flight Center files, National Archives, Washington, D.C., p. 15.

26. This problem, which Pinch and Bijker call "interpretive flexibility," will be discussed extensively in chap. 6. See Trevor J. Pinch and Wiebe E. Bijker, "The Social Construction of Facts and Artefacts: Or How the Sociology of Science and the Sociology of Technology Might Benefit Each Other," *Social Studies of Science* 14 (1984): 399–441.

27. Roger M. Boisjoly, telephone interview by author, 19 March 1993.

28. Coates, interview transcript, 25 March 1986, 12.

29. John Q. Miller, interview transcript, 13 March 1986, Marshall Space Flight Center files, National Archives, Washington, D.C., p. 27.

30. See also Todd R. LaPorte, "Conditions of Strain and Accommodation in Industrial Research Organizations," *Administrative Science Quarterly* 10 (1965): 21–38.

31. See also Robert Zussman, *Mechanics of the Middle Class: Work and Politics among American Engineers* (Berkeley and Los Angeles: University of California Press, 1985), 106.

32. Ray, telephone interview, 27 January 1993.

33. Wear, interview transcript, 12 March 1986, 83.

34. U.S. Congress, House, *Investigation: Report*, 126.

35. Reproduced in Presidential Commission, *Report* 4:188–90.

36. Lawrence O. Wear, personal interview by author, 5 June 1992. 37. Smith, personal interview, 8 June 1992.

38. U.S. Congress, House, *Investigation: Hearings* 1:375.

39. Lawrence O. Wear, personal interview by author, 8 June 1992.

40. Lawrence B. Mulloy, personal interview by author, 8 June 1992.

41. Wear, personal interview, 5 June 1992.

42. U.S. Congress, House, *Investigation: Hearings* 1:394.

43. Alex Roland, personal conversation, 9 June 1995.

44. "Space Shuttle Flight Readiness Reviews," 26 September 1983, 2.

45. Ibid.; Wear, personal interview, 5 June 1992; Lawrence B. Mulloy, telephone interview by author, 5 August 1992.

46. U.S. Congress, House, *Investigation: Report*, 70.

47. Ben Powers, interview transcript, 12 March 1986, Marshall Space Flight Center files, National Archives, Washington, D.C., pp. 75–76.

48. See also H. M. Collins, "The Place of the Core-Set in Modern Science: Social Contingency with Methodological Propriety in Science," *History of Science* 19 (1981): 6–19.

49. Ibid.

50. See, e.g., Alex McCool, interview transcript, 27 March 1986, Marshall Space Flight Center files, National Archives, Washington, D.C., p. 39.

51. Smith, personal interview, 8 June 1992.

52. Bell and Esch, 44.

53. Ibid., 40; Presidential Commission, *Report* 1:121.

54. Bell and Esch, 44; Presidential Commission, *Report* 1:122.

55. Ray, interview transcript, 25 March 1986, 5–9.

56. They had no data indicating rotation occurred on the Titan (ibid., 7). Later, after they discovered joint rotation, they "were able then to get our hands on some Titan experience that we couldn't get before, and we found out that indeed the Titan joint was opening up" (Jack R. Kapp, interview transcript, 12 April 1986, Morton Thiokol, Inc., files, National Archives, Washington, D.C., p. 17). See also Presidential Commission, *Report* 5:1520.

57. Bell and Esch, 42; Presidential Commission, *Report* 1:121–22.

58. W. Leon Ray, telephone interview by author, 21 July 1992; Howard McIntosh, interview transcript, 2 April 1986, Morton Thiokol, Inc., files, National Archives, Washington, D.C., p. 16.

59. Presidential Commission, *Report* 1:123.

60. Ibid.

61. Ray, interview transcript, 25 March 1986, 62.

62. McIntosh, interview transcript, 2 April 1986, 9–16.

63. Ibid., 13.

64. Ibid., 18.

65. Ibid., 27.

66. George B. Hardy, interview transcript, 3 April 1986, Marshall Space Flight Center files, National Archives, Washington, D.C., p. 43.

67. Presidential Commission, *Report* 1:233.

68. Ray, telephone interview, 21 July 1992.

69. Presidential Commission, *Report* 5:1621.

70. Ibid., 1620.

71. Ibid. 1:234–35; 5:1619–20; Ray, interview transcript, 25 March 1986, 9; John Q. Miller, interview transcript, 27 March 1986, Marshall Space Flight Center files, National Archives, Washington, D.C., p. 24.

72. Ray, interview transcript, 25 March 1986, 16–24; see esp. p. 22.

73. Presidential Commission, *Report* 1:234–35.

74. Ray, interview transcript, 25 March 1986, 11, 13.

75. Presidential Commission, *Report* 1:124.

76. Ibid., 236.

77. The primary O-ring was sealing by extruding into the gap that occurred at ignition. Industry and government application practices were that it was to seal by compression, not extrusion (Bell and Esch, 40). The joint was designed to seal by compression. Combustion gas pressure would displace the putty in the space between the motor segments. The process by which the O-rings were to seal the joint was as follows: "The displacement of the putty would act like a piston and compress the air ahead of the primary O-ring, and force it into the gap between the tang and clevis. This process is known as pressure acruation of the O-ring seal. This pressure actuated sealing is required to occur very early during the Solid Rocket Motor ignition transient, because the gap between the tang and clevis increases as pressure loads are applied to the joint during ignition" (Presidential Commission, Report 1:57–58).

78. Ray, telephone interview, 21 July 1992.

79. Arnold R. Thompson, interview transcript, 4 April 1986, Morton Thiokol, Inc., files, National Archives, Washington, D.C., p. 6; McIntosh, interview transcript, 2 April 1986, 23–24.

80. McIntosh, interview transcript, 2 April 1986, 23–24.

81. Presidential Commission, Report 1:238.

82. Ray, interview transcript, 25 March 1986, 26.

83. Ibid.

84. Jack R. Kapp, interview transcript, 2 April 1986, Morton Thiokol, Inc., files, National Archives, Washington, D.C., p. 29.

85. Ben Powers, personal interview by author, 4 June 1992.

86. Ray, interview transcript, 25 March 1986, 21–22.

87. Ibid.

88. Morton Thiokol engineer Arnold R. Thompson reports on the tests that went into his decision to shim with one-size rather than variable-sized shims in Thompson, interview transcript, 4 April 1986, 18–20.

89. McIntosh, interview transcript, 2 April 1986, 48–49.

90. Thompson, interview transcript, 4 April 1986, 11.

91. Kapp, interview transcript, 2 April 1986, 30–31.

92. Ray, telephone interview, 21 July 1992.

93. Ray, interview transcript, 25 March 1986, 61.

94. Maurice Parker, interview transcript, 4 April 1986, Marshall Space Flight Center files, National Archives, Washington, D.C., pp. 26–27.

95. Roger M. Boisjoly, interview transcript, 2 April 1986, Morton Thiokol, Inc., files, National Archives, Washington, D.C., pp. 8–9.

96. Presidential Commission, Report 1:125.

97. Ibid.

98. U.S. Congress, House, Investigation: Report, 142.

99. Ibid., 153.

100. Hammack and Raines, 12.

101. Mulloy, telephone interview, 5 August 1992.

102. Presidential Commission, Report 1:239.

103. The STA-1 tests "verified the seals capability under flight loads and further verified the redundancy of the secondary seal." Recall that

Marshall's Leon Ray did not agree with Thiokol's interpretation of these data. Despite this unresolved dispute, the other tests performed had convinced both camps that the primary provided an effective seal and the secondary an effective backup if the primary failed initially. Since they had no concerns about erosion at this point, they did not envision the possibility that the primary might not seal initially due to erosion.

104. Presidential Commission, *Report* 1:239–40.

105. Henry Petroski, *To Engineer Is Human: The Role of Failure in Successful Design* (New York: St. Martin's, 1985).

106. Gurney and Forte, vii–viii.

107. Robert M. Emerson, "Holistic Effects in Social Control Decision-Making," *Law and Society Review* 17 (1983): 425–55.

108. Gerald R. Salancik, "Commitment and the Control of Organizational Behavior and Belief," in *New Directions in Organizational Behavior,* ed. Barry M. Staw and Gerald R. Salancik (Malabar, Fla.: Krieger, 1977), 33–35; Henry Mintzberg, Duru Raisinghani, and André Théorêt, "The Structure of Unstructured Decision Processes," *Administrative Science Quarterly* 21 (1976): 246–75.

109. Keith B. Coates, interview transcript, 10 April 1986, Marshall Space Flight Center files, National Archives, Washington, D.C., p. 34.

110. Carol A. Heimer, "Doing Your Job *and* Helping Your Friends: Universalistic Norms about Obligations to Particular Others in Networks" (Department of Sociology, Northwestern University, 1992; manuscript).

111. Wear, personal interview, 5 June 1992.

112. Herbert A. Simon, *Models of Man: Social and Rational* (New York: Wiley, 1957).

113. John Schell, interview transcript, 25 March 1986, Marshall Space Flight Center files, National Archives, Washington, D.C., pp. 16–17.

114. Presidential Commission, *Report* 5:1623.

115. Smith, personal interview, 8 June 1992.

CHAPTER 4

1. For a parallel, see also Vaughan, *Uncoupling.*

2. Jack Buchanan, interview transcript, 25 March 1986, Morton Thiokol, Inc., files, National Archives, Washington, D.C., pp. 34–37.

3. Presidential Commission, *Report* 1:125.

4. U.S. Congress, House, *Investigation: Hearings* 1:762.

5. Powers, interview transcript, 12 March 1986, 42.

6. Presidential Commission, *Report* 1:133.

7. Kapp, interview transcript, 2 April 1986, 44–48.

8. Thompson, interview transcript, 4 April 1986, 28.

9. McIntosh, interview transcript, 2 April 1986, 77.

10. Arthur L. Stinchcombe, "Social Structure and Organizations," in *Handbook of Organizations,* ed. James G. March (Chicago: Rand McNally, 1965).

11. Roger M. Boisjoly, telephone interview by author, 18 March 1993.

12. Ibid.

13. Ray, telephone interview, 27 January 1993.

14. Boisjoly, telephone interview, 18 March 1993.

15. George B. Hardy, telephone interview by author, 7 April 1993.

16. Ibid.

17. Morton Thiokol, Inc., "Post-Flight Evaluation of STS-2 SRM Components," Report TRW-13 286, January 1983, National Archives, Washington, D.C.

18. Gurney and Forte, 41.

19. Presidential Commission, *Report* 1:5; Bell and Esch, 45–46.

20. This section on the capture feature is drawn from William J. Broad, "NASA Had Solution to Key Flaw in Rocket When Shuttle Exploded," *New York Times*, 22 April 1986, sec. A, 1.

21. Ibid., sec. B, 8.

22. Kapp, interview transcript, 2 April 1986, 9–12.

23. Ray, telephone interview, 27 January 1993.

24. Ibid.

25. Mulloy joined NASA in 1960, first working as a loads and dynamics analyst and then moving into the Apollo Program. After a leave for postgraduate work in public administration, he became Chief Engineer of the External Tank Project for the Space Shuttle Program, remaining in that position until 1979. Then he became Chief Engineer for NASA on the inertial upper stage in conjunction with the Air Force until taking charge of the SRB in 1982 (Presidential Commission, *Report* 5:826).

26. Boyd C. Brinton, interview transcript, 11 April 1986, Morton Thiokol, Inc., files, National Archives, Washington, D.C., p. 23; Gurney and Forte, 39.

27. Lawrence B. Mulloy, interview transcript, 2 April 1986, Marshall Space Flight Center files, National Archives, Washington, D.C., p. 7.

28. Ibid., 6–10.

29. Presidential Commission, *Report* 1:153.

30. National Aeronautics and Space Administration, "Space Shuttle Program Requirements Document, Level I" (Washington, D.C., 30 June 1977), par. 2.8, cited in Presidential Commission, *Report* 1:127.

31. Presidential Commission, *Report* 1:126–27; 5:1658–60.

32. Mulloy, interview transcript, 2 April 1986; see also Hardy, interview transcript, 3 April 1986.

33. McIntosh, interview transcript, 2 April 1986, 58–71.

34. Ibid., 17–19.

35. Allan J. McDonald, interview transcript, 19 March 1986, Morton Thiokol, Inc., files, National Archives, Washington, D.C., pp. 49–50.

36. Parker, interview transcript, 4 April 1986, 29.

37. Ray, telephone interview, 21 July 1992.

38. McIntosh, interview transcript, 2 April 1986, 63–67.

39. Presidential Commission, *Report* 1:126.

40. Mulloy, interview transcript, 2 April 1986, 21.

41. Wiley C. Bunn, interview transcript, 3 April 1986, Marshall Space Flight Center files, National Archives, Washington, D.C., pp. 50–52.

42. Ray, telephone interview, 21 July 1992.

43. Presidential Commission, *Report* 1:241.

44. See also McIntosh, interview transcript, 2 April 1986, 58–66; Presidential Commission, *Report* 1:123–24, 126.

45. Presidential Commission, *Report* 5:1540.

46. Ibid., 833.

47. Wiley C. Bunn, interview transcript, 17 April 1986, Marshall Space Flight Center files, National Archives, Washington, D.C., p. 23.

48. Presidential Commission, *Report* 1:153.

49. Ibid.

50. NASA, "Space Shuttle Program Requirements Document, Level I," par. 2.8.

51. Presidential Commission, *Report* 5:833; Mulloy, interview transcript, 2 April 1986, 27–28.

52. The ECR for this waiver was written by Marshall engineer W. Trewhitt (Presidential Commission, *Report* 1:243).

53. Ibid.

54. Ibid. 5:1658–62; Mulloy, interview transcript, 2 April 1986, 47–49.

55. The notation reads: "This PRCBD [Program Requirements Control Board Directive] was processed outside the formal Level II PRCB. In the event a discrepancy or impact is identified, implementation of this change shall be held in abeyance pending further disposition by the Level II PRCB. Such discrepancies and/or impacts shall be immediately identified to the Level II PRCB secretary and submitted on a Level II Change Request for appropriate action within sixty days after approval of this directive" (Presidential Commission, *Report* 1:224).

56. Ibid. 5:1660; see also Mulloy, interview transcript, 2 April 1986, 47–48.

57. Mulloy, interview transcript, 2 April 1986, 47–49.

58. Ibid., 48.

59. Presidential Commission, *Report* 5:1662.

60. Ibid. 1:129; 2:H1.

61. Brian Russell, interview transcript, 7 April 1986, Morton Thiokol, Inc., files, National Archives, Washington, D.C., pp. 12–13.

62. "Space Shuttle Flight Readiness Reviews," 26 September 1983, 2.

63. Lawrence B. Mulloy, telephone interview by author, 12 July 1992.

64. For complete FRR report of engineering analysis, see Presidential Commission, *Report* 2:H5–H12.

65. William H. Starbuck and Francis Milliken, "Challenger: Fine-Tuning the Odds until Something Breaks," *Journal of Management Studies* 25 (1988): 319–40.

66. Presidential Commission, *Report* 2:H9.

67. Note the difference in pounds per square inch: since the field joints were more vulnerable to anomalies than the nozzle joint, the field

joint pressure was higher in order to assure sealing. On STS-9, one of the right SRB field joints was pressurized using a 200-psi leak check after a destack. Subsequently, 200 psi was used on all field joints. Because of the distinctive design on nozzle joints, 100 psi was thought to be sufficient for leak check pressure on nozzle joints. STS 41-B was the first flight after STS-9 to be pressurized at 200 psi in the initial field joint leak checks. (Ibid. 1:129.)

68. Ibid., 133–34; Bell and Esch, 43.

69. Presidential Commission, *Report* 1:129–31.

70. Ibid. 2:H6–H7.

71. Ibid., H6, chart 5; H7, chart 9.

72. Ibid., H9, chart 12.

73. Ibid., H7, chart 9; see reference to "short time, localized impingement of hot gas" at ninth bullet. This first-time reference to what later became a well-established standard is admittedly brief but was accompanied by more-detailed oral description in FRR. More complete descriptions of the concept of "self-limitation" appear in testimony and later FRR documents.

74. When pressure was injected through the leak check port after booster assembly to check for leaks, it pushed the secondary downstream, into proper position for sealing, but pushed the primary O-ring upstream to the opposite side of its groove. Before the primary could seal, ignition pressure had to push it "across" its groove to the downstream position. While ignition pressure pushed it downstream into proper position to seal, the potential existed for hot gas to impinge on the primary for about 330 milliseconds. Erosion would stop when (1) pressure equalized in the cavity between the putty and the primary O-ring after the primary O-ring sealed or (2) pressure between the primary and secondary O-rings equaled the motor pressure (ibid., H42).

75. Ibid., H7, chart 9; H11, chart 16. Mulloy, in a typed transcript of a presentation he made at the FRR for STS 51-E, explained the technical aspects of erosion as a self-limiting phenomenon: "The rationale that was developed after observing this erosion on STS-11 (STS-41B) was that it was a limited duration that was self-limiting in that as soon as the pressure in the cavity either between the putty and the primary O-ring after the primary O-ring seats, or the pressure between the primary and the secondary O-ring, equals the motor pressure, the flow stops and the erosion stops" (ibid., H42).

76. Marshall Space Flight Center, Marshall Problem Assessment System, "Problem Title: Segment Joint Primary O-Ring Charred," Record A07934, Contractor Reference DR4-5/30, 7 March 1984, reproduced in Presidential Commission, *Report* 5:1549.

77. U.S. Congress, House, *Investigation: Hearings* 1:361–62.

78. Presidential Commission, *Report* 5:1572.

79. Mulloy, interview transcript, 2 April 1986, 26–27.

80. See charts in Presidential Commission, *Report* 2:H5–H12.

81. Ibid., H1, H5–H14.

82. Ibid., H1.

83. Ibid., H5–H14.

84. Ibid., H5–H9.

85. Ibid., H9.

86. John Q. Miller to George Hardy, memorandum, 28 February 1984, reproduced in Presidential Commission, *Report* 1:245.

87. Presidential Commission, *Report* 1:132.

88. George Morefield, United Technologies, to Larry Mulloy, Marshall Space Flight Center, memorandum, 9 March 1984, reproduced in Presidential Commission, *Report* 1:254.

89. Thompson, interview transcript, 4 April 1986, 23.

90. U.S. Congress, House, *Investigation: Hearings* 1:361.

91. Russell, interview transcript, 7 April 1986, 18.

92. Ibid., 17–18.

93. Boisjoly, interview transcript, 2 April 1986, 15–16.

94. Mulloy, interview transcript, 2 April 1986, 32, 40–41.

95. David Collingridge, *The Social Control of Technology* (Milton Keynes: Open University Press, 1980), 23.

96. Turner, *Man-Made Disasters*.

97. See charts, Presidential Commission, *Report* 1:129–31. I am not including the rings on STS-4, which were never subjected to postflight analysis because the parachutes did not work properly and the boosters were never recovered from the sea.

98. Hardy, interview transcript, 3 April 1986, 44.

99. "Space Shuttle Flight Readiness Reviews," 26 September 1983.

CHAPTER 5

1. Roger M. Boisjoly, "Ethical Decisions: Morton Thiokol and the Space Shuttle *Challenger* Disaster" (paper no. 87-WA/TS-4 presented at winter meeting of American Society of Mechanical Engineers, Boston, 13–18 December 1987), 2. According to Mark Lackey, National Climatic Data Center, U.S. Weather Service, Asheville, N.C., the 30-year official average *low* for the month of January (1950–80) at near-by Titusville, Fla., was 48.6°F (telephone interview by author, 7 November 1994).

2. Gurney and Forte, 160; *Aviation Week and Space Technology*, 28 January 1985, 22, photo caption.

3. Johnson Space Center, "Program Specification Requirements," JSC-07700, vol. 10, appendix 1010.

4. Mark Lackey, U.S. Department of Commerce, National Climatic Data Center, telephone interview by author, 6 June 1993.

5. Presidential Commission, *Report* 1:135.

6. Ibid. 2:H2. The left forward field joint showed "a considerable amount of soot between the primary and the secondary O-rings." On the right center field joint, the blow-by left "black grease (sooted) behind the primary O-ring over a 110° arc and the secondary was affected (but not eroded) by heat over a 29.5 inch span."

7. Ben Powers, interview transcript, 26 March 1986, Marshall Space Flight Center files, National Archives, Washington, D.C., pp. 30–31.

8. Presidential Commission, *Report* 1:135.

9. Ibid., 136; 2:H2, H19.

10. Ibid. 2:H19, charts 33 and 34.

11. Ibid., H20–H32.

12. Roger M. Boisjoly, "Ethical Decisions: Morton Thiokol and the Space Shuttle *Challenger* Disaster" (lecture, Massachusetts Institute of Technology, 7 January 1987; manuscript), 1.

13. Boisjoly, interview transcript, 2 April 1986, 39.

14. McDonald, interview transcript, 19 March 1986, 39–40.

15. Presidential Commission, *Report* 1:147; 2:H2.

16. U.S. Congress, House, *Investigation: Hearings* 1:648.

17. U.S. Congress, House, *Investigation: Report*, 527.

18. Presidential Commission, *Report* 1:136.

19. Marshall Space Flight Center, "Propulsion Trivia," NASA Fact Sheet 21F981 (Huntsville, Ala., March 1980).

20. Roger M. Boisjoly, personal interview by author, 8 February 1990.

21. Boisjoly, interview transcript, 2 April 1986, 24; U.S. Congress, House, *Investigation: Report*, 149.

22. A transcript of Mulloy's videotaped presentation appears in Presidential Commission, *Report* 2:H42.

23. Ibid., H41–H42.

24. Gurney and Forte, 168.

25. Presidential Commission, *Report* 2:H2.

26. Ibid., H3. In the comprehensive review analyzing the O-ring erosion history on all flights, STS 51-D was included. See the charts presented for the August 19, 1985, NASA Headquarters (Level I) briefing (ibid., H68–H95).

27. Reporting in FRR for STS 51-D and 51-B, Mulloy conformed to reporting procedures. A week before STS 51-D flew, the 51-B Marshall FRR briefings on the SRBs had begun. Although the results of the preceding flight were customarily presented, the April 12 launch of STS 51-D and the April 8 SRB Board date at Marshall for STS 51-B made that impossible. The nozzles from STS 51-C, the 53°F January flight, had by this time been disassembled. At the April FRR for STS 51-B, Mulloy presented the findings from the 51-C nozzle joints: no erosion; a small amount of blow-by occurred in both nozzle joints. The conclusion was "observations within data base . . . acceptable for flight" (ibid. 1:130; 2:H42). The completed analysis of the nozzle joints was presented to the Mission Management Team's Launch-Minus-One-Day (L-1) review prior to the launch.

As was customary, the SRB joint performance was reviewed at the Level IV and III FRRs for STS 51-B. When Mulloy made his FRR presentation at Level II and Level I FRR for the 51-B launch, he did not mention O-ring sealing problems. But he was again conforming to the Delta review concept that was used for FRR. After January's 51-C postflight analysis, a full briefing on the SRB joint problems had been given at all FRR levels for STS 51-D, as required by Delta review. Further discussions about O-ring problems in FRRs were contingent on some change that required reanalysis. None had occurred prior to the 51-B launch.

28. Mark Salita, interview transcript, 2 April 1986, Morton Thiokol, Inc., files, National Archives, Washington, D.C., p. 13.

29. McDonald, interview transcript, 19 March 1986, 45.

30. Presidential Commission, *Report* 1:137.

31. Ibid. See Robert E. Lindstrom to Distribution, Marshall Space Flight Center, "Assigning Launch Constraints on Open Problems Submitted to MSFC PAS," memorandum, 15 September 1980, reproduced in Presidential Commission, *Report* 5:1543–45. This constraint was recorded in the MPAS report as a nozzle joint entry. No Launch Constraint ever had been imposed in response to field joint erosion (Presidential Commission, *Report* 1:137). This differential treatment was not an administrative lapse, but occurred because of differences in both design and performance between field joints and nozzle joints (a possible contributing factor to the greater amount of erosion found on this flight was believed to be the greater eccentricity of the nozzle joint compared to the field joint) (Presidential Commission, *Report* 5:1548). No secondary erosion had ever occurred in a field joint. Therefore, based on the engineering rationale and flight experience, a Launch Constraint was not an appropriate response to the field joint erosion (Presidential Commission, *Report* 1:138; see Mulloy testimony quoted at bottom of first column).

32. Presidential Commission, *Report* 5:1531.

33. Johnson Space Center, "PRACA (Problem Reporting and Corrective Action) System Requirements for the Space Shuttle Program," JSC-08126A, May 1985, National Archives, Washington, D.C., p. 4-1. See also Lindstrom to Distribution, "Assigning Launch Constraints."

34. Marshall Space Flight Center, "Problem Assessment System Procedures," MSFC-SE012082TH, March 1981, National Archives, Washington, D.C. The Launch Constraint procedure in this document is described in U.S. Congress, House, *Investigation: Hearings* 1:453.

35. Presidential Commission, *Report* 5:1573.

36. Bunn, interview transcript, 17 April 1986, 61–62.

37. Presidential Commission, *Report* 1:130–31.

38. Ibid. 2:H3, H49–H66.

39. Mark Salita, "Prediction of Pressurization and Erosion of the SRB Primary O-rings during Motor Ignition," part 1, "Motor Development and Validation," Report TRW-14 952, 23 April 1985, National Archives, Washington, D.C; Mark Salita, "Prediction of Pressurization and Erosion of the SRB O-rings during Motor Ignition," part 2, "Parametric Studies of Field and Nozzle Joints," Report TWR-15 186, 29 July 1985, National Archives, Washington, D.C. See also Salita, interview transcript, 2 April 1986, 1–9.

40. Salita, interview transcript, 2 April 1986, 4–6.

41. Ibid., 6–7.

42. Ibid.

43. Presidential Commission, *Report* 5:1591.

44. Ibid. 2:H-3. The "Problem Summary" was one of 10 charts (prepared by Thiokol engineers) that he presented (ibid., H44–H49). This document is significant because it shows that the previously established normative standards for assessing risk were used as a basis for normalizing the deviant nozzle joint performance on STS 51-B just two days after the initial discovery generated alarm (ibid., H44, chart 86).

45. The following discussion is based on two sources: the charts that Thiokol engineers presented in the July 1, 1985, meeting, which gave the data and engineering rationale on which their conclusion was based, and their testimony before the Presidential Commission, which indicated agreement among them (ibid., H49–H66; 5:1570–74).

46. Ibid. 5:1546.

47. Ibid. 2:H66.

48. Ibid. 5:1507.

49. Ibid. 1:148.

50. Ibid. 5:1508, 1513.

51. Ibid., 1531.

52. U.S. Congress, House, *Investigation: Report,* 145.

53. Marshall Space Flight Center, Marshall Problem Assessment System, "Constraining Problems," 25 June 1986, reproduced in U.S. Congress, House, *Investigation: Hearings* 1:455–68.

54. Ibid.

55. U.S. Congress, House, *Investigation: Hearings* 1:454.

56. Presidential Commission, *Report* 1:136–38.

57. Ibid. 5:1546.

58. Ibid., 1513–14.

59. Ibid. 1:143.

60. Ibid., 658. I verified Mulloy's reporting up the hierarchy from the SRB work group by (1) examining the applicable rules, (2) examining FRR documents for all the launches after the constraint was imposed, (3) tracing the presentations on the SRB joint up the hierarchy, and (4) comparing reporting practices for the SRBs with the treatment of other Launch Constraints on the shuttle in those same FRRs. I acquired FRR documents from the FOIA Office, Johnson Space Center. Some of them are reproduced in Presidential Commission, *Report* 2:H49–H97. For supporting statements by Thiokol engineers, see interview transcripts of Roger Boisjoly, Al McDonald, Brian Russell, and Robert Ebeling in the Morton Thiokol files at the National Archives. Further supporting testimony is in Presidential Commission, *Report,* and U.S. Congress, House, *Investigation: Hearings.*

61. Presidential Commission, *Report* 1:139.

62. "Space Shuttle Flight Readiness Reviews," 26 September 1983.

63. Ibid.

64. Ibid.

65. Bunn, interview transcript, 3 April 1986, 42, 44–45.

66. Presidential Commission, *Report* 5:1546; see entry for July 15, 1985.

67. Ibid. 1:148.

68. Ibid., 132.

69. Salita, interview transcript, 2 April 1986, 5–6, 9.

70. Thompson, interview transcript, 4 April 1986, 34–35.

71. Presidential Commission, *Report* 5:1568–69.

72. Mulloy, interview transcript, 2 April 1986, 39–40; Broad, "NASA Had Solution."

73. U.S. Congress, House, *Investigation: Hearings* 1:765–66.

74. Irving Davids, interview transcript, 4 April 1986, NASA Headquarters files, National Archives, Washington, D.C.; Irving Davids, "Case to Case and Nozzle to Case O-ring Seal Erosion Problems" (Washington, D.C.: NASA Headquarters, 17 July 1986), reproduced in U.S. Congress, House, *Investigation: Report,* 376–77.

75. Mulloy, interview transcript, 2 April 1986, 33.

76. Presidential Commission, *Report* 4:689–90.

77. Donald Ketner, interview transcript, 26 March 1986, Morton Thiokol, Inc., files, National Archives, Washington, D.C., p. 3; R. Boisjoly to R. K. Lund, memorandum, 31 July 1985, Morton Thiokol, Inc.; Mulloy, interview transcript, 2 April 1986, 34.

78. John P. McCarty, interview transcript, 19 March 1986, Marshall Space Flight Center files, National Archives, Washington, D.C., pp. 32, 35.

79. Ibid., 26–28.

80. Roger M. Boisjoly, "Progress Report: Applied Mechanics Department," weekly activity report, 22 July 1985, Morton Thiokol, Inc., reproduced in U.S. Congress, House, *Investigation: Report,* appendix V-G, 285.

81. Presidential Commission, *Report* 4:377–99.

82. R. Cook to M. Mann, "Problem with SRB Seals," memorandum, 23 July 1985, National Aeronautics and Space Administration, reproduced in Presidential Commission, *Report* 4:391–92.

83. Presidential Commission, *Report* 4:397.

84. Ibid. 1:251.

85. B. Russell to J. W. Thomas, "Actions Pertaining to SRM Field Joint Secondary Seal," letter, 9 August 1985, reproduced in Presidential Commission, *Report* 5:1568–69.

86. Presidential Commission, *Report* 5:1568.

87. Ibid., 1569.

88. Roger M. Boisjoly, telephone interview by author, 2 August 1989.

89. Michael Weeks, affidavit, 13 November 1986, Barbara J. Webb, Notary Public, District of Columbia, p. 1, reproduced in U.S. Congress, House, *Investigation: Hearings* 2:322–24.

90. Originally, the meeting was scheduled to discuss a different problem NASA was having with Thiokol, but Michael Weeks called Thiokol and, according to McDonald's testimony, said, "While you're here, you

ought to come and address a couple of other issues that have happened recently that we are very interested in." One of them was the nozzle joint secondary erosion on STS 51-B (Presidential Commission, *Report* 5:1592).

91. Ibid., 1587. The charts are reproduced in Presidential Commission, *Report* 2:H68–H95.

92. Ibid. 2:H84.

93. Ibid., H87–H94.

94. Salita, interview transcript, 2 April 1986, 9–10.

95. Michael Weeks, affidavit, 13 November 1986, 2.

96. See, e.g., Mark Maier and Roger Boisjoly, "Roger Boisjoly and the Space Shuttle *Challenger* Disaster" (Binghamton: SUNY-Binghamton, School of Education and Human Development, Career and Interdisciplinary Studies Division, 1988; videotape instructional package).

97. McDonald, interview transcript, 19 March 1986, 52–53.

98. Presidential Commission, *Report* 5:1067–68.

99. Ibid., 1595–96.

100. For Thiokol task force, see U.S. Congress, House, *Investigation: Report*, 58; Presidential Commission, *Report* 1:140. For Marshall task force, see Jerry Peoples, interview transcript, 12 March 1986, Marshall Space Flight Center files, National Archives, Washington, D.C.

101. A. Thompson to S. Stein, memorandum, 22 August 1985, Morton Thiokol, Inc., reproduced in Presidential Commission, *Report* 1:251.

102. U.S. Congress, House, *Investigation: Hearings* 1:491.

103. Ibid., 766.

104. The memo is reproduced in Presidential Commission, *Report* 1:256.

105. Reproduced in Presidential Commission, *Report* 5:1617–18.

106. McDonald, interview transcript, 19 March 1986, 41.

107. William Macbeth, interview transcript, 14 March 1986, Morton Thiokol, Inc., files, National Archives, Washington, D.C., pp. 31–32.

108. S. Stein to R. Lund, A. McDonald, B. Brinton, W. Macbeth, O-Ring Task Force Members, memorandum, 1 October 1985, Morton Thiokol, Inc., p. 1, reproduced in Presidential Commission, *Report* 1:253.

109. Roger M. Boisjoly, weekly activity report, 4 October 1985, Morton Thiokol, Inc., reproduced in Presidential Commission, *Report* 1:254–55.

110. U.S. Congress, House, *Investigation: Hearings* 1:359–60; see also Boisjoly, weekly activity report, 4 October 1985.

111. Presidential Commission, *Report* 5:1577.

112. Robert Ebeling, interview transcript, 19 March 1986, Morton Thiokol, Inc., files, National Archives, Washington, D.C., pp. 8–9.

113. Ibid.

114. Presidential Commission, *Report* 1:142.

115. MPAS tracked open problems in addition to Launch Constraints.

Not all open problems were Launch Constraints. Many were considered "lesser" problems that did not require a Launch Constraint (ibid. 5:1528).

116. Ibid., 1554; the original memo, reproduced here, has no date on it except a stamp of the date on which it was apparently filed (24 December 1985).

117. Ibid. 1:142.

118. Ibid.

119. Ibid. 5:1522.

120. Ibid., 1575.

121. Russell, interview transcript, 7 April 1986, 11. See also McDonald's testimony in Presidential Commission, *Report* 5:1579–80.

122. Russell, interview transcript, 7 April 1986, 16.

123. Ibid., 14–17.

124. Ibid., 12.

125. Presidential Commission, *Report* 5:1611–12.

126. Ibid., 1576.

127. Ibid., 1571.

128. Ibid., 1579–80.

129. Harry Quong, interview transcript, 11 April 1986, NASA Headquarters files, National Archives, Washington, D.C., pp. 23–26.

130. Presidential Commission, *Report* 5:1509, 1526.

131. Ibid. 1:144; U.S. Congress, House, *Investigation: Report*, 217.

132. Presidential Commission, *Report* 5:1526.

133. Wear, personal interview, 2 June 1992.

134. Marshall Space Flight Center, Marshall Problem Assessment System, "Primary O-Ring Charred."

135. Ketner, interview transcript, 26 March 1986, 7.

136. Presidential Commission, *Report* 2:H3.

137. Brian Russell, interview transcript, 19 March 1986, Morton Thiokol, Inc., files, National Archives, Washington, D.C., p. 66.

138. Presidential Commission, *Report* 5:1614.

139. Ibid., 1580–81.

140. Boisjoly, interview transcript, 2 April 1986, 34–35.

141. Boyd Brinton, interview transcript, 13 March 1986, Morton Thiokol, Inc., files, National Archives, Washington, D.C., pp. 57–58, 63.

142. Macbeth, interview transcript, 14 March 1986, 32–33.

143. W. Leon Ray, personal interview by author, 26 January 1993.

144. Powers, interview transcript, 12 March 1986, 42.

145. Coates, interview transcript, 10 April 1986, 46.

146. Smith, personal interview, 8 June 1992.

147. Barbara Levitt and James G. March, "Organizational Learning," *Annual Review of Sociology* 14 (1988): 319–40.

148. McDonald, interview transcript, 19 March 1986, 54; Mulloy, telephone interview, 5 August 1992.

149. Presidential Commission, *Report* 5:1505–83, but see esp. 1527, 1531, 1575.

150. U.S. Congress, House, *Investigation: Report*, 215.

151. Presidential Commission, *Report* 5:1527, 1528.

152. Lindstrom to Distribution, "Assigning Launch Constraints."

153. Mulloy, personal interview, 8 June 1992.

154. Presidential Commission, *Report* 5:1507, 1545–46.

155. Vaughan, *Controlling Unlawful Organizational Behavior*, 102.

156. Anselm Strauss, *Negotiations* (San Francisco: Jossey-Bass, 1978); Gary Alan Fine, "Negotiated Orders and Organization Cultures," *Annual Review of Sociology* 10 (1984): 239–62.

CHAPTER 6

1. Susan Leigh Star, "Scientific Work and Uncertainty," *Social Studies of Science* 15 (1985): 414.

2. Kuhn.

3. Ibid., 10; Latour, *Science in Action*; Latour and Woolgar; Brian Wynne, "Unruly Technology: Practical Rules, Impractical Discourses and Public Understanding," *Social Studies of Science* 18 (1988): 147–67; H. M. Collins, "The TEA Set: Tacit Knowledge," *Science Studies* 4 (1974): 165–86; Pinch and Bijker; Andrew Pickering, ed., *Science as Practice and Culture* (Chicago: University of Chicago Press, 1992); Stephen Cole, *Making Science: Between Nature and Society* (Cambridge: Harvard University Press, 1992).

4. For an application of Kuhn's concept of paradigm to engineering as a disciplinary community, see David A. Bella, "Engineering and Erosion of Trust," *Journal of Professional Issues in Engineering* 113 (April 1987): 117–29.

5. Karin Knorr-Cetina, *The Manufacture of Knowledge* (New York: Pergamon, 1981).

6. Anthony Giddens, *Central Problems in Social Theory: Action, Structure, and Contradiction in Social Analysis* (Berkeley and Los Angeles: University of California Press, 1979); Lynne G. Zucker, "Organizations as Institutions," in *Research in the Sociology of Organizations*, ed. S. B. Bacharach (Greenwich, Conn.: JAI, 1983), 2.

7. Giddens; Pierre Bourdieu, *Outline of a Theory of Practice* (Cambridge: Cambridge University Press, 1977).

8. Paul J. DiMaggio and Walter W. Powell, "The Iron Cage Revisited: Institutional Isomorphism and Collective Rationality in Organizational Fields," *American Sociological Review* 48 (1983): 147–60; James G. March and Johan P. Olsen, "The New Institutionalism: Organizational Factors in Political Life," *American Political Science Review* 78 (1984): 741; Linda Smircich, "Concepts of Culture and Organizational Analysis," *Administrative Science Quarterly* 28 (1983): 350.

9. W. Richard Scott, "The Adolescence of Institutional Theory," *Administrative Science Quarterly* 32 (1987): 507.

10. Ibid.; John W. Meyer, W. Richard Scott, and Terrence E. Deal, "Institutional and Technical Sources of Organizational Structure: Explaining the Structure of Educational Organizations," in *Organization*

and the Human Services: Cross-Disciplinary Reflections, ed. Herman D. Stein (Philadelphia: Temple University Press, 1981).

11. Patricia Riley, "A Structurationist Account of Political Culture," *Administrative Science Quarterly* 28 (1983): 415. For an excellent in-depth discussion of this process in organizations, see Stewart Ranson, Bob Hinings, and Royston Greenwood, "The Structuring of Organizational Structures," *Administrative Science Quarterly* 25 (1980): 1–17.

12. DiMaggio and Powell, "Iron Cage Revisited"; Wildavsky; Mary Douglas, *How Institutions Think* (London: Routledge and Kegan Paul, 1987); Mary Douglas and Aaron Wildavsky, *Risk and Culture* (Berkeley and Los Angeles: University of California Press, 1982).

13. For arguments supporting such a methodological strategy for cultural studies and the sociology of knowledge, see also Henrika Kuklick, "Sociology of Knowledge: Retrospect and Prospect," *Annual Review of Sociology* 9 (1983): 287–310; Stephen R. Barley, "Semiotics and the Study of Occupational and Organizational Cultures," *Administrative Science Quarterly* 28 (1983): 393–413; Ann Swidler, "Culture in Action: Symbols and Strategies," *American Sociological Review* 51 (1986): 280; Maryan S. Schall, "A Communication-Rules Approach to Organizational Culture," *Administrative Science Quarterly* 28 (1983): 557–81; Van Maanen and Barley, 33–35; Edgar H. Schein, *Organizational Culture and Leadership* (San Francisco: Jossey-Bass, 1985).

14. Giddens; Ranson, Hinings, and Greenwood, 3.

15. Lynne G. Zucker, "The Role of Institutionalization in Cultural Persistence," *American Sociological Review* 42 (1977): 741–42; Ronald L. Jepperson, "Institutions, Institutional Effects, and Institutionalism," in *The New Institutionalism in Organizational Analysis*, ed. Walter W. Powell and Paul J. DiMaggio (Chicago: University of Chicago Press, 1991), 151.

16. Jepperson, 153–58; Louis, 79.

17. In his unparalleled history of organization culture, McCurdy carefully and insightfully traces these and other historic changes affecting the space agency. See Howard E. McCurdy, *Inside NASA: High Technology and Organizational Change in the U.S. Space Program* (Baltimore: Johns Hopkins University Press, 1993).

18. Howard S. Becker, "Culture: A Sociological View," *Yale Review* 71 (1982): 513–28.

19. See, e.g., Van Maanen and Barley; Louis; Riley; Vaughan, *Controlling Unlawful Organizational Behavior*; Gregory.

20. For theoretical discussions of differences in worldview produced by differences in structural position, see, e.g., Berger and Luckmann; Bourdieu; DiMaggio and Powell, "Introduction"; Douglas, *How Institutions Think*; Robert K. Merton, *Social Theory and Social Structure* (Glencoe, Ill.: Free Press, 1968), 516–21, 616–24. For an excellent ethnographic portrayal of cultural differences between managers and engineers in the workplace, see Gideon Kunda, *Engineering Culture: Control and Commitment in a High-Tech Corporation* (Philadelphia: Temple University Press, 1992).

21. Geertz, *Interpretation of Cultures*, 312.

22. Ibid., 23.

23. See also Gregory, 361.

24. Zucker, "Organizations as Institutions," 5; Randall Stokes and John P. Hewitt, "Aligning Actions," *American Sociological Review* 41 (1976): 838–49.

25. Joseph Bensman and Robert Lilienfeld, *Craft and Consciousness: Occupational Technique and the Development of World Images* (New York: Aldine de Gruyter, 1991). See also Charles Bosk, *Forgive and Remember: Managing Medical Failure* (Chicago: University of Chicago Press, 1979).

26. Wynne.

27. Robert J. Thomas, *What Machines Can't Do: Politics and Technology in the Industrial Enterprise* (Berkeley and Los Angeles: University of California Press, 1993), 46.

28. Turner, *Man-Made Disasters*, 198.

29. Charles B. Perrow, "Accidents in High Risk Systems," *Technology Studies* 1 (1992): 11–13.

30. Wynne, 149–54.

31. Ibid., 147–67.

32. Ibid., 153.

33. Heimer, "Doing Your Job."

34. Clifford Geertz, *Local Knowledge: Further Essays in Interpretive Anthropology* (New York: Basic Books, 1983).

35. Donald MacKenzie and Graham Spinardi, "Tacit Knowledge, Weapons Design, and the Uninvention of Nuclear Weapons," *American Journal of Sociology* 101 (1995): 44–99.

36. Collins, "The TEA Set"; Susan Leigh Star, "Layered Space, Formal Representations, and Long Distance Control: The Politics of Information" (paper presented at Place of Science Conference, Van Leer Institute, Jerusalem and Tel Aviv, May 1989).

37. Levitt and March.

38. Martin Landau and Russell Stout, Jr., "To Manage Is Not to Control: Or the Folly of Type II Errors," *Public Administration Review* 39 (1979): 153.

39. Richard C. Dorf, *Technology and Society* (San Francisco: Boyd and Fraser, 1974), 26, 34.

40. Petroski, 40.

41. Ibid., 80.

42. Ibid., 44; Thomas, *What Machines Can't Do*, 181.

43. Thomas, *What Machines Can't Do*.

44. Petroski, 108.

45. Starbuck and Milliken, "Challenger: Fine-Tuning the Odds," 335.

46. Pinch and Bijker.

47. Ibid.; Allan Mazur, "Disputes between Experts," *Minerva* 11 (1973): 243–62; Petroski, 112.

48. Perrow, *Normal Accidents*, chaps. 2 and 3.

49. Peter Weingart, "Large Technical Systems, Real-Life Experiments, and the Legitimation Trap of Technology Assessment: The Contribution of Science and Technology to Constituting Risk Perception," in *Social Responses to Large Technical Systems: Control or Anticipation*, ed. Todd R. LaPorte (Boston: Kluwer, 1991), 8–9.

50. Alfred A. Marcus, "Risk, Uncertainty, and Scientific Judgment," *Minerva* 26 (1988): 138–52.

51. Pinch and Bijker.

52. Collins, "The Place of the Core-Set."

53. Robert Perrucci, "Engineering: Professional Servant of Power," *American Behavioral Scientist* 41 (1970): 492–506.

54. Ibid., 301–2.

55. Harry Braverman, *Labor and Monopoly Capital* (New York: Monthly Review, 1974).

56. Arthur L. Stinchcombe, "Authority and the Management of Engineering on Large Projects," in *Organization Theory and Project Management: Administering Uncertainty in Norwegian Offshore Oil*, ed. Arthur L. Stinchcombe and Carol A. Heimer (Oslo: Norwegian University Press, 1985), 235.

57. Carol A. Heimer, "Organizational and Individual Control of Career Development in Engineering Project Work," *Acta Sociologica* 27 (1984): 283–310.

58. Perrucci, 505.

59. For various aspects of this widely held view, see Steven Kerr, Mary Ann Von Glinow, and Janet Schriesheim, "Issues in the Study of 'Professionals' in Organizations: The Case of Scientists and Engineers," *Organizational Behavior and Human Performance* 18 (1977): 329–45; Allan Schnaiberg, "Obstacles to Environmental Research by Scientists and Technologists," *Social Problems* 24 (1977): 500–520; Clovis R. Shepherd, "Orientations of Scientists and Engineers," *Pacific Sociological Review* 4 (1961): 79–83; Zussman; Richard R. Ritti, *The Engineer in the Industrial Corporation* (New York: Columbia University Press, 1971); Fred Goldner and Richard R. Ritti, "Professionalization as Career Immobility," *American Journal of Sociology* 72 (1967): 489–502; David F. Noble, *America by Design* (New York: Knopf, 1977).

60. Peter Mieksins, "The 'Revolt of the Engineers' Reconsidered," *Technology and Culture* 24 (1988): 219–46; Peter Meiksins, "Engineers and Managers: An Historical Perspective on an Uneasy Relationship" (paper presented at annual meetings of American Sociological Association, San Francisco, August 1989).

61. For a discussion of the problem and suggestions for actions to keep engineers from becoming servants of power, see Bella, "Engineering and Erosion of Trust."

62. Heimer, "Organizational and Individual Control," 306–7; Whalley.

63. Lotte Bailyn, "Autonomy in the Industrial R&D Lab," *Human Resource Management* 24 (1985): 129–46; Charles Derber, *Professionals*

as Workers: Mental Labor in Advanced Capitalism (Boston: Hall, 1982); Peter Meiksins and James M. Watson, "Professional Autonomy and Organizational Constraint: The Case of Engineers," *Sociological Quarterly* 30 (1989): 561-85; Zussman, 102-23.

64. Zussman, 106.

65. Dorf, 26, 27, 29, 33.

66. Zussman, 121.

67. Dorf, 25.

68. Ibid., 29.

69. Ibid.

70. Simon, *Models of Man.*

71. David Pye, *The Nature and Aesthetics of Design* (New York: Van Nostrand Reinhold, 1978), quoted in Petroski, 218-19.

72. Petroski, 221.

73. Ibid., 99.

74. Ibid., 98.

75. Merton, *Social Theory and Social Structure,* 250.

76. Heimer, "Organizational and Individual Control."

77. Robert K. Merton, "The Machine, the Worker, and the Engineer," *Science* 105 (1947): 79-81; Howard E. Aldrich, *Organizations and Environment* (Englewood Cliffs, N.J.: Prentice-Hall, 1979), 197-99.

78. Meiksins and Watson, 577-78.

79. Thomas, *What Machines Can't Do,* 81-82.

80. Zussman, 155.

81. Ibid.; Joseph A. Raelin, "An Examination of Deviant/Adaptive Behaviors in the Organizational Careers of Professionals," *Academy of Management Review* 9 (1984): 415; Goldner and Ritti, 491-94.

82. Zussman, 230.

83. McCurdy, *Inside NASA.*

84. Howard E. McCurdy, "The Decay of NASA's Technical Culture," *Space Policy,* November 1989, 302.

85. Smircich, "Concepts of Culture," 348.

86. McDougall, 362.

87. Ibid., 373.

88. These changes are all-encompassing and too numerous for discussion here. For details, see ibid., esp. 376-77; Roger E. Bilstein, *Orders of Magnitude: A History of the NACA and NASA, 1915-1990* (Washington, D.C.: National Aeronautics and Space Administration, Scientific and Technical Information Division, 1989), 42-91; Phillip K. Tompkins, *Organizational Communication Imperatives: Lessons of the Space Program* (Los Angeles: Roxbury, 1993).

89. McCurdy, *Inside NASA.*

90. McCurdy, "The Decay of NASA's Technical Culture," 302.

91. U.S. Congress, House, *Investigation: Report,* 3, 22.

92. Cristy S. Johnsrud, "Conflict, Complementarity, and Consequence: The Administration of Cross-Sector Organizational Linkages"

(paper presented at annual meeting of American Anthropological Association, Phoenix, November 1988); McCurdy, *Inside NASA*.

93. Carol A. Heimer, *Reactive Risk and Rational Action: Managing Moral Hazard in Insurance Contracts* (Berkeley and Los Angeles: University of California Press, 1985); Arthur L. Stinchcombe, "Contracts as Hierarchical Documents," in *Organization Theory and Project Management: Administering Uncertainty in Norwegian Offshore Oil*, ed. Arthur L. Stinchcombe and Carol A. Heimer (Oslo: Norwegian University Press, 1985).

94. Romzek and Dubnick.

95. Ibid.

96. Ibid.

97. I am indebted to Piet Hut for this apt double entendre.

98. John W. Meyer and Brian Rowan, "Institutionalized Organizations: Formal Structure as Myth and Ceremony," *American Journal of Sociology* 83 (1977): 340–63.

99. Diamond, "NASA Wasted Billions."

100. See also John C. Macidull, "Safety Awareness Continuity in Transportation and Space Systems," *Acta Astronautica* 17 (1988): 931–36; Ronald C. Kramer, "State-Corporate Crime: The Space Shuttle Challenger Disaster," in *White-Collar Crime Reconsidered*, ed. Kip Schlegel and David Weisburd (Boston: Northeastern University Press, 1992).

101. U.S. Congress, House, *Investigation: Hearings* 1:565, 569–70.

102. Ranson, Hinings, and Greenwood, 6.

103. Presidential Commission, *Report* 1 and 7; U.S. Congress, House, *Investigation: Report*, chap. 6.

104. See also Murray, 229. Murray notes that neither the Presidential Commission nor the President asked this question, nor did they ask whether the tasks could be done by a few crew members or whether the tasks should be dropped entirely.

105. Schulze, interview transcript, 15 April 1986, 22–26; National Aeronautics and Space Administration, Aerospace Safety Advisory Panel, "National Aeronautics and Space Administration Annual Report: Covering Calendar Year 1984" (Washington, D.C., 1985; mimeograph), 54, 9.

106. Macidull, 931.

107. U.S. Congress, House, *Investigation: Report*, 130.

108. Buchanan, interview transcript, 25 March 1986, 40–41.

109. McCurdy, "The Decay of NASA's Technical Culture"; Romzek and Dubnick.

110. Romzek and Dubnick, 232; U. S. Congress, House, *Investigation: Report*, 166–69.

111. Tompkins, *Organizational Communication Imperatives*, 72.

112. For a listing of delays on each launch and the technical causes, see Gurney and Forte.

113. McDougall; Murray; Tompkins, *Organizational Communication Imperatives.*

114. Michael Wright, personal interview by author, 7 June 1993; Tompkins, *Organizational Communication Imperatives.*

115. For details, see Tompkins, *Organizational Communication Imperatives,* 54–57, 76n.

116. Phillip Tompkins, "Management Qua Communication in Rocket Research and Development," *Communication Monographs* 44 (1977): 1–26, as cited and discussed in Tompkins, *Organizational Communication Imperatives,* 121–22.

117. For accounts of NASA under various directors, see Tompkins, *Organizational Communication Imperatives;* Murray; Bilstein; McDougall.

118. McConnell, 106–11.

119. Ibid., 34; Tompkins, *Organizational Communication Imperatives.*

120. James Smith, interview transcript, 13 March 1986, Marshall Space Flight Center files, National Archives, Washington, D.C., p. 52.

121. Robert J. Schwinghamer, interview transcript, 13 March 1986, Marshall Space Flight Center files, National Archives, Washington, D.C., pp. 57–58.

122. McCurdy, *Inside NASA;* Murray, 213. Under the direction of Administrator James Beggs, who assumed the position in 1981, much of shuttle operations were moved to Kennedy (Murray, 222).

123. Tompkins, *Organizational Communication Imperatives,* 169.

124. Ibid., 107, 108.

125. Ibid., 62–63, 160–63.

126. Ibid., 166.

127. Wear, personal interview, 2 June 1992.

128. Mulloy, telephone interview, 5 August 1992.

129. Leslie F. Adams, interview transcript, 12 March 1986, Marshall Space Flight Center files, National Archives, Washington, D.C., pp. 56–59.

130. Boisjoly, interview transcript, 2 April 1986, 38–40.

131. Stokes and Hewitt, 849.

132. Ibid., 848.

133. Riley, 435.

134. Boisjoly, telephone interview, 19 March 1993.

135. Robert M. Emerson and Sheldon L. Messinger, "The Micro-Politics of Trouble," *Social Problems* 25 (1977): 121–44.

136. See William Lowrance, *On Acceptable Risk: Science and the Determination of Safety* (Los Altos, Calif.: Kaufmann, 1976).

137. Mulloy, interview transcript, 2 April 1986, 27–28.

138. See also Stinchcombe, "Contracts as Hierarchical Documents."

139. See, e.g., Melville Dalton, *Men Who Manage* (New York: Wiley, 1959); James P. Spradley, *The Cocktail Waitress: Woman's Work in a Man's World* (New York: Wiley, 1975); Michael Burawoy, *Manufacturing*

Consent (Chicago: University of Chicago Press, 1979); Joseph Bensman and Israel Gerver, "Crime and Punishment in the Factory: The Function of Deviancy in Maintaining the Social System," *American Sociological Review* 28 (1963): 588–98; Donald Roy, "Efficiency and 'The Fix': Informal Intergroup Relations in a Piecework Machine Shop," *American Journal of Sociology* 60 (1954): 255–67.

140. Wynne.

141. U.S. Congress, House, *Investigation: Report*, 138–39.

142. Ibid., 162.

143. McCurdy, *Inside NASA*.

144. Wear, personal interview, 2 June 1992.

145. Smith, personal interview, 8 June 1992.

146. See also Collins, "The Place of the Core-Set."

147. Ray, telephone interview, 21 July 1992.

148. Powers, interview transcript, 12 March 1986, 59–68. See also Presidential Commission, *Report* 2:16.

149. Ray, interview transcript, 25 March 1986, 52; Miller, interview transcript, 13 March 1986, 31–32.

150. Gurney and Forte, 98.

151. Coates, interview transcript, 10 April 1986, 50.

152. Ray, telephone interview, 13 October 1992.

153. McCarty, interview transcript, 19 March 1986, 37–38.

154. Powers, interview transcript, 12 March 1986, 69.

155. Adams, interview transcript, 12 March 1986, 54–55.

156. Mulloy, telephone interview, 5 August 1992.

157. Boisjoly, telephone interview, 19 March 1993; Roger M. Boisjoly to R. K. Lund, Vice President, Engineering, memorandum, 31 July 1985, Morton Thiokol, Inc., reproduced in Presidential Commission, *Report* 1:249–50.

158. Boisjoly, telephone interview, 19 March 1993.

159. Presidential Commission, *Report* 5:1514–15.

160. Ibid., 1514.

161. Smith, personal interview, 8 June 1992.

162. Mulloy, personal interview, 8 June 1992.

163. Lawrence B. Mulloy, telephone interview by author, 8 February 1993. The program had a basic fixed cost that was the same regardless of the number of flights, but a marginal cost of $60 million had to be added for each additional flight, so that total costs went up with additional flights. For the fallacies built into NASA's idea that cost-per-flight could be reduced by flying more often, see U.S. Congress, House, *Investigation: Report*, 120.

164. Zucker, "Organizations as Institutions," 3.

CHAPTER 7

1. For the importance of material objects and elements in the production of scientific and technical knowledge, see Pickering; Wiebe E. Bijker and John Law, eds., *Shaping Technology, Building Society: Studies in*

Sociotechnical Change (Cambridge: MIT Press, 1992); Adele E. Clarke and Joan H. Fujimura, eds., *The Right Tools for the Job: At Work in Twentieth-Century Life Sciences* (Princeton, N.J.: Princeton University Press, 1992).

2. Presidential Commission, *Report* 1:82.

3. Ibid., 84.

4. Ibid.

5. Ibid.

6. Ibid., 104.

7. Ibid., 152–55.

8. U.S. Congress, House, *Investigation: Report*, 170–71.

9. Ibid.

10. Ibid., 148.

11. Ibid., 4–5.

12. Ibid., 172.

13. In addition to the statements from Wiley Bunn and W. Leon Ray quoted in the text, see Adams, interview transcript, 12 March 1986, 64; Miller, interview transcript, 13 March 1986, 32; Coates, interview transcript, 25 March 1986, 46–47; Smith, interview transcript, 13 March 1986, 52–53, 55; Lawrence O. Wear, telephone interview by author, 23 March 1993.

14. Bunn, interview transcript, 3 April 1986, 64–65.

15. Ray, interview transcript, 25 March 1986, 50–51.

16. George Hardy, telephone interview by author, 10 February 1993.

17. Wear, personal interview, 2 June 1992.

18. Coates, interview transcript, 25 March 1986, 32.

19. Mulloy, personal interview, 8 June 1992.

20. Emerson, "Holistic Effects."

21. "Decision stream" differs from Emerson's concept "case stream," which originated from analysis of specialists processing a number of clients (i.e., caseloads); therefore each decision was made about a different entity. The decision stream notion would apply to the same client returning to the caseworker for additional considerations, in Emerson's example.

22. Emerson, "Holistic Effects," 448–53. The importance of historical process in institutionalization is emphasized in Berger and Luckmann, 54–55.

23. See also U.S. Congress, House, *Investigation: Report*, 161.

24. Emerson, "Holistic Effects," 433.

25. David Sudnow, "Normal Crimes: Sociological Features of the Penal Code in a Public Defender Office," *Social Problems* 12 (1965): 255–76.

26. Emerson, "Holistic Effects," 433.

27. Landau and Stout, 153.

28. Salancik, 34–35; Mintzberg, Raisinghani, and Théorêt; Thomas, *What Machines Can't Do.*

29. Dorothy A. Smith, "Textually-Mediated Social Organization,"

International Social Science Journal 36 (1984): 59–75; Berger and Luckmann.

30. David Matza, *Becoming Deviant* (Englewood Cliffs, N.J.: Prentice-Hall, 1969), 9.

31. Arthur L. Stinchcombe, "Work Institutions and the Sociology of Everyday Life," in *The Nature of Work*, ed. Kai Erikson and Steven Peter Vallas (New Haven, Conn.: Yale University Press, 1990), 103.

32. Salancik, 6; Jerry Ross and Barry M. Staw, "EXPO '86: An Escalation Prototype," *Administrative Science Quarterly* 31 (1986): 276; Vaughan, *Uncoupling*, chap. 8.

33. Salancik, 23.

34. Berger and Luckmann.

35. Salancik, 7.

36. Wear, personal interview, 2 June 1992.

37. Smith, personal interview, 8 June 1992.

38. Vaughan, *Controlling Unlawful Organizational Behavior*, chaps. 5 and 6. See also Weick, "Educational Organizations as Loosely Coupled Systems."

39. Martha S. Feldman, *Order without Design: Information Production and Policy Making* (Stanford, Calif.: Stanford University Press, 1989); Martha S. Feldman and James G. March, "Information in Organization as Signal and Symbol," *Administrative Science Quarterly* 26 (1981): 171–84; Meyer and Rowan, "Institutionalized Organizations."

40. Michael Spence, *Market Signaling* (Cambridge: Harvard University Press, 1974). Spence described how organizations make decisions in conditions of product uncertainty. Although Spence used his model to explain transactions between organizations, it also makes sense of decisions based on transactions *within* organizations. See Vaughan, "Theory Elaboration," 193–95.

41. See also Emmanual Lazega, *The Micropolitics of Knowledge: Communication and Indirect Control in Workgroups* (New York: Aldine de Gruyter, 1992).

42. Magnuson, 14.

43. U.S. Congress, House, *Investigation: Report*, 171.

44. James Beniger, *The Control Revolution: Technological and Economic Origins of the Information Society* (Cambridge: Harvard University Press, 1986).

45. See, e.g., CIL documents reproduced in Presidential Commission, *Report* 1:157, 239.

46. Ibid., 193.

47. Starbuck and Milliken, "Executives' Perceptual Filters."

48. Presidential Commission, *Report* 1:142.

49. McDonald, interview transcript, 19 March 1986, 53.

50. See, e.g., U.S. Congress, House, *Investigation: Hearings* 1:433, 436; Miller, interview transcript, 13 March 1986, 27; Peoples, interview transcript, 12 March 1986.

51. Coates, interview transcript, 25 March 1986, 41.

52. Wear, interview transcript, 12 March 1986, 74–76.

53. These memos deserve close reading because their content was often distorted in postaccident accounts. E.g., the Ray/Miller memorandum of January 1979 typically was reported to have said that the "design" was "unacceptable." In fact, it said, "We find the Thiokol *position* regarding design adequacy of the clevis joint to be completely unacceptable." The meaning of the latter is very different from the former, as the points that follow in the memo demonstrate.

54. Coates, interview transcript, 10 April 1986, 6–7.

55. Ibid.

56. See chap. 3. These memos are reproduced in Presidential Commission, *Report* 1:234–38.

57. Ray, telephone interview, 26 January 1993.1

58. Bunn, interview transcript, 17 April 1986, 15–16.

59. See also Macbeth, interview transcript, 14 March 1986, 32–33. For details of Thompson's memo, see Arnold R. Thompson to S. R. Stein, "SRM Flight Seal Recommendation," memorandum, 22 August 1985, Morton Thiokol, Inc., reproduced in Presidential Commission, *Report* 1:251.

60. Thompson, "SRM Flight Seal Recommendation"; see also Thompson's account of the content of this memo in Thompson, interview transcript, 4 April 1986, 49.

61. Presidential Commission, *Report* 5:1580–81; see also Macbeth, interview transcript, 14 March 1986, 32–33.

62. Boisjoly, interview transcript, 2 April 1986, 19; Russell testified, "My understanding, when he wrote that memo—you'll look and see that I was not on distribution—it was a company private memo and he did not want to make broad distribution. However, I do recall having read it" (Russell, interview transcript, 19 March 1986).

63. Wear, interview transcript, 12 March 1986, 83.

64. Ray, telephone interview, 27 January 1993.

65. Ibid.

66. U.S. Congress, House, *Investigation: Report*, 126.

67. Boisjoly, telephone interview, 19 March 1993.

68. Presidential Commission, *Report* 1:126.

69. Ibid. 5:1507; "Space Shuttle Flight Readiness Reviews," 26 September 1983; Marshall Space Flight Center, Shuttle Projects Office, "Shuttle Project Flight Readiness Review," SPO 8000.1., 29 December 1983, reproduced in U.S. Congress, House, *Investigation: Report*, 387–91.

70. U.S. Congress, House, *Investigation: Report*, 70.

71. Mulloy, telephone interview, 5 August 1992.

72. Boisjoly, interview transcript, 2 April 1986, 41–42. The Delta review principle for O-ring anomalies no longer applied when the Launch Constraint was imposed after the April 1985 primary burn-through. The Launch Constraint mandated Level I problem review each time. Because the performance remained within predictions on subsequent flights, it

was reviewed as a "significant resolved problem" (see chap. 5). Thus, it was "statused" at each review for the rest of 1985. Levels II and I received "Problem Summaries." Extensive reviews and histories mandatory at lower levels were digested into a few "bullets" at Level I.

73. Presidential Commission, *Report* 5:1539; Jesse W. Moore, interview transcript, Johnson Space Center files, National Archives, Washington, D.C., 8 April 1986.

74. "Space Shuttle Flight Readiness Reviews," 26 September 1983.

75. Lawrence B. Mulloy, telephone interview by author, 3 September 1993.

76. Ibid.

77. Presidential Commission, *Report* 5:1539; Mulloy, telephone interview, 3 September 1993.

78. Moore, interview transcript, 8 April 1986, 3–4.

79. Presidential Commission, *Report* 5:1468–78.

80. Powers, personal interview, 4 June 1992.

81. Presidential Commission, *Report* 1:159.

82. Ibid. 2:H84.

83. McDonald, interview transcript, 19 March 1986, 53–54; Boisjoly, interview transcript, 2 April 1986, 38–40.

84. Boisjoly, interview transcript, 2 April 1986, 38–40.

85. Presidential Commission, *Report* 1:148.

86. U.S. Congress, House, *Investigation: Report*, 174.

87. Sudnow, 433.

88. Presidential Commission, *Report* 1:152.

89. Pfeffer and Salancik, 39–61.

90. For the original conceptualization of the problems of autonomy and interdependence, see Vaughan, *Controlling Unlawful Organizational Behavior*, chap. 7. For a deeper analysis of how autonomy and interdependence affected safety regulation at NASA, see Diane Vaughan, "Autonomy, Interdependence, and Social Control: NASA and the Space Shuttle Challenger," *Administrative Science Quarterly* 35 (June 1990): 225–58.

91. For examples of each, see Susan P. Shapiro, *Wayward Capitalists: Targets of the Securities and Exchange Commission* (New Haven, Conn.: Yale University Press, 1984); Nancy Reichman, "Regulating Risky Business," *Law and Policy* 13 (1991): 21–48.

92. U.S. Congress, House, Committee on Science and Astronautics, *Hearings on H.R. 4450, H.R. 6470 (Superseded by H.R. 10340)*, 90th Cong., 1st sess., 1967.

93. *National Aeronautics and Space Act of 1958, U.S. Statutes at Large* 72 (1959): 426.

94. Albert J. Reiss, Jr., "Selecting Strategies of Social Control Over Organizational Life," in *Enforcing Regulation*, ed. Keith Hawkins and John M. Thomas (Boston: Kluwer-Nijhoff, 1984), 23–35.

95. See especially Keith Hawkins, *Environment and Enforcement: Regulation and the Social Definition of Pollution* (New York: Oxford

University Press, 1984); Keith Hawkins, "Bargain and Bluff: Compliance Strategy and Deterrence in the Enforcement of Regulation," *Law and Policy Quarterly* 5 (1983): 35–73; Peter K. Manning, "Managing Uncertainty in the British Nuclear Installations Inspectorate," *Law and Policy* 11 (1989): 350–69.

96. Diamond, "NASA Cut or Delayed Safety Spending"; Presidential Commission, *Report* 1:159–61.

97. U.S. Congress, House, *Investigation: Report*, 176–77.

98. U.S. Congress, House, *Investigation: Hearings* 1:655; Quong, interview transcript, 11 April 1986, 27.

99. J. B. Hammack, "Space Shuttle Crew Safety Panel History," Presidential Commission on the Space Shuttle *Challenger* Accident, National Archives, Washington, D.C., mimeograph, pp. 2–3.

100. Presidential Commission, *Report* 1:161; 5:1411.

101. Hammack, 5, 9–19.

102. Johnson Space Center, "Space Shuttle Program Directive 4A," 14 April 1974, National Archives, Washington, D.C.

103. Hammack.

104. Susan P. Shapiro, "The Social Control of Impersonal Trust," *American Journal of Sociology* 93 (1987): 646.

105. Presidential Commission, *Report* 1:154.

106. Carol A. Heimer, "Substitutes for Experience-based Information: The Case of Off-shore Oil Insurance in the North Sea," Discussion Paper no. 1181 (Bergen: Institute of Industrial Economics, 1980), 28-29.

107. In this section, I draw from Daniel F. Egan, "The Origins of the Aerospace Safety Advisory Panel" (Department of Sociology, Boston College, 1988; manuscript); Gilbert L. Roth, telephone interviews by author, 2 March 1988, 22 July 1988.

108. Presidential Commission, *Report* 1:161.

109. Ibid.

110. U.S. Congress, House, *Investigation: Hearings* 1:655–56. The many transcripts with SR&QA engineers at the National Archives agree.

111. Ibid.

112. Bunn, interview transcript, 3 April 1986, 27–30; Bunn, interview transcript, 17 April 1986, 12–13.

113. Presidential Commission, *Report* 1:155.

114. Ibid., 148.

115. U.S. Congress, House, *Investigation: Report*, 11.

116. Presidential Commission, *Report* 1:152.

117. Ibid., 155.

118. This finding is consistent with Pierre Bourdieu's theory of *habitus*, discussed in *Outline of a Theory of Practice*, in which he asserts that the positions persons occupy in the social structure have shared histories that create regularities in thought, aspirations, dispositions, and strategies of action.

119. David A. Bella, "Organizations and Systematic Distortion of

Information," *Journal of Professional Issues in Engineering* 113 (1987): 360–70.

120. Ibid., 360.

121. Lee Clarke, "The Disqualification Heuristic: When Do Organizations Misperceive Risk?" in *Research in Social Problems and Public Policy*, vol. 5, ed. R. Ted Youn and William F. Freudenberg (Greenwich, Conn.: JAI, 1993); Lee Clarke, *Acceptable Risk? Making Decisions in a Toxic Environment* (Berkeley and Los Angeles: University of California Press, 1989).

122. Clarke, "Disqualification Heuristic."

123. Lee Clarke, "The Wreck of the Exxon *Valdez*," in *Controversies: Politics of Technical Decisions*, ed. Dorothy Nelkin (Beverly Hills, Calif.: Sage, 1992).

124. Feynman, "Personal Observations on Reliability of Shuttle," F1.

125. Ibid.

126. Bourdieu.

127. Lee Clarke and James F. Short, Jr., "Social Organization and Risk: Some Current Controversies," *Annual Review of Sociology* 19 (1993): 381.

128. U.S. Congress, House, *Investigation: Report*, 171–72.

129. Ibid., 70.

130. Ibid., 4–5.

131. Ibid., 148, 70–71.

132. Presidential Commission, *Report* 1:148; U.S. Congress, House, *Investigation: Report*, 4.

133. Presidential Commission, *Report* 2:F1.

CHAPTER 8

1. Presidential Commission, *Report* 5:867.

2. Coates, interview transcript, 25 March 1986, 29.

3. Jack Buchanan, interview transcript, 14 March 1986, Morton Thiokol, Inc., files, National Archives, Washington, D.C., p. 14.

4. Smith, personal interview, 8 June 1992.

5. See, e.g., Russell, interview transcript, 19 March 1986, 67.

6. Boisjoly, 9.

7. Peter L. Berger, *Invitation to Sociology* (New York: Anchor, 1963), 61.

8. Vaughan, *Uncoupling.*

9. Karl Weick, "Sensemaking in Organizations: Small Structures with Large Consequences," in *Social Psychology in Organizations: Advances in Theory and Research*, ed. J. Keith Murnighan (New York: Prentice-Hall, 1993). See also Salancik.

10. "Shuttle Project Flight Readiness Review," 26 September 1983, 2; "Shuttle Project Flight Readiness Review," 29 December 1983, 2; U.S. Congress, House, *Investigation: Report*, 209. Anomalies on STS 61-A and 61-B were not discussed because they had already been dispositioned

as an acceptable risk in the FRRs for STS 61-B and 61-C (U.S. Congress, House, *Investigation: Report*, 209).

11. U.S. Congress, House, *Investigation: Report*, 208.

12. Gurney and Forte, 260.

13. U.S. Congress, House, *Investigation: Report*, 208–10 (documents are reproduced on pp. 407–16). See also Presidential Commission, *Report* 2:H3.

14. U.S. Congress, House, *Investigation: Report*, 412–13.

15. Ibid., 210.

16. Ibid., 409.

17. "Shuttle Project Flight Readiness Review," 29 December 1983 (see esp. p. 389, item d-4).

18. The 51-L FRR charts are not reproduced in the Presidential Commission report. Level I and II 51-L FRR charts are reproduced in U.S. Congress, House, *Investigation: Report*, 395–416. Under the Freedom of Information Act, I filed an application for (and received) 51-L FRR charts from the FOIA Office at Johnson Space Center.

19. Presidential Commission, *Report* 2:H3.

20. U.S. Congress, House, *Investigation: Report*, 210.

21. Ibid.; Presidential Commission, *Report* 5:1525; see also McDonald's testimony, U.S. Congress, House, *Investigation: Report*, 70.

22. I draw this conclusion from my review of documents prescribing FRR procedures, interviews verifying procedural norms, and FRR documents for all elements for STS 51-L. See also U.S. Congress, House, *Investigation: Report*, 208.

23. Which window is designated as primary and which as secondary is determined by flight-specific requirements of mission payloads and by the lighting requirements for TAL site. TAL sites are established in case the Orbiter has postlaunch problems before achieving orbit when it is too far downrange to return to Kennedy. NASA aims for redundancy in TAL sites; if weather or visibility is unacceptable in the primary TAL site (Dakar, Senegal), Dakar's alternative (Casablanca, Morocco) may be used, thus reducing the possibility of a launch delay while waiting for conditions to clear. Although having both alternatives is preferable, launch can proceed if only one TAL site is available.

Challenger's primary launch window was originally in the afternoon, then it was switched to morning. Late afternoon hours would accommodate the Spartan, which was to be deployed from *Challenger*'s payload bay to view Halley's comet with two ultraviolet spectrographs. An afternoon launch window would allow Spartan to view the comet under special conditions, resulting from earth and sun positioning, that existed for only 90 seconds per orbit. However, an afternoon launch window would result in a night landing if a transatlantic abort was necessary, which eliminated alternative TAL site Casablanca because it was not equipped for night landings and could only be used in daylight hours. The decision was made to use the afternoon launch window, accepting risk of launch delays if the weather was unacceptable in Dakar in order to ensure the best possible scientific data on Halley's comet. When the *Challenger* launch date was slipped from the original launch date to January 26, however, the earth/sun positioning for the Spartan's optimal view of Halley's comet no longer existed. At that time, there was no longer any reason to give up the alternative landing site, so the morning launch windows were designated as primary in order to regain Casablanca as a backup landing site, thus minimizing the possibility of delays due to poor weather in Dakar (Presidential Commission, *Report* 2:J10, J17–J19, J23).

24. Ibid.

25. Ibid. 5:913.

26. Wear, interview transcript, 12 March 1986, 7–8; Brinton, interview transcript, 13 March 1986, 4–5.

27. Ebeling, interview transcript, 19 March 1986, 13–15.

28. Russell, interview transcript, 19 March 1986, 4–5.

29. Ibid., 5–6.

30. U.S. Congress, House, *Investigation: Report*, 220.

31. Ebeling, interview transcript, 19 March 1986; Wear, interview transcript, 12 March 1986; Smith, interview transcript, 13 March 1986; Russell, interview transcript, 19 March 1986, 3–5.

32. Lovingood, interview transcript, 23 April 1986.

33. Russell, interview transcript, 19 March 1986, 12.

34. Smith, personal interview, 8 June 1992.

35. Presidential Commission, *Report* 1:88.

36. Ibid. 5:919.

37. Smith, interview transcript, 13 March 1986; Lovingood, interview transcript, 23 April 1986.

38. W. Leon Ray, telephone interview by author, 3 June 1992.

39. Schwinghamer, interview transcript, 13 March 1986; Wilbur Riehl, interview transcript, 13 March 1986, Marshall Space Flight Center files, National Archives, Washington, D.C.

40. Schell, interview transcript, 25 March 1986; Riehl, interview transcript, 13 March 1986; Schwinghamer, interview transcript, 13 March 1986.

41. Schell, interview transcript, 25 March 1986, 14.

42. Riehl, interview transcript, 13 March 1986, 10.

43. Presidential Commission, *Report* 1:129–31; U.S. Congress, House, *Investigation: Report*, 43; McDonald, interview transcript, 19 March 1986, 53–54.

44. Jerald Mason, interview transcript, 2 April 1986, Morton Thiokol, Inc., files, National Archives, Washington, D.C., p. 20.

45. Macbeth, interview transcript, 14 March 1986, 6.

46. Joel Maw, interview transcript, 2 March 1986, Morton Thiokol, Inc., files, National Archives, Washington, D.C., pp. 5–8.

47. Jack Kapp, interview transcript, 19 March 1986, Morton Thiokol, Inc., files, National Archives, Washington, D.C., pp. 15–16.

48. Larry H. Sayer, interview transcript, 2 March 1986, Morton Thiokol, Inc., files, National Archives, Washington, D.C., pp. 11–12.

49. Ketner, interview transcript, 26 March 1986, 16.

50. Presidential Commission, *Report* 4:790.

51. Thompson, interview transcript, 4 April 1986, 57.

52. Wear, interview transcript, 12 March 1986, 17–18.

53. Lawrence O. Wear, interview transcript, 13 March 1986, Marshall Space Flight Center files, National Archives, Washington, D.C., pp. 55–56.

54. Powers, interview transcript, 12 March 1986, 4.

55. Riehl, interview transcript, 13 March 1986, 10.

56. Brinton, interview transcript, 13 March 1986, 19–20.

57. Cecil Houston, interview transcript, 25 March 1986, Marshall Space Flight Center files, National Archives, Washington, D.C., p. 12.

58. Buchanan, interview transcript, 25 March 1986, 12.

59. Macbeth, interview transcript, 14 March 1986, 8–9.

60. Russell, interview transcript, 19 March 1986, 11.

61. Presidential Commission, *Report* 4:790.

62. Ibid., 791.

63. Ibid.

64. Cecil Houston, interview transcript, 10 April 1986, Marshall Space Flight Center files, National Archives, Washington, D.C., pp. 59–62.

65. U.S. Congress, House, *Investigation: Hearings* 1:567.

66. Ibid., 373.

67. Ebeling, interview transcript, 19 March 1986, 25.

68. In the following testimony, Mulloy read from notes he took during the teleconference, which are reproduced in Presidential Commission, *Report* 4:612.

69. Ibid. 5:841. Mulloy's statement is confirmed by notes taken that night by Keith Coates (Coates, interview transcript, 25 March 1986, 16–17). Thiokol's Jack Kapp and Brian Russell recalled still other data that Mulloy pointed that out Thiokol had not taken into account in drawing their conclusions (Russell, interview transcript, 19 March 1986, 18; Kapp, interview transcript, 19 March 1986, 25–26).

70. Powers, interview transcript, 12 March 1986, 13, 20, 23.

71. Russell, interview transcript, 19 March 1986, 18, 21.

72. Sayer, interview transcript, 2 March 1986, 16–17.

73. U.S. Congress, House, *Investigation: Hearings* 1:380.

74. Kapp, interview transcript, 19 March 1986, 25, 28.

75. Presidential Commission, *Report* 4:297.

76. U.S. Congress, House, *Investigation: Report*, appendix VI-A, 292–318.

77. Ibid., 27. For a discussion of ECRs, see chap. 8.

78. Ibid., 148–49.

79. Buchanan, interview transcript, 25 March 1986, 16.

80. Presidential Commission, *Report* 4:291. See also Marshall engineer Ben Powers's statement in Bell and Esch, 47.

81. U. S. Department of Commerce, National Oceanic and Atmos-

pheric Administration, National Weather Service, "Record of River and Climatological Observations," Titusville, Fla., December–January 1985.

82. Ibid.; U.S. Congress, House, *Investigation: Report*, 149–50.

83. Schell, interview transcript, 25 March 1986, 5.

84. Presidential Commission, *Report* 5:840.

85. Ibid., 843.

86. Buchanan, interview transcript, 25 March 1986, 13–14.

87. Wear, interview transcript, 12 March 1986, 21–22.

88. Riehl, interview transcript, 13 March 1986, 13, 18–19.

89. See, e.g., Kapp, interview transcript, 19 March 1986, 26; Brinton, interview transcript, 13 March 1986, 28; Kyle Speas, interview transcript, 20 March 1986, Morton Thiokol, Inc., files, National Archives, Washington, D.C., p. 11.

90. Russell, interview transcript, 19 March 1986, 18.

91. Maw, interview transcript, 2 March 1986, 11.

92. Ebeling, interview transcript, 19 March 1986, 23.

93. Kapp, interview transcript, 19 March 1986, 43.

94. For manager views, see McDonald, interview transcript, 19 March 1986, 13–14; Wear, interview transcript, 12 March 1986, 24–25. For Hardy's explanation, see Presidential Commission, *Report* 5:877, 878.

95. Buchanan, interview transcript, 25 March 1986, 13–14.

96. Brinton, interview transcript, 13 March 1986, 25.

97. Macbeth, interview transcript, 14 March 1986, 8–10, 24.

98. Powers, interview transcript, 12 March 1986, 17–18, 20–23.

99. Coates, interview transcript, 25 March 1986, 14.

100. See also McCarty, interview transcript, 19 March 1986, 18; McDonald, interview transcript, 19 March 1986, 15; Schwinghamer, interview transcript, 13 March 1986, 22.

101. Sayer, interview transcript, 2 March 1986, 17–18, 20.

102. Macbeth, interview transcript, 14 March 1986, 13–14.

103. In addition to the statements quoted here, see, e.g., Reinartz in Presidential Commission, *Report* 5:914; Mason, ibid. 4:779; Mason, interview transcript, 2 April 1986, 32; Calvin Wiggins, interview transcript, 19 March 1986, Morton Thiokol, Inc., files, National Archives, Washington, D.C., pp. 11–12; Kapp, interview transcript, 19 March 1986, 28–29; Miller, interview transcript, 13 March 1986, 11–12; Macbeth, interview transcript, 14 March 1986, 16; McCarty, interview transcript, 19 March 1986, 20.

104. McDonald, interview transcript, 19 March 1986, 16.

105. Only Boisjoly is reported here because the other engineers' interpretations are quoted and discussed extensively in chap. 9.

106. Presidential Commission, *Report* 4:792.

107. Ibid. 5:842–43.

108. Schwinghamer, interview transcript, 13 March 1986, 22–23.

109. Although Mason was grilled extensively by investigators about his "hat" comment to Lund and his assessment of the joint's capability, he was not asked to testify in detail about the way he led the discussion.

My analysis in this section is drawn from testimony from other Thiokol participants, who described in sequence what Mason said and did. See, e.g., Sayer, interview transcript, 2 March 1986, 19; Russell, interview transcript, 19 March 1986, 25-26.

110. Kapp, interview transcript, 19 March 1986, 32.

111. Presidential Commission, *Report* 4:1455; Maw, interview transcript, 2 March 1986, 13.

112. See also Kapp, interview transcript, 19 March 1986, 28-30. The temperature 40°F was derived from both the firing of the qualification motors and static tests of developmental motors at that temperature.

113. Jerry Burn, interview transcript, 25 March 1986, Morton Thiokol, Inc., files, National Archives, Washington, D.C., pp. 23-24; Presidential Commission, *Report* 4:822.

114. Mason, interview transcript, 2 April 1986, 27-29; see also Ebeling, interview transcript, 19 March 1986, 32.

115. U.S. Congress, House, *Investigation: Hearings* 1:383.

116. Macbeth, interview transcript, 14 March 1986, 14.

117. Ebeling, interview transcript, 19 March 1986, 31.

118. Russell, interview transcript, 19 March 1986, 29-30.

119. Sayer, interview transcript, 2 March 1986, 21.

120. Kapp, interview transcript, 19 March 1986, 34, 40.

121. Mason, interview transcript, 2 April 1986, 27-29.

122. Robert K. Lund, interview transcript, 1 April 1986, Morton Thiokol, Inc., files, National Archives, Washington, D.C., p. 45.

123. Presidential Commission, *Report* 4:818-19; Wiggins, interview transcript, 19 March 1986, 19-20; U.S. Congress, House, *Investigation: Hearings* 1:371.

124. Lund, interview transcript, 1 April 1986, 46-47. See also Presidential Commission, *Report* 4:816.

125. Mason, interview transcript, 2 April 1986, 23-24, 27-29.

126. Ebeling, interview transcript, 19 March 1986, 33.

127. Buchanan, interview transcript, 25 March 1986, p. 22.

128. U.S. Congress, House, *Investigation: Hearings* 1:522.

129. Lawrence B. Mulloy, personal interview by author, 9 June 1992.

130. McDonald, interview transcript, 19 March 1986, 20-21.

131. Miller, interview transcript, 13 March 1986, 17.

132. Schwinghamer, interview transcript, 13 March 1986, 31.

133. Brinton, interview transcript, 13 March 1986, 27.

134. Ibid., 28; Powers, interview transcript, 12 March 1986, 16.

135. Brinton, interview transcript, 13 March 1986, 26-27; Speas, interview transcript, 20 March 1986, 14.

136. Speas, interview transcript, 20 March 1986, 14.

137. Presidential Commission, *Report* 5:865; Miller, interview transcript, 13 March 1986, 18; Wear, interview transcript, 13 March 1986, 45-46.

138. Powers, interview transcript, 12 March 1986, 17.

139. Presidential Commission, *Report* 5:848, 852, 865.

140. Miller, interview transcript, 13 March 1986, 19.

141. Presidential Commission, *Report* 4:615; U.S. Congress, House, *Investigation: Hearings* 1:406.

142. Buchanan, interview transcript, 25 March 1986, 27. See also Miller, interview transcript, 13 March 1986, 20–21; Wear, interview transcript, 13 March 1986, 60–61; Russell, interview transcript, 19 March 1986, 38; Kapp, interview transcript, 19 March 1986, 37; Maw, interview transcript, 2 March 1986, 19; U.S. Congress, House, *Investigation: Hearings* 1:525.

143. Powers, interview transcript, 12 March 1986, 34; Brinton, interview transcript, 13 March 1986, 36.

144. Brinton, interview transcript, 13 March 1986, 38, 44.

145. Presidential Commission, *Report* 4:793.

146. Sayer, interview transcript, 2 March 1986, 24.

147. Ketner, interview transcript, 26 March 1986, 22.

148. Burn, interview transcript, 25 March 1986, 31, 46.

149. Larry Sayer, interview transcript, 13 March 1986, Morton Thiokol, Inc., files, National Archives, Washington, D.C., p. 20.

150. Presidential Commission, *Report* 4:822. See also Macbeth, interview transcript, 14 March 1986, 22.

151. Russell, interview transcript, 19 March 1986, 43.

152. Presidential Commission, *Report* 4:793.

153. Maw, interview transcript, 2 March 1986, 20.

154. Ebeling, interview transcript, 19 March 1986, 34.

155. Brinton, interview transcript, 13 March 1986, 39.

156. Russell, interview transcript, 19 March 1986, 49.

157. McDonald, interview transcript, 19 March 1986, 29.

158. Presidential Commission, *Report* 5:849; U.S. Congress, House, *Investigation: Hearings* 1:386. See also Buchanan, interview transcript, 25 March 1986, 21.

159. Houston, interview transcript, 25 March 1986, 71–72.

160. Presidential Commission, *Report* 5:849.

161. McDonald, interview transcript, 19 March 1986, 35.

162. Presidential Commission, *Report* 2:I9–I10.

163. Brinton, interview transcript, 13 March 1986, 43.

164. Ibid., 46.

165. Presidential Commission, *Report* 1:100–101.

166. Ibid. 2:J23; McConnell, 216, 233.

167. Presidential Commission, *Report* 2:J23.

168. Ibid., I9–I10.

169. U.S. Congress, House, *Investigation: Report*, 238.

170. Presidential Commission, *Report* 2:I9–I10.

171. Ibid. 1:110.

172. Ibid.

173. U.S. Congress, House, *Investigation: Report*, 241.

174. Ibid., 243.

175. Presidential Commission, *Report* 1:17.

176. Ibid.

177. Ibid., 115.

178. Ibid., 117.

179. Ibid., 116.

180. Ibid., 117.

181. Ebeling, interview transcript, 19 March 1986, 35.

182. Russell, interview transcript, 19 March 1986, 56.

183. Macbeth, interview transcript, 14 March 1986, 28.

CHAPTER 9

1. The Presidential Commission and the House Committee identified launch-related rule violations by other NASA employees on the eve of the launch (Presidential Commission, *Report* 1:193; U.S. Congress, House, *Investigation: Report*, 146–47). In this chapter, we examine only those attributed to members of the SRB work group, however.

2. Ray, telephone interview, 13 October 1992.

3. Kapp, interview transcript, 12 April 1986, 54.

4. Russell, interview transcript, 19 March 1986, 60.

5. Coates, interview transcript, 25 March 1986, 12.

6. I do not have information from everyone because not all were asked the same questions by official investigators. Some volunteered a view without being asked. My conclusion is based on statements by Ketner, interview transcript, 26 March 1986, 25; Burn, interview transcript, 25 March 1986, 27; Powers, interview transcript, 12 March 1986, 68; Coates, interview transcript, 25 March 1986, 20; Ray, telephone interview, 13 October 1992. Boisjoly's view is discussed later in the chapter.

7. Presidential Commission, *Report* 5:856–57.

8. Wear, interview transcript, 13 March 1986, 59–60.

9. Presidential Commission, *Report* 1:104.

10. Brinton, interview transcript, 13 March 1986, 50–51.

11. Ibid., 50.

12. Powers, interview transcript, 12 March 1986, 46.

13. Lund, interview transcript, 1 April 1986, 50.

14. See, e.g., Brinton, interview transcript, 13 March 1986, 25; Buchanan, interview transcript, 25 March 1986, 13–14; Macbeth, interview transcript, 14 March 1986, 8–10, 24; Wear, interview transcript, 12 March 1986, 17–20, 24, 27; Presidential Commission, *Report* 5:877, 878; Powers, interview transcript, 12 March 1986, 17–18, 20–23; Coates, interview transcript, 25 March 1986, 14.

15. At Thiokol in Utah: Russell, interview transcript, 19 March 1986, 18; Maw, interview transcript, 2 March 1986, 11; Ebeling, interview transcript, 19 March 1986, 23; Kapp, interview transcript, 19 March 1986, 43; Presidential Commission, *Report* 4:793 (Boisjoly). At Kennedy: McDonald, interview transcript, 19 March 1986, 13–14.

16. See also McDonald in U.S. Congress, House, *Investigation: Hearings* 1:524.

17. Presidential Commission, *Report* 4:822.

18. Ibid., 793; U.S. Congress, House, *Investigation: Hearings* 1:531.

19. U.S. Congress, House, *Investigation: Report*, 148.

20. Presidential Commission, *Report* 1:82, 104.

21. Ibid. 2:I17.

22. Salita, interview transcript, 2 April 1986, 9, 14.

23. U.S. Congress, House, *Investigation: Hearings* 1:567.

24. Ibid., 525.

25. Ebeling, interview transcript, 19 March 1986, 25.

26. Bell and Esch, 47.

27. Russell, interview transcript, 19 March 1986, 38.

28. Wear, interview transcript, 12 March 1986, 17–20, 24, 27.

29. Presidential Commission, *Report* 4:793; U.S. Congress, House, *Investigation: Hearings* 1:378, 531.

30. Presidential Commission, *Report* 5:840.

31. Ibid., 867.

32. Powers, interview transcript, 12 March 1986, 34.

33. Presidential Commission, *Report* 4:848, 849, 854; Russell, interview transcript, 19 March 1986, 38; McDonald, interview transcript, 19 March 1986, 32.

34. Presidential Commission, *Report* 1:82.

35. U.S. Congress, House, *Investigation: Hearings* 1:405; Presidential Commission, *Report* 5:849.

36. Presidential Commission, *Report* 1:100–101.

37. Mulloy, telephone interview, 5 August 1992.

38. McDonald, interview transcript, 19 March 1986, 35.

39. Presidential Commission, *Report* 1:101.

40. U.S. Congress, House, *Investigation: Report*, 27.

41. Presidential Commission, *Report* 4:723.

42. Ibid. 5:729–30, 848. See also, e.g., Russell, interview transcript, 19 March 1986, 38.

43. Presidential Commission, *Report* 4:730.

44. Mulloy, personal interview, 8 June 1992.

45. Presidential Commission, *Report* 5:843.

46. McCurdy, "The Decay of NASA's Technical Culture"; Romzek and Dubnick.

47. U.S. Congress, House, *Investigation: Report*, 219.

48. Ray, telephone interview, 13 October 1992.

49. DiMaggio and Powell, "The Iron Cage Revisited"; March and Olsen, "The New Institutionalism," 741; Smircich, 350; Zucker, "The Role of Institutionalization."

50. Other scholars have analyzed the Thiokol engineering presentation, identifying weaknesses from an engineering/statistical point of view. I agree with their conclusions, although I do not agree with all points of their analyses that lead to those conclusions. A factor distinguishing their analyses from mine is that they have examined the data per se, not taking into account information patterns and the social context. See Frederick F. Lighthall, "Launching the Space Shuttle *Chal-*

lenger: Disciplinary Deficiencies in the Analysis of Engineering Data," *IEEE Transactions on Engineering Management* 38 (February 1991): 63–74; Siddhartha R. Dalal, Edward B. Fowlkes, and Bruce Hoadley, "Risk Analysis of the Space Shuttle: Pre-*Challenger* Prediction of Failure," *Journal of the American Statistical Association* 84 (1989): 945–57; Michael Lavine, "Problems in Extrapolation Illustrated with Space Shuttle O-ring Data," *Journal of the American Statistical Association* 86 (1991): 919–22.

51. U.S. Congress, House, *Investigation: Report*, 150–51.

52. Mulloy, personal interview, 8 June 1992.

53. Wear, interview transcript, 13 March 1986, 53.

54. Schell, interview transcript, 25 March 1986, 5.

55. U.S. Congress, House, *Investigation: Report*, 27.

56. Ibid., 150–51.

57. Powers, interview transcript, 12 March 1986, 23.

58. See also Lighthall; Dalal, Fowlkes, and Hoadley; Lavine.

59. Presidential Commission, *Report* 4:297.

60. U.S. Congress, House, *Investigation: Report*, 150–51.

61. Dennis S. Gouran, Randy Y. Hirokawa, and Amy E. Martz, "A Critical Analysis of Factors Related to Decisional Processes Involved in the *Challenger* Disaster," *Central States Speech Journal* 37 (Fall 1986): 127.

62. Presidential Commission, *Report* 4:812.

63. Ibid.

64. Ibid. 5:843.

65. Clarke, "Disqualification Heuristic."

66. Russell, interview transcript, 19 March 1986, 21.

67. See also Lighthall, 72.

68. Riehl, interview transcript, 13 March 1986, 13, 18–19.

69. Presidential Commission, *Report* 1:118; U.S. Congress, House, *Investigation: Report*, 151–52.

70. Cf. Gouran, Hirokawa, and Martz, 130–31; Patrick Moore, "When Politeness Is Fatal: Technical Communication and the *Challenger* Accident," *Journal of Business and Technical Communication* 6 (1992): 269–92.

71. Presidential Commission, *Report* 1:117.

72. U.S. Congress, House, *Investigation: Report*, 151–52.

73. Karl E. Weick, "The Vulnerable System: An Analysis of the Tenerife Air Disaster," *Journal of Management* 16 (1990): 571–93.

74. Giddens; Ranson, Hinings, and Greenwood, 3.

75. See also Lazega.

76. See, e.g., Speas, interview transcript, 20 March 1986, 10; Schell, interview transcript, 25 March 1986, 7.

77. Miller, interview transcript, 27 March 1986, 5.

78. Ibid., 13–14, 15.

79. Powers, interview transcript, 12 March 1986, 4.

80. Ibid., 66.

81. Ibid., 16.

82. Schwinghamer, interview transcript, 13 March 1986, 20–21.

83. Ibid., 69–70.

84. Riehl, interview transcript, 13 March 1986, 21–22.

85. Houston, interview transcript, 10 April 1986, 42, 49; Buchanan, interview transcript, 25 March 1986, 10.

86. Coates, interview transcript, 25 March 1986, 7–8.

87. Ibid., 7, 11.

88. Ibid., 21.

89. Powers, interview transcript, 12 March 1986, 10–11.

90. Ibid., 12, 15.

91. Russell, interview transcript, 19 March 1986, 36.

92. Kapp, interview transcript, 19 March 1986, 34.

93. Jerald Mason, interview transcript, 2 March 1986, Morton Thiokol, Inc., files, National Archives, Washington, D.C., pp. 25–26.

94. Russell, interview transcript, 19 March 1986, 35–36.

95. Mason, interview transcript, 2 April 1986, 4.

96. Presidential Commission, *Report* 1:88, 92.

97. Boisjoly, personal interview, 8 February 1990.

98. Maw, interview transcript, 2 March 1986, 17.

99. Kapp, interview transcript, 19 March 1986, 32.

100. Macbeth, interview transcript, 14 March 1986, 14, 16.

101. Russell, interview transcript, 19 March 1986, 31–32.

102. Sayer, interview transcript, 13 March 1986, 21.

103. Presidential Commission, *Report* 4:822; Russell, interview transcript, 19 March 1986, 29.

104. Sayer, interview transcript, 13 March 1986, 26.

105. Riehl, interview transcript, 13 March 1986, 21–22. See also Russell, interview transcript, 19 March 1986, 33.

106. Presidential Commission, *Report* 5:865.

107. McDonald, interview transcript, 19 March 1986, 22–23.

108. Wear, interview transcript, 13 March 1986, 15.

109. Coates, interview transcript, 25 March 1986; Powers, interview transcript, 12 March 1986.

110. U.S. Congress, House, *Investigation: Hearings* 1:402.

111. See, e.g., Wayne Littles, interview transcript, 19 March 1986, Marshall Space Flight Center files, National Archives, Washington, D.C., pp. 13–14; Lund, interview transcript, 1 April 1986, 41; McCarty, interview transcript, 19 March 1986, 28.

112. Presidential Commission, *Report* 1:93.

113. Russell, interview transcript, 19 March 1986, 41.

114. Kapp, interview transcript, 19 March 1986, 39–40.

115. Powers, interview transcript, 12 March 1986, 28–29.

116. Coates, interview transcript, 25 March 1986, 25.

117. McDonald, interview transcript, 19 March 1986, 23; U.S. Congress, House, *Investigation: Hearings* 1:524.

118. U.S. Congress, House, *Investigation: Hearings* 1:372.

119. Weick, "Vulnerable System," 589.
120. Wear, interview transcript, 13 March 1986, 35–38.
121. Lawrence B. Mulloy, telephone interview by author, 16 June 1994.
122. Erving Goffman, "On Cooling the Mark Out: Some Aspects of Adaptation to Failure," *Psychiatry* 15 (1952): 451–63.
123. Presidential Commission, *Report* 4:822; Russell, interview transcript, 19 March 1986, 29.
124. Miller, interview transcript, 13 March 1986, 5–6, 7.
125. Littles, interview transcript, 19 March 1986, 13–14.
126. Burn, interview transcript, 25 March 1986, 31.
127. Ibid., 31, 40.
128. See chap. 7.
129. Russell, interview transcript, 19 March 1986, 55–56.
130. Ebeling, interview transcript, 19 March 1986, 22.
131. Powers, interview transcript, 12 March 1986, 28–29.
132. See, e.g., Thompson, interview transcript, 4 April 1986, 57.
133. Sayer, interview transcript, 2 March 1986, 16–17.
134. A 40°F floor would have been consistent with their data from two different tests: (1) Qualifying Motor 3 (QM3) was tested at an ambient temperature of 40°F and experienced no O-ring anomalies (see fig. 11.11); and (2) the SRMs were qualified to (i.e., fired at) 40°F–90°F, which also meant that this recommendation would have conformed to an existing rule. The restrictive use of the experience was not called for because their own analysis after the 53°F launch concluded that the joint could tolerate three times the damage that they were likely to experience in the worst case. Although Thiokol's Thompson, Boisjoly, and McDonald testified that both these tests were less than perfect, had they recommended the launch be delayed until the temperature reached 40°F or better, they would have met the technical standard of logical consistency with available data. The recommendation would not have incorporated the 30°F data, but since those data were from bench tests and therefore were not considered in engineering culture to be "as good as" motor firings (more closely approximating flight data), their position would have been strong.
135. The field joint was the main concern on the eve of the launch; therefore these charts do not isolate nozzle joint anomalies. The nozzle joint, because of its design, was not vulnerable to rotation and thus to cold. Nozzle joint anomalies were caused by other factors. Some of the missions plotted in fig. 13.1 that had field joint anomalies also had nozzle joint erosion. In fig. 13.2, those missions with nozzle joint erosion are collapsed into "flights with no incidents," appropriate since the intent is to examine a possible relationship between field joint erosion and temperature.
136. The collective failure of all participants to focus on all data points in order to examine the correlation between temperature and damage was attributed by Lighthall to disciplinary deficiencies in the

professional training of engineers that left teleconference participants uniformly ignorant of simple concepts and methods in covariation. While I agree with much of Lighthall's analysis, the existence of uniform disciplinary deficiencies seems an unlikely explanation. After the disaster, people without engineering training readily came up with a trend analysis using all data points. More logically, the ability of Kehrli and Keel and others (like Lighthall and myself) to put together a trend analysis was enhanced by hindsight, which also obfuscates the ill-structured nature of the problem and the complexity of the task environment that participants faced at the time (Lighthall, 72–73).

137. Boisjoly, telephone interview, 19 March 1993.

138. DiMaggio and Powell, "The Iron Cage Revisited," n. 5.

139. For the importance of making macro-micro connections in institutional arguments, see DiMaggio and Powell, "Introduction." 140. Giddens; Ranson, Hinings, and Greenwood, 3.

141. Becker.

142. Starbuck and Milliken, "Challenger: Fine-Tuning the Odds," 324.

CHAPTER 10

1. David E. Nye, American Technological Sublime (Cambridge: MIT Press, 1994).

2. Presidential Commission, Report 1:ii.

3. U.S. Congress, House, Investigation: Report, 7.

4. Ibid., 3.

5. Presidential Commission, Report 1:148; 2:II7; U.S. Congress, House, Investigation: Report, 4.

6. Russell, interview transcript, 19 March 1986, 60.

7. Presidential Commission, Report 1:60–66; U.S. Congress, House, Investigation: Hearings 1:341.

8. U.S. Congress, House, Investigation: Hearings 1:445.

9. Presidential Commission, Report 1:140; U.S. Congress, House, Investigation: Hearings 1:517; McIntosh, interview transcript, 2 April 1986, 69.

10. Presidential Commission, Report 4:723.

11. Hawkins, "Use of Legal Discretion," 35.

12. Presidential Commission, Report 2:K3.

13. U.S. Congress, House, Investigation: Hearings 1:447–50.

14. Ibid., 449.

15. Wear, personal interview, 2 June 1992; see, e.g., U.S. Congress, House, Investigation: Hearings 1:405, 447–48, 528–29, 534–37.

16. U.S. Congress, House, Investigation: Hearings 1:445.

17. Scott D. Sagan, The Limits of Safety: Organizations, Accidents, and Nuclear Weapons (Princeton, N.J.: Princeton University Press, 1993), 278.

18. See also Perrow, "The President's Commission and the Normal Accident."

19. Starbuck and Milliken, "Executives' Perceptual Filters."

20. Geertz, *Interpretation of Cultures*, 27–28.

21. See, e.g., Daniel Kahneman, Paul Slovic, and Amos Tversky, eds., *Judgment under Uncertainty: Heuristics and Biases* (Cambridge: Cambridge University Press, 1982); Amos Tversky and Daniel Kahneman, "The Framing of Decisions and the Psychology of Choice," *Science* 211 (30 January 1981): 1453–58; Amos Tversky and Daniel Kahneman, "Judgment under Uncertainty: Heuristics and Biases," *Science* 185 (1974): 1124. For an excellent review and analysis, see Heimer, "Social Structure."

22. Mary Douglas, *Risk Acceptability According to the Social Sciences* (London: Routledge and Kegan Paul, 1985); Heimer, "Social Structure"; James F. Short, Jr., "The Social Fabric at Risk: Toward the Social Transformation of Risk Analysis," *American Sociological Review* 49 (1984): 711–25; Freudenburg, "Perceived Risk, Real Risk"; Clarke and Short; James F. Short, Jr., and Lee Clarke, eds., *Organizations, Uncertainties, and Risk* (Boulder: Westview, 1992).

23. The position that these two areas ought to be combined is recent, initiated by and argued in Pinch and Bijker; Latour, *Science in Action*; Bruno Latour, "Some Scholars' Babies Are Other Scholars' Bathwater," *Contemporary Sociology* 22 (July 1993): 487–89; Pickering; Bijker and Law.

24. Mary Douglas observes that in the professional discussion of cognition and choices about risk "the neglect of culture is so systematic and so entrenched that nothing less than a large upheaval in the social sciences would bring about a change" (Douglas, *Risk Acceptability*, 1). For work that also makes culture the central focus see, e.g., Sheila Jasanoff, *Risk Management and Political Culture* (New York: Russell Sage, 1986), and Douglas and Wildavsky.

25. See also Latour and Woolgar; Knorr-Cetina.

26. Latour and Woolgar; Susan Leigh Star, "Simplification in Scientific Work: An Example from Neuroscience Research," *Social Studies of Science* 13 (1983): 205–28; Star, "Scientific Work and Uncertainty."

27. Kuhn, 53.

28. Ibid., 11, 38.

29. Ibid., 7.

30. Ibid., 24.

31. Ibid., 65.

32. Graham T. Allison, *The Essence of Decision: Explaining the Cuban Missile Crisis* (Boston: Little, Brown, 1971); Irving L. Janis, *Groupthink* (Boston: Houghton Mifflin, 1982).

33. For the rationale and examples of the naturalistic perspective, see Peter K. Manning, "The Social Reality and Social Organization of Natural Decision Making," *Washington and Lee Law Review* 43 (1986): 1291–1311; Hawkins, "Use of Legal Discretion"; Emerson, "Holistic Effects."

34. Karl E. Weick, "The Collapse of Sensemaking in Organizations: The Mann Gulch Disaster," *Administrative Science Quarterly* 38 (1993): 628–52.

35. For a foundational work on this perspective, see Weick, *Social Psychology of Organizing*.

36. Mayer N. Zald, "History, Meta-Narratives, and Organizational Theory" (Department of Sociology, University of Michigan, 1995; manuscript).

37. Douglas, *How Institutions Think*.

38. Walter W. Powell and Paul J. DiMaggio, eds., *The New Institutionalism in Organizational Analysis* (Chicago: University of Chicago Press, 1991).

39. DiMaggio and Powell, "Introduction," 11.

40. Janis. See, e.g., James K. Esser and Joanne S. Lindoerfer, "Groupthink and the Space Shuttle *Challenger* Accident: Toward a Quantitative Case Analysis," *Journal of Behavioral Decision Making* 2 (1989): 167–77; Arie W. Kruglanski, "Freeze-think and the *Challenger*," Psychology Today, August 1986, 48–49.

41. Groupthink focuses on "concurrence seeking" and striving for mutual support in decision making. The antecedent conditions that Janis posits were not present. In fact, the constraints on collective thinking that Janis suggests should be present, were present, scripted into the proceedings by organizational rules and norms. Teleconference participants were no "inner circle" consisting of a cohesive small group of decision makers who liked each other and valued membership in the group. Instead, 34 people were present; not all knew each other; and several had not even participated in FRR before. Physically, the group was insulated from others in the organization, but experts outside the SRB work group were present specifically to interject alternative views and information. The discussion did not lack norms requiring methodical procedures for decision making. It was guided by norms and rules of the organization about how technical discussions must be conducted. A critical antecedent condition that was present, not taken into account by Janis's criteria, was the institutionalized cultural construction of risk that was a product of work group interaction in the years preceding the launch.

In the teleconference discussion itself, the pressure toward uniformity that Janis identified was present in group dynamics; however, it originated in culture and structure. Self-censorship was present; however, it did not originate in a desire to preserve the unity of the group, but from the cultural imperatives of the original technical culture, bureaucratic accountability, and political accountability. What Janis identified as "direct pressure on any member who expresses strong arguments or dissent," which engineers experienced that night, was also a function of the original technical culture, bureaucratic accountability, and political accountability. At the end, there was a shared illusion of unanimity, but it was a product of physical separation as well as the triumvirate of cultural imperatives. Finally, what appeared to exemplify the "mindguards" who fail to transmit potentially distressing information to top leaders who might make a painful reevaluation was a consequence of bureaucratic accountability: Level III managers conforming to rules about what was to be forwarded to Level II.

526   NOTES TO PAGES 404–407

42. Steven Brint and Jerome Karabel, "Institutional Origins and Trans-formations: The Case of American Community Colleges," in *The New Institutionalism in Organizational Analysis*, ed. Walter W. Powell and Paul J. DiMaggio (Chicago: University of Chicago Press, 1991), 337.

43. Ibid., 345.

44. Wolf V. Heydebrand, "New Organizational Forms," *Work and Occupations* 16 (1989): 323–57.

45. DiMaggio and Powell suggest that Bourdieu's notion of habitus, with its emphasis on social location of actors, be developed into a microsociology that complements the macrosociological work on which most institutional theorists concentrate. They call for a "theory of practical action" that circumvents the limits of role theory by replacing role with position as the relevant concept. Position encompasses, but is not restricted to, role, extending beyond role to "regularities of thought, aspirations, dispositions, patterns of appreciation, and strategies of action that are linked to the positions persons occupy in the social structure they continually reproduce." See DiMaggio and Powell, "Introduction," 15–27; Bourdieu.

46. Steven Brint's research on a wide variety of professionals demonstrates this conclusion with data that includes multiple organizational forms. Using ingenious methods, Brint examined the political attitudes and class position of professionals. Locating Professionals in layered structures—"spheres of purpose," markets, and different forms of organization within them—he found that worldview varied with social location (Steven Brint, *In an Age of Experts: The Changing Role of Professionals in Politics and Public Life* [Princeton, N.J.: Princeton University Press, 1994]).

47. See appendix C.

48. Vaughan, "The Macro-Micro Connection."

49. Among the classic writings in this tradition are Albert K. Cohen, *Delinquent Boys: The Culture of the Gang* (Glencoe, Ill.: Free Press, 1955); Merton, *Social Theory and Social Structure*; Edwin H. Sutherland, "White-Collar Criminality," *American Sociological Review* 5 (1940): 1–12; Richard E. Quinney, "Occupational Structure and Criminal Behavior: Prescription Violation by Retail Pharmacists," *Social Problems* 11 (1963): 179–85; David M. Ermann and Richard J. Lundman, *Corporate and Governmental Deviance* (New York: Oxford University Press, 1978); Norman K. Denzin, "Notes on the Criminogenic Hypothesis: A Case Study of the American Liquor Industry," *American Sociological Review* 42 (1977): 905–20; Simpson.

50. For a particularly cogent exposition of this view, see Ermann and Lundman.

51. For an excellent review of this argument, see James W. Coleman, "The Theory of White-Collar Crime," in *White-Collar Crime Reconsidered*, ed. Kip Schlegel and David Weisburd (Boston: Northeastern University Press, 1992).

52. Cf. Merton, *Social Theory and Social Structure*, 185–214. See also Diane Vaughan, "Merton's Anomie Theory and Organizational Miscon-

duct" (paper presented at annual meetings of American Society of Criminology, Miami, November 1994).

53. Whether deviance or conformity is learned is a contested issue. See, e.g., Cohen, *Delinquent Boys;* Edwin H. Sutherland, *Principles of Criminology* (Chicago: Lippincott, 1934); Sutherland, *White-Collar Crime;* Walter Miller, "Lower Class Culture as a Generating Milieu of Gang Delinquency," *Journal of Social Issues* 14 (1958): 5–19; Gresham K. Sykes and David Matza, "Techniques of Neutralization: A Theory of Delinquency," *American Sociological Review* 22 (1957): 667–70.

54. Hannah Arendt, *Eichmann in Jerusalem: A Report on the Banality of Evil* (New York: Viking, 1964); Herbert C. Kelman and V. Lee Hamilton, *Crimes of Obedience* (New Haven, Conn.: Yale University Press, 1989).

55. See, e.g., John Braithwaite and Toni Makkae, "Testing an Expected Utility Model of Corporate Deterrence," *Law and Society Review* 25 (1991): 7–9; Raymond Paternoster and Sally S. Simpson, "A Rational Choice Theory of Corporate Crime," in *Routine Activity and Rational Choice,* ed. Ronald V. Clarke and Marcus Felson (New Brunswick, N.J.: Transaction, 1993); Kip Schlegel,2 *Just Desserts for Corporate Criminals* (Boston: Northeastern University Press, 1990); Kirk R. Williams and Richard Hawkins, "Perceptual Research on General Deterrence: A Critical Overview," *Law and Society Review* 20 (1986): 545–72.

56. Stephen J. Pfohl, *Images of Deviance and Social Control* (New York: McGraw-Hill, 1985), 73–74; Stanton Wheeler, "The Problem of White-Collar Crime Motivation," in *White-Collar Crime Reconsidered,* ed. Kip Schlegel and David Weisburd (Boston: Northwestern University Press, 1992), 108–23.

57. Cullen, Maakestad, and Cavender.

58. Sykes and Matza; James W. Coleman, "Toward an Integrated Theory of White-Collar Crime," *American Journal of Sociology* 93 (1987): 406–39.

59. John Braithwaite and Brent Fisse, "Varieties of Responsibility and Organizational Crime," *Law and Policy* 7 (1985): 315–43; Brent Fisse and John Braithwaite, "The Allocation of Responsibility for Corporate Crime: Individualism, Collectivism, and Accountability," *Sidney Law Review* 12 (1988): 468–513.

60. See e.g., Edward Gross, "Organizational Structure and Organizational Crime," in *White-Collar Crime: Theory and Research,* ed. Gilbert Geis and Ezra Stotland (Beverly Hills, Calif.: Sage, 1980); Ronald C. Kramer, "Corporate Crime: An Organizational Perspective," in *White-Collar and Economic Crime,* ed. Peter Wickman and Timothy Daily (Lexington, Mass.: Lexington Books, 1982); Christopher D. Stone, *Where the Law Ends: The Social Control of Corporate Behavior* (New York: Harper and Row, 1975).

61. See, e.g., Geis; Edward Gross, "Organizational Crime: A Theoretical Perspective," in *Studies in Symbolic Interaction,* ed. Norman K. Denzin (Greenwich, Conn.: JAI, 1978); Clinard and Yeager; Cullen,

Maakestad, and Cavender; Coleman, "Toward an Integrated Theory of White-Collar Crime."

62. Marshall B. Clinard, *Corporate Ethics and Crime: The Role of Middle Management* (Beverly Hills, Calif.: Sage, 1983); Jackall.

63. Eric J. Chaisson, *The Hubble Wars* (New York: HarperCollins, 1994); Paul Carroll, *Big Blues: The Unmaking of IBM* (New York: Random House, 1993); Robert S. McNamara, *In Retrospect* (New York: Times Books, 1995).

64. Turner, *Man-Made Disasters*; Turner, "The Organizational and Interorganizational Development of Disasters."

65. Turner, *Man-Made Disasters*, 51.

66. See, e.g., Stephen Jay Gould, *Time's Arrow, Time's Cycle* (Cambridge: Harvard University Press, 1987); Stephen Jay Gould, *Wonderful Life: The Burgess Shale and the Nature of History* (New York: Norton, 1989).

67. Gould, *The Mismeasure of Man*.

68. Ibid., 27.

69. Ibid.

70. Ibid., 62, 66–67, 68–69, 88–107.

71. Ibid., 28.

72. Stephen J. Pfohl, *Predicting Dangerousness: The Social Construction of Psychiatric Reality* (Lexington, Mass.: Lexington Books, 1978).

73. Ibid., 131.

74. Ibid., 143.

75. Ibid., 150.

76. Ibid., 146.

77. Vaughan, *Uncoupling*.

78. Perrow, *Normal Accidents*.

79. The core literature on this approach includes Todd R. LaPorte and Paula M. Consolini, "Working in Practice but Not in Theory: Theoretical Challenges of 'High Reliability Organizations,'" *Journal of Public Administration Research and Theory* 1 (January 1991): 19–47; Karlene H. Roberts, "New Challenges in Organizational Research: High Reliability Organizations," *Industrial Crisis Quarterly* 3 (1989): 111–25; Karl E. Weick and Karlene H. Roberts, "Collective Mind in Organizations: Heedful Interrelating on Flight Decks," *Administrative Science Quarterly* 38 (1993): 357–81. For similarities and differences between high-reliability theory and normal-accidents theory, see Lee Clarke, "Drs. Pangloss and Strangelove Meet Organization Theory: High Reliability Organizations and Nuclear Weapons Accidents," *Sociological Forum* 8 (1993), 675–89; Sagan, *The Limits of Safety*, 11–52; see also a symposium on Sagan's book published in *Journal of Contingencies and Crisis Management* 2 (1994).

80. Here NASA adopted Weick's suggestion that increasing networks and teams of divergent individuals increases reliability in organizations. See Karl E. Weick, "Organizational Culture as a Source of High Reliability," *California Management Review* 29 (1987): 115–16.

81. Presidential Commission, *Report* 4:378–80.

82. After the accident, several analysts suggested ethical training as a remedy. While increasing a sense of moral responsibility in individuals is an important strategy because it has the potential to alter organizational culture, it is insufficient without corresponding attempts to make additional structural changes. See, e.g., Russell P. Boisjoly, Ellen Foster Curtis, and Eugene Mellican, "Roger Boisjoly and the *Challenger* Disaster: The Ethical Dimensions," *Journal of Business Ethics* 8 (1989): 217–30.

83. Karlene H. Roberts, "Some Characteristics of One Type of High Reliability Organization," *Organization Science* 1 (1990): 160–73; Weick, "Organizational Culture"; Karlene H. Roberts, Denise M. Rousseau, and Todd R. LaPorte, "The Culture of High Reliability: Quantitative and Qualitative Assessment aboard Nuclear Powered Aircraft Carriers," *Journal of High Technology Management Research* 5 (Spring 1994): 141–61.

84. Sagan, *The Limits of Safety*, 252.

85. Vaughan, *Controlling Unlawful Organizational Behavior*, 105–12; Scott D. Sagan, "Toward a Political Theory of Organizational Reliability," *Journal of Contingencies and Crisis Management* 2 (1994): 228–40; Perrow, *Normal Accidents*.

86. Sagan, "Toward a Political Theory of Organizational Reliability."

87. Heimer, "Social Structure," 513; Perrow, *Normal Accidents*; Sagan, *The Limits of Safety*.

88. James G. March, Lee S. Sproull, and Michal Tamuz, "Learning from Samples of One or Fewer," *Organization Science* 2 (February 1991): 1–13.

89. Everett C. Hughes, "Mistakes at Work," *Canadian Journal of Economics and Political Science* 17 (1951): 320–27.

90. Marianne Paget, *The Unity of Mistakes: A Phenomenological Interpretation of Medical Work*, (Philadelphia: Temple University Press), 58.

91. See also Perrow, *Normal Accidents*; Perrow, "The Limits of Safety: The Enhancement of a Theory of Accidents," *Journal of Contingencies and Crisis Management* 2 (1994): 212–20; Sagan, *The Limits of Safety*; Sagan, "Toward a Political Theory of Organizational Reliability."

92. William J. Broad, "NASA's New Guide to Space Exploration (On $25 a Day), *New York Times*, 19 March 1995, sec. E, 16; Warren E. Leary, "Space Agency Plans Layoffs, Shrinking to Pre-Apollo Size," *New York Times*, 20 May 1995, sec. A, 1.

APPENDIX A

1. Easterbrook, 53.

2. Ibid.

3. Ibid.; Roland, 39.

4. Presidential Commission on the Space Shuttle *Challenger* Accident, STS 51-L Data and Design Analysis Task Force, "Mission Planning and Operations Team Report," appendix C, "Crew Escape Systems Report" (Washington, D.C., May 1986). See Presidential Commission, *Report* 1:181; 2:J32; Hammack, appendix B, "Representative Examples of

Topics Discussed and Action Items Assigned at the Space Shuttle Crew Safety Panel," 14 April 1986, 15–18.

5. Presidential Commission, *Report* 2:J32.

6. Presidential Commission, "Mission Planning and Operations Team Report," appendix C, C1.

7. Presidential Commission, *Report* 2,;J32.

8. U.S. Congress, House, *Investigation: Report*, 135.

9. Lee Dye, "NASA Rejected Seamless Rocket to Save Money," *Los Angeles Times*, 15 February 1986, pt. 1, 1.

10. NASA, Source Evaluation Board, "Selection of Contractor for SRMs," 12.

11. Richard F. Cottrell, "Considerations Related to Segmented Solid Rockets and the Accident to the Space Shuttle Challenger," 12 March 1986, National Archives, Washington, D.C.

12. Ibid., 6.

13. Aerojet Solid Propulsion Company, "Solid Rocket Motor for Space Shuttle System," Proposal AS 73004099, vol. 3: "Design, Development and Verification Proposal," 27 August 1973, National Archives, Washington, D.C., pp. i–15.

14. Ibid., 1–2.

15. NASA, Source Evaluation Board, "Selection of Contractor for SRMs."

16. General Accounting Office, "Matter of Lockheed Propulsion Company; Thiokol Corporation," File B-173677, 24 June 1974, National Archives, Washington, D.C., p. 19.

17. General Accounting Office, "NASA Procurement: The 1973 Space Shuttle Solid Rocket Booster Contractor Selection," File B-227523 (Washington, D.C., 23 September 1987), 10.

18. NASA, Source Evaluation Board, "Selection of Contractor for SRMs," 20.

19. Fletcher's statement of the logic leading up to this conclusion included the following:

> In considering the results of the Board's evaluation, we first noted that in Mission Suitability scoring the summation resulted essentially in a stand-off amongst the top three scorers (Lockheed, Thiokol and UTC) though with a varying mix of advantages and disadvantages contributing to the total. Within this group, Lockheed's main strengths were in the technical categories of scoring, while they trailed in the management areas. Thiokol led in the management areas but trailed in the technical areas, and UTC fell generally between these two. We noted that Aerojet ranked significantly lower than the other three competitors in the Mission Suitability evaluation, and the proposal offered no cost advantages in relation to the higher ranked firms. Accordingly, we agreed that Aerojet should no longer be considered in contention for selection.
>
> We noted that the Board's analysis of cost factors indicated that Thiokol could do a more economical job than any of the other proposers in both the development and the production phases of the program; and that, accordingly, the cost per flight to be expected from a Thiokol-built motor would be the lowest. We agreed with the Board's

conclusion that this would be the case. We noted also that a choice of Thiokol would give the agency the lowest level of funding requirements for SRM work not only in an overall sense but also in the first few years of the program (the "developmental" stage of the shuttle). We, therefore, concluded that any selection other than Thiokol would give rise to an additional cost of appreciable size. (Ibid., 20–21).

20. Ibid., 12.

21. Cottrell, "Considerations Related to Segmented Solid Rockets," 7.

22. General Accounting Office, "Matter of Lockheed Propulsion Company; Thiokol Corporation," 77.

23. After reviewing the Lockheed and Thiokol proposals, the GAO concluded that "NASA Procurement Regulation 3.805-2, which deemphasizes cost in favor of quality of expected performance, is not violated by selection of contractor for Solid Rocket Motor Project" ("Matter of Lockheed Propulsion Company; Thiokol Corporation," 1) and that "design deficiencies in successful proposal cannot be fairly categorized as major" (6).

24. General Accounting Office, "NASA Procurement: 1973 Space Shuttle Solid Rocket Booster Contractor Selection," 10.

25. Ibid.

26. General Accounting Office, "Matter of Lockheed Propulsion Company; Thiokol Corporation," 1.

27. General Accounting Office, "NASA Procurement: 1973 Space Shuttle Solid Rocket Booster Contractor Selection," 11.

28. See, e.g., McConnell, 52–60.

29. Bell and Esch, 40.

30. Ibid.

31. Ibid., 1–2, 11.

32. Ibid., 1–3, 11–15.

33. Ibid., 3.

34. Mark Thompson, "GenCorp. Unit Warned NASA of O-ring Problems Years Ago," *Akron Beacon Journal*, 2 March 1986, sec. A, 10.

35. Ibid.

36. General Accounting Office, "Matter of Lockheed Propulsion Company; Thiokol Corporation."

37. Roger M. Boisjoly, lecture (Simmons College, Boston, 20 February 1992).

38. Ray, interview transcript, 25 March 1986, 5–9.

39. Bell and Esch, 40.

APPENDIX C

1. For the chronological development of rationale, exegesis, and examples of theory elaboration, see Vaughan, *Controlling Unlawful Organizational Behavior*, 54–104; Diane Vaughan, "Transaction Systems and Unlawful Organizational Behavior," *Social Problems* 29 (1982): 372–79; Diane Vaughan, "Regulating Risk: Implications of the Challenger Accident," *Law and Policy* 11 (1989): 330–49; Vaughan, "Autonomy, Interde-

pendence, and Social Control"; Vaughan, "Theory Elaboration"; Vaughan, "The Macro-Micro Connection"; Diane Vaughan, "Theory Elaboration: From Whistleblowing to a Theory of Organizational Dissent" (paper presented at annual meetings of American Sociological Association, Los Angeles, August 1994); Vaughan, *Theory Elaboration: Social Structure and Organization Theory*, in preparation.

2. Kurt Wolff, trans. and ed., *The Sociology of Georg Simmel* (New York: Free Press, 1950).

3. Alfred R. Lindesmith, *Opiate Addiction* (Bloomington, Ind.: Principia, 1947).

4. Vaughan, *Controlling Unlawful Organizational Behavior*, 54–66.

5. Ibid., 67–87.

6. Ibid., 88–104.

7. Merton, *Social Theory and Social Structure*; Robert K. Merton, "Opportunity Structure: The Emergence, Diffusion, and Differentiation of a Sociological Concept, 1930s–1950s," in *The Legacy of Anomie Theory*, ed. Freda Adler and William S. Laufer (New Brunswick, N.J.: Transaction, 1995), 3–80; Vaughan, *Controlling Unlawful Organizational Behavior*, 55–66, 85–87.

8. Richard A. Cloward and Lloyd E. Ohlin, *Delinquency and Opportunity: A Theory of Delinquent Gangs* (New York: Free Press, 1960); Vaughan, *Controlling Unlawful Organizational Behavior*, 67–68, 84–87.

9. For some preliminary thinking, see Vaughan, "Merton's Anomie Theory."

10. Mayer N. Zald, Comments on Perrow (annual meetings of American Sociological Association, Cincinnati, August 1991); Charles Perrow, "Organizational Theorists in a Society of Organizations" (invited address, section on Organizations and Occupations, annual meetings of American Sociological Association, Cincinnati, August 1991).

11. William F. Whyte, *Street Corner Society: The Social Organization of the Slum* (Chicago: University of Chicago Press, 1955); Burawoy; Timothy J. Diamond, *Making Gray Gold: Nursing Home Narratives* (Chicago: University of Chicago Press, 1992); Susan Ostrander, *Money for Change: Social Movement Philanthropy at Haymarket People's Fund* (Philadelphia: Temple University Press, 1995); Mitchell Duneier, *Slim's Table: Race, Respectability, and Masculinity* (Chicago: University of Chicago Press, 1992); Daniel F. Chambliss, *Beyond Caring: Hospitals, Nurses, and the Social Organization of Ethics* (Chicago: University of Chicago Press, 1996).

12. Diane Vaughan, "Organizations as Research Settings," in *Controlling Unlawful Organizational Behavior* (Chicago: University of Chicago Press, 1983), appendix, 113–36.

13. Mayer N. Zald, "History, Sociology, and Theories of Organization," in *Institutions in American Society: Essays in Market, Political, and Social Organizations*, ed. John E. Jackson (Ann Arbor: University of Michigan Press, 1990), 81–108.

14. Geertz, *Interpretation of Cultures*, 28–29.

Bibliography

DOCUMENTS

Aerojet Solid Propulsion Company. "Solid Rocket Motor for Space Shuttle System." Proposal AS 73004099. Vol. 3, "Design, Development and Verification Proposal." 27 August 1973. National Archives, Washington, D.C.

Boisjoly v. Morton Thiokol, Inc. 706 F.Supp. 795 (D.Utah 1988).

Cottrell, Richard F. "Considerations Related to Segmented Solid Rockets and the Accident to the Space Shuttle *Challenger.*" 12 March 1986. National Archives, Washington, D.C.

General Accounting Office. Mission Analysis and Systems Acquisition Division. Report no. MASAD-82-15:B-202664, 23 February 1982. National Archives, Washington, D.C.

Hammack, J. B. "Space Shuttle Crew Safety Panel History." Presidential Commission on the Space Shuttle *Challenger* Accident. National Archives, Washington, D.C. Mimeograph.

Hammack, J. B., and M. L. Raines. *Space Shuttle Safety Assessment Report.* Johnson Space Center, Safety Division, 5 March 1981. National Archives, Washington, D.C.

Johnson Space Center. "PRACA (Problem Reporting and Corrective Action) System Requirements for the Space Shuttle Program." JSC-08126A, May 1985. National Archives, Washington, D.C.

———. "Space Shuttle Program Directive 4A." 14 April 1974. National Archives, Washington, D.C.

Maier, Mark. "'A Major Malfunction . . .': The Story behind the Space Shuttle *Challenger* Disaster." Albany: Research Foundation of the State University of New York, 1992. Videotape instructional kit.

Maier, Mark, and Roger Boisjoly. "Roger Boisjoly and the Space Shuttle *Challenger* Disaster." Binghamton: SUNY-Binghamton, School of Education and Human Development, Career and Interdisciplinary Studies Division, 1988. Videotape instructional package.

Marshall Space Flight Center. "Problem Assessment System Procedures." MSFC-SE012082TH, March 1981. National Archives, Washington, D.C.

Morton Thiokol, Inc. "Post-Flight Evaluation of STS-2 SRM Components." Report TRW-13286, January 1983. National Archives, Washington, D.C.

Mulloy, Lawrence B. "Level I Flight Readiness Review, 51-E." Reel .098,
21 February 1985. Motion Picture and Video Library, National Archives, Washington, D.C.

National Aeronautics and Space Act of 1958. U.S. Statutes at Large 72
(1959).

National Aeronautics and Space Administration. *NASA Pocket Statistics.* Washington, D.C.: Office of Headquarters Operations, Code DA,
January 1991.

———. Aerospace Safety Advisory Panel. "National Aeronautics and
Space Administration Annual Report: Covering Calender Year 1984."
Washington, D.C., 1985. Mimeograph.

———. News and Information Branch. "Background Information on
Teacher in Space Project." Washington, D.C., 1 December 1985.

———. Source Evaluation Board. "Selection of Contractor for Space
Shuttle Program Solid Rocket Motors." 19 November 1973. National
Archives, Washington, D.C.

Presidential Commission on the Space Shuttle *Challenger* Accident.
*Report to the President by the Presidential Commission on the Space
Shuttle Challenger Accident,* 5 vols. Washington, D.C.: Government
Printing Office, 1986.

———. STS 51-L Data and Analysis Task Force. "Views of *Challenger*
Space Shuttle: Explosion, Tank Failure, etc." 13 June 1986. Motion
Picture and Video Library, National Archives, Washington, D.C.

Salita, Mark. "Prediction of Pressurization and Erosion of the SRB Primary O-rings during Motor Ignition." Part 1, "Motor Development
and Validation." Report TRW-14952, 23 April 1985. National Archives, Washington, D.C.

———. "Prediction of Pressurization and Erosion of the SRB Primary O-
rings during Motor Ignition." Part 2, "Parametric Studies of Field and
Nozzle Joints." Report TRW-15 186, 29 July 1985. National Archives,
Washington, D.C.

U.S. Congress. House. Committee on Science and Astronautics. *Hearings on H.R. 4450, H.R. 6470 (Superseded by H.R. 10340).* 90th Cong.,
1st sess., 1967.

———. *Investigation of the Challenger Accident: Hearings,* 2 vols.
Washington, D.C.: Government Printing Office, 1986.

———. *Investigation of the Challenger Accident: Report.* Washington,
D.C.: Government Printing Office, 1986.

United States ex rel. Boisjoly v. Morton Thiokol, Inc. Civil Action no.
87-209 (D.D.C.).

NEWSPAPERS AND PERIODICALS

Bell, Trudy E., and Karl Esch. "The Fatal Flaw in Flight 51-L." *IEEE Spectrum,* February 1987, 36–51.

Boffey, Philip M. "*Challenger* Crew Knew of Problem, Data Now Suggest; NASA Reverses Its Stand." *New York Times,* 29 July 1986, sec. A, 1.

———. "Engineer Who Opposed Launching *Challenger* Sues Thiokol for $1 Billion." *New York Times,* 28 January 1987, sec. A, 1.

———. "NASA Had Warning of a Disaster Risk Posed by Booster." *New York Times,* 9 February 1986, sec. A, 1.

———. "Panel Asks NASA for All Records on Booster Risks: Space Agency to Comply." *New York Times,* 10 February 1987, sec. B, 10.

Broad, William J. "NASA Chief Might Not Take Part in Decisions on Booster Contracts; Accusations of Bias in '73 Award to Thiokol Prompt His Statement." *New York Times,* 7 December 1986, sec. A, 1.

———. "NASA Had Solution to Key Flaw in Rocket When Shuttle Exploded." *New York Times,* 22 April 1986, sec. A, 1.

———. "NASA's New Guide to Space Exploration (On $25 a Day)." *New York Times,* 19 March 1995, sec. E, 16.

Clark, John F. "NASA Has Been Telling the Truth All Along." Letter, *New York Times,* 18 August 1986, sec. A, 16.

Cook, Richard C. "The Rogers Commission Failed: Questions It Never Asked, Answers It Didn't Listen To." *Washington Monthly,* November 1986, 13–21.

Diamond, Stuart. "NASA Cut or Delayed Safety Spending." *New York Times,* 24 April 1986, sec. A, 1.

———. "NASA Wasted Billions, Federal Audits Disclose." *New York Times,* 23 April 1986, sec. A, 1.

Dye, Lee. "NASA Rejected Seamless Rocket to Save Money." *Los Angeles Times,* 15 February 1986, pt, 1, 1.

Easterbrook, Gregg. "Big Dumb Rockets." *Newsweek,* 17 August 1987, 46–60.

Esch, Karl. "How NASA Prepared to Cope with Disaster." *IEEE Spectrum,* March 1986, 32–36.

"Flight Safety Critical Items Analyzed in Wake of Accident." *Aviation Week and Space Technology,* 17 March 1986, 87–95.

Greenhouse, Linda. "NASA Told to Release *Challenger* Disaster Tape." *New York Times,* 30 July 1988, sec. A, 31.

Greenhouse, Steven. "Thiokol: Contractor at Biggest Risk." *New York Times,* 3 February 1986, sec. A, 11.

Ignatius, David. "Did the Media Goad NASA into the *Challenger* Disaster?" *Washington Post,* 30 March 1986, sec. D, 1.

"Johnson Team Urges Parachutes, Rockets for Shuttle Crew Escape." *Aviation Week and Space Technology,* 14 July 1986, 141.

Jones, Alex S. "Withheld Shuttle Data: A Debate over Privacy." *New York Times,* 27 January 1987, sec. C, 1.

Kruglanski, Arie W. "Freeze-think and the Challenger." *Psychology Today,* August 1986, 48–49.

Leary, Warren E. "Space Agency Plans Layoffs, Shrinking to Pre-Apollo Size." *New York Times,* 20 May 1995, sec. A, 1.

Magnuson, Ed. "Fixing NASA." *Time*, 9 June 1986, 14–25.

"NASA Cover-Up of Challenger Data Alleged." *Boston Globe*, 13 November 1988, sec. A, 1.

Overbye, Dennis. "Success Amid the Snafus." *Discover*, November 1985, 52–67.

Reinhold, Robert. "Astronaut's Chief Says NASA Risked Life for Schedule: 'Awesome' List of Flaws." *New York Times*, 9 March 1986, sec. A, 1.

Robinson, Walter V. "NASA Blamed for Shuttle Disaster." *Boston Globe*, 10 June 1986, sec. A, 1.

Roland, Alex. "The Shuttle: Triumph or Turkey?" *Discover*, November 1985, 29–49.

Rowley, Storer, and Michael Tackett. "Internal Memo Charges NASA Compromised Safety." *Chicago Tribune*, 9 March 1986, sec. A, 8.

Schmidt, William E. "Reporters Use New Technology to Thwart NASA's Secrecy." *New York Times*, 20 March 1986, sec. A, 24.

Shenon, Philip. "NASA Accused of Cover-Up in Shuttle Deaths." *New York Times*, 14 November 1988, sec. B, 6.

"Shuttle Disaster Puts Spotlight on Safety: Trade-off for Performance Is a Critical Question." *Wall Street Journal*, 4 February 1986, 2.

"The Shuttle Inquiry: Business Aspects. Shuttle's Builders Have Varying Stakes in Program's Future." *New York Times*, 30 January 1986, sec. A, 5.

Tackett, Michael. "*Challenger* Cabin, Crew Found: Shuttle's Gear May Yield Clues." *Chicago Tribune*, 10 March 1986, sec. A, 1.

Thomas, Evan. "Painful Legacies of a Lost Mission." *Time*, 24 March 1986, 28–29.

Thompson, Mark. "GenCorp. Unit Warned NASA of O-ring Problems Years Ago." *Akron Beacon Journal*, 2 March 1986, sec. A, 10.

Van Allen, James A. "Space Science, Space Technology, and the Space Station." *Scientific American* 254 (January 1986): 32–39.

BOOKS AND ARTICLES

Allison, Graham T. *The Essence of Decision: Explaining the Cuban Missile Crisis*. Boston: Little, Brown, 1971.

Arendt, Hannah. *Eichmann in Jerusalem: A Report on the Banality of Evil*. New York: Viking, 1964.

Bailyn, Lotte. "Autonomy in the Industrial R&D Lab." *Human Resource Management* 24 (1985): 129–46.

Barley, Stephen R. "Semiotics and the Study of Occupational and Organizational Cultures." *Administrative Science Quarterly* 28 (1983): 393–413.

Becker, Howard S. "Culture: A Sociological View." *Yale Review* 71 (1982): 513–28.

———.*Outsiders: Studies in the Sociology of Deviance*. New York: Free Press, 1963.

Becker, Howard S., and James W. Carper. "The Elements of Identification with an Occupation." *American Sociological Review* 21 (1956): 341–48.

Bella, David A. "Engineering and Erosion of Trust." *Journal of Professional Issues in Engineering* 113 (1987): 117–29.

————. "Organizations and Systematic Distortion of Information." *Journal of Professional Issues in Engineering* 113 (1987): 360–70.

Bensman, Joseph, and Israel Gerver. "Crime and Punishment in the Factory: The Function of Deviancy in Maintaining the Social System." *American Sociological Review* 28 (1963): 588–98.

Bensman, Joseph, and Robert Lilienfeld. *Craft and Consciousness: Occupational Technique and the Development of World Images.* New York: Aldine de Gruyter, 1991.

Berger, Peter. *Invitation to Sociology.* Garden City, N.Y.: Anchor, 1963.

Berger, Peter, and Thomas Luckmann. *The Social Construction of Reality.* New York: Doubleday, 1966.

Bijker, Wiebe E., and John Law, eds. *Shaping Technology, Building Society: Studies in Sociotechnical Change.* Cambridge: MIT Press, 1992.

Bilstein, Roger E. *Orders of Magnitude: A History of the NACA and NASA, 1915–1990.* Washington, D.C.: National Aeronautics and Space Administration, Scientific and Technical Information Division, 1989.

Boisjoly, Roger M. "Ethical Decisions: Morton Thiokol and the Space Shuttle *Challenger* Disaster." Paper no. 87-WA/TS-4 presented at winter meeting of American Society of Mechanical Engineers, Boston, 13–18 December 1987. Mimeograph.

Boisjoly, Russell P., Ellen Foster Curtis, and Eugene Mellican. "Roger Boisjoly and the Challenger Disaster." *Journal of Business Ethics* 8 (1989): 217–30.

Bosk, Charles. *Forgive and Remember: Managing Medical Failure.* Chicago: University of Chicago Press, 1979.

Bourdieu, Pierre. *Outline of a Theory of Practice.* Cambridge: Cambridge University Press, 1977.

Braithwaite, John, and Brent Fisse. "Varieties of Responsibility and Organizational Crime." *Law and Policy* 7 (1985): 315–43.

Braithwaite, John, and Toni Makkae. "Testing an Expected Utility Model of Corporate Deterrence." *Law and Society Review* 25 (1991): 7–39.

Brint, Steven. *In an Age of Experts: The Changing Role of Professionals in Politics and Public Life.* Princeton, N.J.: Princeton University Press, 1994.

Brint, Steven, and Jerome Karabel. "Institutional Origins and Transformations: The Case of American Community Colleges." In *The New Institutionalism in Organizational Analysis,* ed. Walter W. Powell and Paul J. DiMaggio. Chicago: University of Chicago Press, 1991.

Brittain, Jack, and Sim Sitkin. "Facts, Figures, and Organizational Decisions: Carter Racing and Quantitative Analysis in the Organizational Behavior Classroom." *Organanizational Behavior Teaching Review* 14 (1989): 62–81.

Burawoy, Michael. *Manufacturing Consent.* Chicago: University of Chicago Press, 1979.

Carroll, Paul. *Big Blues: The Unmaking of IBM.* New York: Random House, 1993.

Chaisson, Eric J. *The Hubble Wars.* New York: HarperCollins, 1994.

Chambliss, Daniel F. *Beyond Caring: Hospitals, Nurses, and the Social Organization of Ethics.* Chicago: University of Chicago Press, 1996.

Clarke, Adele E., and Joan H. Fujimura, eds. *The Right Tools for the Job: At Work in Twentieth-Century Life Sciences.* Princeton, N.J.: Princeton University Press, 1992.

Clarke, Lee. *Acceptable Risk? Making Decisions in a Toxic Environment.* Berkeley and Los Angeles: University of California Press, 1989.

———. "The Disqualification Heuristic: When Do Organizations Misperceive Risk?" In *Social Problems and Public Policy,* vol. 5, ed. R. Ted Youn and William F. Freudenberg. Greenwich, Conn.: JAI, 1993.

———. "Drs. Pangloss and Strangelove Meet Organization Theory: High Reliability Organizations and Nuclear Weapons Accidents." *Sociological Forum* 8 (1993): 675–89.

———. "The Wreck of the Exxon *Valdez.*" In *Controversies: Politics of Technical Decisions,* ed. Dorothy Nelkin. Beverly Hills, Calif.: Sage, 1992.

Clarke, Lee, and James F. Short, Jr. "Social Organization and Risk: Some Current Controversies." *Annual Review of Sociology* 19 (1993): 375–99.

Clinard, Marshall B. *Corporate Ethics and Crime: The Role of Middle Management.* Beverly Hills, Calif.: Sage, 1983.

Cloward, Richard A., and Lloyd E. Ohlin. *Delinquency and Opportunity: A Theory of Delinquent Gangs.* New York: Free Press, 1960.

Cole, Stephen. *Making Science: Between Nature and Society.* Cambridge: Harvard University Press, 1992.

Coleman, James W. "Toward an Integrated Theory of White-Collar Crime." *American Journal of Sociology* 93 (1987): 406–39.

Collingridge, David. *The Social Control of Technology.* Milton Keynes: Open University Press, 1980.

Collins, H. M. "The Place of the Core-Set in Modern Science: Social Contingency with Methodological Propriety in Science." *History of Science* 19 (1981): 6–19.

———. "The TEA Set: Tacit Knowledge and Scientific Networks." *Science Studies* 4 (1974): 165–86.

Cook, Richard C. "The Challenger Report: A Critical Analysis of the Report to the President by the Presidential Commission on the Space Shuttle *Challenger* Accident." 1986. Mimeograph.

Cullen, Frank T., William J. Maakestad, and Gray Cavender. *Corporate Crime under Attack: The Ford Pinto Case and Beyond.* Cincinnati: Anderson, 1987.

Cyert, Richard M., and James G. March. *A Behavioral Theory of the Firm.* Englewood Cliffs, N.J.: Prentice-Hall, 1963.

Dalal, Siddhartha R., Edward B. Fowlkes, and Bruce Hoadley. "Risk

Analysis of the Space Shuttle: Pre-*Challenger* Prediction of Failure." *Journal of the American Statistical Association* 84 (1989): 945–57.

Dalton, Melville. *Men Who Manage.* New York: Wiley, 1959.

Darnton, Robert. *The Great Cat Massacre and Other Episodes in French Cultural History.* New York: Basic Books, 1984.

Derber, Charles. *Professionals and Workers.* Boston: Hall, 1982.

DiMaggio, Paul J. and Walter W. Powell. "Introduction." In *The New Institutionalism in Organizational Analysis,* ed. Walter W. Powell and Paul J. DiMaggio. Chicago: University of Chicago Press, 1991.

————. "The Iron Cage Revisited: Institutional Isomorphism and Collective Rationality in Organizational Fields." *American Sociological Review* 48 (1983): 147–60.

Dorf, Richard C. *Technology and Society.* San Francisco: Boyd and Fraser, 1974.

Douglas, Mary. *How Institutions Think.* London: Routledge and Kegan Paul, 1987.

————. *Risk Acceptability According to the Social Sciences.* London: Routledge and Kegan Paul, 1985.

Douglas, Mary, and Aaron Wildavsky. *Risk and Culture.* Berkeley and Los Angeles: University of California Press, 1982.

Egan, Daniel F. "The Origins of the Aerospace Safety Advisory Panel." Department of Sociology, Boston College, 1988. Manuscript.

Emerson, Robert M. "Holistic Effects in Social Control Decision-Making." *Law and Society Review* 17 (1983): 425–55.

Emerson, Robert M., and Sheldon L. Messinger. "The Micro-Politics of Trouble." *Social Problems* 25 (1977): 121–34.

Erikson, Kai T. *Everything in Its Path: Destruction of Community in the Buffalo Creek Flood.* New York: Simon and Schuster, 1976.

Esser, James K., and Joanne S. Lindoerfer. "Groupthink and the Space Shuttle *Challenger* Accident: Toward a Quantitative Case Analysis." *Journal of Behavioral Decision Making* 2 (1989): 167–77.

Feldman, Martha S. *Order without Design: Information Production and Policy Making.* Stanford, Calif.: Stanford University Press, 1989.

Feldman, Martha S., and James G. March. "Information in Organizations as Signal and Symbol." *Administrative Science Quarterly* 26 (1981): 171–84.

Feynman, Richard P. *What Do You Care What Other People Think? Further Adventures of a Curious Character.* New York: Norton, 1988.

Fine, Gary Alan. "Negotiated Orders and Organization Cultures." *Annual Review of Sociology* 10 (1984): 239–62.

————. *With the Boys.* Chicago: University of Chicago Press, 1987.

Fisse, Brent, and John Braithwaite. "The Allocation of Responsibility for Corporate Crime: Individualism, Collectivism, and Accountability." *Sidney Law Review* 12 (1988): 468–513.

Freudenberg, William R. "Nothing Recedes Like Success? Risk Analysis and the Organizational Amplification of Risk." *Risk* 3 (1992): 1–35.

————. "Perceived Risk, Real Risk: Social Science and the Art of Probabilistic Risk Assessment." *Science* 242 (7 October 1988): 44–49.

Geertz, Clifford. *The Interpretation of Cultures.* New York: Basic Books, 1973.

————. *Local Knowledge: Further Essays in Interpretive Anthropology.* New York: Basic Books, 1983.

Geis, Gilbert. "The Heavy Electrical Equipment Antitrust Cases of 1961." In *Criminal Behavior Systems,* ed. Marshall B. Clinard and Richard Quinney. New York: Holt, Rinehart and Winston, 1967.

Giddens, Anthony. *Central Problems in Social Theory: Action, Structure, and Contradiction in Social Analysis.* Berkeley and Los Angeles: University of California Press, 1979.

Gieryn, Thomas F., and Anne E. Figert. "Ingredients for a Theory of Science in Society: O-Rings, Ice Water, C-Clamps, Richard Feynman and the *New York Times.*" In *Theories of Science in Society,* ed. Susan Cozzens and Thomas Gieryn. Bloomington: Indiana University Press, 1990.

Goffman, Erving. "On Cooling the Mark Out: Some Aspects of Adaptation to Failure." *Psychiatry* 15 (1952): 451–63.

————. *The Presentation of Self in Everyday Life.* Garden City, N.Y.: Anchor, 1959.

Goldner, Fred, and Richard R. Ritti. "Professionalization as Career Immobility." *American Journal of Sociology* 72 (1967): 489–502.

Gould, Stephen Jay. *The Mismeasure of Man.* New York: Norton, 1981.

————. *Time's Arrow, Time's Cycle.* Cambridge: Harvard University Press, 1987.

————. *Wonderful Life: The Burgess Shale and the Nature of History.* New York: Norton, 1989.

Gouran, Dennis S., Randy Y. Hirokawa, and Amy E. Martz. "A Critical Analysis of Factors Related to Decisional Processes Involved in the Challenger Disaster." *Central States Speech Journal* 37 (Fall 1986): 119–35.

Grabosky, Peter N. *Wayward Governance: Illegality and its Control in the Public Sector.* Woden: Australian Institute of Criminology, 1989.

Gregory, Kathleen L. "Native-View Paradigms: Multiple Cultures and Culture Conflicts in Organizations." *Administrative Science Quarterly* 28 (1983): 359–76.

Gross, Edward. "Organizational Crime: A Theoretical Perspective." In *Studies in Symbolic Interaction,* ed. Norman K. Denzin. Greenwich, Conn.: JAI, 1978.

Gurney, Gene, and Jeff Forte. *Space Shuttle Log: The First Twenty-Five Flights.* Blue Ridge Summit, Pa.: AERO, 1988.

Hawkins, Keith. *Environment and Enforcement: Regulation and the Social Definition of Pollution.* New York: Oxford University Press, 1984.

————. ed. *The Uses of Discretion.* Oxford: Clarendon, 1992.

Hechter, Michael, ed. *The Microfoundations of Macrosociology.* Philadelphia: Temple University Press, 1983.

Heimer, Carol A. "Organizational and Individual Control of Career Development in Engineering Project Work." *Acta Sociologica* 27 (1984): 283–310.

———. *Reactive Risk and Rational Action: Managing Moral Hazard in Insurance Contracts.* Berkeley and Los Angeles: University of California Press, 1985.

———. "Social Structure, Psychology, and the Estimation of Risk." *Annual Review of Sociology* 14 (1988): 491–519.

Heydebrand, Wolf V. "New Organizational Forms." *Work and Occupations* 16 (1987): 323–57.

Hughes, Everett C. "Good People and Dirty Work." *Social Problems* 10 (1962): 3–11.

———. "Mistakes at Work." *Canadian Journal of Economics and Political Science* 17 (1951): 320–27.

Hughes, Thomas P. "The Evolution of Large Scale Technical Systems." In *The Social Construction of Technological Systems: New Directions in the Sociology and History of Technology,* ed. Wiebe Bijker, Thomas P. Hughes, and Trevor J. Pinch. Cambridge: MIT Press, 1987.

Jackall, Robert. *Moral Mazes: The World of Corporate Managers.* New York: Oxford University Press, 1988.

Janis, Irving L. *Groupthink.* Boston: Houghton Mifflin, 1982.

Jasanoff, Sheila. *Risk Management and Political Culture.* New York: Russell Sage, 1986.

Jepperson, Ronald L. "Institutions, Institutional Effects, and Institutionalism." In *The New Institutionalism in Organizational Analysis,* ed. Walter W. Powell and Paul J. DiMaggio. Chicago: University of Chicago Press, 1991.

Johnsrud, Cristy S. "Conflict, Complementarity, and Consequence: The Administration of Cross-Sector Organizational Linkages." Paper presented at annual meeting of American Anthropological Association, Phoenix, November 1988.

Kagan, Robert A., and John T. Scholz. "The 'Criminology of the Corporation' and Regulatory Enforcement Strategies." In *Enforcing Regulation,* ed. Keith Hawkins and John M. Thomas. Boston: Kluwer-Nijhoff, 1984.

Kahneman, Daniel, Paul Slovic, and Amos Tversky, eds. *Judgment under Uncertainty: Heuristics and Biases.* Cambridge: Cambridge University Press, 1982.

Katz, Jack. "Cover-Up and Collective Integrity: On the Natural Antagonisms to Authority Internal and External to Organizations." *Social Problems* 25 (1979): 3–17.

Kelman, Herbert C., and V. Lee Hamilton. *Crimes of Obedience.* New Haven, Conn.: Yale University Press, 1989.

Knorr-Cetina, Karin. *The Manufacture of Knowledge.* New York: Pergamon, 1981.

Kramer, Ronald C. "State-Corporate Crime: The Space Shuttle *Challenger* Disaster." In *White-Collar Crime Reconsidered,* ed. Kip

Schlegel and David Weisburd. Boston: Northeastern University Press, 1992.

Kuhn, Thomas. *The Structure of Scientific Revolutions.* Chicago: University of Chicago Press, 1962.

Kuklick, Henrika. "Sociology of Knowledge: Retrospect and Prospect." *Annual Review of Sociology* 9 (1983): 287–310.

Kunda, Gideon. *Engineering Culture: Control and Commitment in a High-Tech Corporation.* Philadelphia: Temple University Press, 1992.

Ladwig, Alan. "The Space Flight Participant Program: Taking the Teacher and Classroom into Space." Paper IAF-85-438 presented at 36th Congress of International Astronautical Federation. Stockholm: Pergamon, 1985.

Landau, Martin, and Russell Stout, Jr. "To Manage Is Not To Control: Or, the Folly of Type II Errors." *Public Administration Review* 39 (1979): 148–56.

LaPorte, Todd R. "Conditions of Strain and Accommodation in Industrial Research Organizations." *Administrative Science Quarterly* 10 (1965): 21–38.

LaPorte, Todd R., and Paula M. Consolini. "Working in Practice but Not in Theory: Theoretical Challenges of 'High Reliability Organizations.'" *Journal of Public Administration Research and Theory* 1 (January 1991): 19–47.

Latour, Bruno. *Science in Action.* Cambridge: Harvard University Press, 1987.

———. "Some Scholars' Babies Are Other Scholars' Bathwater." *Contemporary Sociology* 22 (July 1993): 487–89.

Latour, Bruno, and Steve Woolgar. *Laboratory Life: The Social Construction of Scientific Facts.* Beverly Hills, Calif.: Sage, 1979.

Lavine, Michael. "Problems in Extrapolation Illustrated with Space Shuttle O-Ring Data." *Journal of the American Statistical Association* 86 (1991): 919–22.

Lazega, Emmanuel. *The Micropolitics of Knowledge: Communication and Indirect Control in Workgroups.* New York: Aldine de Gruyter, 1992.

Lewis, Richard S. *Challenger: The Final Voyage.* New York: Columbia University Press, 1988.

Levitt, Barbara, and James G. March. "Organizational Learning." *Annual Review of Sociology* 14 (1988): 319–40.

Lighthall, Frederick F. "Launching the Space Shuttle *Challenger:* Disciplinary Deficiencies in the Analysis of Engineering Data." *IEEE Transactions on Engineering Management* 38 (February 1991): 63–74.

Lindblom, Charles E. "The Science of Muddling Through." *Public Administration Review* 19 (1959): 79–88.

Logsdon, John M. "The Space Shuttle Program: A Policy Failure?" *Science* 232 (30 May 1986): 1099–1105.

Louis, Meryl Ries. "An Investigator's Guide to Workplace Culture." In *Organizational Culture,* ed. Peter J. Frost, Larry F. Moore, Meryl Ries

Louis, Craig C. Lundberg, and Joanne Martin. Beverly Hills, Calif.: Sage, 1985.

Lowrance, William. *On Acceptable Risk: Science and the Determination of Safety.* Los Altos, Calif.: Kaufmann, 1976.

Macidull, John C. "Safety Awareness Continuity in Transportation and Space Systems." *Acta Astronautica* 17 (1988): 931–36.

MacKenzie, Donald, and Graham Spinardi. "Tacit Knowledge, Weapons Design, and the Uninvention of Nuclear Weapons." *American Journal of Sociology* 101 (1995): 44–99.

MacKenzie, Donald, and Judy Wajcman, eds. *The Social Shaping of Technology: How the Refrigerator Got Its Hum.* Milton Keynes: Open University Press, 1985.

Manning, Peter K. "'Big Bang' Decisions: Notes on a Naturalistic Approach." In *The Uses of Discretion,* ed. Keith Hawkins. Oxford: Clarendon Press, 1992.

———. "Managing Uncertainty in the British Nuclear Installations Inspectorate." *Law and Policy* 11 (1989): 350–69.

———. *Organizational Communication.* New York: Aldine de Gruyter, 1992.

———. "The Social Reality and Social Organization of Natural Decision Making." *Washington and Lee Law Review* 43 (1986): 1291–1311.

March, James G., and Johan P. Olsen. *Ambiguity and Choice.* Bergen: Universitetsforlaget, 1979.

———. "The New Institutionalism: Organizational Factors in Political Life." *American Political Science Review* 78 (1984): 734–49.

March, James G., and Zur Shapira. "Managerial Perspectives on Risk and Risk Taking." *Management Science* 33 (1987): 1404–18.

March, James G., and Herbert A. Simon. *Organizations.* New York: Wiley, 1958.

March, James G., Lee S. Sproull, and Michal Tamuz. "Learning from Samples of One or Fewer." *Organization Science* 2 (February 1991): 1–13.

Marcus, Alfred A. "Risk, Uncertainty, and Scientific Judgment." *Minerva* 26 (1988): 138–52.

Marshall, Eliot. "DOE's Way-Out Reactors." *Science* 231 (21 March 1986): 1357–59.

———. "The Shuttle Record: Risks, Achievements." *Science* 231 (14 February 1986): 664–66.

Martin, Joanne. *Cultures in Organizations: Three Perspectives.* New York: Oxford University Press, 1992.

Matza, David. *Becoming Deviant.* Englewood Cliffs, N.J.: Prentice-Hall, 1969.

Mazur, Allan. "Disputes between Experts." *Minerva* 11 (1973): 243–62.

McConnell, Malcolm. *Challenger: A Major Malfunction.* Garden City, N.Y.: Doubleday, 1987.

McCurdy, Howard E. "The Decay of NASA's Technical Culture." *Space Policy* (November 1989): 301–10.

————. *Inside NASA: High Technology and Organizational Change in the U.S. Space Program.* Baltimore: Johns Hopkins University Press, 1993.

McDougall, Walter A. *And the Heavens and the Earth: A Political History of the Space Age.* New York: Basic Books, 1985.

McNamara, Robert. *In Retrospect.* New York: Times Books, 1995.

Meiksins, Peter. "The 'Revolt of the Engineers' Reconsidered." *Technology and Culture* 29 (1988): 219–46.

Meiksins, Peter, and James M. Watson. "Professional Autonomy and Organizational Constraint: The Case of Engineers." *Sociological Quarterly* 30 (1989): 561–85.

Merton, Robert K. "Bureaucratic Structure and Personality." *Social Forces* 18 (1940): 560–68.

————."The Machine, the Worker, and the Engineer." *Science* 105 (1947): 78–81.

————. "Opportunity Structure: The Emergence, Diffusion, and Differentiation of a Sociological Concept, 1930s–1950s." In *The Legacy of Anomie Theory,* ed. Freda Adler and William S. Laufer. New Brunswick, N.J.: Transaction, 1995.

————. *Social Theory and Social Structure.* New York: Free Press, 1968.

Mileti, Dennis, John Sorenson, and William Bogard. *Evacuation Decision-Making: Process and Uncertainty.* Oak Ridge, Tenn.: Oak Ridge National Laboratory, 1985.

Mintzberg, Henry, Duru Raisinghani, and André Théorêt. "The Structure of Unstructured Decision Processes." *Administrative Science Quarterly* 21 (1976): 246–75.

Moore, Patrick. "When Politeness Is Fatal: Technical Communication and the *Challenger* Accident." *Journal of Business and Technical Communication* 6 (1992): 269–92.

Murray, Bruce. *Journey into Space: The First Thirty Years of Space Exploration.* New York: Norton, 1989.

Nelkin, Dorothy, and Michael Brown. *Workers at Risk: Voices from the Workplace.* Chicago: University of Chicago Press, 1984.

Noble, David F. *America by Design.* New York: Knopf, 1977.

Nye, David E. *American Technological Sublime.* Cambridge: MIT Press, 1994.

Ostrander, Susan. *Money for Change: Social Movement Philanthropy at Haymarket People's Fund.* Philadelphia: Temple University Press, 1995.

Paget, Marianne A. *The Unity of Mistakes: A Phenomenological Interpretation of Medical Work.* Philadelphia: Temple University Press, 1988.

Passas, Nikos E., and Robert S. Agnew, eds. *The Future of Anomie Theory.* Boston: Northeastern University Press, 1996.

Paternoster, Raymond, and Sally S. Simpson. "A Rational Choice Theory of Corporate Crime." In *Routine Activity and Rational Choice,* ed. Ronald V. Clarke and Marcus Felson. New Brunswick, N.J.: Transaction, 1993.

Perin, Constance. "Organizations as Contexts: Implications for Safety Science and Practice." *Industrial and Environmental Crisis Quarterly* 8. In press.

Perrow, Charles. "Accidents in High Risk Systems." *Technology Studies* 1 (1992): 1–13.

———. "The Limits of Safety: The Enhancement of a Theory of Accidents." *Journal of Contingencies and Crisis Management* 2 (1994): 212–20.

———. *Normal Accidents: Living with High Risk Technologies.* New York: Basic Books, 1984.

———. "The President's Commission and the Normal Accident." In *The Accident at Three Mile Island: The Human Dimensions.* Boulder, Colo.: Westview Press, 1981.

Perucci, Robert. "Engineering: Professional Servant of Power." *American Behavioral Scientist* 41 (1970): 492–506.

Petroski, Henry. *To Engineer Is Human: The Role of Failure in Successful Design.* New York: St. Martin's, 1985.

Pfeffer, Jeffrey, and Gerald R. Salancik. *The External Control of Organizations: A Resource Dependence Perspective.* New York: Harper and Row, 1978.

Pfohl, Stephen J. *Images of Deviance and Social Control.* New York: McGraw-Hill, 1985.

———. *Predicting Dangerousness: The Social Construction of Psychiatric Reality.* Lexington, Mass.: Lexington Books, 1978.

Pickering, Andrew, ed. *Science as Practice and Culture.* Chicago: University of Chicago Press, 1992.

Pinch, Trevor J., and Wiebe E. Bijker. "The Social Construction of Facts and Artefacts: Or How the Sociology of Science and the Sociology of Technology Might Benefit Each Other." *Social Studies of Science* 14 (1984): 399–441.

Powell, Walter W., and Paul J. DiMaggio, eds. *The New Institutionalism in Organizational Analysis.* Chicago: University of Chicago Press, 1991.

Quinney, Richard E. "Occupational Structure and Criminal Behavior: Prescription Violation by Retail Pharmacists." *Social Problems* 11 (1963): 179–85.

Raelin, Joseph A. "An Examination of Deviant/Adaptive Behaviors in the Organizational Careers of Professionals." *Academy of Management Review* 9 (1984): 413–27.

Ragin, Charles C. and Howard S. Becker, eds. *What is a Case? Exploring the Foundations of Social Inquiry.* Cambridge: Cambridge University Press, 1992

Ranson, Stewart, Bob Hinings, and Royston Greenwood. "The Structuring of Organizational Structure." *Administrative Science Quarterly* 25 (March 1980): 1–17.

Reichman, Nancy. "Regulating Risky Business." *Law and Policy* 13 (1991): 21–48.

Reiss, Albert J. "Selecting Strategies of Social Control Over Organizational Life." In *Enforcing Regulation,* ed. Keith Hawkins and John M. Thomas. Boston: Kluwer-Nijhoff, 1984.

Riley, Patricia. "A Structurationist Account of Political Culture." *Administrative Science Quarterly* 28 (1983): 414–37.

Ritti, Richard R. *The Engineer in the Industrial Corporation.* New York: Columbia University Press, 1971.

Roberts, Karlene H. "New Challenges in Organizational Research: High Reliability Organizations." *Industrial Crisis Quarterly* 3 (1989): 111–25.

———. "Some Characteristics of One Type of High Reliability Organization." *Organization Science* 1 (1990): 160–73.

Roberts, Karlene H., Denise M. Rousseau, and Todd R. LaPorte. "The Culture of High Reliability: Quantitative and Qualitative Assessment aboard Nuclear Powered Aircraft Carriers." *Journal of High Technology Management Research* 5 (Spring 1994): 141–61.

Romzek, Barbara S., and Melvin J. Dubnick. "Accountability in the Public Sector: Lessons from the *Challenger* Tragedy." *Public Administration Review* 47 (1987): 227–38.

Ross, Jerry, and Barry M. Staw. "EXPO '86: An Escalation Prototype." *Administrative Science Quarterly* 31 (1986): 274–97.

Sagan, Scott D. *The Limits of Safety: Organizations, Accidents, and Nuclear Weapons.* Princeton, N.J.: Princeton University Press, 1993.

———. "Toward a Political Theory of Organizational Reliability." *Journal of Contingencies and Crisis Management* 2 (1994): 228–40.

Salancik, Gerald R. "Commitment and the Control of Organizational Behavior and Belief." In *New Directions in Organizational Behavior,* ed. Barry M. Staw and Gerald R. Salancik. Malabar, Fla: Krieger, 1977.

Saussois, Jean-Michel, and Herve Laroche. "The Politics of Labeling Organizational Problems: An Analysis of the Challenger Case." *Knowledge and Policy: The International Journal of Knowledge Transfer* 4 (1991): 89–106.

Schall, Maryan S. "A Communication-Rules Approach to Organizational Culture." *Administrative Science Quarterly* 28 (1983): 557–81.

Schein, Edgar H. *Organizational Culture and Leadership.* San Francisco: Jossey-Bass, 1985.

Scott, W. Richard. "The Adolescence of Institutional Theory." *Administrative Science Quarterly* 32 (1987): 493–511.

———. *Institutions and Organizations.* Beverly Hills, Calif.: Sage, 1995

Sewell, William H., Jr. "A Theory of Structure: Duality, Agency, and Transformation." *American Journal of Sociology* 98 (1992): 1–29.

Shapiro, Susan P. "The Social Control of Impersonal Trust." *American Journal of Sociology* 93 (1987): 646.

———. *Wayward Capitalists: Targets of the Securities and Exchange Commission.* New Haven, Conn.: Yale University Press, 1984.

Shepherd, Clovis R. "Orientations of Scientists and Engineers." *Pacific Sociological Review* 4 (1961): 79–83.

Short, James F., Jr. "The Social Fabric at Risk: Toward the Social Transformation of Risk Analysis." *American Sociological Review* 49 (1984): 711–25.

Short, James F., Jr., and Lee Clarke, eds. *Organizations, Uncertainties, and Risk.* Boulder, Colo.: Westview, 1992.

Silver, Maury, and Daniel Geller. "On the Irrelevance of Evil: The Organization and Individual Action." *Journal of Social Issues* 34 (1978): 125–35.

Simon, Herbert A. *Administrative Behavior: A Study of Decision-Making Processes in Administrative Organizations,* 3d ed. New York: Free Press, 1976.

———. *Models of Man: Social and Rational.* New York: Wiley, 1957.

Simpson, Sally S. "The Decomposition of Antitrust: Testing a Multi-Level Longitudinal Model of Profit-Squeeze." *American Sociological Review* 51 (1986): 859–75.

Smircich, Linda. "Concepts of Culture and Organizational Analysis." *Administrative Science Quarterly* 28 (1983): 339–58.

Smith, Dorothy E. "The Social Construction of Documentary Reality." *Sociological Inquiry* 44 (1974): 257–67.

———. "Textually-Mediated Social Organization." *International Social Science Journal* 36 (1984): 59–75.

Smith, R. Jeffrey. "Experts Ponder Effect of Pressures on Shuttle Blowup." *Science* 231 (28 March 1986): 1495–98.

Spence, Michael. *Market Signaling.* Cambridge: Harvard University Press, 1974.

Star, Susan Leigh. "Scientific Work and Uncertainty." *Social Studies of Science* 15 (1985): 391–427.

———. "Simplification in Scientific Work: An Example from Neuroscience Research," *Social Studies of Science* 13 (1983): 205–28.

Starbuck, William H., and Frances J. Milliken. "*Challenger:* Fine-Tuning the Odds until Something Breaks." *Journal of Management Studies* 25 (1988): 319–40.

———. "Executives' Perceptual Filters: What They Notice and How They Make Sense." In *The Executive Effect,* ed. Donald C. Hambrick. Greenwich, Conn.: JAI, 1988.

Staw, Barry M., and Eugene Swajkowski. "The Scarcity Munificence Component of Organizational Environments and the Commission of Illegal Acts." *Administrative Science Quarterly* 20 (1975): 345–54.

Stinchcombe, Arthur L. "Authority and the Management of Engineering on Large Projects." In *Organization Theory and Project Management: Administering Uncertainty in Norwegian Offshore Oil,* ed. Arthur L. Stinchcombe and Carol A. Heimer. Oslo: Norwegian University Press, 1985.

———. "Contracts as Hierarchical Documents." In *Organization Theory and Project Management: Administering Uncertainty in Norwegian Offshore Oil,* ed. Arthur L. Stinchcombe and Carol A. Heimer. Oslo: Norwegian University Press, 1985.

————. "Social Structure and Organizations." In *Handbook of Organizations*, ed. James G. March. Chicago: Rand McNally, 1965.

————. "Work Institutions and the Sociology of Everyday Life." In *The Nature of Work: Sociological Perspectives*, ed. Kai Erikson and Steven Peter Vallas. New Haven, Conn.: Yale University Press, 1990.

Stokes, Randall, and John P. Hewitt. "Aligning Actions." *American Sociological Review* 41 (1976): 838–49.

Stone, Christopher D. *Where the Law Ends: The Social Control of Corporate Behavior.* New York: Harper and Row, 1975.

Strauss, Anselm. *Negotiations.* San Francisco: Jossey-Bass, 1978.

Sudnow, David. "Normal Crimes: Sociological Features of the Penal Code in a Public Defender Office." *Social Problems* 12 (1965): 255–76.

Sutherland, Edwin H. *White-Collar Crime.* New York: Dryden, 1949.

Swidler, Ann. "Culture in Action: Symbols and Strategies." *American Sociological Review* 51 (1986): 273–86.

Sykes, Gresham K., and David Matza. "Techniques of Neutralization: A Theory of Delinquency." *American Sociological Review* 22 (1957): 667–70.

Thomas, Robert M. *What Machines Can't Do: Politics and Technology in the Industrial Enterprise.* Berkeley and Los Angeles: University of California Press, 1993.

Tompkins, Phillip. "Management Qua Communication in Rocket Research and Development." *Communication Monographs* 44 (1977): 1–26.

————. *Organizational Communication Imperatives: Lessons of the Space Program.* Los Angeles: Roxbury, 1993.

Trento, Joseph J. *Prescription for Disaster: From the Glory Days of Apollo to the Betrayal of the Shuttle.* New York: Crown, 1987.

Trice, Harrison M., and Janice M. Beyer. *The Cultures of Work Organizations.* Englewood Cliffs, N.J.: Prentice-Hall, 1993.

Tuchman, Barbara W. *Practicing History: Selected Essays.* New York: Ballantine, 1981.

Turner, Barry. *Man-Made Disasters.* London: Wykeham, 1978.

————. "The Organizational and Interorganizational Development of Disasters." *Administrative Science Quarterly* 21 (1976): 378–97.

Tversky, Amos, and Daniel Kahneman. "The Framing of Decisions and the Psychology of Choice." *Science* 211 (30 January 1981): 1453–58.

Van Allen, James A. "Myths and Realities of Space Flight." *Science* 232 (30 May 1986): 1075–76.

Van Maanen, John, and Steve Barley. "Cultural Organization: Fragments of a Theory." In *Organizational Culture*, ed. Peter J. Frost, Larry F. Moore, Meryl Ries Louis, Craig C. Lundberg, and Joanne Martin. Beverly Hills, Calif.: Sage, 1985.

Van Maanen, John, and Edgar H. Schein. "Toward a Theory of Organizational Socialization." In *Research in Organizational Behavior*, vol. 1, ed. Barry Staw and L. L. Cummings. Greenwich, Conn.: JAI, 1979.

Vaughan, Diane. "Autonomy, Interdependence, and Social Control:

NASA and the Space Shuttle *Challenger*." *Administrative Science Quarterly* 35 (1990): 225–58.

———. *Controlling Unlawful Organizational Behavior: Social Structure and Corporate Misconduct*. Chicago: University of Chicago Press, 1983.

———. "The Macro-Micro Connection in 'White-Collar Crime' Theory." In *White-Collar Crime Reconsidered*, ed. Kip Schlegel and David Weisburd. Boston: Northeastern University Press, 1992.

———. "Merton's Anomie Theory and Organizational Misconduct." Paper presented at annual meetings of American Society of Criminology, Miami, November 1994.

———. "Theory Elaboration: The Heuristics of Case Analysis." In *What Is a Case? Exploring the Foundations of Social Inquiry*, ed. Charles C. Ragin and Howard S. Becker. Cambridge: Cambridge University Press, 1992.

———. *Uncoupling: Turning Points in Intimate Relationships*. New York: Oxford University Press, 1986.

Walton, John. "Making the Theoretical Case." In *What Is a Case? Exploring the Foundations of Social Inquiry*, ed. Charles C. Ragin and Howard S. Becker. Cambridge: Cambridge University Press, 1992.

Weick, Karl E. "Educational Organizations as Loosely Coupled Systems." *Administrative Science Quarterly* 21 (1976): 1–19.

———. "Organizational Culture as a Source of High Reliability." *California Management Review* 29 (1987): 116–36.

———. "Sensemaking in Organizations: Small Structures with Large Consequences." In *Social Psychology in Organizations: Advances in Theory and Research*, ed. J. Keith Murnighan. New York: Prentice-Hall, 1993.

———. "The Collapse of Sensemaking in Organizations: The Mann Gulch Disaster." *Administrative Science Quarterly* 38 (1993): 628-52.

———. *The Social Psychology of Organizing*. Reading, Mass.: Addison-Wesley, 1979.

———. "The Vulnerable System: An Analysis of the Tenerife Air Disaster." *Journal of Management* 6 (1990): 571–93.

Weick, Karl E., and Karlene H. Roberts. "Collective Mind in Organizations: Heedful Interrelating on Flight Decks." *Administrative Science Quarterly* 38 (1993): 357–81.

Whalley, Peter. *The Social Production of Technical Work*. Albany: State University of New York Press, 1986.

Wildavsky, Aaron. "Choosing Preferences by Constructing Institutions: A Cultural Theory of Preference Formation." *American Political Science Review* 81 (1987): 3–21.

Williams, Kirk R., and Richard Hawkins. "Perceptual Research on General Deterrence: A Critical Overview." *Law and Society Review* 20 (1986): 545–72.

Woods, Brian. "A Sociotechnical History of the United States Space Shuttle." Masters thesis, University of Edinburgh, 1995.

Wynne, Brian. "Unruly Technology: Practical Rules, Impractical Discourses, and Public Understanding." *Social Studies of Science* 18 (1988): 147–67.

Zald, Mayer N. "History, Meta-Narratives, and Organizational Theory." Ann Arbor: University of Michigan, Department of Sociology. Manuscript.

———. "History, Sociology, and Theories of Organization." In *Institutions in American Society: Essays in Market, Political, and Social Organizations*, ed. John E. Jackson. Ann Arbor: University of Michigan Press, 1990.

Zaltman, Gerald. "Knowledge Disavowal." Paper presented at conference on Producing Useful Knowledge for Organizations, Graduate School of Business, University of Pittsburgh, October 1982.

Zucker, Lynne G. "Organizations as Institutions." In *Research in the Sociology of Organizations*, ed. S. B. Bacharach. Greenwich, Conn.: JAI, 1983.

———. "The Role of Institutionalization in Cultural Persistence." *American Sociological Review* 42 (1977): 726–43.

Zussman, Robert. *Mechanics of the Middle Class: Work and Politics among American Engineers*. Berkeley and Los Angeles: University of California Press, 1985.

Index

acceptable risk: boundaries of expanded, 124, 143, 154, 170; Critical Items List detailing, 107; as engineering concept, 224; Feynman on, 82; in NASA vocabulary, 252; as normative for NASA, 82; O-ring problems as, 111, 144, 185–90; reinterpreting technical deviance as, 61, 120; SRB design as, 106, 109, 111, 112, 121, 125, 130, 244, 283; SRB work group construction of, 191–92; SRB work groups' acceptance of risk approved in FRRs, 247–50; STS 51-B damage as, 166–67, 170–71; three-factor technical rationale for, 61, 66, 120, 138, 140, 141, 156, 161, 176, 186, 350, 351, 481n.102

Acceptable Risk Process, 81–82; Acceptable Risk defined, 81; as aligned with accepted engineering practice, 224; anomalies in, 81, 243; applied to O-ring erosion, 121, 122, 124, 150; Feynman on, 82; negotiating routine decisions, 85–90; no-fly recommendations having to emerge from, 249; risk rationale in, 81. See also acceptable risk

accidents: compliance strategy in prevention of, 266; fault tree analysis of, 53, 479n.73; as normal, 200, 415, 528n.79; organizational design for avoiding, 415–17

Action Items, 91, 92, 95, 261, 277

Adams, Leslie, 221, 231

Aerojet Solid Propulsion Company, 104, 115, 425–31, 530n.19

Aerospace Safety Advisory Panel (ASAP), 266, 268–69, 417

aggregate risk, 82, 95, 162

Aldrich, Arnold: on cold temperature as not a problem, 469n.6; on cold temperature's significance, 9; delays launch until 11:38 A.M., 331; launching despite the ice, 355–56; in NASA launch decision chain, 1, 285; not advised of second teleconference, 288; not informed of Thiokol engineering concerns, 6, 323, 327–28, 343–45

Allison, Graham T., 402–3

amoral calculation, 38–46, 336–39; accusation of not made in Presidential Commission, 55–56; *Challenger* disaster not explained by, 68; detection as factor in, 47; difficulties with hypothesis of, 47–48; failure to redesign booster seen as, 113; in managerial decision making, 35, 36, 37; NASA waiving launch requirements suggesting, 57; as raising question of institutional context, 408; segmented booster use as, 426; in theories of *Challenger* disaster, 72, 278, 334

analytic induction, method of, 457, 459, 462

anomalies: in Acceptable Risk Process, 81, 243; correlation between cold and field joint anomalies, 382–85; at Flight Readiness Review, 92, 160, 259; NASA definition of, 80, 252; as normative in engineering, 223; in STS-4, 128; resolutions in paradigms, 401, 402; in STS 41-B, 138; as taken for granted at NASA, 114, 223. See also nozzle joint anomalies

anthropology of organizations. See ethnography; organizations

Apollo Program: final two missions canceled, 217; German engineers and,

551

mising belief system of NASA, 198;
production of culture within structure
of, 197–98; safety regulators sharing
in, 271; in Space Shuttle work groups,
238; SRB work group affected by, 67,
198, 236, 396–97

data dumps, 86, 228
Davids, Irving, 173, 445
Dean, Bunda L., 54
decision making: amoral calculation in
managerial, 35, 36, 37; banality of
daily, 119, 410; commitment to group
stance generated by, 248; conformity
in NASA, 407; decision stream in, 243,
247, 278, 506n.21; democratic processes in, 419; in early shuttle
development, 112; economic strain
linked to, 36–38, 46–47, 335; engineering decisions biased toward making
existing systems work, 116; first decision's importance, 112, 244–45, 247;
holistic effects in, 243; under
ignorance, 148–49; information influencing, 67; in-house standards for,
115–16; micro-macro connection in,
38; NASA's four-tiered launch decision
chain, 1, 285; in organizations, 37–38,
67, 116–17, 119, 197, 251, 402–5,
507n.40; patterns of, 66; production
pressures institutionalized in NASA,
xiii–xiv, 68, 231–32, 335, 372; psychiatric, 412–13; rationality in organizational, 37–38, 197, 403; relying on signals for, 251, 507n.40; residual risk
and work group, 79–83; social context
of, 430; SRB joint history of, 61–62, 64,
74, 77; in SRB work group, 61–62, 114,
483n.12. See also launch decision for
Challenger mission
decision stream, 243, 247, 278, 506n.21
Delta review concept: blow-by in STS
41-D reported due to, 143; in Flight
Readiness Review, 91, 92, 94–95, 259,
341; Lucas not informed of teleconference results due to, 345; as no longer
applying to O-rings after Launch Constraint imposed, 508n.72; O-ring erosion in STS-2 not reported due to, 136,
140; second teleconference not reported due to, 344; SRB problems not
reported in Challenger FRR due to,
283
deviance: as common in innovative technologies, 223; conformity and in engi-

neers, 225; as controllable, not eliminable, in the shuttle, 223; culture as
cause of organizational, 406; defined,
58; flying with known flaws as not
deviant, 114, 151, 185–90, 406; from
industry standards, 114–16; Mulloy's
lifting Launch Constraints seen as,
167–70; NASA actions seen by
outsiders as, 58, 72–73, 113, 114; as
negative, 480n.87; norms, rules, and
deviance in the work group, 112–18,
148–52, 190–95; organizational, 405–9;
request to close-out O-ring problem
seen as, 183, 194; shifting the burden
of proof seen as, 339, 340; Thiokol's
original launch recommendation as,
349. See also anomalies;
normalization of deviance
Diamond, Stuart, 480n.91
DiMaggio, Paul J., 37, 526n.45
disqualification heuristic, 273–74,
275–77, 354
documents: as asserting consensus, 247,
248; Challenger documents scattered
throughout NASA, 54; commitment
generated by, 248, 395; early documentation of O-ring problems, xi–xii;
House Committee report, 11, 72; initial decisions becoming encoded in,
112; NASA procedural requirements
described by, 57; organizational structure and culture constructed from, 61;
Presidential Commission's report, 10,
39, 60, 70–72; reduced during Flight
Readiness Review, 258; reliance on in
this study, 59, 279, 459–60; Smith on
construction of documentary reality,
68–69
Dorf, Richard C., 206
Douglas, Mary, 400, 524n.24
Dubnick, Melvin J., 211, 348

Ebeling, Robert: on close-out of O-ring
problem, 183; on complete burn-
through as possible, 380; on end of
Thiokol's off-line caucus, 320; as feeling that shuttle flights should stop,
253–54; on Hardy's feeling appalled at
Thiokol recommendation, 312; on
how he should have responded to Mulloy's "April" remark, 381; on lack of
support for O-ring task force, 180,
450–51; as not getting to vote on
launch decision, 325; as not sleeping
well after second teleconference, 326;

Marshall Center Board FRR, 94, 219–20, 284; official organizational practices of, 258; operating from the bottom up, 91, 93–95; O-ring erosion problem in, 122–24, 125, 140; problem resolution in, 258–59; reporting practices of, 259; requiring most recent flight experience to be assessed, 113; scientific positivist standards in, 221; second teleconference compared to, 340–47, 357, 420–21; signals relied on in, 262–64; specialization in, 260–62; structural secrecy in, 257, 416; systematic censorship in, 257–64; time constraints on, 260. *See also* Delta review concept

Ford Motor Company, 34–35, 408

frame of reference. *See* worldview

FRR. *See* Flight Readiness Review

Freudenberg, William R., 400

full disclosure principle, 369

Garn, Jake, 27, 162, 213

Geertz, Clifford, 79, 199, 393, 462

Giddens, Anthony, 357

Glaysher, Bob, 331, 332

global assessments of risk, 274–75, 276

Gore, Albert, Jr., 429

Gould, Stephen Jay, 411–12

Government Accounting Office (GAO), 428–29, 531n.23

Grabosky, Peter, 36

gradualism. *See* incrementalism

Graham, William, 54

Green, Shirley, 14–15

Greenwood, Royston, 213

groupthink, 404, 525n.41

habitus, 510n.118, 526n.45

Hamilton, Lee, 407

Hardy, George: as appalled at Thiokol recommendation, 6, 40, 280, 305, 311–13, 318, 334, 373, 376–78; Boisjoly on, 56; conversation with Brinton on morning of launch, 329; on difficulty remembering second teleconference, 280; discussing delay of launch, 322, 334; as enacting his usual role in second teleconference, 341; in first prelaunch teleconference, 288; on the hydroburst test, 100; on increasing risk for a few hours of schedule, 49, 56; influence of remarks at second teleconference, 362–63; Mulloy succeeds, 127–28; as not launching

against contractor recommendation, 312, 334, 377; on not wanting to make a dumb mistake, 314, 334, 377; on openness of Flight Readiness Review, 241; on O-ring erosion in STS-2, 123; performance pressure on, 374; requesting further discussion at end of second teleconference, 323, 368–69; shifting the burden of proof denied by, 342

Heimer, Carol, 114, 268, 400

Hercules, Inc., 126, 127

high-reliability theorists, 415–16, 422

Hinings, Bob, 213

historical ethnography, xv, 61, 459–63

history: organizational decision making affected by, xiv; reconstruction of in daily life, 281; resting on being able to step into the past, 73; retrospection and, 68–72, 393; revisionist account of *Challenger* disaster, 72–76, 198; social construction of, 393; systematic distortion of, 70

holistic effects in decision making, 243

Hollings, Ernest F., 13

Hotz, Robert B., 54, 310

House Committee on Science and Technology: on August 1985 meeting, 263–64; on communication at NASA, 240; on escape rockets being scrapped, 425; on Flight Readiness Review, 240; historically accepted account of disaster and, xii; on information flow in NASA, 252, 276; Launch Constraints misunderstood by, 193; on McDonald's protest of launch decision, 371; on NASA and Thiokol not understanding joint operation, 390; NASA personnel blamed for disaster by, 11; on need to get space program going again, 389; oversight of NASA, 470n.18; on political elites as source of disaster, 390; Presidential Commission usurps role of, 388; on production pressure as cause of *Challenger* disaster, 31–32; on reporting rules not violated at second teleconference, 345; report of, 11, 72; SR&QA criticized by, 271; on Thiokol raising its concerns on eve of launch, 351–52

Houston, Cecil, 301, 327, 345, 361, 385

Hughes, Everett C., 421, 475n.12

hydroburst test, 98–100

Ice/Frost Inspection Team, 7, 328, 329, 330, 331

ing a management decision, 317; Mulloy requests fax of Thiokol's launch recommendation, 6, 322–23, 346–47; Mulloy requests Thiokol management recommendation on launch, 4, 305–6; as obstacle to O-ring task force, 180–81, 454–55; requesting off-line caucus during second teleconference, 6, 314; signing Thiokol's launch recommendation, 323–24, 347; at SRM Board, 92; in Thiokol management's decision to launch, 317–20

Kingsbury, James: informed of Thiokol's concerns about temperature, 7, 288, 328, 329; requests briefing on O-ring problem, 179

Kuhn, Thomas, 196, 348, 400–402

Lamberth, Horace, 331, 332

Landau, Martin, 202

language, technical, 57, 200, 252–53

Launch Constraints: House Committee and Presidential Commission misunderstanding, 193; as ineffective indicators of problems, 253; lifted for *Challenger*, 168; as a NASA process, 167; not all problems as, 496n.115; on nozzle joints for flights after STS 51-B, 164, 493n.31, 508n.72

Launch Constraint waivers, 56–58, 153–54, 167–70, 192

launch decision for *Challenger* mission, 1–32, 278–333; amoral calculator hypothesis of, 38–48, 278, 336–39; as committed action, 283; communications problems as inadequate explanation of, 11; conventional explanation of, 32, 389–90; as cost/safety trade-off, 47, 50; data existing to cause launch delay, 381–85; discussion subsequent to second teleconference, 324–28; engineers fearing increased damage, not disaster, 380; events between the two teleconferences, 288–99; final Thiokol launch recommendation, 326; first concern about cold, 2, 286; first teleconference on eve of launch, 2, 287–88; as first time Thiokol recommended no-launch, 305, 339; groupthink in, 404; House Committee on Science and Technology report on, 11; launch schedule pressure behind, 45; launch to correlate with State of the Union message, 14–15, 18; lessons learned from, 387–422; Level II and I

FRRs for, 284; Level IV FRR for, 283; Marshall and Kennedy staff unaware of dissent at Thiokol, 323; Marshall Center Board FRR for, 284; media pressures behind, 12; NASA challenges 53°F minimum temperature recommendation, 6, 40, 306–7, 310, 318, 354, 376; NASA decision to launch despite Thiokol misgivings, xii, 11, 40–41; NASA managers on eve of launch, 47–49; NASA managers taking calculated risk to launch, 32; NASA personnel during Thiokol off-line caucus, 321–22; NASA requests Thiokol management recommendation on launch, 4, 305–6, 352; NASA's four-tiered launch decision chain, 1, 285; as one decision in a decision stream, 278; political pressures behind, 12–14; Presidential Commission on, 8; pressures on Shuttle Program as pressure behind, 16–17; revisionist account of, 72–76; as rule-based, 339–47, 386; secondary launch window, 334; Teacher in Space Project as pressure behind, 16; TAL sites as factor in, 285, 330, 512n.23; Thiokol management reverses recommendation against launch, 6, 40–41, 317–20, 478n.59; Thiokol's initial recommendation against launch, 6, 9, 305–6; Thiokol's midday meeting on cold, 287; Thiokol's off-line caucus during second teleconference, 6, 314–22, 363–68, 378. *See also* second prelaunch teleconference

"lying down in the bucket," 349, 356, 379

learning by doing, 191, 394

Levitt, Barbara, 191

liability of newness, 123, 125, 140

Lighthall, Frederick E., 522n.136

Lilienfeld, Robert, 200

Lindesmith, Alfred R., 457

Littles, Wayne, 301, 360, 379

local knowledge, 201–2, 235

Lockheed Propulsion Company, 426, 427, 428–29, 430, 530n.19

logical consistency, 354

Logsdon, John, 21

Lovingood, Jud: discussing delay of launch, 322, 334; downplays role of O-rings and cold temperature, 8, 9; locating people for second teleconference, 289; on a no-launch recommendation,

Program Requirements Control Board, 134, 135, 489n.55

Project Managers, 80; as being "the long poles," 242; competition among, 218–19; Engineering Change Request approval bypassing, 86, 89–90; in engineering disputes, 87–88; in Flight Readiness Review, 84, 93; as links between engineers and administrators, 215; as "neck of the hour glass," 93, 359; problem resolution as responsibility of, 258; S&E's differences in orientation, 88–89

putty. *See* zinc chromate putty

Pye, David, 206

Quinones, John, 12

Quong, Harry, 176, 184

Ranson, Stewart, 213

Rather, Dan, 12

rationality: bounded rationality, 37; *Challenger* launch decision contradicting rational choice theory, 394, 403; engineers as purveyors of, 200; in organizational decision making, 37–38, 197, 403

Ray, W. Leon: on blow-by in STS 51-C, 155; on change in Criticality status of SRB joint, 130; on communication channels in NASA, 241; on continuing to fly despite O-ring problem, 189; on deviating from formal industry standards, 115; on ECRs as attention getting, 256; on extrusion by O-rings, 102; on joint rotation, 40, 97, 99–106; on "lying down in the bucket," 349, 356; on managers driven by costs and schedule, 230; memos of, 101, 102, 113, 255; on no small mistakes in aerospace, 228; as not present at second teleconference, 289, 421; on O-ring erosion in STS-2, 123; on postdisaster corrections to SRB, 421; Ray/Miller memorandum, 101, 225, 508n.53; on redundancy of secondary O-ring, 106, 487n.103; on safety hazards and real safety hazards, 256; on safety regulators, 270; on S&E winning disputes over risk, 89; on schedule as not a factor in launch decision, 336; on Source Evaluation Board, 97, 428, 479n.67; in SRM joint corrections, 60, 104, 105, 111; on SRM redesign, 105, 113, 127; on Thiokol's booster design,

97, 100, 430; Thiokol's dislike of, 87; visiting O-ring manufacturers, 102–3, 115–16

Reagan, Ronald: administration's alleged pressure on NASA, 12–14; announcing that a teacher would fly on the shuttle, 27; *Challenger* flight to give proeducation image to, 388; policies making NASA act like a business, 211; Presidential Commission appointed by, 7; speech at fourth orbital test flight, 24, 125, 151

real safety hazards, 252, 257

Redstone Arsenal, 215–16

redundancy: in Criticality categories, 107, 129, 133–34; fail-safe requirement, 129, 133; in NASA reporting procedures, 419; Powers on, 104; in Space Shuttle Program information flow, 251; of TAL sites, 512n.23. *See also* secondary O-ring as redundant

regulation, safety. *See* safety regulation

regulatory ineffectiveness, 34, 458

Reinartz, Stanley: Aldrich not informed of Thiokol concerns by, 6, 323, 343–45; in first prelaunch teleconference, 288; on 40°F as legitimate data point, 309; Lucas and Kingsbury informed of Thiokol concerns, 7, 288, 328, 329, 345; Lucas not awakened to be informed of final launch decision, 328, 345; McDonald protests launch decision to, 371; in NASA launch decision chain, 1, 285; not advising Aldrich of second teleconference, 288; poll on feasibility of January 28 launch, 286; requesting further discussion at end of second teleconference, 323, 368–69; requests Hardy's opinion of Thiokol recommendation, 6, 305, 311; in teleconference on booster recovery, 327

residual risk, 79–83, 114, 162, 247

Resnick, Judith, 1, 283

Ride, Sally, 353

Riehl, Bill, 289–90, 300–301, 309, 311, 359–60, 368

risk: aggregate risk, 82, 95, 162; assessment by technical workers, 399–402; assessment in NASA work groups, 80–84, 400; collective construction of, 272–77; construction of changed in operational stage, 125, 128; construction of formalized for SRBs, 106–10;

Challenger, 7, 328; Launch Constraint on waived, 56–58, 153–54, 167–70, 192; normalization of technical deviations of, xiii; as not tested at 31°F lower limit, 391; parachute problem, 234–35, 284; photographs of flame in aft segment of *Challenger*'s right, 8–9, 53; reusability of, 21, 97; right SRB breaks loose during *Challenger* launch, 10; in Space Shuttle system, 3; in STS-1, 109–10, 112; surface temperature of right SRB on morning of launch, 330; Thiokol design as lowest ranked but cheapest, 42, 427; Thiokol refusing to acknowledge flaw in, 77; used to reduce Shuttle Program development costs, 21, 423; wind shear's effect on, 391. *See also* Solid Rocket Motor; SRB work group

Solid Rocket Motor (SRM): as acceptable risk, 106, 112, 121, 125, 130, 244, 283; capture feature incorporated in, 126–27, 173, 176, 179, 191; Criticality status changed, 129–32; diagram of, 4; dimensions of joints, 40; factory and field joints in, 4, 5; field joint cross-section, 5; history of decision making about the joints, 61–62, 64, 74, 77; hydroburst test of, 98–100; joint failure, 50, 132, 139; joint problem as well-structured after the disaster, 69–70, 149; joint rotation, 40, 97–106, 110–11, 126–27, 128, 244, 485n.56; joint rotation distinguished from erosion, 131–32, 139; joints classified as Criticality 1R, 108, 111, 253; joints tinkered with rather than redesigned, 42–43; Mark's call for review of joint problems, 140; Marshall Space Flight Center's oversight of, 85; neither NASA nor Thiokol understanding the joint, 277, 390, 470n.19; normalization of deviance in development of, 96–111, 114; normalization of deviant performance of joints, 65–68, 78, 106, 113, 119, 122, 124, 394; open problems with, 184; Ray's corrections to the joints, 60, 104, 105; redesign considered, 105–6, 113, 117–18, 127, 147–48, 232–33, 235; risk in negotiated by SRB work group, 65, 78; segmented versus monolithic, 425–31; shimming the joint, 100, 104, 111; signals of danger in, 110, 124, 141, 170, 395, 397; SRB work group learning the dynamics

of the joints, 119–20, 192; technical deviance of joints redefined as acceptable risk, 61; Thiokol reviews cold's effects on, 2–7, 286–87; three-factor technical rationale on joints, 61, 66, 120, 138, 140, 141, 156, 161, 176, 186, 350, 351, 481n.102. *See also* nozzle joint anomalies; O-rings; zinc chromate putty

Source Evaluation Board (SEB), 97, 425–30

Space Shuttle Crew Safety Panel (SSCSP), 266–67, 269, 417

Space Shuttle Program (Space Transportation System): Air Force endorsement of, 19, 20, 26; anomalies expected in, 80; begun in midst of decline of appropriations, 18; blizzard of paperwork in, 251; as business, 24–28; civilians in flight crews, 26–28; commercial goals of, 24–26; competition and scarce resources for, 17–29; competitors of, 25–26; compromise in design of, 20–22; conversion from development to operational stage, 24, 28, 119, 125–28, 150–51, 162; Cook's list of dangerous conditions, 45, 478n.53; cost/safety trade-offs in, 17, 19–24, 42–43, 47, 49–50, 117–18, 423–31; Critical Items List, 107–8, 419, 438–41; design as unprecedented, 79, 80; development authorized, 22; development costs required to be low, 21; deviations as controllable not eliminable in, 223; divided managerial responsibility for, 23; as done on the cheap, 42; early decision making in, 112; economic and safety interests competing in, 47; fixed cost of, 25; fleet reduced from five to four, 2; flight manifest changes, 28–29; flight schedule of, 22, 24–25, 28, 474n.138; as fundamental to manned space program, 19; information flow in, 251; initial orbital tests delayed, 23; launch decisions as continuous feedback loop, 248; launch delay history, 51–52, 215; launch delays occurring at the Cape, 95, 340; launching cost of, 25; launch schedule as important to, 214; launch turnaround time, 25; launch windows, 285–86, 512n.23; management structure and Flight Readiness Review, 83; military links of, 19–20; NASA argues for compromise design, 22–24; NASA's

work group culture: organizational culture compared to, 64–65; work group defined, 64; worldview of, 65. *See also* Space Shuttle work groups

worldview (frame of reference), 62–63; differences in engineers and managers, 199; dominant worldview of a culture, 66; in interpreting information, 75–76, 409; managerial, 275; occupational technique and, 200; of readers of this study, 73; societal culture affecting, 411; surviving despite persistent challenges, 62–63, 196; in work group culture, 65. *See also* paradigms

Wynne, Brian, 200–201, 223

Young, John W.: management responsibilities of, 261; on NASA compromising flight safety, 44–45; on NASA teleconferences, 305; no accusation of individual wrongdoing by, 55; in STS-1, 109

Zald, Mayer N., 403, 460

zinc chromate putty: blowholes in, 121, 122, 136; function of, 121, 486n.77; leak check change adding to the blowholes, 137, 138–39, 142, 165; more required than in the Titan, 97; in Presidential Commission hearings, 9; propellant gases penetrate during *Challenger* launch, 10; in SRM field joints, 5; use as controversial, 141

Zucker, Lynne G., 236

Zussman, Robert, 205, 206